D0220547

The Evolutionary Biology of the Threespine Stickleback

The Evolutionary Biology of the Threespine Stickleback

Edited by

MICHAEL A. BELL
Department of Ecology and Evolution
State University of New York at Stony Brook
Stony Brook, New York, USA

and

SUSAN A. FOSTER
Department of Biological Sciences
University of Arkansas
Fayetteville, Arkansas, USA

Oxford New York Tokyo
OXFORD UNIVERSITY PRESS
1994

Oxford University Press, Walton Street, Oxford OX2 6DP

Oxford New York Toronto
Delhi Bombay Calcutta Madras Karachi
Kuala Lumpur Singapore Hong Kong Tokyo
Nairobi Dar es Salaam Cape Town
Melbourne Auckland Madrid

and associated companies in
Berlin Ibadan

Published in the United States
by Oxford University Press Inc., New York

A catalogue record for this book is available from the British Library

Library of Congress Cataloging in Publication Data
The Evolutionary biology of the threespine stickleback / edited by Michael A. Bell, Susan A. Foster
Includes bibliographical references and indexes.
1. Threespine stickleback. 2. Threespine stickleback—Evolution. 3. Threespine stickleback—Ecology. I. Bell, Michael A. II. Foster, Susan Adlai.
QL638.G27E96 1993 597′.53—dc20 92-38415
ISBN 0 19 857728 1

Typeset by Colset Pte Ltd, Singapore
Printed in Great Britain by
St Edmundsbury Press, Bury St Edmunds

Preface

The threespine stickleback, *Gasterosteus aculeatus*, has been used to address diverse problems in evolutionary biology and other subdisciplines in biology. Its wide use in biology largely reflects its favourable biological properties: its wide distribution, ecological versatility, extreme phenotypic variation, small size, and adaptability to the laboratory. However, as information about its biology has accumulated, the value of its intrinsic properties has been complemented by the accumulated knowledge which has facilitated subsequent research. Progress in our understanding of evolutionary processes is strongly dependent upon appreciation of the influences of multiple simultaneous effects, and consequently this accumulation of knowledge is crucial for the stickleback's use in evolutionary biology.

The breadth and volume of research on the threespine stickleback has stimulated previous comprehensive works on this species complex. Wootton's (1976) excellent first book has been an indispensable introduction for students and a valuable general reference for those interested in stickleback biology. A stickleback bibliography by Coad (1981) and later books by Paepke (1983, in German) and another, more narrowly focused book by Wootton (1984) have further increased accessibility of information on stickleback biology. There is even a delightful (and largely accurate) children's book on the life cycle of the threespine stickleback (Lane 1981). The importance of the threespine stickleback for biological research as well as for an early introduction to natural history has become widely recognized.

It has already been eight years since the most recent review of stickleback biology was published. There have been major developments during this time, and continued progress in the use of the threespine stickleback for evolutionary studies requires a new synthesis of our knowledge. We believe that the explosive growth of knowledge on stickleback behaviour, ecology, physiology, and evolution would make it very difficult for a single author or even a small group of coauthors to undertake a book that would adequately treat the major areas of research on evolution of the threespine stickleback. However, we also did not believe that a disconnected collection of research papers on stickleback biology would meet this need.

Accordingly, we set out to develop a volume that would combine the consistency, integration, and coverage of a single-author book with the currency, depth of knowledge, and critical judgement that specialists could bring to their subjects. Most chapters review major research areas to which the chapter authors have contributed. Although many chapters include unpublished data or information on the authors' research that will soon appear in the journal literature, most of the chapters take a broader perspective. The individual chapters of this book are the intellectual products of the chapter authors, but we have tried to assemble a cohesive set of topics that centre on evolutionary biology and include the major developments in threespine stickleback biology. Author and subject indexes at the back of the volume represent a further attempt to facilitate use of the volume as a general introduction to the evolutionary biology of the threespine stickleback. We have also tried to impose a reasonably uniform format, style, and terminology throughout the book, and to resolve factual inconsistencies or conflicts of interpretation between chapters by different authors. Indeed, we have taken the liberty to intrude deeply into the usual prerogatives of contributors to multiauthored books. We hope that this intrusion has resulted in a reasonable level of integration and consistency of presentation.

The diversity of information bearing on the evolutionary biology of the threespine stickleback would make it difficult even to render informed editorial judgement of the chapters in this book without the assistance of numerous reviewers. Therefore, we have exploited the generosity of many colleagues who agreed to write critiques of the chapters and to make suggestions for their improvement. We are most grateful for the help of these external reviewers, who were John A. Baker, Theo C.M. Bakker, George W. Barlow, Jeffrey L. Beacham, Mark Bevelhimer, Bertil Borg, Brian W. Coad, Richard G. Coss, John P. Ebersole, Harry W. Greene, Helga E. Guderley, Anne E. Houde, G.J. Kenagy, Manfred Milinski, Gary G. Mittelbach, Guillermo Ortí, Donald H. Owings, Mark S. Ridgway, William J. Rowland, David L. Soltz, David W. Stephens, Nikki C. Tousley, David B. Wake, Jeffrey A. Walker, George C. Williams, David Sloan Wilson, and three anonymous reviewers. The reviewers often contributed extensive comments, and we are indebted to them for the numerous improvements to the book that resulted from their advice. However, their advice was not always followed by authors, and they are not responsible for the content of chapters. We are also most grateful to Nikki C. Tousley and Simon C. Nemtzov, who did most of the work of verifying, organizing, and integrating citations from individual chapters into a single list of references. Editorial work on this book was completed while M.A.B. was on sabbatical leave at St Francis Xavier University, and thanks are due to the University for supporting this work and to D. Max Blouw for his hospitality.

Stony Brook M.A.B.
Fayetteville S.A.F.
December 1993

Contents

Contributors

John A. Baker Department of Biological Sciences, University of Arkansas, Fayetteville, Arkansas 72701, USA

Theo C.M. Bakker Universität Bern, Zoologisches Institut, Abteilung Verhaltensökologie, Wohlenstrasse 50a, CH-3032 Hinterkappelen, Switzerland

Michael A. Bell Department of Ecology and Evolution, State University of New York at Stony Brook, Stony Brook, New York 11794-5245, USA

Patricia S. Bowne Department of Zoology, University of Alberta, Edmonton, Alberta T6G 2E9, Canada (Present address: Biology Department, Alverno College, 3401 South 39th Street, Milwaukee, Wisconsin 53234-3922, USA)

Donald G. Buth Department of Biology, University of California, Los Angeles, California 90024-1606, USA

Gerard J. FitzGerald Département de biologie, Université Laval, Ste Foy, Québec G1K 7P4, Canada

Susan A. Foster Department of Biological Sciences, University of Arkansas, Fayetteville, Arkansas 72701, USA

Andrew B. Gill Department of Zoology, University of Leicester, University Road, Leicester LE1 7RH, UK

Helga E. Guderley Département de biologie, Université Laval, Ste Foy, Québec G1K 7P4, Canada

Thomas R. Haglund Department of Biology, University of California, Los Angeles, California 90024-1606, USA

Paul J.B. Hart Department of Zoology, University of Leicester, University Road, Leicester LE1 7RH, UK

F.A. Huntingford Department of Zoology, University of Glasgow, Glasgow G12 8QQ, UK

J.D. McPhail Ecology Group, Department of Zoology, University of British Columbia, 6270 University Boulevard, Vancouver, British Columbia V6T 1Z4, Canada

Thomas E. Reimchen Department of Biology, University of Victoria, P.O. Box 1700, Victoria, British Columbia V8W 2Y2, Canada

William J. Rowland Department of Biology, Indiana University, Bloomington, Indiana 47405, USA

J.F. Tierney Department of Zoology, University of Glasgow, Glasgow G12 8QQ, UK

Frederick G. Whoriskey Department of Renewable Resources, Macdonald

College of McGill University, 21,111 Lakeshore Road, Ste Anne de Bellevue, Quebec H9X 1CO, Canada

R.J. Wootton Department of Biological Sciences, University College of Wales, Aberystwyth, Dyfed SY23 3DA, Wales, UK

P.J. Wright Department of Zoology, University of Glasgow, Glasgow G12 8QQ, UK (Present address: SOAFD Marine Laboratory, P.O. Box 101, Aberdeen AB9 8DB, UK)

1

Introduction to the evolutionary biology of the threespine stickleback

Michael A. Bell and Susan A. Foster

The threespine stickleback, *Gasterosteus aculeatus*, is a species complex that comprises thousands of phenotypically diverse populations. It is widely distributed in boreal and temperate regions of the northern hemisphere and inhabits coastal marine waters, brackish waters, and a wide array of freshwater habitats. Its broad geographical and ecological distribution and the fragmentation of its gene pool into many thousands of isolated or semi-isolated demes in freshwater habitats have generated an extraordinary range of phenotypic diversity. In some instances, divergent populations have acquired the isolating mechanisms that characterize biological species. This phenotypic diversity offers an exceptional opportunity for analysis of evolutionary mechanisms.

The highly ritualized behaviour of the threespine stickleback has also established it as a frequent subject for behavioural research. In the 1950s Tinbergen (1951) established it as a model system in ethology, and threespine stickleback continue to play a prominent role in ethology and behavioural ecology (van den Assem and Sevenster 1985; various chapters this volume). The development of ethology in western Europe, where threespine stickleback are abundant and easy to obtain, led to nearly exclusive use of populations from this area in early ethological research. Use of populations from this restricted area, combined with the imposition of uniform laboratory environments during ethological observations, created an impression of extreme uniformity of threespine stickleback behaviour.

In contrast, systematic ichthyologists, sampling widely throughout the Northern Hemisphere, were confronted with a bewildering array of morphological variation among threespine stickleback. This variation proved intractable from a taxonomic perspective, resulting in description of more than 40 junior synonyms to *G. aculeatus* after the original description by Linnaeus (1758). Regan (1909) argued that most populations in both the Pacific and Atlantic basins should be treated as a single variable species. Since 1925, when Bertin synonymized most nominal species of *Gasterosteus* with *G. aculeatus*, it has become common practice not to recognize phenotypic differentiation of threespine stickleback populations taxonomically, or to recognize only a few variable subspecies (Miller and Hubbs 1969; Buth *et al.* 1984; Buth and Haglund page 76 this volume).

The striking contrast between morphological diversity and seeming behavioural uniformity of threespine stickleback seems in retrospect to have demanded immediate explanation. However, the resolution of this paradox did not emerge until the late 1960s, when Hagen (1967) and McPhail (1969) began to study speciation and isolating mechanisms between biological species of threespine sticklebacks. It was immediately clear to Hagen and McPhail (1970) that the morphological diversity that had overwhelmed systematic ichthyologists was only the tip of the iceberg. Hagen and McPhail's focus on speciation also demanded integrated studies of morphological and genetic divergence, of the ecological significance of this divergence, and of behavioural and life history differences responsible for reproductive isolation. In addition to providing fundamental insights into the difficult problem of the nature of isolating mechanisms, this work borrowed methodology from the ethological school and provided the first glimpse of behavioural variation among threespine stickleback populations. Subsequent work by Wilz (1973) and by Huntingford and co-workers (e.g. Huntingford 1982; Giles and Huntingford 1984; Huntingford *et al.* Chapter 10 this volume; see also Foster Chapter 13 this volume) has brought the remarkable extent of interpopulation behavioural diversity in the threespine stickleback complex into clear view. A goal of this book is to clarify the interrelationships of stickleback ethology, systematics, and population biology.

GEOGRAPHICAL DISTRIBUTION

Threespine stickleback are restricted to the Northern Hemisphere and occur around the margins of the Atlantic and Pacific oceans (Fig. 1.1; reviewed in Wootton 1976, 1984*a*; Bell 1984*a*; Paepke 1983). They are distributed across the northern rim of the Atlantic, from the Iberian Peninsula through the British Isles to Iceland and southern Greenland, and south along the coast of North America to Chesapeake Bay (Hardy 1978; Lee *et al.* 1980). Freshwater populations occur throughout most of this range, but are usually absent south of Maine, USA, the northern limit of many primary freshwater fishes (Lee *et al.* 1980).

Marine *G. aculeatus* must have penetrated the Mediterranean Sea during the Pleistocene (Bell page 444 this volume) but is largely (Crivelli and Britton 1987) restricted to fresh water south of the English Channel (Münzing 1963). Freshwater populations are widely distributed along the European coast of the Mediterranean, and isolates occur in south-western Turkey, Syria, and North Africa (e.g. Regan 1909; Krupp and Coad 1985; Arnoult 1986). *Gasterosteus aculeatus* is present in the Black Sea and penetrates inland waters across eastern Europe north-west to the Baltic Sea. Both marine and freshwater populations also extend from the North

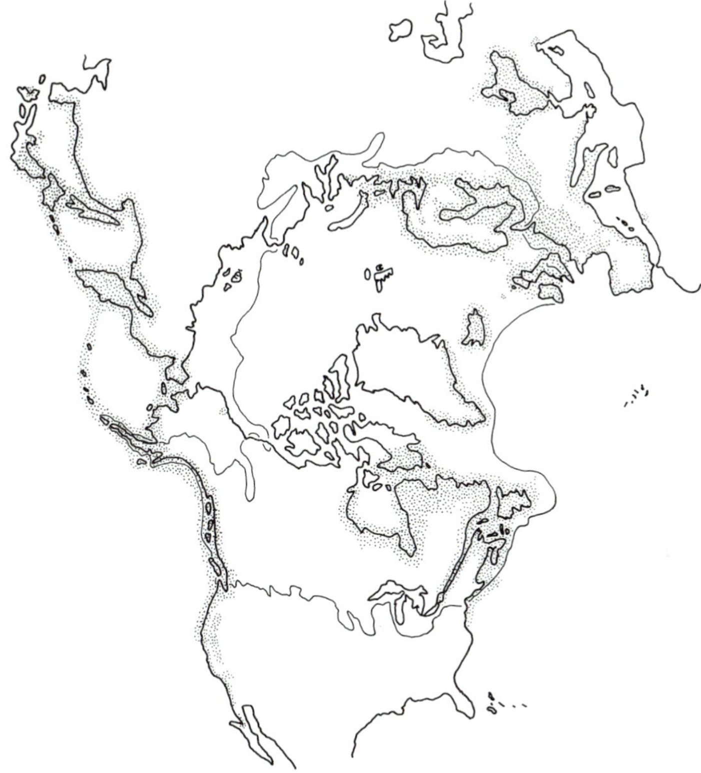

Fig. 1.1 North polar projection of the world showing the distribution (stippled) of the *Gasterosteus aculeatus* species complex. Ranges are based on maps in Berg (1965), Wootton (1976, 1984), Paepke (1983), Bell (1984), and Lelek (1987), and information from other papers cited in the text. The fine line is the approximate southern limit of contiguous continental glaciation and permanent sea ice during the last glacial maximum.

Atlantic, east into the Arctic Ocean as far as Novaya Zemlya, and west to south-eastern Baffin Island and most of Hudson Bay.

In the Pacific basin, threespine stickleback occur from Baja California, Mexico (Miller and Hubbs 1969), northward along the coast of North America, across the Bering Strait and south-west along the coast of mainland Asia and Japan to the south-west coast of Korea (Berg 1965). In contrast to the Atlantic basin, the northern Pacific range of *G. aculeatus* seems to end abruptly at the Bering Strait, except for an isolated population on the Arctic coast of Alaska (McPhail and Lindsey 1970; Morrow 1980; Lindsey and McPhail 1986). Marine populations are absent in the eastern Pacific south of about San Francisco Bay, California, USA (Miller and Hubbs 1969).

Marine and freshwater populations are present in Japan, but the southern limit of marine populations in Asia is unclear. Unlike in central Europe, freshwater populations do not appear to penetrate significantly into interior fresh waters of Asia and North America.

ECOLOGICAL DISTRIBUTION

Threespine stickleback occur in a wide range of aquatic habitats. The complex includes marine, anadromous, and resident freshwater populations. *Gasterosteus aculeatus* has been collected in the open ocean throughout much of its range (Jones and John 1978; Quinn and Light 1989), and has been reported up to 110 km from the Atlantic coast of New York, USA (Cowen *et al.* 1991) and 800 km from shore in the Gulf of Alaska (Morrow 1980). It is not clear whether these records represent strays that are lost to the breeding population or are part of a regular migration pattern. However, threespine stickleback are absent from open ocean habitats near the southern extremes of their distribution in western Europe (Münzing 1963; Crivelli and Britton 1987) and western North America (Miller and Hubbs 1969; Bell 1976*a*, 1979). In contrast, their most southerly records in eastern North America are marine (Hardy 1978; Lee *et al.* 1980; Bell and Baumgartner 1984). *Gasterosteus* evidently always constructs a benthic nest (Wootton 1984*a*), even if the eggs are not brooded (Blouw and Hagen 1990), and populations that spend most of their life cycle in the open ocean presumably must return to coastal habitat or fresh water, where they spawn and the young spend the first weeks of life. However, only Picard *et al.* (1990) have actually observed movement of adult *G. aculeatus* onshore from open estuarine waters to river and salt-marsh breeding grounds.

Threespine stickleback inhabit diverse fluvial environments. At one extreme, they occupy small floodplain potholes and tiny ephemeral streams in arid southern California, USA (Miller and Hubbs 1969; Baskin 1975; Bell unpubl. data). They have been reported from permanent flowing waters with a wide range of sizes (e.g. Heuts 1947*a*; Hagen 1967; McPhail 1969; Miller and Hubbs 1969; Baumgartner and Bell 1984). Picard *et al.* (1990) sampled threespine stickleback well off shore in the St Lawrence Estuary, but there seem to be no other studies that included systematic sampling for *G. aculeatus* in offshore habitats of large rivers. Thus, the extent to which they occupy midstream portions of more rapidly flowing inland reaches of rivers is unclear. Threespine stickleback evidently do not tolerate high-gradient streams, and rarely occur in fluvial habitats more than a few hundred metres above sea level (Hardy 1978; Bell 1982).

Gasterosteus aculeatus occurs in many thousands of separate lentic habitats, over a wide range of sizes. For example, Bell and Baumgartner (1984) reported a population from a spring pool measuring 10 × 26 m

(0.026 ha), and at the other extreme, Hubbs and Lagler (1970) reported *G. aculeatus* from Lake Ontario, North America, with an area of 196 840 ha. Threespine stickleback use a variety of lacustrine habitats and resources (e.g. McPhail 1984, page 418 this volume; Lavin and McPhail 1985; Schluter and McPhail 1992).

GENERAL CHARACTERIZATION OF THE THREESPINE STICKLEBACK

The variability that makes the threespine stickleback such a fascinating subject for evolutionary studies dooms to failure any attempt to characterize it briefly. Although several phenotypic features are shared by the vast majority of populations in the *G. aculeatus* complex (e.g. Wootton 1976, 1984*a*; Bell 1984*a*; Bowne pages 29–36 this volume), there are major exceptions to any generalization in at least a few populations (Fig. 1.2). Nevertheless, it is important to give a general description to introduce the organism.

The 'typical' threespine stickleback is a small streamlined fish, about 5 cm from the tip of the snout to the end of the vertebral column (i.e. standard length, SL, *sensu* Hubbs and Lagler 1970), but SL ranges up to about 110 mm. The pectoral fin is fairly large compared with that in most other teleosts, but the general body form is not otherwise unusual. The threespine stickleback is distinguished most strongly by its bony armour. It has three dorsal spines, the first two of which are large, serrated, and isolated from the dorsal fin. The smaller anal and third dorsal spines are located immediately in front of the dorsal and anal fins. When maximally expressed, bony lateral plates form a single row along each side of the body, beginning as small ossicles above the pectoral girdle (cleithrum), broadening dorsoventrally to cover most of the lateral abdominal surface, and then tapering progressively toward the caudal peduncle, where they expand laterally to form a keel (Fig. 1.3(c)). The distribution and number of plates varies extensively within and among populations (reviewed in Bell 1984*a*; Wootton 1984*a*; Reimchen Chapter 9 this volume; see also below). The pelvic girdle is large and complex compared with that of most teleost fishes, and the ventral (pelvic) fin includes one large, serrated spine and one fin ray (Nelson 1971; Bell 1987; Bowne page 34 this volume). The large dorsal and pelvic spines lock in an erect position (Hoogland 1951). Together, the dorsal spines and their supports, the lateral plates, and the pelvic girdle enclose the abdomen in a flexible, segmented armour complex that protects the stickleback from vertebrate predation (Hoogland *et al.* 1957; Reimchen 1983 page 248 this volume).

General body colour is cryptic and varies according to habitat. As males come into reproductive condition, typically the eye and body take on a blue tinge, and the ventral surface of the head and trunk becomes red (Wootton 1976; Reimchen 1989). Expression of male reproductive coloration varies

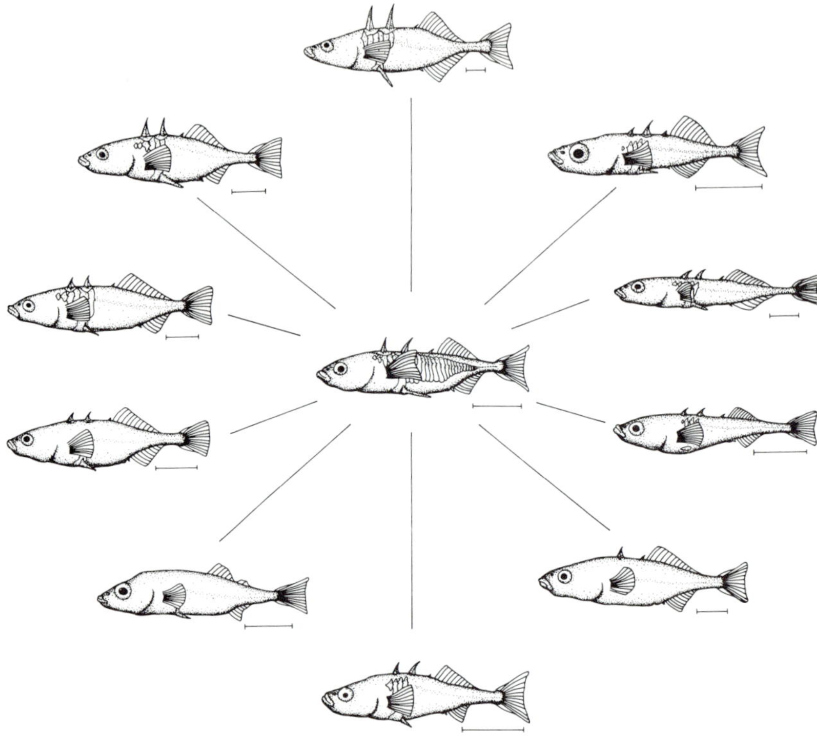

Fig. 1.2 Variation of body form and external features among North American populations of *Gasterosteus aculeatus*. Forms around the periphery have probably been derived independently from the marine ancestors, represented by the specimen shown in the middle. The specimen in the middle is an anadromous fish from Birch Cove, Cook Inlet, Alaska (AK), USA, and those around the periphery are from (clockwise from the top): Mayer Lake, Queen Charlotte Islands, British Columbia, Canada (BC); Enos Lake, Vancouver Island, BC (limnetic species); Twin Island Lake, Cook Inlet, AK; Bear Paw Lake, Cook Inlet, AK; Paxton Lake, Texada Island, BC (benthic species); Enos Lake, Vancouver Island, BC (benthic species); Rouge Lake, Queen Charlotte Islands, BC; upper Santa Clara River (*G. aculeatus williamsoni*) and lower Santa Clara River, California, USA; and Garcia River, California. The scale bars are 1 cm.

during the reproductive cycle (McLennan and McPhail 1989*a*). Reproductive females may experience less dramatic but biologically significant colour change (Rowland *et al.* 1991), and distension of the pale abdomen by a massive brood of eggs distinguishes reproductive females.

Threespine stickleback also exhibit a wide array of conspicuous, often ritualized behaviours. Aggression, territoriality, courtship, and male parental behaviour have long been employed to address basic problems in

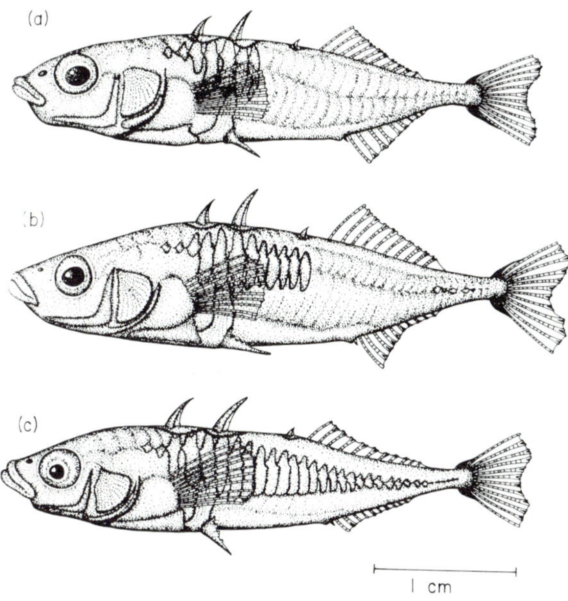

Fig. 1.3 Lateral plate morphs of *Gasterosteus aculeatus* (Sports Lake, Alaska, USA): (a) low, (b) partial, (c) complete.

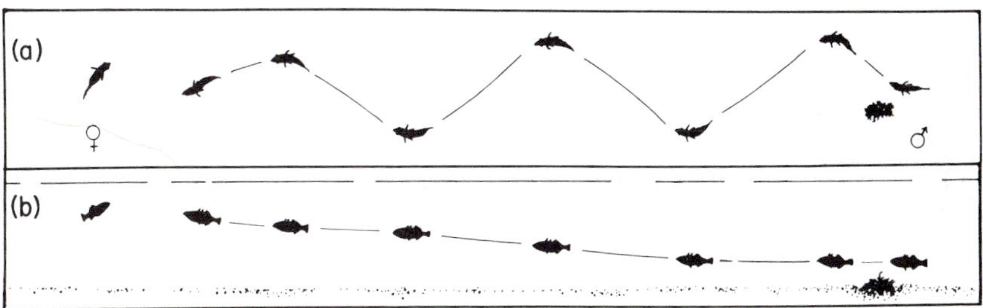

Fig. 1.4 Diagram of the zigzag dance of *Gasterosteus aculeatus* in (a) plan and (b) side view. The sequence begins with the male at the right near his nest and the female at far left. Each image of the male represents a brief pause in his progress toward the female. Details of the dance (e.g. total distance travelled, distance between turns, and height in the water column) vary among populations.

behaviour (Rowland page 313 this volume). Best known are male courtship and parental behaviours, which include biting, zigzagging, dorsal pricking, ritual and paternal nest fanning, showing, and fry retrieval. The male zigzag dance (Fig. 1.4) is a highly ritualized courtship behaviour, elements of which are shared with several stickleback species, whereas dorsal pricking (Fig. 1.5)

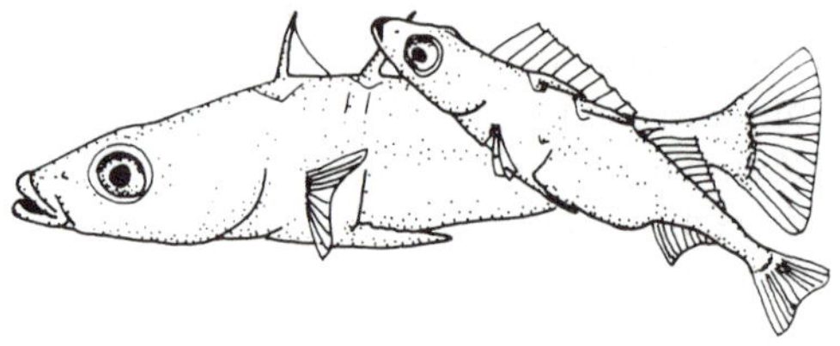

Fig. 1.5 Dorsal pricking of a female (*smaller*) by a male.

appears to be unique to *G. aculeatus* (McLennan *et al.* 1988; Rowland page 315 this volume). Female behaviour plays an integral role in courtship but is less well studied (Rowland Chapter 11 this volume). Threespine stickleback foraging (Hart and Gill Chapter 8 this volume) and predator avoidance behaviour (Huntingford *et al.* Chapter 10 this volume) also have been used to study basic problems in behaviour, ecology, and evolution.

MAJOR DIMENSIONS OF PHENOTYPIC VARIATION IN THREESPINE STICKLEBACK

Although it is simple to give a superficial description of the threespine stickleback, many populations deviate significantly in some respect. In this section we note the major dimensions of phenotypic differentiation among stickleback populations and establish terminology for some of the variable features. We focus on morphological phenotypes because they are most easily studied and have attracted the most attention to date. However, behavioural, physiological, and ecological phenotypes also exhibit interesting patterns of variation and form the subject of several chapters in this volume.

General body form and fin variation

General body form and fin morphology vary greatly in *G. aculeatus* (Fig. 1.2, 8.1) but have only been superficially characterized. Most studies only present body depth, which is expressed either as a mean ratio with SL (e.g. Moodie and Reimchen 1976*a*) or as a mean length adjusted to some arbitrary size using a regression equation (e.g. McPhail Chapter 14 this volume). The only exhaustive analysis of body form in *G. aculeatus* (Baumgartner *et al.* 1988) confirmed the utility of body depth as the major dimension of variation in lacustrine threespine stickleback populations (McPhail page 419 this volume) but also revealed other aspects of body form variation.

The pectoral fin is the primary propulsive organ for routine locomotion (labriform swimming) in threespine stickleback, and it is also used by males to fan water through the nest. It almost always has ten soft rays, but its relative size and shape vary in relation to the locomotor demands of foraging, migration, and the hydrodynamic properties of the environment (Baumgartner 1986*a*). The pectoral fin and musculature are conspicuously enlarged and the fin is more falcate in anadromous stickleback than in resident freshwater populations (Taylor and McPhail 1986).

Analysis of body form in threespine stickleback has been limited, and simple methods have generally been used. Use of ratios between structures should be avoided, and 'size correction' using SL should be employed with caution (Bookstein *et al.* 1985; Baumgartner *et al.* 1988). Recent development of powerful morphometric methods (Rohlf and Bookstein 1990) offers exciting opportunities for more precise and informative analysis of body form evolution in the threespine stickleback species complex.

Variation in armour structure

The conspicuous bony armour of threespine stickleback has received more attention from evolutionary biologists than have other aspects of the stickleback phenotype. Armour was employed initially to diagnose species (reviewed in Wootton 1984*a*: 252), next in zoogeography (Münzing 1963; Miller and Hubbs 1969), and finally as a tool in ecological genetics. In particular, a series of studies by Hagen and his colleagues (Hagen and Gilbertson 1972, 1973*a*,*b*; Kynard 1972, 1979*a*; Hagen 1973; Moodie *et al.* 1973) on geographic variation, genetics, and fitness effects of lateral plate number phenotypes in low plate morphs (see below) yielded important insights into their evolution and established the explanatory power of ecological genetics for armour variation in this group. Ecological genetics (e.g. Creed 1971; Ford 1975) has become a major theme in threespine stickleback biology but has seen limited use in other sticklebacks (e.g. Nelson 1977; Reist 1980*a*,*b*, 1981; Blouw and Hagen 1981, 1984*a*,*b*,*c*; Hagen and Blouw 1983).

Lateral plate morphs

The lateral plate series almost always (Ziuganov 1983; Francis *et al.* 1985) falls into three major morphs (Fig. 1.3), within each of which plate number also varies. Hagen and Gilbertson (1972) introduced the terms complete, partial, and low morphs, which are used uniformly throughout this volume for the lateral plate trimorphism. The terms trachura, semiarmata, and leiura pre-dated Hagen and Gilbertson's terminology for plates (Bakker and Sevenster 1988), but the utility of these terms has been compromised by their inconsistent application to plate morphology and life history modes.

A series of about 36 plates runs along the length on each side of the body in complete morphs (Bell 1981). Partial morphs have an unplated region separating an abdominal plate row and plates on the caudal peduncle, where the keel may be weakly expressed (Hagen and Gilbertson 1973*a*). Low morphs tend to have fewer (usually ≤ 7) abdominal plates than partial morphs, and they lack plates behind the abdomen. Hagen and Gilbertson (1973*a*) presented a model for lateral plate morph inheritance, and plate morphs clearly have a strong genetic basis. However, the general applicability of Hagen and Gilbertson's (1973*a*) model has been questioned (Ziuganov 1983; Francis *et al.* 1985), and plate morph inheritance clearly varies among populations (Avise 1976). Marine, anadromous, and estuarine populations usually are monomorphic or are dominated by the complete morph (but see Bell 1979; Crivelli and Britton 1987). Plate morph frequencies vary in fresh water in relation to climate (Hagen and Moodie 1982; but see Coad 1983) and more locally with stream velocity (Baumgartner and Bell 1984), but selection of plate morphs is poorly understood (MacLean 1980).

Low morph lateral plate number

The number of lateral plates also varies within each plate morph (e.g. Heuts 1947*a*; Bell 1981), but variation in the number of plates of low morphs has received the most attention (reviewed in Bell 1984*a*; Reimchen Chapter 9 this volume). Low morph plate number phenotypes vary extensively within and among populations. Individual plate counts range from zero to nine per side, but usually range between four and seven. Most workers report the number of plates on one side; the sum for both sides is less frequently reported. The series of studies by Hagen and his colleagues cited above shows that lateral plate phenotypes have a strong genetic basis and vary geographically in relation to the distribution of predatory fishes, which are selective predators on plate number phenotypes.

Reimchen (1983) proposed recognition of individual plates of low morphs according to their position on the abdomen. He demonstrated functional differences among plates and proposed a numbering system for abdominal plates. This approach facilitates establishment of relationships between plate number variation, functional morphology, and selection in natural populations by different types of predators.

Dorsal spines

Dorsal spines vary in number, placement on pterygiophores, length, and degree of serration. Each of these properties has been studied (e.g. Gross 1978), but only spine length has been studied extensively (Hoogland 1951; Hagen and Gilbertson 1972; Moodie 1972*a*; Moodie and Reimchen 1976*a*; Gross 1978). Dorsal and pelvic spines tend to be longer in populations sympatric with predatory fishes, and they appear to interfere with ingestion by fishes and birds (Hoogland *et al.* 1957; Reimchen 1983, 1991*a*, Chapter 9 this volume). Until recently, the length of individual spines adjusted for fish size (SL) has been studied, but the distance between the base of the pelvic spines and body depth above the pelvis vary, influencing the effectiveness of the spines against predatory vertebrates. The functionally significant measure of spine size is the distance between the tips of the pelvic spines and the dorsal spine most nearly above the pelvic spines (Reimchen 1991*a*, Chapter 9 this volume).

Pelvic structure

Bowne (page 34 this volume) provides a detailed description of the complex pelvic structure of *G. aculeatus* (see also Nelson 1971; Bell 1987). It is a bilateral structure consisting of the pelvic plate, a spine, and a soft ray on each side (Fig. 2.4). The pelvic plate is formed by anterior and posterior processes in the ventral plane and the ascending branch, which extends dorsally along the lateral sides. The pelvic plates on either side are joined by a midventral suture. As noted above, the pelvic girdle, lateral plates, and dorsal spines form an armour complex that protects the abdomen against predatory vertebrates.

The pelvic girdle and spines are robust in most threespine stickleback populations (Bell 1987, 1988). Minor variations in the pelvic girdle, including its width, the number of distal forks on the ascending branch (usually one to three), the depth of the notch (if any) separating the anterior processes, and length and serration of spines, are ubiquitous (e.g. Regan 1909; Myers 1930; Penczak 1965; Hagen and Gilbertson 1972; Gross 1978). Extreme pelvic reduction (including complete loss) is relatively rare and largely restricted to lake populations in recently deglaciated regions (Bell 1987). Elements of the pelvic girdle are almost always lost in sequence:

(1) spines

(2) posterior process

(3) ascending branch

(4) anterior process.

The uniformity of this sequence is remarkable, considering the wide distribution of populations that exhibit it. Although there have been no detailed studies of the genetics of pelvic reduction in *G. aculeatus*, Blouw and Boyd (1992) recently showed that in the ninespine stickleback (*Pungitius pungitius*) it is heritable and not strongly influenced by variation of pH, dissolved calcium, or salinity. Pelvic reduction in natural populations of *G. aculeatus* is associated with absence of predatory fishes and low calcium ion concentration (Giles 1983*a*; Bell *et al.* 1985*a*, in press; Bell 1987).

Variation in trophic morphology

The snout and gill rakers vary in relation to food type. The snout is long and narrow in planktivorous populations (limnetics, *sensu* McPhail 1984, page 418 this volume) and relatively short and broad in stickleback that consume larger benthic prey (McPhail 1984 Chapter 14 this volume; Hart and Gill page 210 this volume). Gill rakers (Fig. 1.6) project from the branchial arches, forming a sieve structure between the pharynx and gill chamber. The

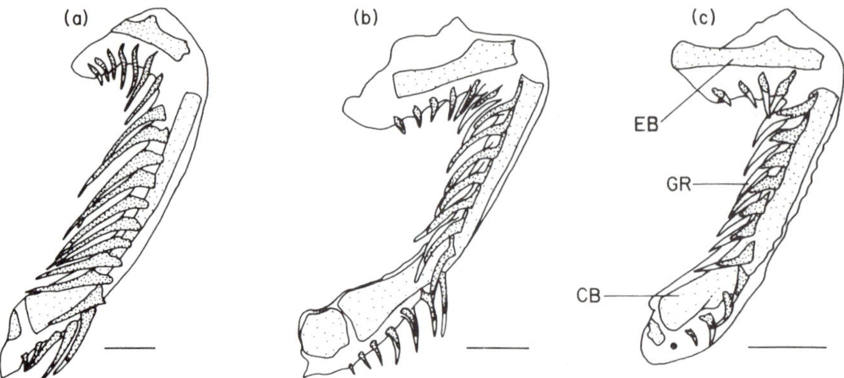

Fig. 1.6 Contrasting gill raker morphology on the first left branchial arch of *Gasterosteus aculeatus* from (a) marine (Birch Cove, Cook Inlet, Alaska, USA; 71.3 mm SL), (b) lacustrine (Twin Island Lake, near Big Lake, Alaska, USA; 64.7 mm SL), and (c) stream populations (Santa Clara River, near Acton, California, USA; 43.8 mm SL). Note the differences in number, length, and thickness of gill rakers. Anterior is to the left. The gill rakers rotate from the anterior to the posterior edge of the gill arch near the ventral end of the ceratobranchial bone. Gill filaments are not shown; bone is stippled. Abbreviations: CB, ceratobranchial; EB, epibranchial; GR, gill raker. The scale bars are 1 mm. See Fig. 2.3 for the location of these structures in the branchial skeleton.

size of spaces between gill rakers is inversely correlated with the number of gill rakers, and populations with high numbers of gill rakers eat smaller food items (Bentzen and McPhail 1984; Gross and Anderson 1984; Hart and Gill page 211 this volume). The mean number of gill rakers varies from a minimum of 14 in stream populations, which tend to eat larger benthic prey, to a maximum of about 24 in planktivorous lacustrine and marine populations, which also have longer gill rakers (e.g. Hagen and Gilbertson 1972; Gross and Anderson 1984; McPhail Chapter 14 this volume). Feeding habits have pervasive effects on phenotype that go beyond trophic structures, and we return to this topic at length later in this chapter.

Variation in body colour

Threespine stickleback are generally cryptic, with brown-to-green barring above and a more uniform pale colour below. In well-lighted, open-water habitats they are more silvery, and melanism is associated with deeply stained waters, either as year-round or as male nuptial coloration (McPhail 1969; Moodie 1972*a*; Hagen and Moodie 1979; Hagen *et al.* 1980; Reimchen 1989; McPhail page 412 this volume).

As males attain reproductive condition, they become less cryptic. The eye becomes iridescent blue, the dorsolateral surface turns iridescent blue to black, and red colour spreads progressively from the mouth to ventrolateral surfaces of the head, shoulder, and pelvic region (McLennan and McPhail 1989*b*). In some populations, red coloration may expand onto the flanks behind the pectoral fin or along the base of the anal fin (Bell 1984*a*; McLennan and McPhail 1989*a*). There is substantial population differentiation in the intensity, amount, and distribution of red nuptial coloration of males. Factors as diverse as parasite load, sexual selection, predation regime, heterospecific convergence for threat display, and water colour have all been implicated as causes of this variation in male nuptial coloration (reviewed in Reimchen 1989; Rowland Chapter 11 this volume). Existence of dorsolateral mottling or barring in sexually receptive females was reported for one population (Rowland *et al.* 1991). More research is needed to determine the functional implications of variation in nuptial coloration in male *G. aculeatus* and whether female reproductive coloration varies among populations.

PHYLOGENY OF THE THREESPINE STICKLEBACK SPECIES COMPLEX

Phylogenetic trees, bushes, and racemes

The extraordinary amount of phenotypic variation among freshwater populations of *G. aculeatus* is a recurrent theme in this volume. Comparison of this rampant intraspecific differentiation with the larger-scale pattern evident in supraspecific phylogeny reveals a curious paradox. The distinctive properties of freshwater populations evolve rapidly and reach remarkable extremes, and there is ample evidence that such population differentiation is an ancient phenomenon in *G. aculeatus*. Except for *G. wheatlandi*, however, no *Gasterosteus* lineage has undergone sustained divergence from the *G. aculeatus* species complex to become recognizable as a phenotypically distinctive species (Nelson 1971; Wootton 1976; Buth and Haglund page 81 this volume). Population differentiation is rapid and common, but does not lead to sustained divergence that culminates in formation of widespread, phenotypically distinctive species. The phylogenetic topology produced by this process does not resemble a conventional phylogenetic tree.

Use of botanical metaphors to illustrate phylogenetic topology is an old practice; phylogenetic trees, bushes, and even lawns are used to express evolutionary history. Williams (1992) recently extended the botanical metaphor for phylogeny, referring to 'phylogenetic racemes.' A raceme is an inflorescence with an elongated central axis about which flowers project symmetrically. A phylogenetic raceme comprises a phenotypically stable, temporally persistent lineage (the central axis) from which divergent lineages (the flowers) project symmetrically and rapidly terminate in extinction. Brown (1957, 1958) argued that a generalized species may give rise to numerous specialized isolates that are prone to extinction, a process that would produce a raceme. Endemic non-parasitic 'satellite species' of lampreys bear this relationship to the parasitic species from which they have been derived (Vladykov and Kott 1979). The *G. aculeatus* complex appears to form a huge phylogenetic raceme.

The phylogenetic raceme of the threespine stickleback species complex

Intraspecific phylogeny of the *G. aculeatus* species complex appears to fit the phylogenetic raceme model (Fig. 1.7), but several criteria must be met to demonstrate this fit. The central axis of the raceme must be formed by a persistent and conservative common ancestor. Numerous populations must be derived from this common ancestor, evolve their apomorphic states at a high rate, and have a very high probability of rapid extinction. In the remainder of this section, we discuss this model for the evolution of intraspecific diversification within the *G. aculeatus* complex.

Evolutionary polarity of the marine–fresh water transition

We must infer whether *G. aculeatus* is primitively a marine or a freshwater fish. Outgroup analysis is generally accepted as an effective method to infer

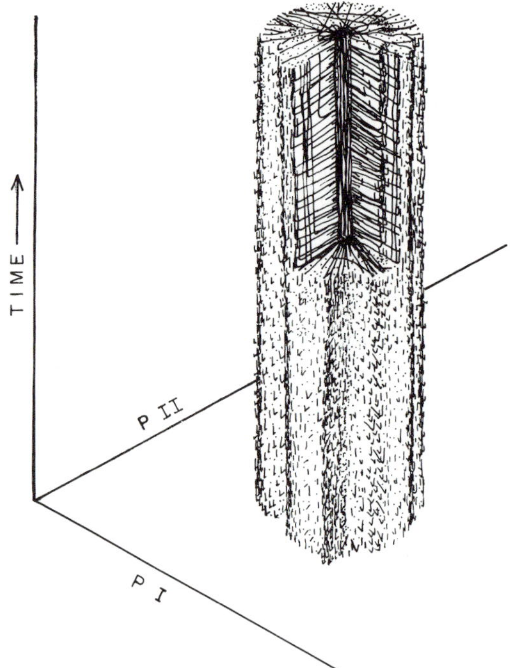

Fig. 1.7 The phylogenetic raceme (simplified) of the *Gasterosteus aculeatus* species complex, showing the process of iterative diversification in a hypothetical two-dimensional phenotypic space (abscissas: PI, PII) through time (ordinate). A section in the upper part of the raceme is cut away to reveal its internal structure, including the central stem of marine ancestors and the diverging freshwater isolates. See text for additional discussion.

the ancestral state of a character (reviewed in e.g. Wiley 1981; Maddison *et al.* 1984). Outgroup analysis assumes that the primitive state of a character for members of a group is the state shared with related species outside the group in question.

Moving from closer to more distant relatives of *G. aculeatus* (see Maddison *et al.* 1984), outgroup analysis indicates that the species complex is primitively marine. The blackspotted stickleback (*G. wheatlandi*), a phenotypically (Hubbs 1929; McInerney 1969; Nelson 1971; McLennan *et al.* 1988) and genetically distinct (Buth and Haglund page 82 this volume) but closely related stickleback (Wootton 1976; McLennan *et al.* 1988), is almost strictly marine (McAllister 1960; Sargent *et al.* 1984; Wootton 1984*a*). Among the other stickleback genera (Fig. 2.12), the brook stickleback (*Culaea inconstans*) is unique in being restricted to fresh water, but the ninespine (*Pungitius pungitius*) and the fourspine (*Apeltes quadracus*) sticklebacks are frequently marine; the most distantly related stickleback, the fifteenspine (*Spinachia spinachia*), is exclusively marine (Wootton 1984*a*). Similarly, two species of the family Aulorhynchidae, the sister group of the Gasterosteidae (Nelson 1971; McLennan *et al.* 1988), are also strictly marine, and other closely related families (Bowne this volume) are primarily marine (Nelson 1984). With the exception of the brook stickleback, several successive outgroups are either strictly or frequently marine, indicating that *G. aculeatus* is primitively marine. However, a point we must return to below is whether permanent freshwater residence evolved once (or a few times) and spread from a central point by dispersal through fresh water, or has evolved many times locally from marine populations.

Phenotypic stability and persistence of marine populations

Threespine stickleback appear in marine fossil deposits about 10 Ma (1 Ma = 1 million years) old (Bell page 444 this volume). Despite claims to the contrary (Sychevskaya and Grechina 1981), the earliest marine threespine stickleback do not appear to differ significantly from their extant counterparts and certainly do not approach the magnitude of divergence seen among freshwater populations (Bell 1977, page 446 this volume). Extant marine populations also appear to exhibit little geographical variation in morphological features. Although more work is needed on other characters, the armour structures, general body form, and trophic morphology appear to be relatively homogeneous among marine populations, except that there is lateral plate morph variation near their southern range limit (Münzing 1963; Black 1977). It is reasonable to conclude that the vast majority of freshwater populations are derived from a marine ancestor that closely resembled the specimen shown at the centre of Fig 1.1. Although more evidence concerning non-morphological phenotypes is needed, morphological evidence suggests that the common marine ancestor of freshwater populations forms a persistent and phenotypically stable central axis in the raceme.

Evidence of repeated derivation of freshwater populations from marine ancestors

Now we can return to the question of whether freshwater *Gasterosteus* evolved once (or a few times) and spread by dispersal through fresh water, or has evolved locally innumerable times. Three lines of evidence support the latter view:

1. Marine populations enter coastal marine or freshwater habitats to breed, suggesting the capacity for frequent colonization of fresh water;
2. The geographical distribution of freshwater populations is inconsistent with dispersal only through fresh water;
3. Genetic variation in marine and freshwater populations is consistent with the repeated derivation of freshwater populations from marine ones.

The natural history of marine and anadromous threespine stickleback populations favours repeated colonization of fresh water. Marine populations migrate from offshore habitats in spring to breed in tidal pools (Weeks 1985*a*,*b*; Picard *et al.* 1990), and a gradual decrease in sea level could transform these intertidal breeding areas into lowland lakes and freshwater wetlands to which marine populations would have ready access. Similarly, anadromous populations breed in streams (e.g. Hagen 1967; reviewed in Wootton 1976) and lakes (Ziuganov *et al.* 1987), and could easily become isolated in fresh water. Recently, Bell and Havens (unpubl. data) sampled a *G. aculeatus* population in a lake near Cook Inlet, Alaska, USA, that had been poisoned eight years before to eliminate stickleback. The new population closely resembled anadromous stickleback (i.e. complete lateral plate morphs, high gill raker number), suggesting that they had colonized the lake from adjacent streams in the interim. Thus, it is plausible that both marine and anadromous populations could frequently give rise to resident freshwater populations.

Occurrence of numerous freshwater populations of *G. aculeatus* on islands, fiords, and peninsulas that must have been colonized postglacially from the sea provides the most compelling evidence for their recent and frequent local origin. Freshwater stickleback habitats in many glaciated regions were totally covered by glaciers or submerged by isostatic depression under seawater during glaciation. Recolonization of these habitats via freshwater dispersal routes in the brief time since deglaciation is extremely unlikely (e.g. Münzing 1963; McPhail and Lindsey 1970; Bell 1976*a*, 1984*a*; Moodie and Reimchen 1976*b*; Campbell 1979, 1984; Ziuganov 1983; Bell *et al.* 1985*a*; McPhail page 401 this volume). Rather, the ubiquity of threespine stickleback in lowlands along the glacially defaunated coasts of north-western Europe (e.g. Münzing 1963; Gross 1978), Scotland (Giles 1983*a*; Campbell 1984), and north-western North America, and on many small islands in each of these regions (e.g. Hagen and Gilbertson 1972; Moodie and

Reimchen 1976*a*,*b*; Bell *et al.* 1985*a*; Reimchen Chapter 9, McPhail Chapter 14 this volume) is compelling evidence that fresh water was colonized independently in each area. Phenotypic divergence of these populations from marine ancestors and their frequent occurrence in land-locked lakes leaves no doubt that they are resident freshwater populations. If such distant regions were invaded independently, there is no reason to doubt that many adjacent drainages within recently deglaciated regions have been colonized from the sea independently of each other. The wide distribution of freshwater threespine stickleback in recently deglaciated habitats can be explained only by polyphyletic origin of resident freshwater habits in *G. aculeatus*. Populations in continuously unglaciated habitats to the south probably have similar phylogenetic histories, but they may have been in fresh water longer and may have spread more widely from the point of colonization.

Genetic variation among freshwater and marine populations in south-western British Columbia, Canada, is also consistent with recurrent colonization of fresh water. If freshwater populations have been derived from marine ancestors numerous times recently, they should be less polymorphic than marine populations, and individual populations should contain a random sample of the polymorphic loci present in adjacent marine populations (Bell 1976*a*). Withler and McPhail (1985) compared allozyme variation among samples from freshwater and anadromous populations at 56 sites in south-western British Columbia. Anadromous populations are more polymorphic and less spatially heterogeneous than freshwater populations, which are strongly differentiated. Although other explanations are possible (Withler and McPhail 1985), these results are consistent with polyphyletic derivation of freshwater populations from marine ones. Similar results have been obtained from an analysis of allozyme variation in anadromous sockeye salmon and kokanee (*Oncorhynchus nerka*), their resident freshwater descendants (Foote *et al.* 1989).

Rapid divergence and extinction of freshwater populations

Both phenotypic diversity of threespine stickleback in habitats that must have been colonized postglacially (reviewed in e.g. Bell 1984*a*, 1987; Campbell 1984; Chapters by Reimchen, Huntingford *et al.*, Foster, and McPhail this volume) and the fossil record (Bell *et al.* 1985*b*; Bell page 470 this volume) demonstrate rapid divergence in this form. Morphological, behavioural, life history, and physiological traits are all affected. The most strongly divergent freshwater phenotypes tend to occur in lakes, which are insular, ecologically unstable, and short-lived compared with the marine environment (Bell 1987, 1988). The difficulty of moving between lakes, combined with their tendency to change rapidly or disappear, makes extinction likely (Pease *et al.* 1989). The most divergent populations in periglacial regions face the additional hazard of extirpation by glacial advance. Thus, freshwater populations of *G. aculeatus* diverge rapidly and have a high probability of extinction.

The antiquity of invasion, divergence, and extinction of freshwater populations

Even if the marine populations are old, and postglacial freshwater populations have arisen often and diverged rapidly, the phylogenetic raceme requires that this process of differentiation be ancient. Fossil freshwater stickleback exhibit preservable apomorphic phenotypes that are restricted to freshwater populations today (Bell page 446 this volume). For example, pelvic girdle reduction occurred in two or three Miocene lineages in south-western USA and has evolved postglacially dozens of times (Bell 1987, 1988; Bell *et al.* in press). Periglacial environments, which appear to be the habitats most conducive to stickleback diversification, have existed since early to middle Miocene times in western North America (Plafker and Addicott 1976). The potential for invasion of fresh water and rapid evolution of pelvic reduction are both ancient, and it appears that differentiation of many other traits in freshwater populations is ancient.

Fit of threespine stickleback phylogeny to a raceme

The phylogenetic raceme shown in Fig. 1.7 should be a good general rendition of threespine stickleback phylogeny. The horizontal axes summarize phenotypic variation in a two-dimensional 'phenotypic space' and the vertical axis is time, beginning at least 10 Ma ago (possibly 16 Ma; Bell page 441 this volume). The central stem of the raceme represents the phenotypically stable marine populations from which freshwater populations, the flowers of the raceme, diverge rapidly after invasion of fresh water. The nearly perpendicular angle of the flowers relative to the axis represents extremely rapid divergence in freshwater lineages, which is indicated by the abundance of highly divergent young populations in recently deglaciated regions. The flowers of the raceme fill the surrounding phenotypic space *more or less* symmetrically because most traits can increase or decrease in value after colonization of fresh water. For example, the strength of armour structures, number of gill rakers, body size, and relative body depth can increase or decrease in freshwater populations relative to values in the marine ancestor. At any time since the *G. aculeatus* complex arose, its occupation of phenotypic space should be *relatively* constant, not because a large set of population lineages have diverged and then persisted at fixed positions in the phenotypic space for a long time, but because the two-dimensional phenotypic space has been occupied continuously through a *steady-state process of divergence and extinction*.

The *G. aculeatus* species complex exhibits some deviations from the strict topology of a raceme. Many flowers split, reflecting divergence of populations after their common ancestor colonized fresh water. Other flowers wrap around the stem, representing evolutionary reversals in response to environmental change or dispersal into a different freshwater habitat. The density of lineages radiating in different directions from the axis and the distance they diverge through the phenotypic space may vary. However, this variation in distribution about the axis should be uniform along the length of the raceme if biases in the direction and extent of divergence persist through time. Finally, even the central axis of the raceme is not a fully integrated unit, instead resembling a cable with strands of wire (Dobzhansky 1951; Anderson 1936) that represent the separate marine lineages (Haglund *et al.* 1992*a*; Buth and Haglund page 78 this volume) whose phenotypic stability and cohesion are maintained by stabilizing selection in a relatively stable and homogeneous marine environment (Ehrlich and Raven 1969). The phylogenetic topology of the *G. aculeatus* species complex deviates strikingly from the classical arborial rendition of phylogeny but conforms well to the structure of a raceme.

The paradox of raceme formation without taxonomic diversification

In view of the frequency and magnitude of postglacial diversification today, and the existence of the *G. aculeatus* species complex and conditions that would favour its diversification since the middle Miocene, numerous old, well-defined stickleback lineages ought to exist. Instead, there are only five well-defined stickleback genera, and *G. wheatlandi* appears to be the only old, widespread species (Buth and Haglund page 78 this volume) to have diverged from the *G. aculeatus* complex. The 'white stickleback' (Blouw and Hagen 1990) is widely distributed in Nova Scotia, Canada, but is so similar genetically to *G. aculeatus* (Haglund *et al.* 1990; Buth and Haglund page 76 this volume) that it probably evolved from it postglacially and may be an ephemeral species (Brown 1957, 1958). After at least 10 (possibly 16) million years of population differentiation, only one morphologically distinctive, geographically widespread species has emerged from the raceme and persisted to the present!

This situation probably reflects the ecological distribution of threespine stickleback. Because marine populations are free to shift their range in response to environmental change, they need not experience environmental change. By moving in response to environmental change, they experience an ecologically homogeneous, spatially continuous, temporally stable and persistent environment. In contrast, freshwater populations occupy ecologically diverse habitat patches that are unstable and ephemeral, and between which dispersal is severely limited. Theoretical results (Pease *et al.* 1989) indicate that evolutionary stasis and phyletic persistence ought to occur in marine populations, but that specialization followed by extinction are probable in freshwater populations. In the long run, only the evolutionarily conservative marine populations are viable, a situation that represents stabilizing selection among populations. Thus, the threespine stickleback complex provides one of the few clear cases of selection among populations (Bell 1987, 1988).

Epistemological implications of the phylogenetic raceme for comparative studies of freshwater threespine stickleback populations

Similarities among related groups may reflect either retention of the phenotypic state of their common ancestor (i.e. symplesiomorphy) or independent evolution of this state subsequent to their divergence from that ancestor (i.e. homoplasy of autapomorphies). Distinguishing between these alternatives is difficult but crucial in intraspecific comparisons (e.g. Ridley 1983; Felsenstein 1985; Sillen-Tullberg 1988; Donoghue 1989; Maddison 1990; Baum and Larson 1991; Brooks and McLennan 1991; Harvey and Pagel 1991). If the similarity reflects symplesiomorphy, it evolved only once and its distribution among habitats or associations with other traits may be an accident of phylogeny. In contrast, if the similarity is due to homoplasy, such distributions and associations are phylogenetically independent and must reflect some cause other than possession by the common ancestor. The latter situation permits tests for association of character states (phenotypes) with other characters or environmental variables to infer evolutionary causation.

Demonstrating independent origin of character states in freshwater threespine stickleback is relatively simple and does not require formal cladistic analysis. As marine populations are relatively homogeneous and freshwater populations are diverse, many derived (apomorphic) character states of freshwater populations are absent in the marine ancestor. Post-glacial freshwater populations have had little time for dispersal through fresh water and are often sufficiently isolated by distance or geographic barriers that they must have evolved independently of each other from marine ancestors. Any character states they share that are absent in the marine ancestor must be homoplasies and can be treated as statistically independent events. The degree of isolation necessary to assume independent derivation from marine ancestors varies directly with geographical relief and tectonic stability, and inversely with the age of the terrain (Bell *et al.* 1985*a*; Bell 1988; Gach and Reimchen 1989). An added advantage of using *G. aculeatus* is that the ancestral character states still occur in marine populations, and thus genetics, development, function, and fitness effects of ancestral states can be studied. The magnitude and polarity of character transformation between the common marine ancestor and freshwater stickleback can be inferred for many characters in numerous independently evolving freshwater populations. Such analyses would be impossible or difficult in a group whose phylogeny does not form a raceme.

Implications of the phylogenetic raceme for interpretation of the fossil record

The topology of a raceme, with iterative evolution of very similar character states at different times in the history of the group, may be a serious complication for interpretation of fossil sequences (Bell 1987, 1988). Similar fossil samples from adjacent stratigraphic levels are commonly interpreted to have had ancestor–descendant relationships (Gingerich 1979, 1990). A sequence of fossil stickleback samples that accumulated in isolated, ecologically similar freshwater environments might have very similar phenotypes, even if their similarities evolved independently from marine ancestors. Consequently, this sequence of lacustrine sticklebacks would be interpreted to represent evolutionary stasis instead of the rapid succession of divergence and extinction events that had actually occurred. The generality of this problem in fossil taxa is unclear, but it is a cause for concern.

OTHER IMPORTANT ATTRIBUTES OF *GASTEROSTEUS ACULEATUS* FOR EVOLUTIONARY STUDIES

The threespine stickleback's abundance and wide distribution in the Northern Hemisphere makes it readily available for research. Within limited areas it may occupy diverse habitats and be strikingly differentiated, providing excellent opportunities for comparative studies. Because it is small, large samples of threespine stickleback can be collected and maintained live in the laboratory, and stickleback husbandry is well developed (e.g. Leiner 1960; Lindsey 1962; Hagen 1967, 1973; McPhail 1977, 1984). The high densities at which threespine stickleback frequently occur also allow field observation of large numbers of individuals within a restricted area over a short period of time. Extensive research on threespine stickleback morphology, behaviour, physiology, and ecology provides a source of evolutionary hypotheses and guidance for the design of laboratory and field studies, and experimental methodology. Books by Wootton (1976, 1984*a*) and Paepke (1983, in German) summarize this research, and Coad's (1981) bibliography provides access to the older literature.

RESEARCH PROGRAMMES USING THE THREESPINE STICKLEBACK

The research summarized throughout this volume was usually performed to address basic biological problems, and *G. aculeatus* will continue to play an important role in this kind of research. However, the threespine stickleback is suitable for research that draws on diverse aspects of an organism's biology to develop insights that would be impossible using most species. Below we discuss two research programmes that depend on a rich background of biological information for one group. 'Multidimensional analysis' (Wake and Larson 1987; Bell 1987) focuses on a single character, and may be thought of as a vertical approach that incorporates information ranging from genetic mechanisms to fitness effects, in order to reveal how processes at various levels interact in character evolution. 'Phenotypic integration' is orthogonal to multidimensional analysis in that it focuses on a single level, that of the phenotype, and is concerned with how different phenotypic traits interact in evolution. Thus, these research programmes are complementary.

Multidimensional analysis of evolution

Multidimensional analysis is an amalgamation of structuralist and neo-Darwinian traditions in evolutionary biology and is aimed at understanding interactions among processes that determine the evolution of a specific morphological character (Wake and Larson 1987). The character is compared among species of a monophyletic group using an explicit phylogeny. Variation of the character within the group is explained by the interactions of genetic and developmental processes, adult structure, function, and natural selection. Tinbergen (1963) advocated a similar approach for analysis of behaviour. Relying on published information and new data on geographical variation, Bell (1987, 1988) performed a multidimensional analysis of vestigial pelvic girdle phenotypes in *Gasterosteus*. The existence of post-glacial populations with pelvic reduction and of a fossil record of pelvic reduction adds a temporal perspective to this multidimensional analysis. This research programme could be applied to other traits of threespine stickleback to develop a clearer understanding of the interactions of diverse biological processes in evolution.

Phenotypic integration

'Phenotypic integration' emphasizes the interactions of multiple characters in response to contrasting selection regimes of separate populations (see also Kingsolver and Schemske 1991). The term phenotypic integration is adopted from Olson and Miller's (1958) *Morphological integration* and is a conceptual extension from its focus on morphology to encompass other aspects of the phenotype. When similar coordinated responses to specific selection regimes are observed for multiple characters among several populations in which the character complex has evolved independently, the similarity among populations may reflect linkage or pleiotropy, possibly mediated by shared developmental pathways (Maynard Smith *et al.* 1985), functional interactions of the characters, shared patterns of phenotypic plasticity, or independent responses of the characters to selection. Thorpe (1976) and Sokal (1978) proposed comparison of character correlations within and among populations to distinguish natural selection from other causes for interpopulation correlation (see also Francis *et al.* 1986 for an application to *G. aculeatus*). Statistical procedures from evolutionary genetics (e.g. reviewed in Endler 1986; Barton and Turelli 1989) and morphometrics (Rohlf and Bookstein 1990) can be used to analyse character correlations, to detect factors underlying multicharacter variation, and to control individual variables to minimize confounding effects. These procedures are necessary for analysis of phenotypic integration.

McPhail and his collaborators (McPhail page 418 this volume) have developed clear evidence for phenotypic integration in lacustrine threespine stickleback from British Columbia, Canada. *G. aculeatus* is strictly carnivorous, but lacustrine populations vary greatly in the proportion of plankton

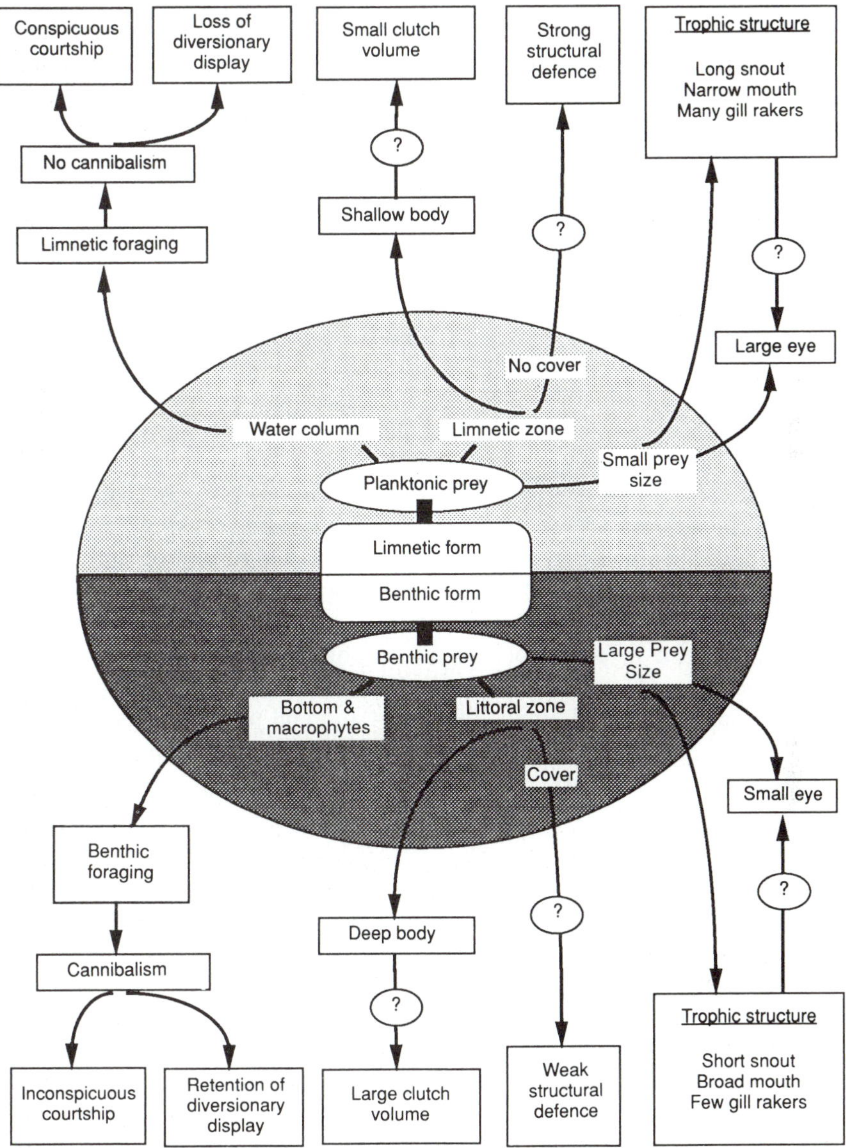

Fig. 1.8 Evolutionary consequences of planktivory (limnetic form) and benthic feeding (benthic form) for phenotypes of lacustrine *Gasterosteus aculeatus*. Arrows marked by question marks are hypothetical or poorly studied, and unmarked arrows have more extensive empirical support (see text).

and benthos they eat. Although there is a continuum of prey use, 'limnetic' populations specialize on plankton whereas 'benthic' populations prefer benthic prey (Bentzen and McPhail 1984; Schluter and McPhail 1992); McPhail page 422 this volume). Limnetic and benthic populations are divergent for traits that are directly related to feeding, but several other traits appear to diverge as an indirect consequence of trophic differentiation (Fig. 1.8).

Primary adaptations of lacustrine threespine stickleback

The primary adaptive differences between limnetic and benthic stickleback have to do with vision and trophic structures. Limnetic populations feed on small planktonic prey, and the ability to resolve prey is limited by the properties of the eye in small fish (Endler 1978; Hairston *et al.* 1982). In bluegill sunfish, for which there is good information, visual acuity increases with body size. Although the spacing of receptor cells remains constant during growth in bluegill, increased eye size increases the acuity of larger fish (Hairston *et al.* 1982), and the rate of increase with size is higher in smaller fish (Breck and Gitter 1983). Thus, slightly larger relative eye size of limnetic stickleback (McPhail 1984; Baumgartner *et al.* 1988) can significantly increase the volume of water searched or ability to detect small prey. The larger prey of benthics may reduce demand for visual resolution and favour evolution of larger mandibular musculature, which might compete with the eye for space in the head (Barel 1984).

Prey size also affects evolution of trophic morphology in threespine stickleback (e.g. Hagen and Gilbertson 1972; McPhail 1984, Chapter 14 this volume; Schluter and McPhail 1992; Hart and Gill page 211 this volume). In limnetics, the mouth is relatively narrow and the snout is long for suction feeding, and the gill rakers are long and numerous (Fig. 1.6) to prevent escape of small prey. Benthic populations have relatively broad mouths and short snouts, which are suitable for capture of relatively large benthic prey, and their short and widely spaced gill rakers suffice to prevent escape of large benthic prey without unnecessarily impeding flow of water for respiration into the gill chamber. Similar trophic divergence has been recognized in a wide variety of teleost fishes.

Secondary adaptations of lacustrine threespine stickleback

Limnetic stickleback feed in open water on vulnerable, dispersed prey, and streamlining should increase efficiency of locomotion in this environment. In contrast, benthic stickleback occupy a structurally complex environment and prey on larger, potentially more active prey. Manœuvrability should be important for benthic stickleback. Greater body depth, which increases manœuvrability, characterizes benthic populations. Of course, other factors, such as size-selective predation (Reimchen 1991*a*, page 271 this volume) and reproductive success (see below), may interact with selection of body form based on feeding.

Limnetics and benthics forage in entirely different contexts. Limnetics search mid-water for planktonic prey but benthics search the bottom. Because of their searching habit, groups of benthic stickleback frequently encounter nests of conspecifics. These groups often attack the nests, feeding on the eggs and fry. Cannibalism may be a significant source of mortality in benthic populations (Hyatt and Ringler 1989*a*). Limnetics do not seem to engage in such cannibalism. Occurrence of cannibalism seems to have selected for evolution of less conspicuous courtship in cannibalistic than in non-cannibalistic populations, and for maintenance of the diversionary displays by which nest-guarding males attempt to lure schools of would-be cannibals away from the young in their nests (reviewed in Foster page 386 this volume).

A potential consequence of benthic habits for life history traits is that the deeper-bodied benthics may produce a relatively larger mass of eggs per spawning than limnetics. Our preliminary observations suggest that clutch volume may be more limited in limnetic females, because greater clutch volume conflicts with their swimming performance. Vitt and Congdon (1978) observed a similar limitation of clutch mass in lizard species that are active either to pursue prey or to flee from predators. However, the advantage of greater relative clutch masses may be offset by longer interclutch intervals, if larger clutches take longer to produce.

Adaptation to fish predation may also be modulated by the feeding ecology of lacustrine stickleback. The microhabitat within which predation occurs may be characteristically different in benthic and limnetic populations. Benthic stickleback generally forage closer to cover and the bottom than limnetics, and it is more feasible for them to sprint to cover than it is for limnetics. Thus, adaptations to fish predation, including boldness and flight behaviour (Huntingford *et al.* Chapter 10 this volume), expression of spines and plates (Reimchen Chapter 9 this volume), and possibly body size (Moodie 1972*a*; McPhail 1977), may all be contingent upon foraging habits.

Conversely, it seems possible that limnetics would be less likely to evolve in lakes in which fish predation is a major source of mortality. Reimchen (1991) observed that juvenile stickleback often took refuge at the bottom after evading the initial attack by trout (*Oncorhynchus clarki*), and this means of escape would be unavailable to limnetics feeding in deep water. In effect, the fitness value of plankton as a food resource might be discounted by natural selection in proportion to the increased risk of predation in open water.

Overview of phenotypic integration

Moving from the more obvious biomechanical adaptations for feeding and locomotion, with numerous precedents in other fishes (e.g. Alexander 1967;

Webb 1975; Lauder and Liem 1982; Webb and Weihs 1983), to more remote associations of feeding habits with life history, behaviour, and armour, the phenotypic dichotomy between extreme benthic and limnetic forms of threespine stickleback is less well documented and, in some cases, only hypothetical at present. However, recognizing the unitary function and fitness of the phenotype (Olson and Miller 1958; Gould and Lewontin 1979; Mayr 1983), one expects to observe a web of interacting (reinforcing and conflicting) phenotypic states which may be potentially free to evolve independently but are bound by selection to evolve as a correlated set. The degree to which such networks exist and the extent to which their distinguishable phenotypic components have become correlated by developmental canalization and genetic correlation represent interesting problems for future analysis (Francis *et al.* 1986).

CONCLUSIONS

In this chapter we have attempted briefly to characterize the threespine stickleback species complex and to summarize the known dimensions of phenotypic variation among populations. This summary is intended to provide a foundation for the remaining chapters of the book, but it also indicates the ecological diversity of threespine stickleback and phenotypic variables that lend themselves to evolutionary analysis. These variables and the conceptual frameworks within which they have been studied form the basis for this book.

The threespine stickleback has several important properties that distinguish it as an outstanding model system for evolutionary studies. These attributes include accessibility and abundance, extreme and conspicuous variation within and among populations, and existence of an ample literature on the species complex. However, its most useful attribute may be its phylogeny, which allows treatment of separate freshwater populations as statistically independent, natural, evolutionary experiments. Freshwater populations are analogous to replicate inoculations taken from a common bacterial culture and allowed to grow for several thousand generations under diverse conditions. However, this analogy pertains only when the phenotypic states studied are absent in marine populations and the populations compared were derived independently from marine ancestors. Compared with other species used in evolutionary biology, it is relatively easy in the threespine stickleback to satisfy these criteria for large numbers of populations. Our emphasis on the value of the intraspecific phylogeny of *G. aculeatus* reflects the necessity of making statistically independent observations to test hypotheses.

Finally, we argue that several favourable attributes of threespine stickleback allow one to undertake evolutionary analyses that are difficult or impossible in most species. Multidimensional analysis allows assembly of information from different biological subdisciplines to develop a clearer understanding of the causes of phenotypic diversity. Phenotypic integration views individual phenotypic traits, which traditionally have been atomized and studied in isolation, as part of an integrated phenotypic unit. Although the threespine stickleback will continue to serve an important role as a model system for narrowly focused mechanistic studies, it is unusually well suited to investigations of interactions among mechanisms. We emphasize the feasibility of these broad-based approaches to evolutionary studies because we believe that they will yield insights that could not be gained from more narrowly focused studies.

ACKNOWLEDGEMENTS

We are deeply indebted to the contributors to this volume for sharing their ideas about stickleback biology with us. We thank J.A. Baker, S.C. Nemtzov, N.C. Tousley, and J.A. Walker for criticizing the manuscript. Most illustrations were prepared by D. Chandler and G. Ortí. Our research has been supported primarily by USA National Science Foundation grants (BSR 86-00114, BSR 89-05758 to M.A.B.; BSR91-08132 to S.A.F.), and a USA National Institute of Mental Health National Research Service Award F32 MH09244 (to S.A.F.).

2

Systematics and morphology of the Gasterosteiformes

Patricia S. Bowne

One goal in studying the evolution of sticklebacks, or any other group, is to suggest causes for the evolution of traits. This is commonly done by identifying environmental or other factors correlated with the presence of the trait in question, and hypothesizing that these factors promote evolution of the trait. The mere identification of a correlation between supposed causative factors and a trait's presence, however, leaves some important questions unanswered. Did the factors actually predate the trait's appearance? Did organisms exposed to the factors actually evolve the trait more often than those not exposed, or does their joint possession of it reflect a single evolutionary event? Such questions can only be answered if the history of the group is known, or reconstructed through phylogenetic analysis.

Ideally, a causative factor should be associated with the historical appearance of a trait, rather than its presence in extant forms. If a phylogeny of the group is available, we can infer the points in the group's history at which traits appeared. As Donoghue (1989) demonstrated, a phylogenetic context allows correlations to be made between supposed causative factors and individual evolutionary events (reviewed in Harvey and Pagel 1991).

While phylogenies are potentially useful in evolutionary analysis, several drawbacks restrict their actual use. The phylogeny of many groups is unclear, and may be based on the very characters and hypotheses it might be used to evaluate. For a phylogeny to serve as a credible test of evolutionary hypotheses, it must have at least as much credibility as the hypotheses themselves, and should be supported by a more varied data set. To date, such a phylogeny has not been available for the Gasterosteiformes.

The order Gasterosteiformes (*sensu* Nelson 1971) contains two monotypic genera of tubesnouts (Aulorhynchidae) and five genera of sticklebacks (Gasterosteidae). Within Gasterosteidae, three genera, *Spinachia*, *Culaea*, and *Apeltes*, are also monotypic. The remaining genera, *Gasterosteus* and *Pungitius*, are widely variable and have been divided into numerous species and subspecies; currently, two species are recognized within each genus (Nelson 1984). This chapter will present the major osteological data on which hypotheses of gasterosteid relationships have been based, a cladistic analysis based on these data, and an evaluation of the proposed hypotheses about phylogenetic relationships, both within the Gasterosteidae and between the Gasterosteiformes and other teleost groups. Numerous errors in my previous osteological descriptions (Bowne 1985) are corrected in the following descriptions.

OSTEOLOGY OF *GASTEROSTEUS*

Osteological methods

Specimens of Gasterosteiformes examined in detail (listed in Appendix 2.1) were from the Milwaukee Public Museum (MPM) and University of Alberta Museum of Zoology (UAMZ). Specimens were cleared and stained according to Taylor (1967). Measurements and counts were made according to Hubbs and Lagler's (1970) specifications with the following exceptions: the last two elements of the dorsal and anal fins were counted as two rays, regardless of whether they shared a common base, and the head lengths were measured from the anteriormost point of the snout to the posterior border of the occipital condyle. Bone lengths were measured and figures drawn with a camera lucida, except for Fig. 2.5, which was based on a photograph, and Fig. 2.11, which was redrawn from Nelson (1971).

Neurocranium

The gasterosteiform neurocranium, represented by *Gasterosteus aculeatus* (Fig. 2.1), can be most easily visualized as a box, the braincase proper, connected by dorsal and ventral anteriorly extending struts to a pyramid-shaped ethmoid complex. The top of the braincase is formed by the frontals, supraoccipital, and parietals, its lateral walls by the sphenotic and pterotic bones, and its floor by the parasphenoid, basisphenoid, and prootics. The posterior corners and wall of the braincase are formed by the dorsal epiotics and the ventral exoccipitals. Intercalaries lie over the exoccipital–pterotic sutures.

The braincase is attached to the ethmoid region dorsally by the frontal–nasal and ventrally by the parasphenoid–vomer strut. The lateral and median ethmoids connect the anterior termini of these struts, forming the posterior and medial walls of the nasal capsules. There is no supraethmoid. The nasal bears a distinctive flat process (NP, Fig. 2.1 b,c) which extends medially from its lateral edge to meet the median ethmoid and rostral cartilage, separating the lateral nasal capsule from a median groove within which the premaxillary ascending process slides during jaw protrusion and retraction. The median ethmoid is composed of two vertical laminae, connected to each other by bony struts and solid dorsal and ventral plates.

The major feature in which gasterosteiform neurocrania differ from other acanthomorph neurocrania is in the presence of direct contact between the parasphenoid and frontal at the posterior end of the orbit (Fig. 2.1*c*). In most other acanthomorphs, this contact is accomplished by a separate pterosphe-

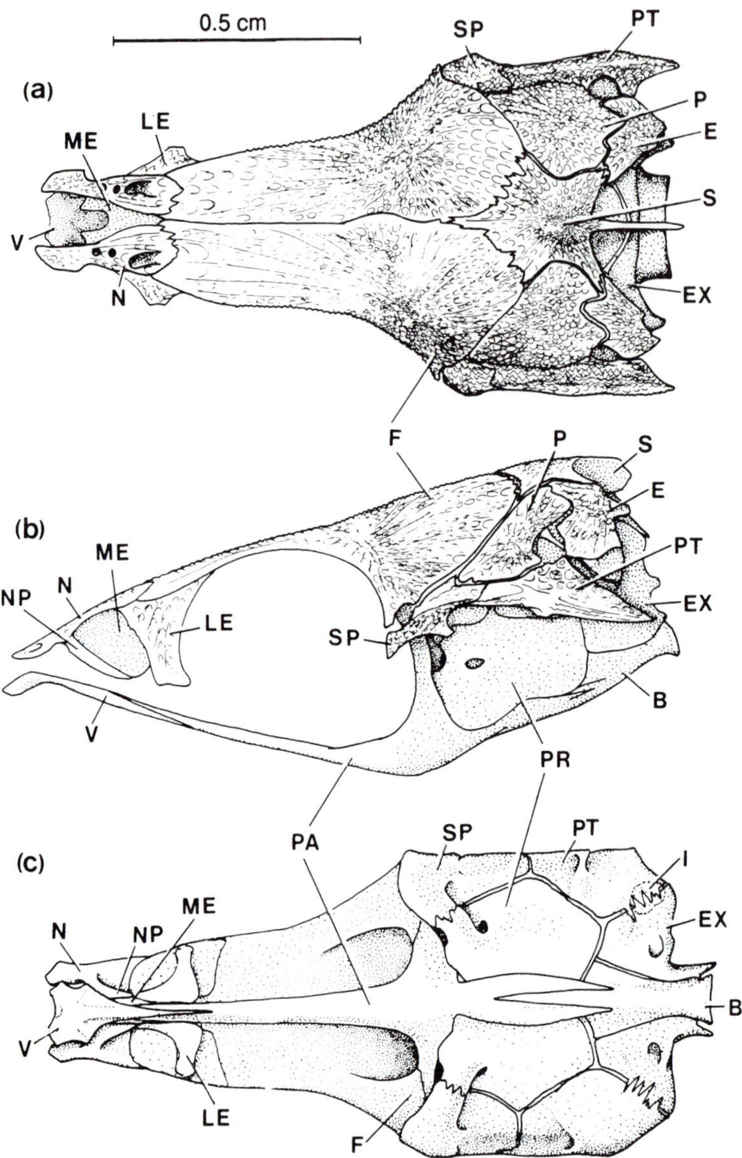

Fig. 2.1 Neurocranium of *Gasterosteus aculeatus* (composite of three specimens of lot UAMZ 5512, anterior to left): (a) dorsal view, (b) lateral view (vomer displaced ventrally), (c) ventral view. Abbreviations: B, basioccipital; E, epiotic; EX, exoccipital; F, frontal; I, position of intercalary (shown on right side of ventral view only; the intercalary was not visible on these specimens); LE, lateral ethmoid; ME, median ethmoid; N, nasal; NP, nasal ventrolateral process; P, parietal; PA, parasphenoid; PR, prootic; PT, pterotic; S, supraoccipital; SP, sphenotic; V, vomer.

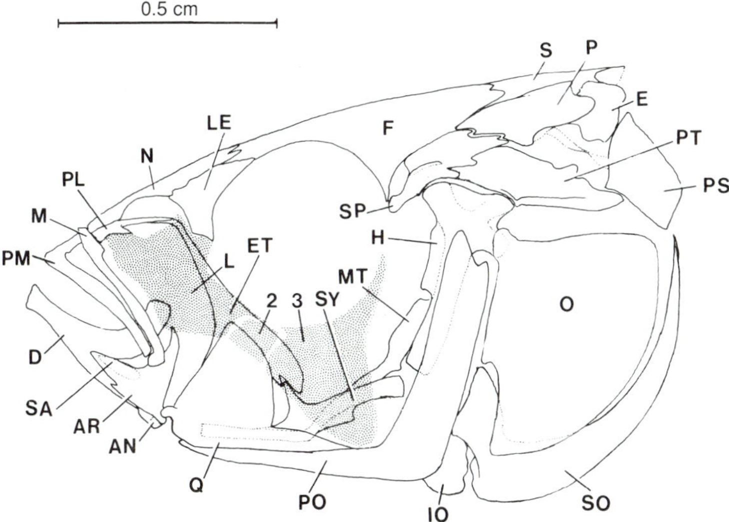

Fig. 2.2 Left lateral view of *Gasterosteus aculeatus* head based on a composite of three specimens of lot UAMZ 5512 (suborbital series stippled). Abbreviations: AN, angular; AR, articular; D, dentary; E, epiotic; ET, fused ectopterygoid and endopterygoid; F, frontal; H, hyomandibular; IO, interoperculum; L, lachrymal (or first suborbital); LE, lateral ethmoid; M, maxilla; MT, metapterygoid; N, nasal; O, operculum; P, parietal; PL, palatine; PM, premaxilla; PO, preoperculum; PS, posttemporal; PT, pterotic; Q, quadrate; S, supraoccipital; SA, sesamoid articular; SO, suboperculum; SP, sphenotic; SY, symplectic; 2, 3, second and third suborbital.

noid. Although Banister (1967) observed a separate pterosphenoid in some specimens of *Aulorhynchus*, as a general rule no separate pterosphenoid is visible in the gasterosteiforms. The other bones of the interorbital septum, the orbitosphenoid and basisphenoid, are also absent. The posterior suborbitals are lacking (Fig. 2.2) and the anterior suborbitals are platelike rather than tubular, the third suborbital extending ventrally to meet the preoperculum. The supra- and infraorbital sensory canals are reduced, and the bony canal bearing the interorbital commissures is absent.

Branchiocranium

Development of the neurocranium and branchiocranium has been described by Swinnerton (1902). The branchiocranium consists of the jaws and suspensoria (Fig. 2.2), the opercular series, the hyoid arch (Fig. 2.3a), and the branchial arches (Fig. 2.3b). The upper jaw is supported both by ligamentous connections to the ethmoid region (Anker 1974) and by a bony palatine–pterygoid strut extending ventrally to the quadrate. The premaxilla bears an ascending process and lacks both articular and postmaxillary processes. The anterior end of the palatine bears a hooklike process to the lachrymal, and

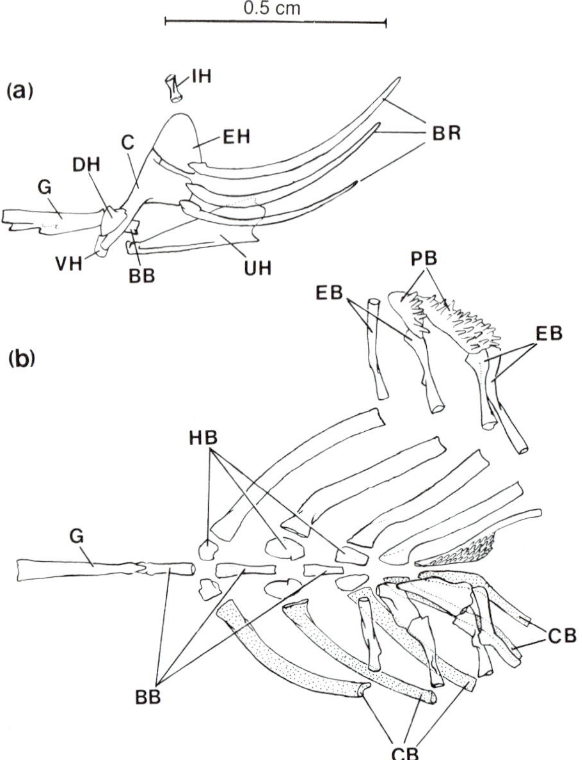

Fig. 2.3 *Gasterosteus aculeatus* hyoid and branchial arches (UAMZ 5512; anterior to left): (a) lateral view of left hyoid arch, (b) dorsal view of branchial arches (right epibranchials and pharyngobranchials reflected, left ceratobranchials stippled). Abbreviations: BB, basibranchials; BR, branchiostegal rays; C, ceratohyal; CB, ceratobranchials; DH, dorsal hypohyal; EB, epibranchials; EH, epihyal; G, glossohyal; HB, hypobranchials; IH, interhyal; PB, pharyngobranchials; UH, urohyal; VH, ventral hypohyal.

a single triradiate pterygoid extends from the palatine to the quadrate, filling the position held in many other fishes by the ecto- and endopterygoids.

The metapterygoid is reduced into a bony splint and the symplectic bears a dorsally extending flange which curves anteriorly toward the posterior portion of the pterygoid.

The gills are supported by the hyoid arch–branchial arch complex. Three basibranchials support the branchial arches (Fig. 2.3b). Three hypobranchials are present on each side, supporting the first three ceratobranchials. The first four of the five ceratobranchials are long tubular bones, connected distally to the four epibranchials. The fifth ceratobranchial is shorter and is anteriorly expanded into a toothed plate. There are two pairs of pharyngobranchials, both toothed. The first is associated with the second epibranchial. The second is a large plate supported by both the third and the fourth epibranchials, and probably represents fused third and fourth pharyngobranchials.

The ceratohyal and epihyal (Fig. 2.3a) are connected dorsally by a strut which originates on the ceratohyal. The urohyal consists of a horizontal plate with a vertical flange arising from its dorsal midline.

Paired fins

Pectoral girdles and fins

Nelson (1971) described the gasterosteiform paired fin girdles in detail and published illustrations of pectoral and pelvic skeletons of all gasterosteiform

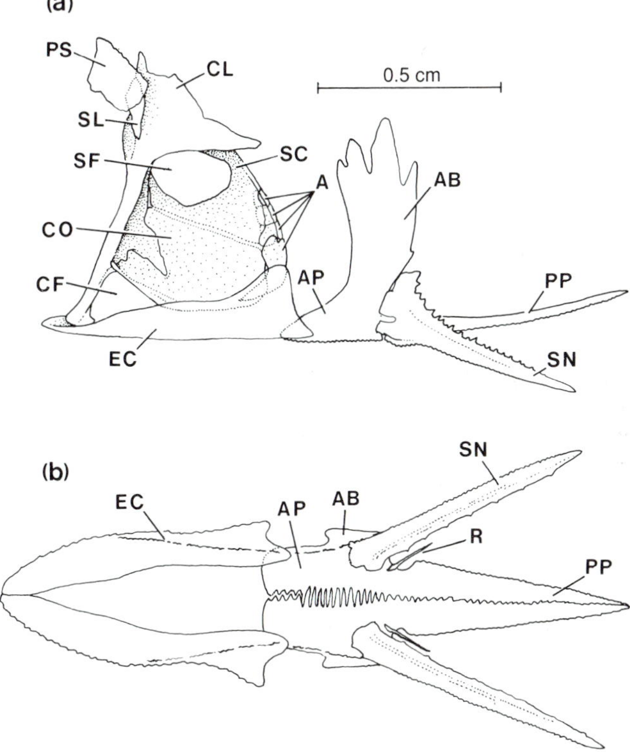

Fig. 2.4 Pectoral and pelvic girdles of *Gasterosteus aculeatus* (UAMZ 5512; pectoral fin rays not shown; anterior to left): (a) lateral view, (b) ventral view. Abbreviations: A, actinosts; AB, ascending branch of pelvic plate; AP, anterior process of pelvic plate; CF, coracoid foramen; CL, cleithrum; CO, coracoid; EC, ectocoracoid; PP, posterior process of pelvic plate; PS, posttemporal; R, pelvic soft ray; SC, scapula; SF, scapular foramen; SL, supracleithrum; SN, pelvic spine.

genera. His terminology will be followed here. The anterior border of each pectoral girdle (Fig. 2.4a) is formed by the vertically orientated cleithrum, which is suspended from the cranium by the posttemporal in *Gasterosteus aculeatus*. The posttemporal articulates with the lateral face of the epiotic and bears a cylindrical medially extending process to the posterior face of the exoccipital. A single supracleithrum is present in *G. aculeatus*, but it and the postcleithrum are absent in *G. wheatlandi*. The shafts of the right and left cleithra meet in a ventral symphysis.

The scapula is pierced by a large foramen which is usually open anteriorly (Starks 1902; Nelson 1971). A closed scapular foramen is found in some specimens of *G. aculeatus*. The sculptured ectocoracoid lies below the coracoid and meets its counterpart below the cleithral symphysis (Fig. 2.4b).

The flat, bilaminar actinosts are supported by the posterior edge of the scapula–coracoid plate, which is notched to receive them. The posterior margins of actinosts and scapula are thickened to form a continuous posterior border. The anteroventral corner of the fourth actinost extends to the ectocoracoid.

Most specimens of *Gasterosteus* have ten pectoral rays. Nelson (1971) records a range of eight to eleven rays for *G. aculeatus*.

Pelvic skeleton and fins

The gasterosteiform pelvic girdle (Fig. 2.4) consists of a bilateral pair of plates, each bearing anterior and posterior processes and an ascending branch located lateral to the base of the pelvic spine. The ontogeny of these structures in *G. aculeatus* has been described by Bell and Harris (1985). The anterior process is relatively short and is usually suturally connected to its counterpart. The posterior process is long and the ascending branch extends dorsally up the side. Presence of a dorsally extending ascending branch seems to function to resist deflection of the spines by predatory vertebrates (Reimchen 1983). Another factor in the spines' ability to resist deflection is the extent of overlap between the pelvic ascending branch and the lateral plates, which suggests that the ascending branch may be functionally less important in specimens lacking lateral plates.

One to three small pelvic rays are present. Serrations appear on the pelvic spine in *G. wheatlandi* and in most specimens of *G. aculeatus* (Gross 1978).

In several populations of *Gasterosteus* the pelvic fins and lateral plates have seemingly undergone independent reduction, perhaps associated with repeated events of lacustrine invasion (Bell 1987). Bell (1987) distinguished two separate modes of pelvic reduction in *Gasterosteus*: paedomorphosis, in which those elements which appear later in ontogeny are more reduced and the anterior and posterior processes may be present as separated rudiments, and distal truncation, in which reduction involves a decrease in the lengths of the ascending branch and posterior process.

Median skeleton and fins

The median skeleton and fins (Fig. 2.5) consist of the vertebrae and associated ribs, pterygiophores, dorsal, anal, and caudal fin elements, and lateral plates.

G. aculeatus has 29–33 vertebrae, 13–14 of them precaudal, and *G. wheatlandi* has 26–8 vertebrae, 11–12 of them precaudal. The neural spine of the first vertebra and the neural and haemal spines of the last two or three vertebrae are expanded. A small postzygapophysis is present on each side of the posterior precaudal vertebrae. On the second caudal vertebra it fuses distally with the haemal arch, forming the posterior margin of a large foramen.

Epipleural ribs begin on the first vertebra and the first epipleural rib is expanded into a flat blade. *G. aculeatus* has nine to fifteen epipleural ribs, while the specimens of *G. wheatlandi* examined had only four to five. The nine to eleven pairs of pleural ribs usually begin on vertebra three (2–4). They originate medially to the epipleural ribs on the transverse processes.

The free dorsal spines and their associated pterygiophores are among the distinguishing characteristics of the Gasterosteiformes. Each of the free dorsal spines is supported by a pterygiophore series in which the proximal, medial, and distal pterygiophores are fused. The spines are supported by large dorsal plates perforated by foramena which receive the proximal ends of the spines (for a detailed discussion see Hoogland 1951).

The configuration of pterygiophores and free dorsal spines anterior to the dorsal fin varies somewhat, but that shown in Fig. 2.5 appears to be common (Bell and Baumgartner 1984). Six pterygiophores usually precede the dorsal fin, and spines generally occur on the third, fourth, and sixth pterygiophore.

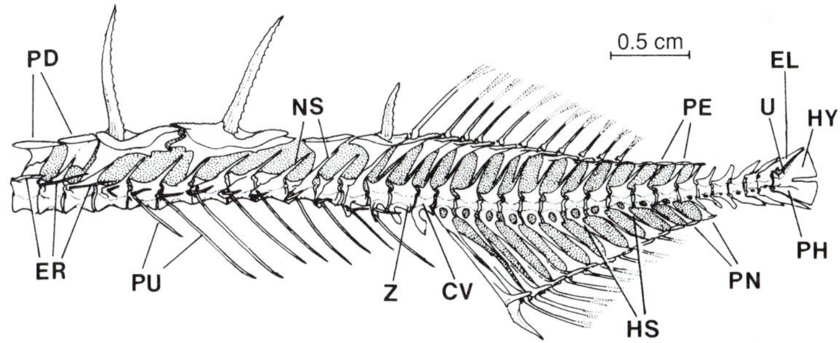

Fig. 2.5 Left lateral view of median skeleton of *Gasterosteus aculeatus* (UAMZ 5512, 6.8 cm SL). Abbreviations: CV, first caudal vertebra; EL, epurals (fused to preural neural spine of this specimen); ER, epipleural ribs; HS, haemal arches and spines; HY, fused hypural fan; NS, neural spines; PD, predorsal pterygiophores; PE, postdorsal pterygiophores; PH, parhypural (fused to lower hypural fan); PN, postanal pterygiophores; PU, pleural ribs; U, uroneural; Z, postzygapophyses.

Each pterygiophore has a proximal process associated (from anterior to sixth pterygiophore) with the first, second, third, sixth, tenth, and twelfth neural spine. Pterygiophores supporting soft dorsal fin rays follow posteriorly in association with subsequent vertebrae. They are less highly fused than those anterior to the dorsal fin, small, cap-shaped distal pterygiophores being visible at the bases of the rays.

The last dorsal ray is followed by two to four pterygiophores in *G. aculeatus*, but by none or one in *G. wheatlandi*. The first and second spines are generally about as long as the anterior dorsal soft rays, and the third spine is shorter.

The first haemal spine supports two pterygiophores in most specimens, the pterygiophore supporting the anal spine lying along its anterior margin and that supporting the first anal ray along its posterior margin. The anal pterygiophores are similar to the dorsal pterygiophores, although their medial components are longer. A series of postanal pterygiophores is present, its development parallelling that of the postdorsal pterygiophores.

The lateral plates (Fig. 1.2), when present, form a single row of vertically elongated dermal ossifications, and their extensive variation has been well documented (e.g. Hagen and Moodie 1982; Bell 1984*a*; Bell and Foster page 9; Reimchen page 251 this volume). Roth (1920) and Igarishi (1964, 1970*a*) have described the ontogeny of the lateral plates.

The caudal skeleton is highly fused, with the hypurals and parhypural forming a deeply notched hypural plate. The uroneural lies along the dorsal margin of this plate, and is only visible as a separate element at its distal tip (Monod 1968). One or two epurals lie between the last neural spine and the uroneural, and are fused to the last neural spine in some adult specimens (Fig. 2.5). Twelve principal caudal rays and a variable number of procurrent rays are present.

VARIATION AMONG THE GASTEROSTEIFORMES

Some of the variation observed among the Gasterosteiformes seems related to elongation, as reflected in characters of the snout, jaws, and numbers of medial skeletal elements. Many osteological characters, however, show no clear association with elongation. The descriptions below emphasize features which serve as characters in my phylogenetic analysis. They are listed in Appendix 2.2, with their distribution among genera in Appendix 2.3.

Neurocranium

Most of the variation in the gasterosteiform ethmoid region seems to be associated with differences in snout length. The nasals, median ethmoid, and lateral ethmoids are longer in Aulorhynchidae and *Spinachia* than in other gasterosteids. Median ethmoid length, however, shows variation even when standardized for snout length.

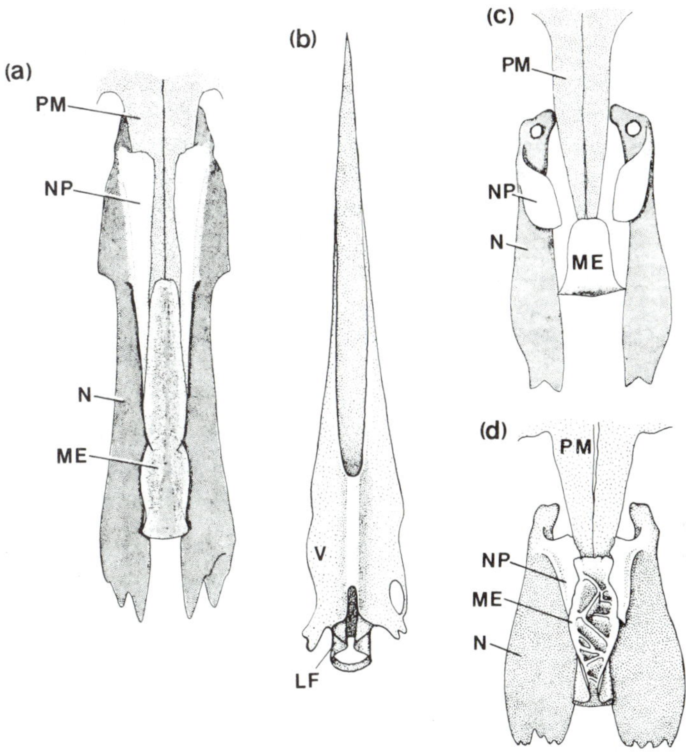

Fig. 2.6 Ventral view of gasterosteiform median ethmoids and adjacent bones (anterior up): (a) *Spinachia spinachia* (UAMZ 5548, × 9.4); (b) *Aulorhynchus flavidus* (median ethmoid only; UAMZ 1694, × 10.8); (c) *Pungitius pungitius* (UAMZ 4754, × 18.7); (d) *Gasterosteus aculeatus* (UAMZ 5512, × 6.4). Abbreviations: LF, lateral flange; ME, median ethmoid; N, nasal; NP, nasal ventromedial process; PM, premaxilla ascending process; V, ventral plate.

In *Spinachia* (Fig. 2.6a) and *Aulichthys* the median ethmoid is a long, flat, ventrally concave plate. The median ethmoid of *Apeltes* is similar except for length. In *Aulorhynchus* (Fig. 2.6b) its posterior edges bend ventrally, forming deep lateral flanges (LF) which are connected by a ventral plate (V). The median ethmoid of *Pungitius* (Fig. 2.6c) is a short plate which lies ventral to the premaxillary ascending processes and is curved dorsally posterior to them. In *Gasterosteus* (Fig. 2.6d) and *Culaea* the basic pattern seen in *Pungitius* is supplemented by ventral flexion of the plate's lateral edges. In *Gasterosteus* this flexion is so extreme that the plate's lateral edges approach each other in the midline. They are connected to each other by bony struts and a ventral plate (removed from Fig. 2.6d).

The medially extending process from the nasal bone is fan-shaped in

Pungitius and *Apeltes*, and antero-posteriorly elongated in *Gasterosteus* (Fig. 2.1) and the other two gasterosteid genera. The Aulorhynchidae lack the nasal process.

The aulorhynchids (Fig. 2.7a) possess a complete series of tubular suborbitals and an elongated lachrymal, a state which is probably associated with increased snout length, as it also occurs in *Spinachia*. The infraorbital canal of the cephalic lateral line system passes through the suborbital series and continues posteriorly from it into a bony tube through the dermopterotic, an elongate bone which overlies much of the sphenotic and almost reaches the orbit margin. The frontal bears the supraorbital sensory canal and the interorbital commissure in bony canals.

In the Gasterosteidae, on the other hand (Fig. 2.2 and 2.7b,c), the suborbital series is incomplete and the suborbitals are flattened rather than tubular. In most specimens the third suborbital extends posteroventrally to touch the preoperculum. In *Gasterosteus* (Fig. 2.2) and *Pungitius* its dorsal margin also extends posteriorly, filling some of the space occupied by the fourth and fifth suborbitals of the aulorhynchids. The dermopterotic is smaller and does not bear sensory canals. The bony canal bearing the inter-

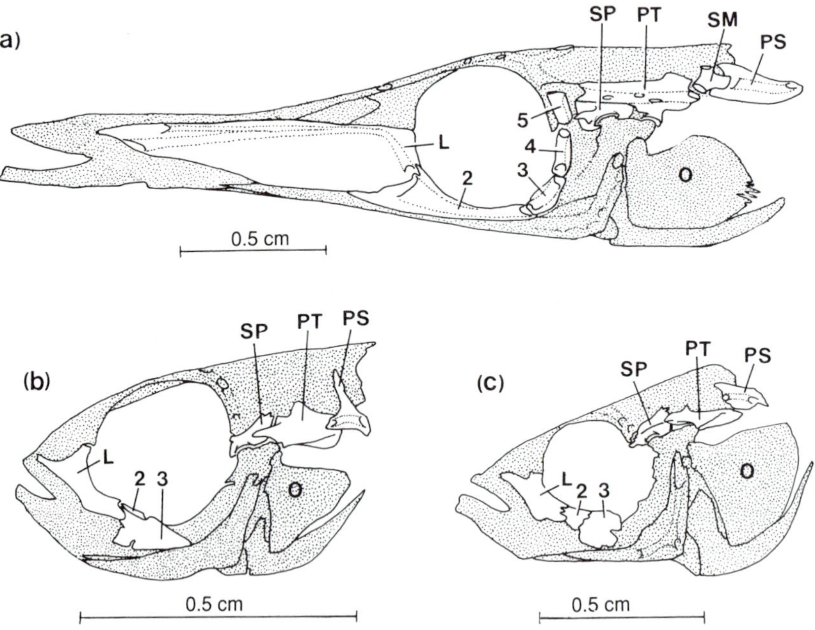

Fig. 2.7 Left lateral views of gasterosteiform heads, showing suborbital series and relationship between sphenotic and pterotic: (a) *Aulorhynchus flavidus* (UAMZ 1694); (b) *Apeltes quadracus* (UAMZ 5540); (c) *Culaea inconstans* (UAMZ uncatalogued). Abbreviations: L, lachrymal; O, operculum; PS, posttemporal; PT, pterotic; SM, supratemporal; SP, sphenotic, 2–5, second through fifth suborbitals.

orbital commissure is present only as a medially opening pore (in *Spinachia*, some specimens of *Apeltes*) or is absent (in *Gasterosteus*, *Pungitius*, *Culaea*, some specimens of *Apeltes*).

Branchiocranium

Some characters from the branchiocranium also reflect snout length. The premaxillary ascending process is longer than the toothed alveolar process in the long-snouted forms and shorter in the short-snouted forms (Fig. 2.8). Only *Aulichthys* (Fig. 2.8a) possesses a marked postmaxillary process. This process is reduced in *Spinachia* (Fig. 2.8b) and *Pungitius* (Fig. 2.8c) and is absent in all the other Gasterosteiformes. The palatines are also proportionately longer in aulorhynchids and *Apeltes*. The symplectic flanges (Fig. 2.9) are longest in aulorhynchids and *Spinachia*, perhaps owing to their elongated suspensoria, but they are also relatively well developed in *Gasterosteus* and some specimens of *Apeltes*.

Dentary sensory canals are present in the Aulorhynchidae and absent in the Gasterosteidae.

In most gasterosteids the operculum is shaped like a quarter-circle, its dorsal margin lying parallel to the dorsal surface of the head. The exception is *Apeltes* (Fig. 2.7b), in which the posterior margin of the operculum is straight or concave. In contrast, aulorhynchid opercula are antero-posteriorly elongated and their posterior margins are extended. In *Aulorhynchus* (Fig. 2.7a) the dorsal margin of the operculum extends dorsally, giving this genus the largest opercular angle (i.e. the angle between the dorsal and anterior margins).

The number of branchial elements is the same in all of the gasterosteiforms except *Aulorhynchus*, which has three toothed pharyngobranchials. The

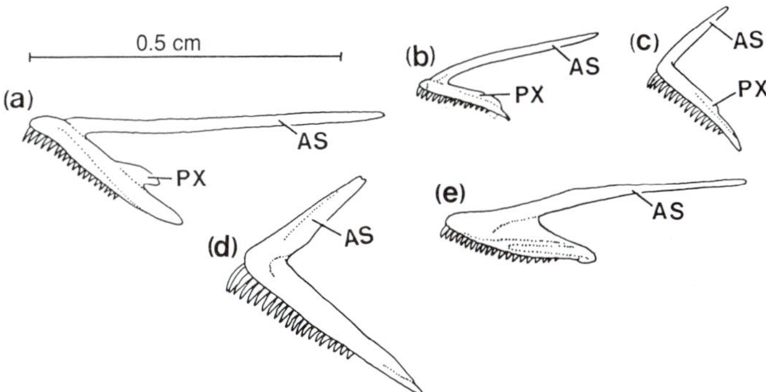

Fig. 2.8 Left lateral views of gasterosteiform premaxillae: (a) *Aulichthys japonicus* (UAMZ 5542); (b) *Spinachia spinachia* (UAMZ 5548); (c) *Pungitius pungitius* (UAMZ 4754); (d) *Gasterosteus aculeatus* (UAMZ 5512); (e) *Aulorhynchus flavidus* (UAMZ 1694). Abbreviations: AS, ascending process of premaxilla; PX, postmaxillary process.

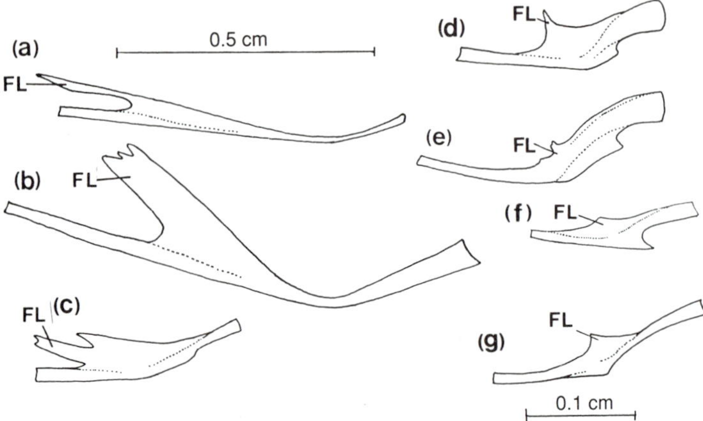

Fig. 2.9 Left lateral views of gasterosteiform symplectics: (a) *Aulorhynchus flavidus* (UAMZ 1694); (b) *Aulichthys japonicus* (UAMZ 5542); (c) *Spinachia spinachia* (UAMZ 5548); (d) *Gasterosteus aculeatus* (UAMZ 5512); (e) *Apeltes quadracus* (UAMZ 5540); (f) *Culaea inconstans* (UAMZ uncatalogued); (g) *Pungitius pungitius* (UAMZ 4754). Abbreviation: FL, symplectic flange. The 0.5 cm scale bar applies to (a) to (f).

first is supported by the first epibranchial and the second and third by the second through fourth epibranchials. The hyoid arches are more elongate in the Aulorhynchidae and *Spinachia* than in the short-bodied gasterosteids, perhaps reflecting snout elongation. The Aulorhynchidae have four branchiostegal rays, three arising from the ceratohyal; the Gasterosteidae have three, two arising from the ceratohyal.

Paired fins

Pectoral girdles and fins

All the Gasterosteiformes except *Gasterosteus wheatlandi*, *Aulorhynchus*, and *Spinachia* possess a single supracleithrum. *G. wheatlandi* also lacks the posttemporal. The posttemporal medial process is absent in *Aulorhynchus*, relatively low in *Culaea*, *Spinachia*, and *Apeltes*, and best developed in *Aulichthys*, *Gasterosteus*, and *Pungitius*. The lateral surface of the posttemporal bears a closed sensory canal in all genera except *Gasterosteus* and *Culaea*.

The cleithrum symphysis is expanded in the Aulorhynchidae, and a flange extends laterally from the entire length of the vertical cleithrum shaft (Fig. 2.10a). In other gasterosteids the lateral flange does not extend to the ventral end of the cleithrum, and the cleithrum symphysis is narrower. *Spinachia* (Fig. 2.10b) has a slightly expanded symphysis. A marked posterior flange (Fig. 2.4, 2.10c) extends from the cleithrum shaft to the coracoid in all the short-bodied gasterosteids except *Apeltes*, and in one specimen of *Spinachia* from UAMZ 5548.

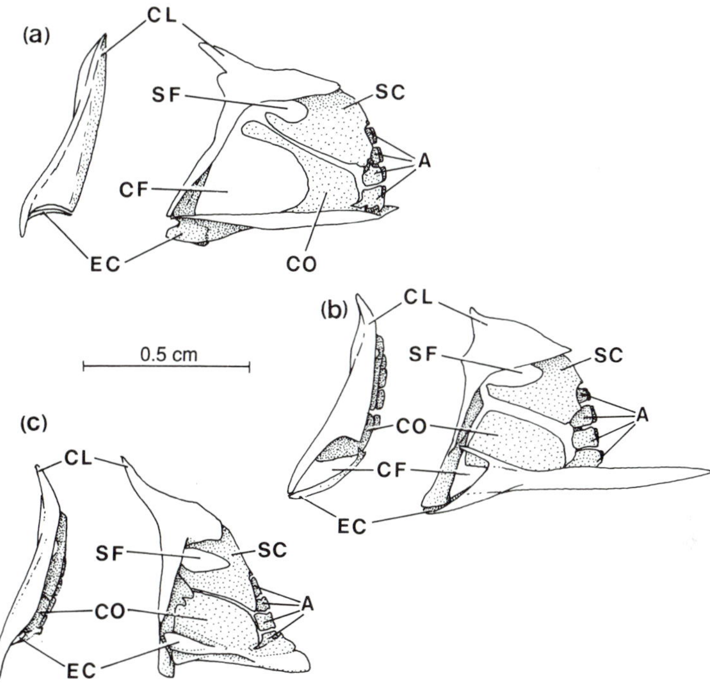

Fig. 2.10 Anterior and left lateral views of left gasterosteiform pectoral skeletons: (a) *Aulorhynchus flavidus* (UAMZ 1694); (b) *Spinachia spinachia* (UAMZ 5548); (c) *Culaea inconstans* (UAMZ uncatalogued). Abbreviations: A, actinosts, CF, coracoid formaen; CL, cleithrum; CO, coracoid; EC, ectocoracoid; SC, scapula; SF, scapular foramen.

The scapular foramen is smaller in *Apeltes* and the aulorhynchids than in the other gasterosteiforms. Closed scapular foramena are found in some specimens of *G. aculeatus*, *Pungitius*, and *Aulichthys*.

The aulorhynchid coracoid–ectocoracoid complex (Fig. 2.10a) is a wedge-shaped plate of bone lying with its open ends directed toward the cleithrum shaft, enclosing the coracoid foramen. The anterior end of the ventral limb, which lies against the lateral side of the cleithrum symphysis, is dorsoventrally expanded and bears a narrow lateral flange along its dorsal margin. Posteriorly, this lateral flange expands into the sculptured lateral face of the ectocoracoid.

In *Gasterosteus* (Fig. 2.4), *Pungitius*, and *Spinachia* (Fig. 2.10b), the ventral limb of the coracoid is not apparent, its position being filled by the ectocoracoid. The dorsal limb is expanded, arising from the ectocoracoid and extending anteriorly to meet the cleithrum posterior flange. The ectocoracoid meets its counterpart below the cleithrum symphysis.

In *Culaea* (Fig. 2.10c) and *Apeltes* the ectocoracoid is small, and in *Culaea* it is very difficult to distinguish from the coracoid. The cleithrum shaft extends ventrally below the ectocoracoid. The coracoid is a roughly quadrilateral plate of bone broadly connected to the cleithrum posterior flange, which is larger in *Culaea* than in *Apeltes*. There is no coracoid foramen in most specimens, though in some specimens of *Culaea* there is a shallow notch in the anterior margin of the coracoid.

The interdigitating processes that connect the actinosts are most clearly visible in the Aulorhynchidae. The connections between actinosts, and between the actinosts and the scapulae, are strongest in *Gasterosteus* (Fig. 2.4), although in *Aulichthys* no suture line can be found separating the fourth actinost from the ectocoracoid.

Aulichthys has 11 pectoral rays. All other Gasterosteiformes have 10 rays, though infrequent specimens of *Gasterosteus aculeatus* and *Apeltes* have 11 rays.

Pelvic skeleton and fins

In the Aulorhynchidae (Fig. 2.11a,b) the pelvic plate does not reach the pectoral girdle. The pelvic plate has short posterior and long anterior processes, and the ascending branch is small and extends posteriorly rather than dorsally.

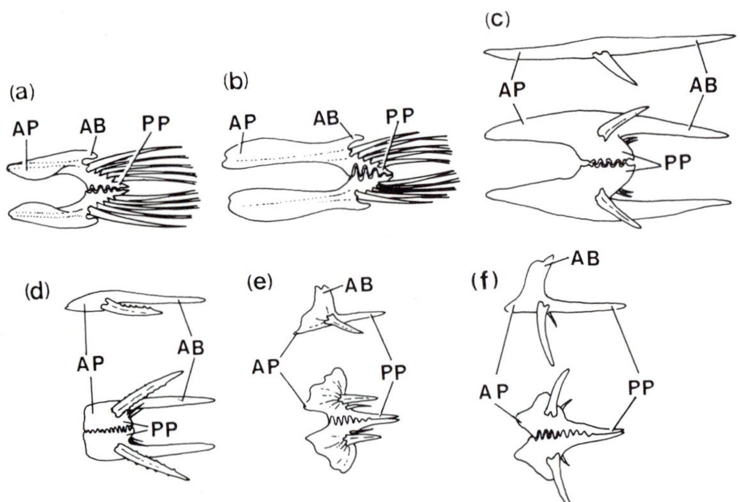

Fig. 2.11 Left lateral (except (a) and (b)) and ventral views of gasterosteiform pelvic skeletons: (a) *Aulorhynchus flavidus* (UAMZ 1694, × 5.7); (b) *Aulichthys japonicus* (UAMZ 5542, × 3.5); (c) *Spinachia spinachia* (UAMZ 5548; × 2.2); (d) *Apeltes quadracus* (UAMZ 5540; × 5); (e) *Culaea inconstans* (UAMZ uncatalogued; × 2.4); (f) *Pungitius pungitius* (UAMZ 4754; × 2.7). Abbreviations: AB, ascending branch of the pelvic plate; AP, anterior process of pelvic plate; PP, posterior process of pelvic plate.

The pelvic skeleton in the gasterosteids reaches anteriorly to the ectocoracoids, although it may be separated from them by a fold of soft tissue (Starks 1902). *Spinachia* (Fig. 2.11c) and *Apeltes* (Fig. 2.11d) are unique among the gasterosteids in having short pelvic posterior processes and ascending branches which bend posteriorly, like those of the Aulorhynchidae, rather than projecting dorsally. The ascending branches in *Spinachia* and *Apeltes* are larger than those in the Aulorhynchidae. In *Spinachia* the anterior process is elongated; in *Apeltes*, as in all the other short-bodied gasterosteids, the anterior process is relatively short. In the other gasterosteids (Fig. 2.4 and 2.11e,f) the ascending branch extends dorsally up the side of the body.

The suture between the two pelvic plates varies. In *Spinachia*, *Culaea*, and the Aulorhynchidae it involves the body of the plate and the posterior processes, but leaves the anterior processes separated from each other. In the other species, while the suture connects the posterior processes, its anterior extension and the portion of the posterior processes which are separated from each other may vary.

In the Aulorhynchidae each pelvic fin has a long, slender spine and four long rays. *Spinachia* has a relatively short, unserrated pelvic spine and two to three pelvic rays. *Apeltes* also has two to three pelvic rays. One pelvic ray is commonly found in *Culaea*, *G. aculeatus*, and *P. pungitius*, but *G. wheatlandi* generally has two. Serrated pelvic spines occur in *Gasterosteus*, *Pungitius*, and *Apeltes*.

All three of the freshwater genera of Gasterosteidae possess populations in which the pelvic fins and plates seem to have undergone independent reduction, perhaps associated with repeated events of lacustrine invasion (Bell 1987). The repeated independent evolution of the reduced pelvics makes the homology of pelvic vestiges in different populations questionable, to say nothing of their homology among genera. Of the two separate modes of pelvic reduction distinguished in *Gasterosteus* (Bell 1987), distal truncation has been tentatively identified in *Pungitius*, but initial examination indicates that the mode of reduction in *Culaea* is highly variable (Bell pers. comm.).

Median skeleton and fins

The numbers of most elements of the median skeleton vary with body elongation, the aulorhynchids and *Spinachia* having more vertebrae, pleural ribs (with the exception of *Aulorhynchus*), lateral plates, predorsal spines, and postdorsal and postanal pterygiophores than the shorter-bodied gasterosteids. Some characters, however, seem independent of body elongation. Notable among these are the numbers of dorsal and anal soft rays, the number of predorsal pterygiophores that do not bear spines, and the number of epipleural ribs. *Aulorhynchus* is distinguished by its reduced number of ribs; their supportive function may be performed by direct connections between lateral processes from the vertebra and medial processes from the lateral plates (Nelson 1971).

Aulorhynchid lateral plates resemble the posttemporals, being rectangular shields which bear the lateral line in a closed canal. The plates are small and rhomboid in *Spinachia*, bearing processes on their posterior margins, vertically elongated plates in *Gasterosteus* and some morphs of *Pungitius*, and small, isolated circles with central pores in *Culaea*. *Apeltes* lacks lateral plates. *Gasterosteus* and *Pungitius*, the two genera with enlarged plates, also show parallel variation in plate position and development, with each genus possessing complete, low, and partial morphs. While in *Gasterosteus* the low morphs possess a cuirass (i.e. abdominal plates) and the partial morphs a cuirass and keel, *Pungitius* also displays morphs in which the plates are restricted to the keel (illustrated in Ziuganov 1981) or reduced to small, circular plates (Nelson 1971). Like the pelvic reduction discussed above, variation in lateral plates has seemingly evolved independently on numerous occasions (Bell 1981). As is the case for pelvic reduction, any postulated synapomorphy must not be between identical states of plate development, but between the patterns of plate reduction, which may reflect shared genetic or developmental constraints. As the other genera of Gasterosteidae lack either the plates themselves (*Apeltes*) or the variation in habitat from fresh to salt water which is thought to partially underlie the evolution of different morphs (*Culaea* and *Spinachia*), it is impossible to compare their patterns of plate reduction with those of *Gasterosteus* and *Pungitius*, making this characteristic, even if homologous, of limited systematic value.

In most gasterosteiforms the dorsal spines are of equal lengths. The spines of Gasterosteidae are proportionately larger than those of the Aulorhynchidae, being largest in *Gasterosteus* and *Apeltes*, in which the first and second spines are most often about as long as the dorsal soft rays and the third spine is much shorter. In *Spinachia*, *Pungitius*, and *Apeltes*, the spines originate along the median axis and diverge laterally, successive spines pointing to opposite sides.

In all genera except *Culaea* the dorsal spines are supported by flat dorsal plates. In *Culaea*, the dorsal spines articulate with tubular medial pterygiophores. The predorsal pterygiophores preceding the first spine are similar in structure to those bearing spines in all genera.

The pterygiophore series supporting the soft dorsal rays are less highly fused than those supporting the spines, separate distal pterygiophores being visible at the bases of the rays in all genera except *Pungitius*. The distal pterygiophores are small and cap-shaped except in *Aulichthys*, in which each distal pterygiophore is forked, forming two lateral processes behind the ray it supports.

In *Spinachia*, the pterygiophore series continues posteriorly from the soft dorsal onto the caudal peduncle, where the dorsal pterygiophores and lateral plates meet. In the Aulorhynchidae and *Pungitius* the postdorsal pterygiophores likewise extend to the caudal peduncle, but are too narrow to meet the lateral plates. In *Gasterosteus*, *Apeltes*, and *Culaea*, the postdorsal pterygiophores do not extend as far on to the caudal peduncle. A series of postanal pterygiophores is present, its development parallelling that of the postdorsal pterygiophores in all genera.

The size of the anal spine is correlated with the size of the last dorsal spine, being relatively small in the Aulorhynchidae, *Spinachia*, *Culaea*, and *Gasterosteus*, and relatively large in *Pungitius* and *Apeltes*.

The caudal skeleton is simple in the Gasterosteiformes, consisting only of a hypural plate and its associated procurrent and principal caudal rays, with one or two epurals visible in all genera except *Aulorhynchus*. Most gasterosteids have two epurals which may be fused to each other, although *Apeltes* has a single epural. *Gasterosteus* is unique among the Gasterosteiformes in having a split hypural plate. Mural (1973) reported ten to twelve hypurals in some specimens of *Spinachia*.

INTERRELATIONSHIPS OF THE GASTEROSTEIFORMES

The Gasterosteiformes have been divided into the two families Aulorhynchidae and Gasterosteidae by many authors (e.g. Nelson 1971; Wootton 1976), and the two families are easily distinguished (Table 2.1). Relationships among the five gasterosteid genera, however, are not clearly understood. While several studies have addressed this problem, only one (McLennan *et al.* 1988) has applied phylogenetic systematics to a data set containing all five genera, and the data set used contained only characters related to reproduction. In this section I present a phylogenetic analysis based on parsimony cladistics, using the morphological data from Appendices 2.2 and 2.3.

Systematic methods

The method used here, parsimony cladistics, assumes that character states shared by two or more taxa are more likely to have evolved once, in their common ancestor, than to have appeared independently in each taxon. Therefore, hypotheses about the evolutionary relationships between taxa may be compared by evaluating the number of character state changes, or evolutionary steps, implied by each. The hypothesis implying the fewest character state changes is preferred as the most parsimonious.

Competing phylogenetic hypotheses, presented as cladograms, are compared by evaluating how many character state changes each implies for the data set used. The number of state changes implied by a cladogram can be inferred by first reconstructing the probable character states of the common ancestors (plesiomorphic states) proposed by a particular cladogram, and then counting the number of character state changes which must have occurred within the cladogram. Because only 105 fully resolved (bifurcated)

Table 2.1 Characters distinguishing the two families of Gasterosteiformes.

Aulorhynchidae	Gasterosteidae
Nasal process absent	Nasal process present
Complete series of tubular suborbitals	Suborbitals flat, posterior suborbitals absent
Canal-bearing dermopterotic overlies most of sphenotic	Dermopterotic short, without canal
Operculum elongated	Operculum not elongated
Cleithral symphysis expanded, cleithra bear wide lateral flange	Cleithral symphysis not widely expanded, lateral flange short or narrowed ventrally
Coracoid foramen large, ectocoracoid closely applied to coracoid	Coracoid foramen small or absent, ectocoracoid partially free from coracoid
Pelvic plate does not touch pectoral girdle, anterior process longer than posterior process, ascending branch very small; spine weak, four rays	Pelvic plate usually touches pectoral girdle, anterior process not longer than posterior process, ascending branch well developed; spine usually stout, three or fewer rays
52 or more vertebrae	42 or fewer vertebrae
24–26 dorsal spines	16 or fewer dorsal spines
Epipleural ribs absent	Epipleural ribs usually present
Four branchiostegal rays	Three branchiostegal rays
Lateral plates small, bearing canals	Lateral plates without canals
Caudal fin forked, 13–14 principal rays	Caudal fin usually rounded, 12 principal rays
Eggs attached to plants (Limbaugh 1962) or deposited in the peribranchial cavity of an ascidean (Uchida 1934)	Eggs deposited in a nest built by the male; elaborate mating behaviours

cladograms are possible for a five-taxon group like the Gasterosteidae (Felsenstein 1978), it was possible to compare all of them using the computer program MacClade (Maddison and Maddison 1987). MacClade reconstructs ancestral character states, and the number of evolutionary steps required to explain reversible binary and ordered multistate characters, by an algorithm described in Maddison and Maddison (1987). For unordered multistate characters it uses the algorithms of Fitch (1971) and Hartigan (1973).

As I assumed that the Gasterosteidae and the Aulorhynchidae had evolved from a common ancestor, the Aulorhynchidae were added to each cladogram as a basal sister group. This ensured that the analysis compared not only the number of evolutionary steps within the Gasterosteidae, but also the number of steps required to derive both families from a common ancestor.

Assumptions about the possible evolutionary pathways between character states affect the number of steps calculated for a cladogram. All binary characters were assumed to be reversible. Morphometric and meristic multistate characters (characters 1, 6, 12, 15, 32, 34, 35, 39–44, and 47–50) were assumed to be ordered such that evolution from state 0 to state 2 must pass through the intermediate state 1, this process counting as two steps. Those multistate characters which appeared to show several stages in their development or loss (characters 3, 7, 10, 13, 19–21, and 38) were also assumed to be ordered. Other multistate characters were assumed to be unordered, change between any two states counting as one step.

If all characters were assigned the same weight, multistate characters would have a disproportionate effect on the final cladogram, as a single multistate character can add as many steps to the cladogram as several binary characters. To avoid this bias, I weighted the characters so that the product of the weight and the minimum number of steps a character could add to the cladogram was constant for all characters (Chappill 1989). The weighting factor, W, was computed using $W = 6 / (N - 1)$, where N is the number of states of the character. Binary characters, for example, were assigned a weight of six, while seven-state characters were assigned a weight of one.

Cladograms were compared according to the number of weighted evolutionary steps they implied and their consistency indices. The consistency index indicates the congruence of the character state distributions among taxa with the cladogram. MacClade computes it by dividing the minimum possible number of steps for a cladogram which was completely congruent with the data set by the number of steps implied by the cladogram being evaluated (for further discussion see Maddison and Maddison 1987).

Results of cladistic analysis of the Gasterosteidae

Nelson (1971) stated that osteological data did not clearly resolve the relationships within Gasterosteidae. My analysis largely supports his conclusion. The 105 possible cladograms for the Gasterosteidae had lengths ranging from 583 to 668 weighted steps, indicating a high level of homoplasy in the data. The shortest cladogram, with 583 steps, had a consistency index of only 0.51.

The basal position of *Spinachia*

Only three cladograms had fewer than 597 weighted steps (Fig. 2.12). In all three, the short-snouted gasterosteids (*Apeltes*, *Gasterosteus*, *Pungitius*, and *Culaea*) were separated from *Spinachia*, forming a group distinguished by character states which are probably associated with decreased snout length (characters 6, 8, and 11) and less elongated bodies (characters 39, 40, 43, 44, and 50). The short-snouted gasterosteids also have less well-developed symplectic flanges and narrower cleithral symphyses, and tend to have reduced interorbital commissures and pelvic girdles with short anterior

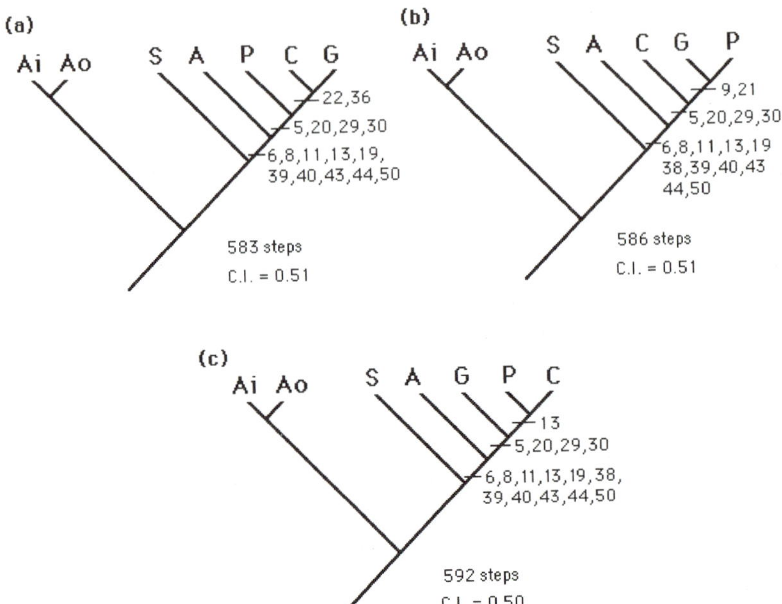

Fig. 2.12 The three shortest cladograms ((a)–(c)) for the Gasterosteidae. Numbers along segments of the cladograms indicate those characters from Appendices 2.2 and 2.3 that change state along the segment indicated. Abbreviations: A, *Apeltes*, Ai, *Aulichthys*; Ao, *Aulorhynchus*; C, *Culaea*; C.I., consistency Index (see text for explanation); G, *Gasterosteus*; P, *Pungitius*; S, *Spinachia*.

processes. In most of these characters, *Spinachia* is intermediate between the Aulorhynchidae and the short-snouted gasterosteids.

Behavioural data (Hall 1956; McLennan *et al*. 1988) also support a basal position for *Spinachia* within the Gasterosteidae. *Spinachia* has relatively simple mating behaviours in which there is no courtship dance, the male simply lunging toward the female and biting her caudal peduncle and the female simply turning toward the male. In the short-snouted gasterosteids, on the other hand, the male's lunge toward the female is supplemented with a pummelling or zigzag ritual, the female responds with a head-up display, and a courtship dance involving a spiral dance, zigzagging, or tail-flagging ensues. *Spinachia* also has relatively simple agonistic behaviours, with neither a head-up nor a broadside threat display.

Appellöf (1894) crossed *Spinachia* and *Gasterosteus*, and Leiner (1934) crossed *Spinachia* with *Gasterosteus* and *Pungitius*; both found that the hybrids did not hatch and had many abnormalities of development.

Interrelationships of the short-snouted gasterosteids

Morphology does little to resolve relationships among the remaining gasterosteids. The four shortest cladograms separate *Apeltes* from *Gasterosteus*, *Pungitius*, and *Culaea* (Fig. 2.12), placing the last three genera in a group with a median ethmoid that does not bend dorsally behind the premaxillary processes, a marked flange extending posteriorly from the cleithrum shaft to the coracoid, and pelvic plates with dorsally directed ascending branches and large posterior processes. Two 598-step cladograms, however, unite *Apeltes*, *Gasterosteus*, and *Culaea* in a group in which the postmaxillary process is absent, the dorsal spines are of different lengths (a character state absent in *Culaea*), and the postdorsal pterygiophores do not form a complete series extending to the uroneural.

Hudon and Guderley (1984) presented an electrophoretic analysis of fourteen proteins from *Apeltes*, *Gasterosteus*, and *Pungitius*. Their calculations of mean genetic identities and differences between the genera indicate that *Gasterosteus* and *Pungitius* are more similar to each other than either is to *Apeltes*. However, Buth and Haglund (Chapter 3 this volume) presented additional allozyme comparisons among gasterosteid genera, and they caution that stickleback genera are too highly divergent for allozymes to be very useful for resolution of their phylogeny. Chen and Reisman (1970) placed *Culaea*, *Gasterosteus*, and *Pungitius* together because the *Gasterosteus–Pungitius* karyotype could be more simply derived from the *Culaea* karyotype than from that of *Apeltes*. *Culaea*, *Gasterosteus*, and *Pungitius* have fewer submetacentric chromosomes than *Apeltes*, but *Apeltes*, *Culaea*, and *Gasterosteus* have fewer metacentric chromosomes than *Pungitius*.

Behavioural data (Hall 1956; McLennan *et al.* 1988) also support cladograms that separate *Apeltes* from the other short-snouted gasterosteids. Like *Spinachia*, *Apeltes* has relatively simple agonistic behaviours. Hall (1956) noted that it has a simple threat display and lacks aggressive colour changes, and that its nest construction involves long bouts of gluing. McLennan *et al.* (1988) also noted that *Apeltes* lacks circle fighting and that no tunnel entrance into the nest is formed. Wootton (1976), however, reported the presence of a tunnel entrance. The male's nuptial colour is confined to his pelvic fins and head (Wootton 1976) and only tail-biting, rather than quivering at the female's tail, is employed by the male to induce spawning. In addition, the male neither builds a nursery nor retrieves the fry. *Apeltes* and *Gasterosteus* were hybridized by Moenkhaus (1911) with very low fertilization rates, and hybrids lived only a few days after hatching.

The unresolved trichotomy of *Gasterosteus*, *Pungitius*, and *Culaea*

The proposed clade containing *Gasterosteus*, *Pungitius*, and *Culaea* cannot be clearly resolved using morphological data. In the shortest cladogram (Fig. 2.12a), *Gasterosteus* and *Culaea* are placed together by their joint possession of posttemporals without closed sensory canals and dorsal spines which do not diverge from the midline. In one (Fig. 2.12b), *Gasterosteus* is placed with *Pungitius* on the basis of their dorsally extended third suborbitals and spurlike posttemporal medial processes. In the other cladogram (Fig. 2.12c), *Pungitius* and *Culaea* are placed together. This placement is supported by only one character, a further reduction in the size of the symplectic dorsal flanges.

Chen and Reisman (1970) placed *Gasterosteus* and *Pungitius* together, as they shared a reduced diploid chromosome number of 42 and possessed four distinctive large metacentric chromosomes and four large submetacentrics. Some behavioural data support this placement in that *Gasterosteus* and *Pungitius* share several behavioural traits, among them short gluing bouts (Hall 1956), creeping through the nest, and the zigzag dance (McLennan *et al.* 1988). McLennan *et al.*, however, grouped *Culaea* with *Pungitius* on the basis of another suite of reproductive characteristics: the presence of insertion gluing, nest showing with the male's snout above the nest, fanning during nest showing, black nuptial coloration, and nursery construction. Sexual coloration of *Pungitius* and *Culaea* is black, while that of *Gasterosteus* is most commonly red. Black populations of *Gasterosteus* are known (Wootton 1976), suggesting that black may be a plesiomorphic state for *Gasterosteus*, *Pungitius*, and *Culaea*, or subject to homoplasy.

Leiner (1940) successfully reared hybrids of *Gasterosteus aculeatus* and *Pungitius pungitius* to maturity, although most were sterile and those which did develop sexually did not develop the complex mating and nest-building behaviours of their parents. Kobayasi (1962) reared larvae of similar crosses to maturity, but found no gametogenesis in the adult hybrids.

Many systematists (e.g. Wiley 1981) hold the view that interfertility is a primitive characteristic which cannot imply close phylogenetic relationships. Leiner's (1940) success in rearing *G. aculeatus* × *P. pungitius* hybrids, then, would indicate that these two species retain the primitive interfertility, while *Spinachia* and *Apeltes* have developed advanced characteristics which make them intersterile with other gasterosteids. Preliminary crosses between *Gasterosteus aculeatus* and *G. wheatlandi* also failed to develop (Bell pers. comm.), indicating that *G. wheatlandi* may also have developed traits causing intersterility.

Uncertainty of the gasterosteid cladogram

The results of osteological analysis of the Gasterosteidae are highly ambiguous. The distributions of osteological characters are not consistent with one another, as reflected in the low consistency indices of the cladograms, and several different cladograms have similar numbers of steps. The basal position of *Spinachia* is supported mainly by characters related to snout and body elongation, which are likely to be functionally interdependent. McLennan *et al.*'s (1988) analysis of reproductive characters yielded clado-

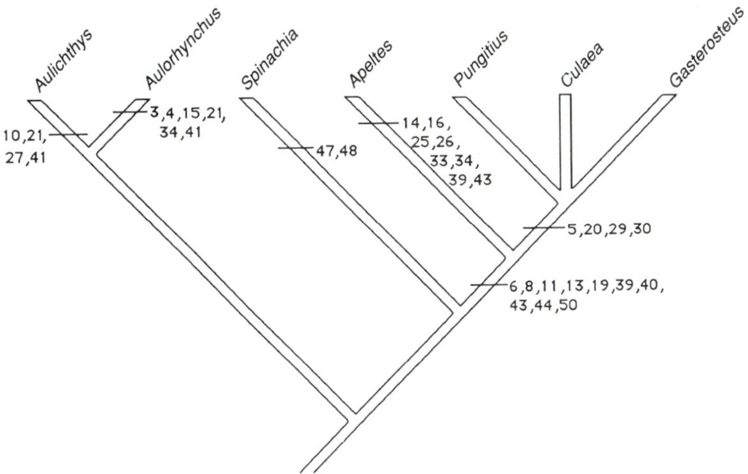

Fig. 2.13 Phylogenetic tree for the Gasterosteiforms. Numbers along segments of the cladogram refer to characters from Appendices 2.2 and 2.3.

grams with higher consistency indices, but these characters also may be functionally correlated (Foster page 384 this volume).

The cladogram presented in Fig. 2.13 is the best-supported cladistic analysis of Gasterosteidae, consistent with the majority of the behavioural and karyotypic data and with the results of electrophoretic studies. The position of *Culaea* remains questionable, because karyotypic and behavioural data disagree and it was not included in the electrophoretic studies. I therefore feel it prudent at this point to regard the *Gasterosteus–Pungitius–Culaea* group as unresolved. Analyses based on other data sets, such as DNA sequences, may resolve this trichotomy.

PROPOSED RELATIVES OF THE GASTEROSTEIFORMES

The Gasterosteiformes traditionally have been placed with the Syngnathiformes (e.g. Cope 1872; Gill 1884; Bridge and Boulenger 1904; Goodrich 1909; Pietsch 1978; Lee *et al.* 1980), a widespread, predominantly marine order characterized by elongate tubular snouts and rigidly armoured bodies. The Syngnathiformes are divided into three lineages: one contains the pipefishes (Syngnathidae) and ghost pipefishes (Solenostomidae), another contains the trumpetfishes (Aulostomidae) and flutemouths (Fistularidae), and the last contains the snipefishes (Macrorhamphosidae) and shrimpfishes (Centriscidae). The synganthid–solenostomid lineage is regarded as the sister group of the other two lineages (Pietsch 1978).

Pietsch (1978), in the most recent review of gasterosteiform relationships, linked the gasterosteiform–syngnathiform assemblage with the Pegasidae (sea-moths) and *Indostomus*, a similarly elongate and armoured fish from Lake Indawgi, upper Myanmar (Burma). He also discussed evidence for including the Dactylopteridae (flying gurnards) in this assemblage, and Ida's (1976) argument for placing the monotypic genus *Hypoptychus* as the sister group of Aulorhynchidae.

The question of whether these groups constitute a monophyletic assemblage is complicated by several factors. Many of the fishes involved have highly modified and reduced skeletons, and it is difficult to homologize their osteological characters with those of other, less specialized fishes. Proposed synapomorphies, once discovered, are likely to be losses whose underlying genetic homology is questionable. The Syngnathiformes are a diverse group and many of the character states that have been used to link them to the Gasterosteiformes are found in only a minority of syngnathiform families. Finally, none of the studies addressing the question has suggested an outgroup by comparison with which the character states uniting these taxa could be polarized. Is the presence of a restricted gill opening, for example, a synapomorphy uniting the pegasids and syngnathiforms, or is it a symplesiomorphy indicating that they are both allied with some other group? Hypotheses about the interrelationships of these fishes can be phylogenetically evaluated only in the context of a broader hypothesis about their position among the Teleostei.

The position of the Gasterosteiformes among teleosts

The gasterosteiforms and their allies have been assigned three different positions in the pantheon of teleost orders. Swinnerton (1902) viewed *Gasterosteus* as a derivative of *Belone* (Series Atherinomorpha). An alternative position was given the Gasterosteiformes by Regan (1913), who placed them with the scorpaenoid fishes because they have the suborbital stay, a connection between the third suborbital and the preoperculum. Nikol'skii (1961) placed them in the same area, regarding them as allies of the cottoid fishes.

Bridge and Boulenger (1904) placed the Gasterosteiformes and Syngnathiformes with the Lamprididae; Lauder and Liem (1983) retain a similar grouping, placing the Gasterosteiformes, Dactylopteridae, and Lampridiformes as separate offshoots of the lineage leading to Percomorpha. They point out, however, that the diagnosis of Percomorpha is extremely unclear. I will therefore concentrate in this section on evaluating evidence for placing the Gasterosteiformes and their allies as derived offshoots of the Atherinomorpha and the Scorpaeniformes, both groups for which several defining character states have been suggested.

The Gasterosteiformes share one of the apomorphies Rosen and Parenti (1981) identified as distinguishing the Atherinomorpha, absence of the fourth

Table 2.2 Character states distinguishing scorpaeniform groups which are also found in Gasterosteiformes.

Scorpaeniformes (Nelson 1984)
1. Suborbital stay present

Hexagrammid–Zaniolepidid–Cottoid lineage (Quast 1965)
1. Hypurals fused into upper and lower plates
2. Single dorsal pterygiophore in each interneural space
3. Gill membranes united

Cottoidea (Yabe 1985)
1. Metapterygoid lamina absent
2. Scapular foramen open
3. Scapula not attached to coracoid
4. Haemal spine on second preural is fused to the centrum
5. Basisphenoid absent
6. Opisthotic small, not reaching prootic
7. Fourth pharyngobranchial absent

Cottidae (Quast 1965)
1. One or two pairs of pharyngobranchials
2. First neural spine often reduced
3. Fewer than five pelvic rays
4. Actinosts square, flat, and platelike
5. First actinost not fused to scapula

pharyngobranchial, and one of the apomorphies uniting Cyprinodontiformes and Beloniformes, the absence of the first pharyngobranchial.

Fewer synapomorphies have been proposed for the Scorpaeniformes, but some of the Gasterosteiformes possess a connection between the third suborbital and the preoperculum. They also possess many of the character states Quast (1965) discussed as defining the Cottidae and the Hexagrammid–Zaniolepidid–Cottoid lineage within the Scorpaeniformes, and several of the cottoid apomorphies identified by Yabe (1985) (Table 2.2). If the Gasterosteiformes are viewed as derived from cottids or the line leading to cottids, what happens to the arguments used to unite them with the Syngnathiformes, *Indostomus*, the Pegasidae, the Dactylopteridae, and *Hypoptychus*?

The proposed gasterosteiform assemblage

Evidence for placing the Gasterosteiformes with the Syngnathiformes

Many of the character states suggested as uniting the Gasterosteiformes and Syngnathiformes are losses or reductions. Some, however, do appear to distinguish these groups from other cottoids. Their body and snout elongation, dorsal connection between the ceratohyal and epihyal, enlarged supraoccipital separating the parietals, and symplectic flange are potential synapomorphies.

The symplectic flange was identified by Banister (1967) as the single character state which most strongly indicated a relationship between these two groups. In the Gasterosteiformes (Fig. 2.9), it is most fully developed in the long-snouted forms. All syngnathiform families investigated possess the flange (Fig. 2.14), but it most closely resembles the gasterosteiform flange in Syngnathidae (Fig. 2.14c). If the presence of a symplectic flange is viewed as a synapomorphy of the Gasterosteiformes and Syngnathiformes, the form of it shared by the Syngnathidae and Aulorhynchidae would be viewed as plesiomorphic, and the expanded flange in other Syngnathiformes might be regarded as a derivation associated with loss or reduction of the anterior suborbitals and expansion of the pterygoid and symplectic to fill their position along the walls of the snout.

Other potential synapomorphies are difficult to reconcile with one another, as each appears in only one syngnathiform lineage. The Gasterosteiformes share the connection between the coracoid and an anteroventral spur from the fourth actinost with the macrorhamphosid–centriscid lineage, and the presence of ectocoracoids with the aulostomid–fistularid lineage. If either of these is taken as a synapomorphy uniting the Gasterosteiformes with all Syngnathiformes, it must be viewed as having undergone reversal in two of the three syngnathiform lineages.

Some syngnathids also resemble the Gasterosteiformes in the direct connection between frontal and parasphenoid behind the orbit (Fig. 2.1c). This connection may have resulted from fusion of the pterosphenoid, which occupies this position in most Percomorpha, with either the frontal or the parasphenoid. In some syngnathids the pterosphenoid is absent, and in others the parasphenoid and frontal contact one another anteriorly to it in a manner very similar to that observed by Yabe (1985) in some specimens of cottoids. This character state may prove to be plesiomorphic for the Gasterosteiform–Syngnathiform lineage, providing more information about their position among cottoids than about their relationship to one another.

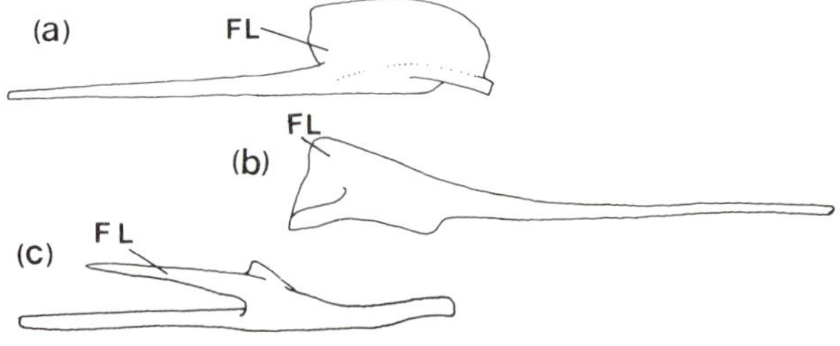

Fig. 2.14 Left lateral view of syngnathiform symplectics: (a) *Macrorhamphosus gracilis* (CAS 47118; × 6.3); (b) *Aulostomus chinensis* (CAS 47120; 5.9); (c) *Syngnathus griseolineatus* (UAMZ 3190, × 4.5). Abbreviation: FL, symplectic flange.

The position of the Pegasidae

The Pegasidae have been placed with the Gasterosteiformes and Syngnathiformes by numerous authors (e.g. Cope 1872; Goodrich 1909). The most recent, Pietsch (1978), advocated placing the pegasids within Syngnathiformes as a sister group to the solenostomid–syngnathid lineage.

Like scorpaeniforms, pegasids have a suborbital stay. Like the Gasterosteiformes and Syngnathiformes, they have a bony strut connecting the ceratohyal and epihyal. In pegasids and syngnathids the body is encased in bony armour, the posttemporals are co-ossified with the cranium, the metapterygoid and supracleithrum are absent, the hyoid arch is short, bearing long filamentous branchiostegals, the gills are lobed, and the gill chamber itself is almost entirely closed, its opening reduced to a dorsal pore on each side. The formation of this restricted gill opening, however, differs greatly in the two groups. The syngnathid opercular series retains a relatively normal teleost morphology, with an operculum forming most of each gill cover and the suboperculum closely applied to its margin. In Pegasidae the operculum and suboperculum are minute, separated bones restricted to the posterodorsal corners of the flattened gill chamber, and most of the gill cover is formed by the expanded and thickened preoperculum. The ventral portion of the preoperculum bends medially to form the floor of the gill chamber. Alexander (1967) pointed out that constricted, dorsally positioned opercular openings are commonly found in bottom-dwelling fishes like the pegasids. A similar pattern of opercular reduction with a large, strong preoperculum whose ventral portion bends medially is found in several scorpaeniform fishes with depressed heads, notably *Peristedion cataphractum* (Allis 1909). Lobed gills may be either synapomorphies of Pegasidae and Syngnathidae or independently associated with the extremely constricted gill opening.

The anterior vertebrae of Pegasidae, like those of Aulostomidae and Fistularidae, are elongated and suturally united, and the neural arches are fused into a solid bony partition along the dorsal midline. The fossil pegasiform *Rhamphosus* resembles Macrorhamphosidae in its possession of a large serrated dorsal spine, the support for which is retained in living pegasiforms, and of anterior scutes which are replaced posteriorly by small scales (Pietsch 1978).

The position of *Indostomus*

Indostomus paradoxus was placed with the Syngnathiformes upon its description by Prashad and Mukerji (1929) and with the Aulorhynchidae and Aulostomidae by Bolin (1936) and Greenwood *et al.* (1966). This placement has been based on general body shape and armature, a reduced lateral line system, and a sutural connection between the cranium and the posttemporal. It is also supported by the presence of free anterior dorsal spines and ossified intramuscular tendons in both *Indostomus* and *Aulostomus*. *Indostomus* also has a dorsally connected ceratohyal and epihyal, and direct connection between the coracoid and an anteroventral spur from the fourth actinost. Its anterior vertebrae are connected by interdigitating processes, their neural spines and arches fused into a solid bony partition on the dorsal midline. *Indostomus* might belong with either the syngnathid–solenostomid or aulostomid–fistularid lineages. Pietsch (1978) suggested that it belonged with the syngnathid–solenostomid lineage, along a line leading to Pegasidae. He based this placement on the presence of a large median bone roofing the snout, which he homologized with the expanded, fused nasals in the Pegasidae.

The position of the Dactylopteridae

Dactylopterids are superficially similar to triglids, and traditionally have been placed with them in Scorpaeniformes (e.g. Allis 1909; Gosline 1971). They possess fused nasals, circumorbital bones attached to the preoperculum by a structure similar to the suborbital stay, and a reduced opercular opening in which the preoperculum forms most of the gill cover, the operculum being reduced and the suboperculum a relatively small, splint-like bone lying at an angle to the opercular margin. The anterior vertebrae and their neural spines are suturally connected to one another. The caudal skeleton in dactylopterids is less fused than in any of the other groups examined, and they have a distinctive pectoral fin with free anterior rays, similar to the triglid pectoral fin.

The character states of the opercular apparatus, as discussed above, are similar to those found in some triglids with depressed heads (although dactylopterids themselves do not have depressed heads). The state of the anterior vertebrae may support placing the Pegasidae and Dactylopteridae with the Syngnathiformes, but as Pietsch (1978) pointed out, in the Dactylopteridae these characteristics may be part of a sound-receiving complex also involving the swim bladder and its associated musculature. If dactylopterid fused vertebrae are in fact derived from such a complex, this raises the question of whether fused vertebrae in the pegasids are similarly derived. The fused vertebrae might provide a synapomorphy uniting the Pegasidae and Dactylopteridae not with Syngnathiformes, but with one another in a triglid lineage.

The position of *Hypoptychus*

Hypoptychus dybowskii, an elongated fish related to Ammodytidae by most authors, was removed to the Gasterosteiformes by Ida (1976). Several of the character states on which he based this placement (i.e. absence of the ectopterygoid, small supracleithrum, large coracoid, large scapular foramen, reduced numbers of circumorbitals and branchiostegal rays, and a caudal skeleton with relatively few rays, fused parahypural, hypural, and urostyle, and non-autologous haemal spines on the last two preural centra) are found in members of various teleost groups, including the Cottoidei (Yabe 1985). Others, the premaxilla with an elongate ascending process and reduced articular process, and a direct connection between the fourth actinost and the coracoid, appear in members of Atherinomorpha. These character states could be viewed either as synapomorphies uniting *Hypoptychus* with the Gasterosteiformes or as indicators that *Hypoptychus* does not belong with the Gasterosteiformes at all, but is a derived atherinomorph. I find the latter suggestion more plausible, as *Hypoptychus* possesses a well-developed maxillary hook very similar to that found in several atherinomorphs.

Conclusion

I accept the traditional grouping of the Gasterosteiformes with the Syngnathiformes on the basis of the symplectic flange. *Indostomus* shares apparent synapomorphies with all three syngnathiform lineages, making it a probable member of this assemblage. Although the Pegasidae share many characters found in Syngnathidae, the opercular series is similar in pegasids and dactylopterids, and resembles that found in triglids with depressed heads. There is a possibility that these two groups belong together in some other scorpaeniform lineage.

ACKNOWLEDGEMENTS

Parts of the research on which this chapter is based were done at the University of Alberta under the supervision of J.S. Nelson as partial requirement for the Ph.D., and were supported by a University of Alberta Dissertation Fellowship. Access to specimens was also provided by W. Roberts, University of Alberta, and R. Henderson, Milwaukee Public Museum. The Computer Center staff at Alverno College, Milwaukee, provided advice in the preparation of the manuscript. D. Engelmann kindly reviewed the osteological descriptions and figures. Comments from an anonymous reviewer were invaluable.

APPENDIX 2.1

Specimens of Gasterosteiformes examined.

Apeltes quadracus, UAMZ 5540, eight specimens, 2.49, 2.63, 2.47, 2.32, 2.43, 2.37, 2.58, and 2.21 cm SL.

Aulichthys japonicus, UAMZ 5542, four specimens, 9.34, 11.25, 11.88, and 9.61 cm SL.

Aulorhynchus flavidus, UAMZ 5541, three specimens, 8.37, 7.9, and 5.85 cm SL; UAMZ 1694, four specimens, 9.92, 10.56, 10.84, and 12.13 cm SL.

Culaea inconstans, UAMZ 5023, two specimens, 5.38 and 6.09 cm SL; UAMZ 5564, three specimens, 4.84, 4.29, and 4.12 cm SL; MPM 11778, one specimen, 4.63 cm SL.

Gasterosteus aculeatus, UAMZ 5512, three specimens, 6.97, 4.37, and 2.8 cm SL; UAMZ 4727, two specimens, 4.7 and 5.32 cm SL.

Gasterosteus wheatlandi, UAMZ uncatalogued specimens, five specimens, 3.5, 3.6, 3.5, 3.6, and 3.8 cm SL.

Pungitius pungitius, UAMZ 4437, two specimens, 5.10 and 5.52 cm SL; UAMZ 4754, three specimens, 3.05, 3.3, and 3.65 cm SL.

Spinachia spinachia, UAMZ 5548, six specimens, 7.84, 5.47, 3.00, 6.14, 10.58, and 5.41 cm SL.

APPENDIX 2.2

Characters used in cladistic analyses of Gasterosteidae.

Ethmoid region

1. *Median ethmoid length/snout length (range 0.15–0.54)
2. Nasal process: (0) absent, (1) elongated, (2) fan-like
3. Median ethmoid lateral flanges: (0) absent, (1) shallow, (2) deep
4. Median ethmoid lateral flanges: (0) separated, (1) close
5. Median ethmoid dorsal plate: (0) flat, (1) anterior end bent ventrally

Orbital region

6. *Eye length/head length (range 0.14–0.42)
7. Interorbital commissure: (0) bony canal, (1) medially opening pore, (2) not apparent
8. Lachrymal: (0) elongated, (1) short
9. Third suborbital posterodorsal extension: (0) absent, (1) present

Suspensorium and branchial skeleton

10. Postmaxillary process: (0) absent, (1) slight, (2) marked
11. Premaxillary ascending process: (0) long, (1) short
12. *Palatine length/head length (range 0.093–0.37)
13. Symplectic dorsal flanges: (0) small, (1) medium, (2) large
14. Operculum posterior margin: (0) concave, (1) rounded, (2) extended
15. *Opercular angle (range 68°–117°)
16. Gill membranes: (0) united, (1) joined to isthmus (Nelson 1971)

Pectoral skeleton

17. Supracleithrum: (0) present, (1) absent
18. Cleithra lateral flange: (0) full length, (1) narrow ventrally, (2) short
19. Cleithral symphysis: (0) narrow, (1) slightly expanded, (2) wide
20. Cleithral extension to coracoid fan: (0) small or absent, (1) low crest, (2) large
21. Posttemporal medial process: (0) absent, (1) low, (2) spurlike
22. Posttemporal canal: (0) closed, (1) open or absent
23. Scapular foramen: (0) small, (1) large
24. Scapular foramen: (0) open, (1) closed
25. Ectocoracoids: (0) meet below cleithrum, (1) touch cleithrum symphysis, (2) touch cleithrum shafts
26. Coracoid foramen: (0) absent, (1) present
27. Number of pectoral rays: (0) 10, (1) 11

Pelvic skeleton

28. Pelvic anterior processes: (0) united, (1) short and separated, (2) long and separated
29. Pelvic posterior process: (0) short, (1) long
30. Pelvic ascending branch direction: (0) dorsal, (1) posterior
31. Pelvic ascending branch size: (0) short, (1) long
32. Number of pelvic rays
33. Pelvic spine: (0) smooth, (1) serrated
34. Pelvic spine origin: (0) anterior, (1) mid-P1, (2) posterior

Median skeleton

35. Number of predorsal pterygiophores
36. Anterior dorsal spines: (0) divergent, (1) straight
37. Dorsal spines: (0) of equal length, (1) different lengths
38. Post-dorsal pteryglophores: (0) absent, (1) incomplete series, (2) complete series to uroneural
39. *Number of precaudal vertebrae (range 10–33)
40. *Number of caudal vertebrae (range 14–30)
41. *Number of pleural ribs (range 0–24)
42. *Number of epipleural ribs (range 0–23)
43. *Number of lateral plates (range 0–59)
44. *Number of dorsal spines (range 1–26)
45. Lateral plates: (0) absent, (1) small, (2) large
46. Lateral plate position: (0) full length, (1) cuirass and keel, (2) cuirass, (3) keel
47. *Number of dorsal rays (range 5–13)
48. *Number of anal rays (range 5–12)
49. Number of caudal rays: (0) 11, (1) 12, (2) 13, (3) 14
50. *Number of post-dorsal pterygiophores (range 2–15)

*These characters were coded by dividing the range into seven equal segments corresponding to character states 0–6.

APPENDIX 2.3

Coded data used in cladistic analyses of Gasterosteidae; characters and codes are explained in Appendix 2.2.

Char.	Genus[a]						
	Auli	Aulo	Spin	Gast	Cula	Apel	Pung
Ethmoid region							
1	4–5	5–6	2–4	1–2	0–1	1–4	2
2	0	0	1	1	1	2	2
3	1	2	1	2	1	1	0
4	0	1	0	1	0	0	?
5	0	0	0	1	1	0	1
Orbital region							
6	0–1	1–2	2	2–3	3–5	4–6	4
7	0	0	1	2	2	1–2	2
8	0	0	0	1	1	1	1
9	0	0	0	1	0	0	1

Appendix 2.3 *cont.*

Char.	Genus[a]						
	Auli	Aulo	Spin	Gast	Cula	Apel	Pung
Branchiocranium							
10	2	0	1	0	0	0	1
11	0	0	0	1	1	1	1
12	3–4	1–3	0	0–1	0	3–6	0–1
13	2	2	2	1	0	0–1	0
14	2	2	1	1	1	0	1
15	2–4	5–6	1–3	4–5	1–3	0–4	1–4
16	0	0	0	1	0	1	0
Pectoral girdles and fins							
17	0	1	1	0–1	0	0	0
18	0	0	2	1	1–2	1–2	1
19	2	2	1	0	0	0	0
20	0	0	0–1	2	2	1	2
21	2	0	1	2	1	1	2
22	0	0	0	1	1	0	0
23	0	0	1	1	1	0	1
24	0–1	0	0	0–1	0	0	0–1
25	1	1	0	0	2	2	0
26	1	1	1	1	0–1	0	1
27	1	0	0	0–1	0	0–1	0
Pelvic skeleton and fins							
28	2	2	2	0–1	1	0	1
29	0	0	0	1	1	0	1
30	1	1	1	0	0	1	0
31	0	0	1	0–1	0	1	0–1
32	4	4	2–3	1–3	1	2–3	1–2
33	0	0	0	0–1	0	1	0–1
34	2	0	2	1	1	0	1
Median skeleton							
35	0	0–2	2–3	2	2	1	2
36	1	1	0	1	1	0	0
37	0	0	0	1	0	1	0
38	2	2	2	1	0–1	1	2
39	3–4	4–6	2–3	1	1	0	1
40	5–6	5–6	3	1–2	0–1	1–2	1–2
41	5–6	0–1	4–5	2	2–3	2	1–2
42	0	0	4–6	2–4	4	2	2–4
43	6	6	4–5	0–4	3–4	0	0–3
44	6	6	3–4	0	0–1	0–1	1–4
45	1	1	1	2	1	0	1–2
46	0	0	0	0–2	0	?	0–3
47	4	4	1	0–5	2–6	4–6	0–5
48	4–5	4–5	1–2	0–6	2–6	3–5	1–6
49	2	2–3	1	0–1	0–1	1	1
50	5–6	6	5–6	0–1	0–1	0–2	1–2

[a] Genera: Apel, *Apeltes*; Auli, *Aulichthys*; Aulo, *Aulorhynchus*; Cula, *Culaea*; Gast, *Gasterosteus*; Pung, *Pungitius*; Spin, *Spinachia*.

3

Allozyme variation in the *Gasterosteus aculeatus* complex

Donald G. Buth and Thomas R. Haglund

Biologists from many disciplines have chosen to study *Gasterosteus aculeatus* because of its highly variable behavioural, ecological, physiological, and morphological characters, which have provided the raw materials with which to examine correlations and have challenged researchers to explain the observed diversity through analytical and experimental approaches. However, the very characters that have made *G. aculeatus* so attractive to researchers are incompletely understood in terms of genetic control and vulnerability to environmental influence. Questions about population divergence and gene flow have been answered ambiguously, or only in localized contexts.

The electrophoresis of allozymes has been, and continues to be, a valuable and economical method for analysis of the evolutionary processes of divergence and gene flow (Avise 1974; Buth 1984, 1990; Morizot and Schmidt 1990; Murphy *et al.* 1990). Yet despite the circumglobal distribution, popularity as an experimental organism, and intriguing patterns of variation of *G. aculeatus*, its allozymes did not begin to receive rigorous attention until the mid 1980s.

Electrophoretic studies of allozymes and other proteins of *G. aculeatus* are summarized in Table 3.1. Hagen (1967) was the first to apply electrophoretic technology to address the divergence of anadromous and freshwater populations in North America. Soon comparable situations were studied in Asia (Muramoto *et al.* 1969) and Europe (Raunich *et al.* 1971, 1972). However, these early studies examined the products of only one or two variable loci.

A second generation of allozyme studies addressed specific problems of dimorphic variation in sympatry (Avise 1976; McPhail 1984), allopatric or parapatric divergence within a single drainage (Bell and Richkind 1981; Buth *et al.* 1984; Baumgartner 1986*b*), or variation in a limited region (Haglund and Buth 1988; Haglund *et al.* 1990; Yang and Min 1990; Buth *et al.* unpubl. data). These studies usually employed an increased number of allozyme loci but depended upon identifying enough variable loci with which to make statistical comparisons. Only recently has this design been expanded to

Table 3.1 Selected studies of allozymes and (nonspecific) proteins of *Gasterosteus aculeatus*.

Nature of comparison (Location)	Number of geographic samples	Number of loci		Findings[a]	Reference
		Total	Polyallelic		
Inheritance study (Canada)	1	6	6	Confirmed Mendelian interpretation for five loci; sexual dimorphism in isocitrate dehydrogenase expression	Withler *et al*. (1986)
Anadromous v. freshwater (Canada)	5+	1	1	Identified a diagnostic general muscle protein	Hagen (1967)
Anadromous v. freshwater *G. a. aculeatus* v. *G. a. microcephalus* (Japan)	3?	2	2	Additional electromorphs in anadromous sample; freshwater samples identical	Muramoto *et al*. (1969)
Anadromous v. freshwater (Italy, Germany, Sweden)	11 + 6	1	1	Second haemoglobin allele limited to anadromous samples	Raunich *et al*. (1971, 1972)
Anadromous v. freshwater (Poland)	2 + 4	5	1	Second sMdh-A allele limited to anadromous samples	Zietara (1989)
Anadromous v. freshwater (Poland)	6 + 13	13	7	$D = 0\text{–}0.03$	Rafinski *et al*. (1989)
Anadromous v. freshwater; temporal resampling (Canada, USA)	16 + 40	8	6	$F_{ST} = 0.046$ among anadromous, 0.293 among freshwater; no temporal changes	Withler and McPhail (1985)
Complete v. low plate morphs in sympatry (USA)	1	15	4	$I = 0.999$	Avise (1976)

'Benthic' v. 'limnetic' morphs in sympatry (Canada)	1	6	4	Represent separate genomes	McPhail (1984)
Clinal variation (plates) (USA)	7	12	1	D = 0–0.003; no Pgm-A cline	Bell and Richkind (1981)
Clinal variation (allozymes); temporal resampling (USA)	7	25	4	Stable clines; no temporal changes	Baumgartner (1986*b*)
Regional study (Korea)	7	25	15	Mean S = 0.963	Yang and Min (1990)
G. a. aculeatus v. *G. a. microcephalus* (USA)	2	19	9	I = 0.97	Avise (1976)
G. a. microcephalus v. *G. a. williamsoni* (USA)	2	45	6	D = 0.034	Buth *et al.* (1984)
G. a. microcephalus v. *G. a. williamsoni* (USA)	17	7	7	Regional divergence with minimal intergrade zone	Buth *et al.* (unpubl. data)
G. a. williamsoni v. '*Shay Creek stickleback*' (USA)	3	44	11	Represent separate genomes	Haglund and Buth (1988)
G. aculeatus v. '*white stickleback*' (Canada)	3	19	6	I = 0.995–1.00	Haglund *et al.* (1990)
Intra- and intercontinental (Japan, USA, Canada, UK, Italy, Sweden)	16	18	13	Some Japanese samples highly divergent from all others	Haglund *et al.* (1992*a*)
Intra- and intergeneric (Canada)	1	'14'	'14'	Four gasterosteid species different; *G. aculeatus* sister to *G. wheatlandi*	Hudon and Guderley (1984)

[a] Includes Nei's (1972, 1978) coefficients of genetic distance (D) and genetic identity (I), and Rogers' (1972) coefficient of genetic similarity (S).

examine variation among non-problematic populations of *G. aculeatus* on a global level (Haglund *et al.* 1992*a*).

Direct taxonomic comparisons have always been tempting, especially among named subspecies (Avise 1976; Buth *et al.* 1984). Comparisons with other gasterosteids are now becoming popular (Hudon and Guderley 1984; Yang and Min 1990; Haglund *et al.* 1992*a*,*b*, unpubl. data).

This chapter reviews known patterns of allozyme divergence within the *G. aculeatus* complex. Beginning with what is known of patterns of intra-population variation, the review discusses polyallelic loci, heterozygosity estimates, and concordance with Hardy–Weinberg expectations. The scope of studies is then expanded to examine local or regional interpopulation allozyme divergence with respect to differences in lateral plate morphology, colour polymorphisms, and among western North American subspecies. This examination culminates with a discussion of what is known of the global pattern of divergence and evidence for the recognition of species within the complex. The patterns of allozyme variation within the *G. aculeatus* complex are compared with those known for other gasterosteids. The chapter concludes with a discussion of potential directions for future research.

METHODS

Current enzymatic technology is sufficient to resolve the gene products of more than 100 loci in teleosts like *G. aculeatus* (e.g. the products of both single locus and multilocus systems stained according to the formulae listed in Buth and Murphy 1990 and in Morizot and Schmidt 1990). However, it is unlikely that all such products could be stained from a single specimen, owing to the limited amount of tissue available in such a small species. Most limiting, of course, would be liver tissue, which is often the only or predominating source of many enzymes (e.g. Adh-A, sAat-B, Gcdh-A, G3pdh-B, Iddh-A, Xdh-A; see Table 3.2). Another limiting factor is that of 'shelf-life' of the enzymes. It has been our experience with cypriniform and perciform fishes that the activity of many enzymes in frozen storage can be maintained for many years if

(1) intact specimens are frozen in ice blocks upon capture and maintained this way until dissection to prevent desiccation;
(2) tissue extracts are prepared only for immediate use, not as a means of long-term storage; and
(3) freezing and thawing of specimens or extracts are kept to a minimum.

Even under these conditions, many enzymes of *G. aculeatus* (e.g. adenosine deaminase) do not remain active in frozen storage for as long as they do in most other teleosts. Some enzymes (e.g. lactate dehydrogenase, malate

Table 3.2 Enzymes, loci, tissue sources, and buffer conditions in selected studies of *Gasterosteus aculeatus*.

Enzyme	Enzyme Commission number	Locus	Multiple alleles reported	Tissue source(s)	Buffer[a] and reference[b]
Acid phosphatase	3.1.3.2	Acp-A	+	Liver	A[1] B[2] C[3]
Aconitate hydratase	4.2.1.3	mAcoh-A	+	Muscle	D[2]
		sAcoh-A	–	Liver	D[2]
Adenosine deaminase	3.5.4.4	Ada-A	+	Muscle	B[2] E[1] F[2]
Adenylate kinase	2.7.4.3	Ak-A	+	Muscle	B[1] D[4]
Alcohol dehydrogenase	1.1.1.1	Adh-A	+	Liver	C[3] D[1]
Aspartate aminotransferase	2.6.1.1	mAat-A	–	Liver, muscle	C[3,5] F[1]
		sAat-A	+	Liver, muscle	C[3,5] F[1]
		sAat-B	–	Liver	F[2]
Calcium binding proteins	–	Cbp-1[c]	–	Muscle	D[2]
		Cbp-2[c]	–	Muscle	D[2]
Creatine kinase	2.7.3.2	Ck-A	+	Muscle	D[1,4] G[3]
		Ck-B	+	Brain + eye	D[1,4] G[3]
Dihydrolipoamide dehydrogenase	1.8.1.4	Ddh-1	–	Brain + eye	E[1]
Esterase[d]	3.1.1.–	'Est-1'	+	Muscle	C[5]
		'Est-2'	–	Muscle	C[5]
		'Est-3'	–	Muscle	C[5]
		Est-3	–	Brain + eye	E[1]
		Est-4	–	Brain + eye	E[1]
		Est-5	+	Brain + eye	E[1]
Fructose-bisphosphatase	3.1.3.11	Fbp-1	–	Liver	D[1]
		Fbp-2	–	Liver	D[2]
Fructose-bisphosphate aldolase	4.1.2.13	Fba-A	–	Muscle	E[1]
		Fba-B	–	Liver	E[1]
		Fba-C	–	Brain + eye	E[1]

Table 3.2 *cont.*

Enzyme	Enzyme Commission number	Locus	Multiple alleles reported	Tissue source(s)	Buffer[a] and reference[b]
Fumarate hydratase	4.2.1.2	Fumh-A	–	Liver	E^{1}
General protein	-	'Pt-0'	–	Liver	C^{5}
		'Pt-1'	+	Muscle	C^{5}
		'Pt-2'	–	Muscle	C^{5}
		'Pt-3'	+	Muscle	C^{5}
		Gp-1[c]	–	Muscle	D^{1}
Glucokinase	2.7.1.2	Gk-1	–	Muscle	C^{3}
Glucose dehydrogenase	1.1.1.47	Gcdh-A	+	Liver	E^{1}
Glucose-6-phosphate isomerase	5.3.1.9	Gpi-A	+	Muscle, brain	$C^{1,5}D^{4}G^{3}$
		Gpi-B	+	Muscle	$C^{1,5}D^{4}G^{3}$
Glucose-6-phosphate dehydrogenase	1.1.1.49	G6pdh-A	–	Muscle	$E^{1}G^{3}$
Glutamate dehydrogenase	1.4.1.2	Gtdh-A	–	Liver	E^{1}
Glyceraldehyde-3-phosphate dehydrogenase	1.2.1.12	Gapdh-A	–	Muscle	$D^{1}F^{2}$
Glycerate dehydrogenase	1.1.1.29	Glydh-A	–	Liver	E^{2}
Glycerol-3-phosphate dehydrogenase	1.1.1.8	G3pdh-A	+	Muscle	$A^{2}B^{1}E^{2}H^{3}J^{5}$
		G3pdh-B	–	Liver	A^{1}
L-iditol dehydrogenase	1.1.1.14	Iddh-A	+	Liver	D^{1}
Isocitrate dehydrogenase[e]	1.1.1.42	mIcdh-A	–	Muscle	E^{1}
		sIcdh-A	–	Muscle	$D^{4}E^{1}H^{3}J^{5}$
L-lactate dehydrogenase	1.1.1.27	Ldh-A	+	Muscle	$C^{1,3,5}I^{6}J^{4}$
		Ldh-B	–	Heart, brain + eye	$C^{1,3}I^{6}J^{4}$
		Ldh-C	–	Brain + eye	$C^{1,3}I^{6}J^{4}$
Malate dehydrogenase (NAD-dependent)	1.1.1.37	mMdh-A	+	Muscle	$F^{1}H^{2}J^{4,5}$
		sMdh-A	+	Muscle	$F^{1}H^{2}J^{4}I^{6}$
		sMdh-B	+	Muscle	$F^{1}H^{2}J^{4}I^{6}$

Malate dehydrogenase (NADP-dependent)	1.1.1.40	mMdhp-A	+	Muscle	F[2]
		sMdhp-A	+	Muscle	F[1] H[3]
Mannose-6-phosphate dehydrogenase	5.3.1.8	Mpi-A	+	Muscle	C[1,3]
Peptidase[f]	3.4.-.-	Pep-A	+	Muscle	D[1]
	3.4.-.-	Pep-B	−	Muscle	D[2]
	3.4.-.-	Pep-C	+	Brain + eye	E[2]
	3.4.13.9	Pep-D	−	Liver, muscle	D[2]
	3.4.11.1	Pep-E	−	Muscle	E[1]
	3.4.-.-	Pep-F	−	Muscle	C[2] D[2]
	3.4.-.-	Pep-S	+	Muscle	D[1]
Phosphoglucomutase	5.4.2.2	Pgm-A	+	Muscle	C[1,5] J[4] K[3]
Phosphogluconate dehydrogenase	1.1.1.44	Pgdh-A	+	Liver, muscle	D[1] H[3] J[5]
Purine-nucleoside phosphorylase	2.4.2.1	Pnp-A	+	Muscle	F[1]
Pyruvate kinase	2.7.1.40	Pk-A	−	Muscle	F[1]
Superoxide dismutase	1.15.1.1	sSod-A	−	Liver	B[2] C[3] D[1]
Triose-phosphate isomerase	5.3.1.1	Tpi-B	+	Liver	C[3,5]
Xanthine dehydrogenase	1.1.1.204	Xdh-A	−	Liver	D[1]

[a] A, histidine-citrate, pH 7.0 (Brewer 1970); B, histidine-citrate, pH 8.0 (Brewer 1970); C, Poulik system (Selander *et al*. 1971); D, tris-citrate, pH 8.0 (Selander *et al*. 1971); E, phosphate-citrate (Selander *et al*. 1971); F, tris-citrate, pH 7.0 (Whitt 1970); G, lithium hydroxide-borate (Selander *et al*. 1971); H, tris-citrate pH 7.0 (Siciliano and Shaw 1976); I, tris-citrate pH 7.0 (Zietara 1989); J. tris-citrate-EDTA, pH 7.0 (Avise *et al*. 1975); K, tris-maleate-EDTA, pH 7.6 (Brewer 1970).

[b] 1, Buth *et al*. (1984); 2, Buth and Haglund (unpubl. data); 3, Baumgartner (1986*b*); 4, Rafinski *et al*. (1989); 5, Avise (1976); 6, Zietara (1989).

[c] Follows Buth (1982).

[d] Five esterase products (numbered from cathode to anode) were resolved from brain + eye extracts by Buth *et al*. (1984), however only the three most anodal products were consistently scorable. These products may or may not be homologous to any of the esterase products scored from muscle by Avise (1976).

[e] If only a single ICDH locus was scored in a given study, it was assumed that it was the sIcdh-A locus whose products predominate.

[f] Peptidase substrates followed the recommendations of Buth and Murphy (1990).

dehydrogenase), however, will remain active for several years. These variables place constraints on the experimental design of studies of *G. aculeatus*. If the gene products of a large number of loci are to be studied, the electrophoresis should be done soon after the specimens are obtained. If many geographic samples are compared, their collection must be coordinated so that all can be examined while the tissues are comparably fresh. To date, most investigators have scored nearly every enzyme examined from every specimen available. However, if the number of loci studied is to be increased much more, investigators may have to settle for scoring 'genomes per sample', not specimens per sample, because the limited tissue available will require that different sets of individuals contribute different enzyme systems to the database.

Table 3.2 lists the enzymes, loci, tissue sources, and buffer conditions used in six electrophoretic studies of *G. aculeatus*. With the exception of the peptidases (which follow Buth and Murphy 1990) and calcium-binding proteins (which follow Buth 1982), the enzyme nomenclature used herein is that recommended by the International Union of Biochemistry (1984). The system of locus nomenclature followed the recommendations of Buth (1983) for teleost studies, with minor modifications. Not all studies cited in Table 3.1 followed these recommendations. Several enzyme names were updated, including aspartate aminotransferase (formerly 'glutamic-oxaloacetate transaminase' = GOT), fructose-bisphosphatase aldolase (formerly 'aldolase' = ALD), glucose-6-phosphate isomerase (formerly 'phosphoglucoisomerase' = PGI), glycerol-3-phosphate dehydrogenase (formerly 'a-glycerophosphate dehydrogenase' = aGPD), NADP-dependent malate dehydrogenase (formerly 'malic enzyme' = ME), peptidase-E (formerly 'aminopeptidase' = AP, or 'leucine aminopeptidase' = LAP), and superoxide dismutase (formerly 'tetrazolium oxidase' = TO). Locus abbreviations include the prefixes m or s to identify mitochondrial and 'supernatant'/cytosolic subcellular localization of expression, if relevant, and locus symbols of letters (if vertebrate homology can be inferred) or numbers (if such homology is unknown). In most cases, the identity of a locus could be inferred from the tissue source of its product scored in a given study (e.g. Ldh-A and Ck-A from muscle, Ck-B and Ldh-C from eye or brain). Inferring homologies for nonspecific esterases and general proteins proved to be the most difficult problem across studies, and their original published locus designations were retained in Table 3.2.

We have omitted some loci from Table 3.1 because the numbers reported are at odds with expression reported for diploid teleosts (e.g. Fisher *et al.* 1980) or our own investigations. For example, Hudon and Guderley (1984) reported the products of five malate dehydrogenase loci, whereas only three MDH loci are expressed in diploid teleosts (Rainboth and Whitt 1974; pers. obs.). Baumgarter (1986*b*) reported products of three glycerol-3-phosphate dehydrogenase loci (as Gpd-1, Gpd-2, Gpd-3) from muscle extracts, whereas diploid teleosts express only two G3PDH loci (Fisher *et al.* 1980) and only the products of G3pdh-A in muscle tissue (pers. obs.). Yang and Min (1990) reported the products of three phosphoglucomutase loci, but we have conservatively listed only Pgm-A from muscle tissue. The previously listed studies did not provide zymograms of these problematic expressions of alleged multilocus products, so the resolution of some of these inconsistencies remains to be investigated.

The buffers listed in Table 3.2 are simply those reported by the investigators. In the few studies in which the same enzymes were scored by different investigators, it is usually the case that different buffer systems were used. Whereas in cyprinid fishes, a large number of buffers can often yield comparable results for many enzymes (Buth *et al.* 1991), the same is not necessarily true for *G. aculeatus*. Most investigators have arrived at 'optimal' buffers by a trial-and-error method, instead of a rigorous but expensive and time-consuming screen of all systems using many buffers, and what may be deemed optimal by one investigator may not be so for another. Improved resolution may be possible for many enzymes listed in Table 3.1 (Lawson pers. comm.). We urge that additional buffer experimentation be conducted and that zymograms depicting optimal resolution be published.

PATTERNS OF ALLOZYME VARIATION

Variation within populations

Polyallelic loci

Nearly half the loci listed in Table 3.2 have been reported to have multiple allelic forms in *G. aculeatus* by the references therein and in Table 3.1. Additional variants have been reported at a general protein locus (Hagen 1967), galactose-6-phosphate dehydrogenase and glucose-6-phosphate dehydrogenase (Muramoto *et al.* 1969), and haemoglobin (Raunich *et al.* 1971, 1972). However, the genetic control of the differences apparent in these early studies was usually incompletely known. For many of the loci listed above, allelic variation was limited to a few uncommon heterozygotes and/or polyallelism reported from a single geographic locality. Several variables, including locations sampled, array of loci studied, and electrophoretic conditions used, preclude the discussion of general trends in allozyme variation.

Certain better-known enzyme systems, for example glucose-6-phosphate isomerase, lactate dehydrogenase, malate dehydrogenase, and phosphoglucomutase, examined routinely in most allozyme studies of fishes, were included in most of the studies listed in Table 3.1. Of these four systems, malate dehydrogenase was most frequently reported as having polyallelic loci in *G. aculeatus*. However, these studies rarely reported the allelic composition of all three known loci, or used an informative system of locus nomenclature that would distinguish the non-interactive mitochondrial product (mMdh-A) from the interactive supernatant products (sMdh-A, sMdh-B), which form an AB interlocus heterodimer (Buth 1990). Studies that scored all three loci include those of Baumgartner (1986*b*), Withler *et al.* (1986), and Rafinski *et al.* (1989), who simply numbered the loci (i.e. Mdh-1, Mdh-2, Mdh-3), evidently in order of increasing anodal migration of their products. However, this numbering scheme was not necessarily followed in other studies that scored only two MDH loci. McPhail (1984) and Withler and McPhail (1985) had previously made reference to 'Mdh-1' and 'Mdh-3' products. Identification of the three loci as mMdh-A, sMdh-A, and sMdh-B in *G. aculeatus* was first made by Buth *et al.* (1984), although no zymogram was depicted because these loci were invariant in the southern California populations they examined. Zymograms expressing MDH variation have been published by Withler *et al.* (1986), Zietara (1989), and Haglund *et al.* (1992*a*). The identification of the loci by Buth *et al.* (1984) and Haglund *et al.* (1992*a*), as well as the scoring of all three loci by Withler and McPhail (1985) and Baumgartner (1986*b*), was facilitated by the clear separation of the products in populations from the west coast of North America (Fig. 3.1). In eastern North America and in Europe, different MDH alleles yield products in *G. aculeatus* that do not separate as well electrophoretically. For example, Zietara (1989) scored only 'Mdh-A' and 'Mdh-B' products but not a mitochondrial product. Judging from his published zymogram, we suspect that the mMdh-A product is superimposed with the AB interlocus heterodimer in his samples from Poland. We also suspect that his 'A' and 'B' products are reversed, because sMdh-B products predominate in muscle tissue. Given the predominance of mMdh-A and sMdh-B products in muscle tissue, the interaction of sMdh-A and sMdh-B products, and the expected intermediate mobility of their AB heterodimer (e.g. Fig. 3.1), we

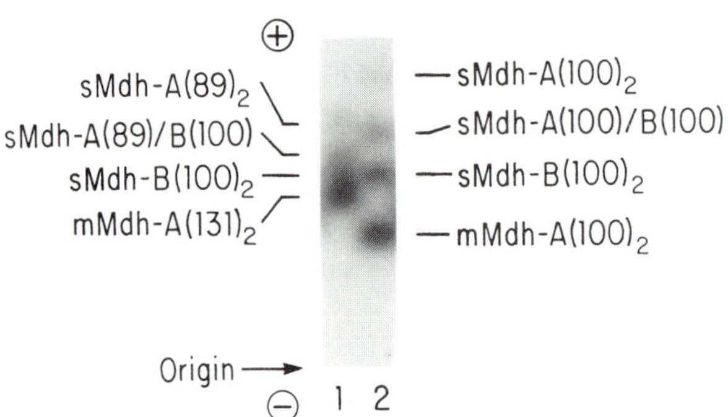

Fig. 3.1 Zymogram of NAD-dependent malate dehydrogenase expression from muscle extracts, showing the typical Atlantic basin pattern (specimen 1) and the typical western North American pattern (specimen 2). The composition of all electromorphs is indicated. (From Haglund *et al.* 1992*a*.)

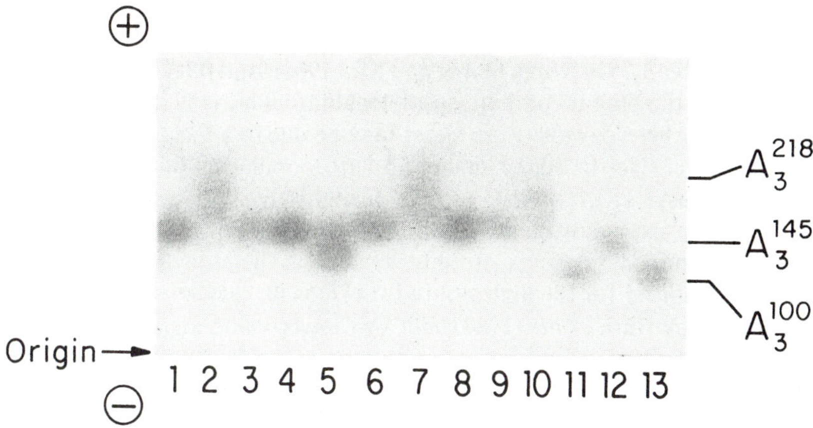

Fig. 3.2 Zymogram of trimeric purine-nucleoside phosphorylase (Pnp-A) variation in a southern California, USA, sample of *G. aculeatus*. Products of four genotypes are shown: specimens 1, 3, 4, 6, 8, 9, and 12 express the 145/145 homozygous condition; specimens 11 and 13 express the 100/100 homozygous condition; specimen 5 expresses the 100/145 heterozygous condition; specimens 2, 7, and 10 express the 145/218 heterozygous condition. Only the homotrimeric products are labelled.

suspect that the MDH zymogram of Withler *et al.* (1986), with covarying 'Mdh-1' and 'Mdh-2', has been printed upside down. Variation in MDH can be quite useful in population studies of *G. aculeatus*, but specific identification of the gene products and publication of zymograms is necessary to facilitate comparisons among studies.

Purine-nucleoside phosphorylase (Pnp-A) is scorable from muscle tissue in *G. aculeatus* and has been shown to vary in virtually every population in which it has been studied (Buth *et al.* 1984; Haglund and Buth 1988; Haglund *et al.* 1992*a*). It was the only locus to vary in every sample in the global study of Haglund *et al.* (1992*a*), and was often triallelic (Fig. 3.2). Unfortunately, other investigators to date have not chosen to score Pnp-A products. We urge that Pnp-A be added to future studies.

Heterozygosity estimates

The percentage of polyallelic loci in *G. aculeatus* is quite high in most populations studied: 21.0–42.1 per cent (Avise 1976), 16.0–48.0 per cent (Yang and Min 1990), and 16.7–38.9 per cent (Haglund *et al.* 1992*a*). It follows that heterozygosity levels in *G. aculeatus* might be higher than the mean value of 0.051 reported for teleosts (Nevo 1978). Although direct-count heterozygosity varies from place to place (Table 3.3), the mean values are higher than the teleost average in most studies that scored 18 or more loci: 0.115 (two locations) in central California, USA (Avise 1976), 0.070 (three locations) in Nova Scotia, Canada (Haglund *et al.* 1990), 0.086 (seven locations; Table 3.3) in Korea (Yang and Min 1990), and 0.065 (16 locations; Table 3.3) in the global comparison of Haglund *et al.* (1992*a*). Exceptions yielding lower heterozygosity estimates may be due to a bias from a smaller number of loci (Nei 1978), or may be due to population bottlenecks (Bell 1976*a*). Rafinski *et al.* (1989) found mean heterozygosities to be 0.036 and 0.028 in six anadromous and 13 freshwater samples, respectively, but they examined products of only 13 loci. A heterozygosity value of 0.029 was reported for the unarmoured stickleback, *Gasterosteus aculeatus williamsoni*, by Buth *et al.* (1984), but this lower value might be expected in inland, upstream, freshwater populations, as has been shown for *Poeciliopsis* by Vrijenhoek *et al.* (1977).

Haglund *et al.* (1992*a*; Table 3.3) noted that in each of 16 geographic samples, the observed (direct-count) heterozygosity was nominally below the value expected from Hardy–Weinberg estimates. The consistency of the observed heterozygote deficiency within the samples indicates a significant tendency toward heterozygote deficiency (exact probability that all results would be lower than the Hardy–Weinberg expectation is $(1/2)^{N-1} = 0.000030517$). This result could be attributed to a low level of inbreeding in finite natural populations. However, this relationship has not been corroborated in other studies (Avise 1976; Yang and Min 1990; Table 3.3).

The evolutionary significance of heterozygosity has been debated. However, at the very least, heterozygosity can serve as an important marker for calculations of certain population phenomena (see below). The levels of variation throughout the range of *G. aculeatus* suggest that severe bottlenecks that drastically reduce variation must be rare in this complex.

Hardy–Weinberg equilibrium

Experimental crosses have verified that allozymes of *G. aculeatus* behave as Mendelian characters (Withler *et al.* 1986). High levels of variation provide ample material with which to test whether the genotypic distributions correspond with Hardy–Weinberg expectations. However, with an alpha-level of 0.05, one should expect some deviation by chance alone.

Most studies of *G. aculeatus* that tested for correspondence with Hardy-Weinberg expectations did find it. Reported departures from equilibrium were always due to heterozygote deficiencies. Avise (1976) found no significant departure from equilibrium expectations for four loci scored from pooled samples from the dimorphic Friant population in California, USA. All loci were in equilibrium for seven intramorph comparisons in the dimorphic Enos Lake (British Columbia, Canada) population, but Ck-A and Pgm-A exhibited significant departures if the samples were pooled, suggesting a lack of gene flow between the two forms (McPhail 1984, page 419 this volume). Baumgartner (1986*b*) reported complete correspondence with

Table 3.3 Estimates of genetic variation in 16 geographic samples (nos 1–16) of *Gasterosteus aculeatus* based on the gene products of 18 loci from Haglund *et al*. (1992*a*), and 7 Korean samples (nos 17–23) based on the gene products of 25 loci from Yang and Min (1990).

Sample number and location	Mean sample size per locus (mean ± SE)	Mean number of alleles per locus (mean ± SE)	Percentage of loci polyallelic[a]	Proportion of heterozygotes	
				Observed[b] (mean ± SE)	Expected[c] (mean ± SE)
1. Japan: Gifu	15.9 ± 0.9	1.2 ± 0.1	22.2	0.045 ± 0.027	0.058 ± 0.035
2. Japan: Shiga	12.4 ± 0.5	1.3 ± 0.1	27.8	0.048 ± 0.023	0.051 ± 0.023
3. Japan: Ishikawa	15.8 ± 1.0	1.4 ± 0.1	38.9	0.087 ± 0.030	0.096 ± 0.032
4. Japan: Aomori-1	19.2 ± 0.6	1.4 ± 0.2	27.8	0.050 ± 0.020	0.065 ± 0.026
5. Japan: Aomori-2	6.9 ± 0.1	1.4 ± 0.1	33.3	0.073 ± 0.031	0.082 ± 0.030
6. Japan: Nanae	19.6 ± 0.2	1.2 ± 0.1	22.2	0.042 ± 0.026	0.054 ± 0.030
7. USA: Alaska	19.8 ± 0.2	1.2 ± 0.1	16.7	0.032 ± 0.023	0.056 ± 0.037
8. USA: Oregon	20.0 ± 0.0	1.4 ± 0.1	33.3	0.097 ± 0.042	0.110 ± 0.047
9. USA: North California	20.0 ± 0.0	1.5 ± 0.2	38.9	0.100 ± 0.042	0.109 ± 0.040
10. USA: South California	20.0 ± 0.0	1.2 ± 0.1	16.7	0.053 ± 0.030	0.055 ± 0.032
11. USA: New York	20.0 ± 0.0	1.3 ± 0.1	27.8	0.083 ± 0.038	0.090 ± 0.042
12. Canada: Nova Scotia	20.0 ± 0.0	1.3 ± 0.1	22.2	0.061 ± 0.034	0.072 ± 0.038
13. Sweden: Äsko Lab	20.0 ± 0.0	1.2 ± 0.1	22.2	0.064 ± 0.033	0.070 ± 0.036
14. UK: South Yorkshire	20.0 ± 0.0	1.2 ± 0.1	16.7	0.058 ± 0.040	0.063 ± 0.039
15. UK: Humberside	20.0 ± 0.0	1.4 ± 0.1	33.3	0.100 ± 0.042	0.101 ± 0.043
16. Italy: Lombardy	20.0 ± 0.0	1.2 ± 0.1	22.2	0.050 ± 0.027	0.055 ± 0.028
17. Korea: Sokcho	2.0 ± 0.0	1.2	16.0	0.080	0.080
18. Korea: Opong-ri	13.0 ± 0.0	1.6	40.0	0.083	0.078
19. Korea: Oho-ri	20.0 ± 0.0	1.7	48.0	0.098	0.098
20. Korea: Kyongpo	20.0 ± 0.0	1.4	32.0	0.100	0.093
21. Korea: Pyonggok	20.0 ± 0.0	1.6	40.0	0.066	0.074
22. Korea: Kanggu	20.0 ± 0.0	1.4	32.0	0.094	0.087
23. Korea: Koje	13.0 ± 0.0	1.5	44.0	0.080	0.081

[a] A locus was considered to be polyallelic if the frequency of the common allele was less than 0.95.
[b] Based on a direct count of heterozygotes.
[c] An unbiased estimate based on Hardy–Weinberg expectations (Nei 1978).

Hardy–Weinberg expectations (21 comparisons) in the Brush Creek drainage, California. In 56 samples from British Columbia and Washington scored for gene products of five loci, Withler and McPhail (1985) found equilibrium conditions in 194 of 197 relevant comparisons; three cases of heterozygote deficiency were noted at 'Mdh-1'. Buth *et al.* (1984) reported equilibrium conditions in seven of nine comparisons in the Santa Clara drainage, California; significant deficiencies of heterozygotes were noted at Est-5 and Pgm-A. Haglund *et al.* (1990) found correspondence with equilibrium expectations in 13 of 15 cases in Nova Scotia; heterozygote deficiency occurred at sMdh-A, and a rare Ldh-A homozygote was found (no heterozygotes; see Fig. 2 of Haglund *et al.* 1990). Equilibrium conditions were found in 71 of 76 comparisons in the global study of Haglund *et al.* (1992*a*); heterozygote deficiencies were found in single cases for sMdh-A, mIcdh-A, and Gpi-A, and twice for Pnp-A. The greatest number of departures from equilibrium was noted in a southern California study (Buth *et al.* unpubl. data) in which only 55 of 72 comparisons corresponded to Hardy–Weinberg expectations (departures involved six of seven polyallelic loci). Buth *et al.* (unpubl. data) recognized three regional population assemblages, and genetic interaction among them may account for the high number of local departures from equilibrium.

With the exception of pooled data for a pair of sympatric species in Enos Lake (McPhail 1984) and the parapatric exceptions in southern California (Buth *et al.* unpubl. data), local populations of *G. aculeatus* are in Hardy–Weinberg equilibrium. Other than MDH loci (scored in most studies), no locus exhibits consistent departure from equilibrium. Departures by chance should vary in both directions of heterozygote deficiency and excess, yet only the former is observed in *G. aculeatus*. We have no explanation for this pattern, other than to suggest that it may be the result of a low level of inbreeding.

Divergence among populations

Anadromous and freshwater populations

The difference between completely plated, anadromous populations of *G. aculeatus* (form 'trachurus'; but see discussion by Bakker and Sevenster 1988; Bell and Foster page 9 this volume) and variably plated, resident freshwater populations (form 'leiurus') was the first to attract electrophoretic comparisons. The general muscle protein difference reported by Hagen (1967) may actually involve two covarying characters. Examination of Hagen's Fig. 4 shows the 'diagnostic band' for 'leiurus' to be part of a one-banded and, less frequently, two-banded polymorphism that is monomorphic for the less anodal electromorph in 'trachurus'. The 'diagnostic band' for 'trachurus' found near the origin has a less discrete appearance than the more anodal products, and appears to have a faintly resolved counterpart in 'leiurus'. These products may be part of a different protein character that varies in regulation of expression in which 'trachurus' > 'leiurus'. Both characters presumably intergrade in specimens obtained from hybrid zones. Other studies reported electromorphs or formally recognized allelic products present in anadromous populations but absent from neighbouring freshwater populations (Muramoto *et al.* 1969; Raunich *et al.* 1971, 1972; Rafinski *et al.* 1989; Zietara 1989). This distribution of products is consistent with the hypothesis that freshwater populations have been established from a small number of individuals drawn from anadromous populations (founder effect; Bell 1976*a*).

In general, anadromous populations express a greater number of alleles per locus (Avise 1976; Rafinski *et al.* 1989), but less geographic divergence than freshwater populations. Withler and McPhail (1985) noted F_{ST} values of 0.046 and 0.293 among anadromous and freshwater populations, respectively, indicating greater heterogeneity among the latter. Rafinski *et al.* (1989) expressed this relationship in terms of genetic distances, which were significantly smaller among anadromous populations. Interestingly, Avise (1976), Withler and McPhail (1985), and Rafinski *et al.* (1989) all noted that the increased number of alleles in anadromous populations did not necessarily mean that these populations exhibited significantly more heterozygosity than freshwater populations.

Colour forms

Attention to variation in male nuptial coloration began in earnest with McPhail's (1969) study of the 'black' stickleback, which occurs in parapatry with normally coloured (red) males in the Chehalis River system, Washington, USA. Since that time, parapatric or sympatric cases of black and normal males have been described from other localities including Mayer Lake, Queen Charlotte Islands, British Columbia, Canada (Moodie 1972*a*,*b*; McPhail page 411 this volume), Holcomb Creek, Mojave River drainage, California (Bell 1982), and Enos Lake, Vancouver Island, British Columbia (McPhail 1984). Additional colour forms have been described from Lake Wapato, Washington (Semler 1971), the Baltic Karlskrona Archipelago of south-eastern Sweden (Borg 1985), Nova Scotia, Canada (Blouw and Hagen 1984*d*, 1990), and throughout the Queen Charlotte Islands (Reimchen 1989).

For all the aforementioned situations, allozyme data are available only for the Enos Lake pair (McPhail 1984), the 'white' stickleback of Nova Scotia (Haglund *et al.* 1990), and the 'black' stickleback of Holcomb Creek in southern California (Buth *et al.* unpubl. data). The Enos Lake forms differed in body shape and gill raker architecture in addition to male nuptial coloration. McPhail (1984) referred to these forms as 'benthics' (black males, but otherwise similar to other populations in the region) and 'limnetics' (normal coloration, but highly adapted for plankton feeding). Of four polyallelic loci studied, three ('Mdh-3', 'Ck' = Ck-A, and 'Pgm' = Pgm-A) exhibited statistically significant differences ($P < 0.05$) in their allele frequencies. At 'Mdh-3', the limnetic population expressed an allele ('55'; $N = 152$ specimens; $F(55) = 0.178$) that was absent in the benthic population ($N = 124$ specimens). McPhail (1984, page 419 this volume) argued that these data indicate a restriction of gene flow between these sympatric forms, which nevertheless can produce fertile hybrids in the laboratory. He recommended that these forms be considered as separate species, but stopped short of a formal description. It is not known which, if either, of the Enos Lake forms is *G. aculeatus*. Similar pairs of forms are known from other lakes in the Strait of Georgia region (Schluter and McPhail 1992; McPhail 1992, 1993, page 418 this volume). Comparisons of these additional pairs with the Enos Lake pair, plus comparisons with more distant populations (e.g. Haglund *et al.* 1992*a*), could clarify this situation.

Unlike the Enos Lake pair, the 'white' stickleback of Nova Scotia and the 'black' stickleback of southern California lack morphological correlates to their male nuptial colour differences. The 'white' stickleback expresses even the same minor alleles in similar frequencies (Haglund *et al.* 1990) and is essentially genetically identical to neighbouring populations of normal, red *G. aculeatus* (Table 3.1). There is no allozyme support for the recognition of the 'white' stickleback as a separate species. The 'black' stickleback population of Holcomb Creek in southern California was found to be essentially genetically identical with normal, red *G. a. microcephalus* found in coastal drainages (Buth *et al.* unpubl. data). No taxonomic recognition is warranted in this case either. However, the southern California 'black' stickleback may be exhibiting coloration homoplastic with that of 'black' sticklebacks of the Pacific North-west (e.g. McPhail 1969). Its allozyme identity in southern California may not extend to these other populations.

Western North American subspecies

Miller and Hubbs (1969) recognized three subspecies of *G. aculeatus* along the western coast of North America: *G. a. aculeatus* is a completely plated form that is usually anadromous, *G. a. microcephalus* is a variable, but usually low-plate morph freshwater form, and *G. a. williamsoni* is a low-plate morph form, which often lacks plates entirely and now is limited to the head waters of the Santa Clara drainage in southern California (Buth *et al.* 1984, unpubl. data). The first two subspecific names have been applied to some Japanese populations (e.g. Muramoto *et al.* 1969; Masuda *et al.* 1984), but not to eastern North American or European populations.

No single study has compared the allozymes of all three subspecies. Avise (1976) compared single geographic samples of *G. a. aculeatus* and *G. a. microcephalus* from central California; he found no diagnostic loci (of 19 scored), and a high level of genetic similarity ($I = 0.97$). Buth *et al.* (1984) compared single geographic samples of *G. a. microcephalus* and *G. a. williamsoni* from southern California and found only a single diag-

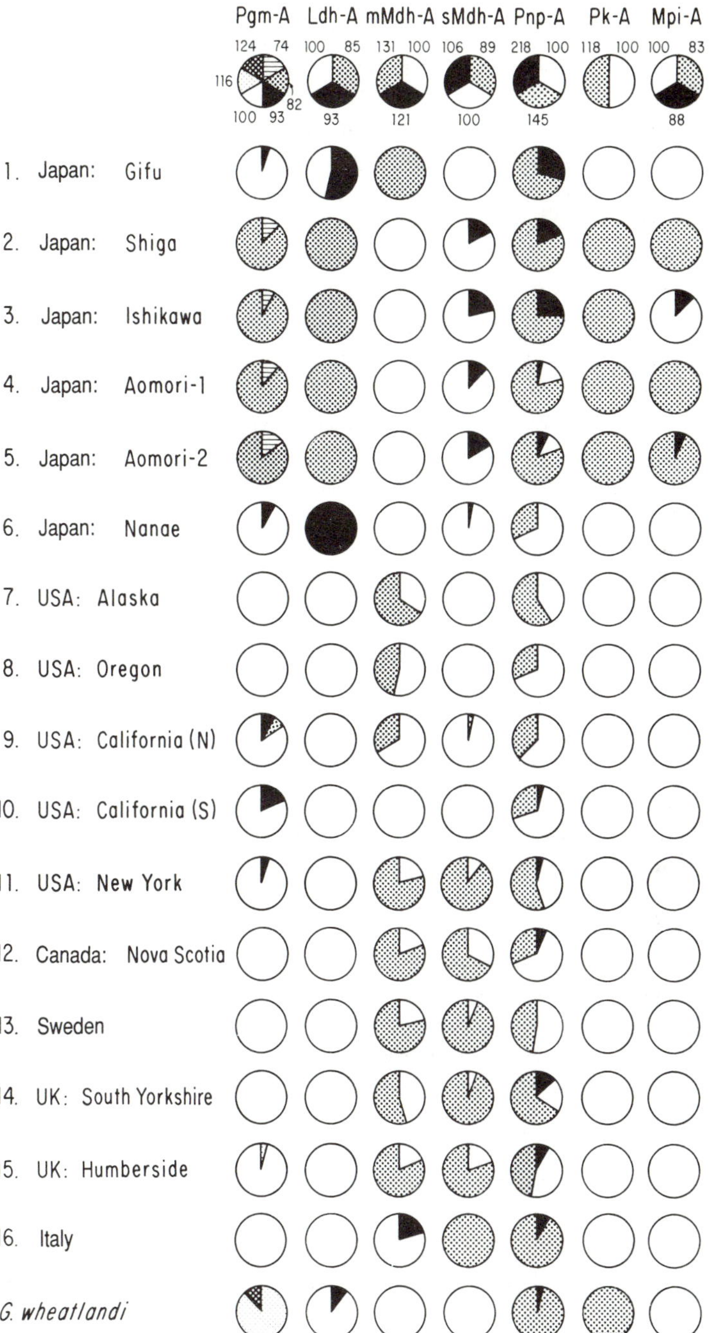

Fig. 3.3 Proportional depiction of allelic composition at seven loci in 16 geographic samples of *Gasterosteus aculeatus* and in *G. wheatlandi* (Haglund *et al.* 1992*a*).

nostic locus (Ada-A; of 45 scored) but a comparably high level of genetic similarity ($I = 0.97$). Expanded geographic sampling in southern California has shown that Ada-A is a useful diagnostic character with a narrow intergrade zone (Buth *et al.* unpubl. data; Haglund, unpubl. data), and can be used to recognize *G. a. williamsoni*.

Divergence within the complex

Global pattern

The only study to date that addressed the question of allozyme divergence in more than a local or regional context is that of Haglund *et al.* (1992*a*), who compared 16 Asian, North American, and European populations of *G. aculeatus*. Except for a greater collecting effort in Japan, the locations sampled were chosen simply to represent diverse portions of the broad range of *G. aculeatus*. Problem areas identified in previous studies were not sampled. The sampling design reflected a decision to search for major

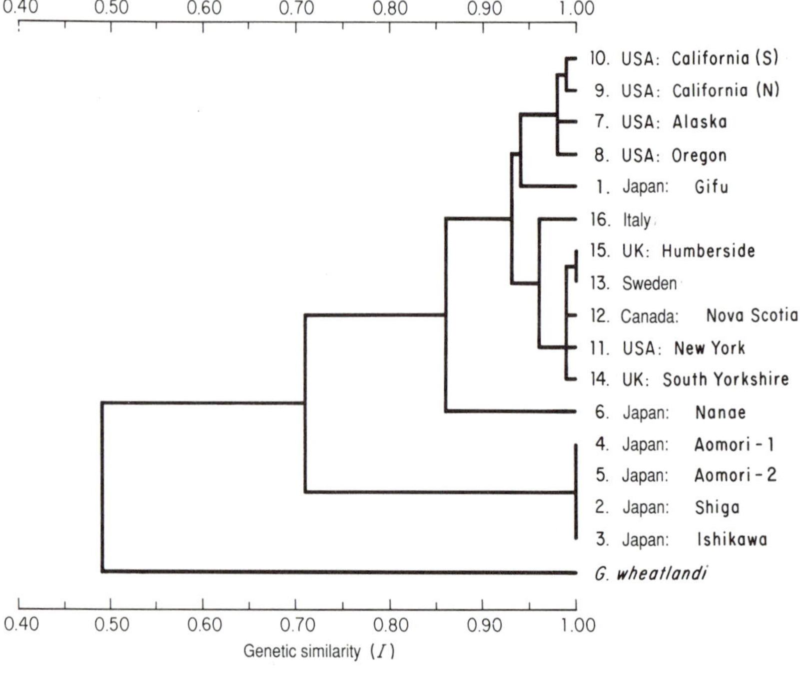

Fig. 3.4 Phenogram of relationships among *Gasterosteus wheatlandi* and 16 geographic samples of *Gasterosteus aculeatus* based on the gene products of 18 loci (Haglund *et al.* 1992*a*). Coefficients of genetic similarity (*I*: Nei 1978) between all pairwise combinations of samples have been clustered using the unweighted pair-group method with arithmetic means (UPGMA). The cophenetic correlation for this phenogram is 0.989.

patterns of divergence, rather than to address known local problems. Multiple alleles were found at 13 of 18 loci. Products of seven loci were especially informative and are summarized in Fig. 3.3. Other loci expressed minor allelic variation of lower information content. An outgroup perspective was gained by comparisons with *Gasterosteus wheatlandi*, a sympatric congener from eastern North America.

Relationships among the populations examined by Haglund *et al.* (1992*a*) can be expressed in terms of overall similarity (phenetics) or phylogenetic history—genealogy (cladistics). Phenetically, the samples from Europe, both coasts of North America, and at least one sample from Japan (Gifu) show a very high degree of genetic similarity; $I > 0.93$ (Fig. 3.4). Four Japanese samples, from three locations, were essentially identical to one another ($I > 0.99$), yet displayed considerable dissimilarity ($I = 0.71$) when compared with the larger assemblage just mentioned. One Japanese sample (Nanae) showed an intermediate level of difference ($I = 0.86$) compared with the two other assemblages.

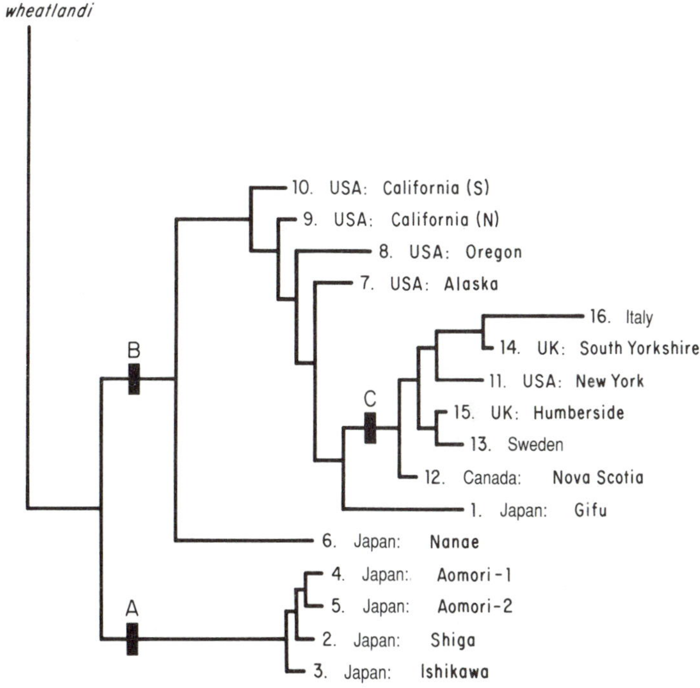

Fig. 3.5 Cladogram of 16 geographic samples of *Gasterosteus aculeatus* (modified from Haglund *et al.* 1992*a*) constructed using the distance Wagner procedure (Farris 1972) to cluster modified Rogers' (1972) coefficient of genetic distance (Wright 1978), and rooted using *Gasterosteus wheatlandi* as the outgroup. The cophenetic correlation of the cladogram is 0.959. Synapomorphies are mapped for the following loci to indicate major clades: A, Ldh-A and Mpi-A; B, Pk-A; and C, sMdh-A.

Table 3.4 Summary of mean *F*-statistics across all polyallelic loci for selected comparisons of samples of *G. aculeatus* from Haglund *et al.* (1992*a*). Sample numbers correspond to those listed in Table 3.3 and used in Fig. 3.3, 3.4, and 3.5.

Assemblages compared	Mean value across all polyallelic loci		
	F_{IS}	F_{IT}	F_{ST}
Divergent Japanese samples (2–5)	0.080	0.111	0.034
Western North American samples (7–10)	0.127	0.269	0.163
Eastern North American and European samples (11–16)	0.053	0.242	0.199
Western and eastern North American, and European samples (7–16)	0.085	0.407	0.352
Western North American and non-divergent Japanese samples (1, 6–10)	0.147	0.495	0.408
All (1, 6–16) except divergent Japanese samples (2–5)	0.100	0.506	0.452
All samples (1–16)	0.095	0.685	0.651

A cladistic treatment of the same data yielded a slightly different arrangement (Fig. 3.5) and permitted an evolutionary interpretation of the allozyme variation. *F*-statistics (Table 3.4) revealed that several coherent genetic units correspond to lineages, or parts thereof. Two major clades can be recognized in the *G. aculeatus* complex: (1) the phenetically most distinctive Japanese samples can be defined by synapomorphies (A in Fig. 3.5) at Ldh-A and Mpi-A (and perhaps Pgm-A), (2) the remainder of the complex, including some Japanese samples, can be defined by a synapomorphy at Pk-A (B in Fig. 3.5). The latter assemblage (2) can be further subdivided by a synapomorphy (C in Fig. 3.5) at sMdh-A (Fig. 3.1) that linked the eastern North American and European samples as an Atlantic basin clade.

Haglund *et al.* (1992*a*) have demonstrated that the major evolutionary pattern in the *G. aculeatus* complex is not a patchwork of regional divergence, but rather a single, widespread form to which a single name, *Gasterosteus aculeatus*, should be applied. A second, highly divergent evolutionary unit exists in Japan, perhaps in parapatry with other, less derived populations that may have North American or continental Asian populations as closest relatives.

Divergent isolates

Whereas Haglund *et al.* (1992*a*) identified a widespread evolutionary unit as *G. aculeatus*, many distinctive locally endemic populations are known. Bell's (1976*a*) interpretation of the complex as 'a superspecies composed of semispecies, some of which are polytypic and polymorphic' provided a cautionary note regarding hasty taxonomic recognition. Previous studies using allozyme data in conjunction with other characters (Table 3.1), and others identifying dimorphic conditions without allozyme data (e.g. Ziuganov *et al.* 1987), have all stopped short of formal description of new taxa. Clearly, not all patterns of variation warrant taxonomic recognition. However, congruent patterns of ecological, morphological, behavioural, and allozyme variation in sympatry make the strongest argument for taxonomic recognition, but this has only been described for the Enos Lake pair (reviewed in McPhail page 419 this volume).

Perhaps another reason for a taxonomically conservative approach is that, until recently, no fixed or complete allelic differences had been demonstrated in any sympatric, parapatric, or allopatric comparisons of populations of *G. aculeatus*. Statistical significance had been demonstrated for allelic frequency differences (e.g. McPhail 1984), but complete differences, as had been expected in traditional taxonomic recognition of species (Avise 1974; Buth and Mayden 1981), were lacking.

The unarmoured threespine stickleback, a southern California isolate recognized as *G. a. williamsoni*, was the first population of *G. aculeatus* to have been shown to express a complete allelic difference at a single locus (Buth *et al.* 1984). Because this population exists in parapatry with *G. a. microcephalus* (Bell 1978; Buth *et al.* unpubl. data) and does have a narrow geographic zone of genetic intergradation (Haglund unpubl. data), the recognition of this difference at the subspecific level is justified. Concern for polyphyletic subspecies is negated here because *G. a. williamsoni* inhabits just one headwater region in southern California (Buth *et al.* unpubl. data).

The Shay Creek isolate is the most genetically distinct population in southern California. It expresses complete allelic differences at two loci (Pep-C and Est-5) relative to other populations in the region and possesses unique alleles, often at high frequency, at six other loci (Haglund and Buth 1988). Taxonomic recognition of the Shay Creek isolate is complicated by its allopatric distribution. Designation of this distinctive form as a subspecies of *G. aculeatus* or as a separate species may be seen as an arbitrary decision based on a 'magnitude of difference' argument (Buth and Mayden 1981) that has no place in modern classification. Additional morphological and behavioural studies of this isolate are in progress.

Other distinctive populations of *G. aculeatus* that exhibit lateral plate dimorphism or unusual coloration patterns have been shown to have nondistinctive allozyme compositions (Avise 1976; Haglund *et al.* 1990; Buth *et al.* unpubl. data). A conservative taxonomic approach is suggested in situations in which the null hypothesis of conspecificity cannot be falsified with characters for which the genetic basis is known.

Comparison with other gasterosteids

Allozyme comparisons of *G. aculeatus* with its congener, the blackspotted stickleback, *G. wheatlandi*, have only recently been conducted. These two species can be readily distinguished on the basis of morphological

(Hubbs 1929; McAllister 1960; Scott and Crossman 1964) and behavioural (McInerney 1969) characters. Ecological separation is not complete, and the two species are found in sympatry throughout much of north-eastern North America (Worgan and FitzGerald 1981*a*; Cowen *et al.* 1991). Hudon and Guderley (1984) obtained both species in sympatry in Quebec and found them to be quite different (genetic identity of $I = 0.37$). Haglund *et al.* (1992*a*) obtained both species in sympatry in New York and also found them to be quite different ($I = 0.49$; Fig. 3.5) based on a slightly different set of allozyme loci. Not only is it surprising to find differences of such magnitude in species that are closest relatives, it is also interesting to note that the populations of *G. aculeatus* that possess the most primitive allozyme states (i.e. those shared with *G. wheatlandi*) are those from the west coast of North America, not the ones in sympatry on the east coast (Fig. 3.4). These data and the derived nature of the eastern North American and European populations of *G. aculeatus* (Fig. 3.4) suggest the following scenario for speciation in the genus: (1) an initial separation and isolation of an eastern North American population from Pacific basin populations yielded speciation as *G. wheatlandi* and ancestral *G. aculeatus* (Pacific basin), respectively; (2) an expansion of range that resulted in derived populations of *G. aculeatus* on the east coast of North America and on into Europe brought the two species into sympatry, but reproductive isolating mechanisms were in place and evidently no introgression occurred. Analysis of the fossil record is consistent with this interpretation; *Gasterosteus* was widely distributed in the Pacific basin in the Miocene but does not appear in the Atlantic basin until the Pliocene (Bell page 444 this volume).

Intergeneric comparisons of allozymes of *G. aculeatus* with gasterosteids are also quite recent. Hudon and Guderley (1984) found that *G. aculeatus* shared only a few allozymes with *Pungitius pungitius* and *Apeltes quadracus* from Quebec (genetic identities of $I = 0.21$ and 0.15, respectively). Yang and Min (1990) found that Korean *G. aculeatus* shared very few allozymes with *Pungitius sinensis* and *Pungitius kaibarae* (Rogers' 1972 average genetic similarity of $S = 0.052$). Haglund *et al.* (1992*b*) confirmed this considerable divergence in a comparison of *G. aculeatus* from southern California and *Pungitius* from Korea; these populations shared alleles only at Ldh-C of 18 loci scored. However, this divergence may be due to considerable apomorphy among Korean *Pungitius*. Haglund *et al.* (1992*b*) found alleles shared at Gpi-A, Gtdh-A, Ldh-C, sMdhp-A, and Pep-C between *G. aculeatus* from southern California and *Pungitius* from Japan, Canada, England, and/or Sweden. Our own comparisons of allozymes of various species of gasterosteids (Haglund and Buth unpubl. data), including *Culaea inconstans* and *Spinachia spinachia*, confirm that all gasterosteid species are highly divergent from one another. With little in common, identification of allozyme synapomorphies will be difficult and the allozyme database may yield a very limited contribution to the resolution of gasterosteid phylogeny.

DIRECTION OF FUTURE RESEARCH

Many of the most recent allozyme studies have revealed interesting patterns of genetic divergence, but call for additional study for a complete evolutionary interpretation and taxonomic resolution. Obvious evolutionary and taxonomic clarification is needed for the Enos Lake pair and others in the region (McPhail 1984, Chapter 14 this volume), the Shay Creek isolate (Haglund and Buth 1988), and the divergent Japanese evolutionary unit (Haglund *et al.* 1992*a*).

Additional regional studies should be integrated with the global framework initiated by Haglund *et al.* (1992*a*). For example, Korean populations of *G. aculeatus* are homogeneous (Yang and Min 1990), but are they equivalent to any of the different forms found in Japan (Haglund *et al.* 1992*a*)? The sampling hiatus from northern Japan to southern Alaska should be filled in to see whether zones of intergradation are present. A finer-grained sampling across Europe should prove especially interesting. *G. aculeatus* exhibits more allozyme variation in Britain (Lawson pers. comm.) than was suggested by the limited sampling of Haglund *et al.* (1992*a*). The complete-morph forms around the Black Sea (Münzing 1963) and the unarmoured forms found on the Outer Hebrides, Scotland (Campbell 1984) deserve special attention.

An allozyme database should be considered an essential element in studies of behavioural or life history differences or those of trophic divergence. If such data were included, we could gain insight into the role of genetic divergence during the process of speciation.

It is clear that allozyme contributions to our understanding of the biology of the *G. aculeatus* complex have barely scratched the surface. Yet we are not encouraged by a pattern of publications that elucidate the potential of allozymes to address key problems in the complex, but fail to pursue these problems in subsequent investigations. The combination of few researchers choosing to study the allozymes of *G. aculeatus*, and the depletion of their ranks to study other taxa or to employ other, 'newer' technologies suggests that the full potential for allozyme data may not be realized. The investigatation of mitochondrial DNA (mtDNA) of *G. aculeatus* is in its preliminary stages (Gach and Reimchen 1989; Ortí *et al.* unpubl. data). Coordinated investigations of allozyme and mtDNA variation would be optimal for the resolution of many of the population-level problems widespread in *G. aculeatus*.

SUMMARY

Allozyme variation in the *G. aculeatus* complex has been shown to be very informative in studies of regional and global population structure and gene flow, and has aided our understanding of variation expressed in other characters. Nevertheless, most studies have been limited to localized problems and/or have underutilized the potential of allozyme technology that is currently available. The very interesting findings to date should continue to be pursued and expanded. Many local problems remain to be studied before a complete global perspective can be gained.

The systematic relationships within the complex, and the taxonomy that should reflect these relationships, are especially problematic areas in which allozyme characters could contribute considerably. Future clarification could possibly end our application of the vague concept of 'complex' to this taxon.

We have discussed a few of many seeming errors that have entered the allozyme literature on *G. aculeatus*, and some of the limitations faced when studies are to be compared. For optimal communication among future researchers, we urge that they

(1) become familiar enough with the enzyme systems being studied to be able to use an informative locus nomenclature that is based on current enzyme nomenclature;

(2) publish zymograms; and

(3) publish primary data as *genotype* arrays, not as less informative allele frequency or genetic distance/similarity matrices.

ACKNOWLEDGEMENTS

This study was supported in part by the University of California (UCLA) Department of Biology Fisheries Program, the UCLA Biomedical Support Grant, and the UCLA Committee on Research (U.R. 3674 to D. G. B).

4

Physiological ecology and evolution of the threespine stickleback

Helga E. Guderley

The life cycle of the threespine stickleback exposes it to diverse physical conditions. The anadromous stickleback, in particular, faces marked changes in the salinity, temperature, and oxygenation of its environment during its annual cycle. Neuroendocrine control mechanisms and modification of tissue function provide the physiological basis of its responses to environmental changes; the physiological preparation for reproduction may influence its response to seasonal changes. For example, mobilization of somatic reserves for gonad production or parental care could accentuate the effect of high temperature on nutritional requirements and on muscle metabolic organization. Such interactions must be considered when interpreting the adaptive significance of the stickleback's physiological responses to the environment. Common environmental signals may coordinate the reproductive cycle and adaptation to changing environmental conditions. For example, as photoperiod modifies both salinity preferences (Baggerman 1957; Audet *et al.* 1986*a*) and reproductive readiness (Baggerman 1957), it is likely that these responses are mediated, at least in part, by common mechanisms.

In examining the physiological ecology of the threespine stickleback, I will focus on the endocrine mechanisms that coordinate the physiological and behavioural responses to environmental conditions, the physiological, metabolic, and biochemical responses themselves, and the environmental and endocrine control of the reproductive cycle. Because Wootton (1976) provides an excellent review of earlier studies concerning these phenomena, I will primarily cover the more recent literature, except when the earlier literature is critical for the understanding of the adaptive mechanisms.

THE EFFECT OF ENVIRONMENTAL SALINITY, TEMPERATURE, AND OXYGENATION ON STICKLEBACK

Freshwater stickleback carry out only short reproductive migrations and live in fairly constant physical conditions, whereas the reproductive migration of anadromous stickleback exposes them to alterations in salinity, temperature, and oxygen availability. Furthermore, depending upon the exact habitat chosen for reproduction, stickleback may move from stable to fluctuating environments. For example, stickleback that reproduce in the tidepools at Isle Verte (see Fig. 7) on the south shore of the St Lawrence Estuary, Quebec, Canada, face temperatures as low as 2 °C or as high as 3 °C. These tidepools can be diluted by rain or dried up by the sun and their dissolved oxygen levels can drop as low as 2.8 mg l^{-1} (Reebs *et al.* 1984). Members of this population also reproduce in the freshwater section of an adjacent tidal river in which temperatures are intermediate and salinities are constantly low (Kedney *et al.* 1987). Clearly, anadromous threespine stickleback are able to reproduce in a variety of environmental conditions. This flexibility may come at a cost. The physiological adjustments required for reproduction in an unstable environment may reduce the fitness of the parental fish or the survivorship of the young. The threespine stickleback may be 'best' suited for stable habitats, but more demanding, unstable environments may also provide numerous, suitable opportunities for reproduction.

Salinity: osmoregulatory mechanisms

The migration from salt to fresh water exposes the stickleback to large shifts in the ionic and osmotic environment. In fresh water, the stickleback must work to keep its ions, whereas in water with a salinity above approximately 14‰, the fish must eliminate excess ions. While generally, low-plated ('leiurus') stickleback are found in fresh water and completely plated ('trachurus') stickleback have an anadromous life history, exceptions to this pattern are found. Plate morph and salinity tolerance are well correlated in regions where this generalization is valid, but low-plated stickleback that live part of the year in brackish water (Crivelli and Britton 1987) or reproduce in seawater (Bell 1979) likely have physiological characteristics different from those of completely or low-plated stickleback living in fresh water. Furthermore, different freshwater populations of low-plated stickleback may have arisen independently from completely plated anadromous stickleback (e.g. Bell and Foster page 16 this volume, McPhail page 401 this volume) and may differ in their adaptation to fresh water. Unfortunately, much of the work on the salinity tolerance and osmoregulatory mechanisms of stickleback only specifies plate morph and not life history. If in the following section, only the plate morph is given, this is because the life history of the fish under study was not specified.

When completely plated stickleback are transferred from salt water to fresh water, the gills show morphological changes consistent with an enhanced ion uptake. Effectively, the gill chloride cells, which are responsible for the absorption of monovalent salts in fresh water and their excretion in seawater, show an increase in apical surface area during the transfer to fresh water (Biether 1970). The chloride cells of completely plated stickleback represent features typical of ion-transporting cells: an abundance of mitochondria, highly developed agranular endoplasmic reticulum, and a functional asymmetry in the cell membranes (Matej 1980). While these morphological features are present both in freshwater- and in seawater-adapted fish (Matej *et al.* 1981), mitochondrial abundance is higher in seawater-adapted stickleback (Biether 1970).

Kidney histology has been interpreted to suggest an increased glomerular filtration rate and an enhanced tubular ion uptake in anadromous completely plated stickleback transferred from seawater to fresh water. During this transfer, the number of membrane particles in the basal labyrinth of tubular cells increases (Wendelaar Bonga and Veenhuis 1974). Because the number of these particles also rises after prolactin is injected into seawater-adapted fish, and because prolactin increases the levels of Na^+/K^+ ATPase in the basal labyrinth of kidney cells in other species, these authors conclude that the particles represent enzyme complexes involved in ion transport. Prolactin accelerates the growth of the tubular epithelium during adaptation of the anadromous stickleback to fresh water, but does not modify the rate of change of the glomeruli (Wendelaar Bonga 1976). Generally, when anadromous, completely plated stickleback are adapted to fresh water, their kidneys have larger capsules, glomeruli, and intratubular lumens than when they are adapted to seawater. Furthermore, in the glomerulus of these stickleback, the filtration slits cover a greater area, endothelial fenestrations are more numerous, and the basal lamina is thinner than in seawater-adapted fish (de Ruiter 1980). Clearly, these modifications require energy, both for their synthesis and for their operation. This cost may, however, be negligible relative to the advantages of exploiting habitats with markedly different salinities.

In sexually mature males, the transformation of the kidney to a glue-secreting structure may interfere with its role in osmoregulation. In this respect, male stickleback are unique among teleosts. As of the production of glue is essential for nest building and thus for the male's reproductive success, a compromise must be reached in which both osmoregulation and glue production are assured. During the transformation, the final two-thirds of the nephron undergoes marked changes. The height of the tubular cells increases dramatically and the ion-transporting cells become mucus-secreting cells. A high-speed autoradiographic study indicates that proximal tubules lengthen and proliferate during this transformation (Mourier 1979). Concomitantly, the glomerulus assumes the appearance typical of seawater-adapted, completely plated stickleback, in that the basal lamina thickens, the endothelial fenestrations are reduced, and the filtration slit membrane area is reduced (de Ruiter 1980). These changes suggest a reduction in glomerular filtration and urine production. The androgen dependence of the renal transformation is so well established (Wai and Hoar 1963; Mourier 1976*a*,*b*; de Ruiter 1981; de Ruiter and Mein 1982) that it has been used to evaluate whether spermatogenesis is subjected to androgen inhibition (Borg 1982*a*).

As sexual maturation brings no changes in the osmotic permeability of the gills (de Ruiter *et al.* 1985), compensation for a reduction in urine formation must occur elsewhere. In salt water, the mucus secreted by the kidney may bind some divalent ions and maintain part of the kidney's normal functions. In fresh water, renal excretion of hypotonic fluid is partly assumed by the intestine. Androgen administration and sexual maturation increase production of hypotonic liquid by the intestine (de Ruiter *et al.* 1984, 1985). Simultaneously, the extent of the basal labyrinth in the enterocytes of the posterior intestine increases, suggesting that a solute-linked transport system allows this intestinal fluid production (de Ruiter *et al.* 1985). While other freshwater teleosts have a similar basal labyrinth in the enterocytes (Yamamoto 1966; Noaillac-Depeyre and Gas 1973; Stroband and Debets 1978), it is more extensive in the stickleback intestine (de Ruiter *et al.* 1985). The intestine of most other freshwater teleosts is thought to contribute little to water elimination. The kidney's transformation to a glue-secreting structure can explain why reproductively mature male stickleback exploit this unusual osmoregulatory mechanism.

Salinity tolerance

Although not all threespine stickleback are faced with variations in salinity during their life cycles, all lateral plate morphs tolerate salinity fluctuations fairly well. Anadromous, completely plated stickleback do some have difficulty maintaining plasma Cl^- at low salinities, while low-plated stickleback have no such difficulty (Heuts 1947*a*). Gutz (1970) compared the osmoregulatory responses of laboratory-raised offspring of completely plated stickleback from the seashore with those of the offspring of low-plated stickleback from inland freshwater ditches. Completely plated, anadromous stickleback increase their oxygen consumption during transfer from fresh to salt water, presumably owing to the cost of ion pumping and of adaptive tissue modifications. Low-plated, freshwater stickleback exposed to salt water show a volume-regulatory response in which tissue concentrations of free amino acids increase, presumably owing to an increase in plasma osmolarity, but oxygen consumption does not rise. As this volume-regulatory response loses its effectiveness at low temperatures, low-plated fish do not tolerate high salinities at low temperature (Gutz 1970).

The response of the low-plated stickleback to changes in salinity is similar to the osmoregulatory response of organisms ranging from eubacteria to vascular plants, marine invertebrates, and cyclostome fishes (Somero 1986). Volume regulation by changes in concentrations of organic solutes such as amino acids and small carbohydrates has the advantage of not perturbing macromolecular structure and function, as would changes in ion concentrations. While cellular volume regulation by changes in solute contents is part of the teleost repertoire of osmoregulation (King and Goldstein 1983), the required response has been reduced through the iono- and osmoregulatory actions of the gills, kidney, and intestine, which maintain plasma hypotonicity in seawater and hypertonicity in fresh water. One of the best osmoregulators, the sheepshead minnow, *Cyprinodon variegatus*, maintains plasma osmolarity at 340 mOsm kg^{-1} in fresh water and at 385 mOsm kg^{-1} in double-strength seawater (Nordlie and Walsh 1989). Muscle solute concentrations are modified by acclimation of teleosts to different salinities (Schmidt-Nielsen 1977; King and Goldstein 1983), but the response is smaller than that in elasmobranchs and hagfish. Presumably, the greater reliance of low-plated stickleback on cellular volume regulation reflects impaired osmoregulatory performance by the gill, kidney, or intestine.

Hedgpeth (1957) has hypothesized that freshwater forms arising from euryhaline ancestors retain characteristics including regulation of plasma osmolarity over a wide range of salinities. The osmoregulatory responses of cyprinodontids from habitats with different degrees of salinity fluctuation support this hypothesis (Nordlie and Walsh 1989). In contrast, the seeming loss of osmoregulatory capacity in the freshwater stickleback (Gutz 1970) does not conform to this hypothesis. Since the closely related low- and completely plated stickleback differ in their osmoregulatory strategies, this system could provide a unique opportunity to elucidate which aspects of cellular physiology are altered to cope with changes in tissue osmolarity.

The salinity tolerance of the different morphs of threespine stickleback varies seasonally. For example, the freshwater, low-plated form can tolerate salinities as high as 45‰ outside the reproductive season, but loses this tolerance when it is reproductively mature (Koch and Heuts 1942). The more euryhaline completely plated form also undergoes seasonal variations in salinity tolerance. In the autumn, anadromous fish lose their ability to survive in low salinities, even though they easily tolerate low salinities during their breeding season (Lam and Hoar 1967).

Seasonal changes in salinity tolerance may well be under hormonal control. Early experiments which suggest that high prolactin levels in spring fish facilitate their adaptation to fresh water (Lam and Hoar 1967; Leatherland and Lam 1969) have been supported by histological studies of the prolactin cells in the freshwater (Benjamin 1974) and anadromous forms (Leatherland 1970*a*; Honma *et al.* 1976). Adrenocorticotropin (ACTH, which stimulates the secretion of interrenal hormones such as cortisol) and thyroid hormones have also been implicated in the adaptation of anadromous stickleback to fresh water (Koch and Heuts 1942; Lam and Leatherland 1970). Taken together, these studies indicate that several pituitary hormones, either directly or indirectly, favour adaptation to fresh water by stickleback. In an evolutionary perspective, it is interesting to ask why so many hormones have similar effects. A simple proximal cause such as structural similarity among the hormones is unlikely, since prolactin, ACTH, and thyrotropin have markedly differing structures. Alternatively, the hormones may be activating different targets, each of which facilitates adaptation to fresh water. Current evidence indicates that in stickleback, as in other teleosts (Hirano and

Mayer-Gostan 1978; Prunet *et al.* 1985), prolactin is the most important hormone assuring survival in fresh water.

As in many teleosts, the prolactin-secreting cells in stickleback modify their activity in response to changes in salinity. Exposure of freshwater-adapted completely plated stickleback to fresh water containing Ca^{2+} or Mg^{2+} at seawater concentrations reduces prolactin cell activity to levels found in seawater-adapted stickleback (Wendelaar Bonga 1978). Prolactin cell activity is also reduced by removal of the Stannius corpuscles, presumably through increases in plasma Ca^{2+} levels (Wendelaar Bonga and Greven 1978). Because adaptation to seawater with low levels of Ca^{2+} and Mg^{2+} stimulates prolactin cells to the same extent as adaptation to fresh water, these ions seem to be the primary chemical signals that determine prolactin cell activity in stickleback. By contrast, in tilapia, *Sarotherodon mossambicus*, homologous radioimmunoassays of prolactin indicate that osmolarity or NaCl concentration is a more important signal for prolactin secretion than Ca^{2+} or Mg^{2+} concentrations (Nicoll *et al.* 1981). In coho salmon, *Oncorhynchus kisutch*, transfer to freshwater increased prolactin cell activity even though plasma Na^+, K^+, Mg^{2+}, and Ca^{2+} concentrations and plasma osmolarity remained stable (Brewer and McKeown 1980).

The positive correlation between prolactin cell activity, epidermal thickness, and mucous cell activity suggests that prolactin facilitates freshwater adaptation by increasing the barriers to ion loss (Wendelaar Bonga 1978). In completely plated, anadromous stickleback, the increase in prolactin secretion in fresh water is probably accompanied by decreases in calcitonin secretion by the ultimobranchial bodies and in hypocalcin secretion by the Stannius corpuscles (Wendelaar Bonga *et al.* 1977; Wendelaar Bonga 1980). These latter hormones decrease plasma Ca^{2+} levels partly through enhanced renal excretion of Ca^{2+}. Benjamin (1980) suggests that the prolactin cells in the freshwater, low-plated stickleback he studied respond more slowly and less markedly to changes in salinity and season than the prolactin cells in the completely plated stickleback studied by Leatherland (1970*a*) and Honma *et al.* (1976). By analogy with the relatively ineffective osmoregulatory mechanism used by low-plated stickleback, their hormonal response to changing salinities may be less extensive than that of the completely plated fish.

The hormonal control of adaptation of stickleback to high salinities has received little study. In other teleosts, cortisol and growth hormone (GH) are the putative seawater-adapting hormones. Cortisol stimulates Na^+/K^+ ATPase activity in the salmonid gill, both *in vivo* (Björnsson *et al.* 1987; Richman and Zaugg 1987) and *in vitro* (McCormick and Bern 1989). Cortisol levels have frequently been shown to increase with transfer from fresh water to salt water (Hirano and Mayer-Gostan 1978) and are known to facilitate increased water absorption by the intestine. In light of these findings, the favourable effects of ACTH on freshwater survival in stickleback are surprising. The recent development of homologous radioimmunoassays for GH have allowed it to be distinguished from prolactin. Despite their considerable structural similarity, these molecules play very distinct roles. During the adaptation of salmonids to seawater, plasma GH levels increase (Sweeting *et al.* 1985; Hasegawa *et al.* 1987), and metabolic clearance rates of GH increase after transfer of trout into 75 per cent sea water (Sakamoto *et al.* 1990). The effects of GH on salt water adaptation seem to occur independently from its growth enhancing capacity (Bolton *et al.* 1987; Collie *et al.* 1989). Unfortunately, the small size of stickleback will hamper the development of homologous radioimmunoassays, and precise understanding of the role of pituitary hormones in adaptation to salinity awaits other technological advances.

Besides its importance in osmoregulation, the levels of prolactin have also been correlated with the regulation of parental behaviour (Molenda and Fiedler 1971). Implantation of a pituitary prolactin lobe into the dorsal musculature of nesting anadromous completely plated males increased the number of epithelial mucocytes as well as increasing fanning behaviour (de Ruiter *et al.* 1986). The rate of synthesis and release by prolactin cells is increased markedly during the period when completely plated males display parental fanning behaviour (Slijkhuis *et al.* 1984). Displacement fanning is positively correlated with prolactin cell activity in several freshwater teleosts, but this correlation is less well established for parental fanning. Given that the sensitivity of prolactin cells to salinity and season seems lower in low-plated than in completely plated stickleback, it would be interesting to determine whether parental behaviour is also correlated with enhanced activity of prolactin cells in the low-plated stickleback.

As well as varying on a seasonal basis, salinity tolerance may vary with developmental stage (Gerking and Lee 1980). Although adult anadromous stickleback easily accept low salinities during the reproductive season, the eggs and fry have a higher surface-to-volume ratio and may not have as highly developed osmoregulatory capacities. To test whether exposure to low salinities results in high mortality in fry of anadromous stickleback, Campeau *et al.* (1984) allowed males to hatch eggs at 20‰. At different intervals after hatching, the salinity tolerance of the fry was tested. One-week-old fry had a significantly higher mortality at 0.7 and 28‰ than at the control salinity (21‰). Clearly, fry of anadromous stickleback that hatch at fairly high salinities cannot tolerate a transfer to low salinities. However, the fry were euryhaline at 5 weeks. The osmotic fragility of young fry would only decrease their survivorship if they were to hatch in brackish water tidal pools that were markedly diluted by rain or concentrated by intense sun.

To evaluate whether the fry of anadromous completely plated stickleback that hatch in fresh water also have a low freshwater tolerance, Belanger *et al.* (1987) compared the salinity tolerance of fry that were hatched in fresh and in brackish water (20‰). The prediction that fry that had developed in fresh water would have a lower mortality in fresh water than fry that had developed in brackish water was supported. Despite the indications in Campeau *et al.* (1984), a high juvenile mortality does not accompany reproduction of anadromous stickleback in fresh water. Fry that hatched in fresh water maintained similar growth rates at salinities of 0, 7, 14, and 21‰ and reduced growth at high salinties (28‰). Fry that hatched in brackish water had higher mortalities at 0 and 28‰ than at intermediate (7,14, and 21‰) salinities. These fry grew little in fresh water and had their highest growth rates at 28‰ (Belanger *et al.* 1987). As in other fish (Gerking and Lee 1980), the salinity experienced during development influences the subsequent salinity tolerance of the stickleback. The mechanisms responsible for this response are not clear, but its adaptive value is a key element in the reproductive success of completely plated stickleback in fresh water.

The salinity and pH experienced during embryonic development influence the pH sensitivity of eggs and fry of the freshwater low-plated morph (Faris and Wootton 1987). Increasing salinity to 10 per cent seawater enhanced hatching of eggs and survival of fry at low pH. Furthermore, females kept in 10 per cent seawater produced eggs with greater hatching success at low pH than females maintained in fresh water. As larval survival was not affected by the salinity at which the females were kept, the maternal effect may reflect an increased proportion of high-quality, viable eggs.

Some classic, and unfortunately unique, studies on the adaptive significance of plate number and associated physiological characteristics (Heuts 1947*a*) indicate that hatching success at low salinities is greatest for low-plated stickleback. Within a population of low-plated stickleback, hatching success at low salinity is greatest for the fish with the lowest plate number. The completely and low-plated stickleback have complementary profiles of hatching success with respect to salinity and temperature. Under conditions in which the low-plated stickleback has low hatching success, the completely plated morph thrives. The salinity dependence of hatching success is maternally inherited. Heuts (1947*a*) postulated that genetic linkage of physiological and morphological characters explains the distribution of plate number in different populations. While this hypothesis remains to be tested, it could provide a general explanation of the distributional pattern of plate morphs in the threespine stickleback.

Environmental and hormonal control of salinity preferences

As migration exposes anadromous stickleback to changes in salinity, it is possible that salinity tolerance and migratory behaviour are controlled by similar environmental signals and mediated by common neuroendocrine mechanisms. Furthermore, seasonal changes in salinity preferences could coordinate the migratory behaviour of anadromous stickleback. Baggerman's monumental work has documented both the seasonal changes in, and the photoperiod control of, salinity preference, as well as possible hormonal controls of the shifts in salinity preference (Baggerman 1957). In these studies, only anadromous stickleback were used. Generally, salinity preferences change over an annual cycle. In late winter and spring, fish prefer low salinities, whereas at the end of the breeding season, the preference shifts to high salinities. This pattern would facilitate migration to the breeding grounds in the spring and to the ocean in the autumn.

Changes in daylength may mediate seasonal changes in salinity preferences (Baggerman 1957). When anadromous completely plated stickleback collected in early spring in the St Lawrence Estuary are acclimated to long daylengths, they also prefer low salinities (7 and 14‰) (Audet *et al.* 1985*a*, 1986*a*). As would be expected if daylength plays an important role in regulating salinity preferences, acclimation to short daylengths shifts the preference to high salinities. However, Baggerman (1957) found that the salinity preference shifted from fresh to salt water at the end of the breeding season, even though daylengths were still long. Termination of reproduction may somehow shift salinity preferences, through shifts in the production of gonadal steroids (see below) or of the pituitary hormones that regulate reproductive activities.

Temperature has also been implicated in the seasonal regulation of salinity preference. Baggerman (1957) found that increases in temperature in late winter rapidly stimulate a preference for low salinities. In autumn and early winter, such temperature increases are without effect. As gonadectomized and intact fish show similar shifts in salinity preferences, these changes do not require gonadal steroids. However, the changes in salinity preference coincide with reproductive maturation in intact fish.

In many populations, juvenile stickleback remain on the breeding grounds considerably longer than the adults. Juvenile fish prefer high salinities about two months after hatching. Perhaps only photoperiodic control of salinity preference is operative in juveniles, and the long summer daylengths may maintain a preference for low salinities. Interestingly, photoperiod strongly influences salinity preferences in the fourspine stickleback, *Apeltes quadracus*, but has little influence on salinity preferences of the blackspotted stickleback, *Gasterosteus wheatlandi*, and none at all for the ninespine stickleback, *Pungitius pungitius* (Audet *et al.* 1986*a*). Possibly the shorter life span of the two latter species in the St Lawrence Estuary can explain these differences in the environmental control of salinity preferences.

In her studies on the hormonal control of the seasonal changes in salinity preferences, Baggerman (1957) concentrated on thyroid hormones as these were known to enhance the ability of low-plated stickleback to osmoregulate in fresh water. In her experiments, thyroxine rapidly induced a preference for fresh water in both intact and gonadectomized anadromous stickleback. When the thyroid inhibitor, thiourea, was administered to fish preferring fresh water, their preference shifted to higher salinities. Mammalian gonadotropins also induced preferences for low salinities, but much more slowly than thyroxine. These responses argued for a major role of thyroid hormones in controlling the seasonal migrations of stickleback. Thyroid hormones are also thought to control salinity preferences in juvenile salmon (Baggerman 1960). In anadromous stickleback, histological data indicate effects of photoperiod on the epithelium of the thyroid gland (Ahsan and Hoar 1963) and an activation of thyroglobulin production in late winter and early spring (Leatherland 1970*b*), but seasonal variations in the levels of thyroid hormones, their binding proteins, and/or their receptors have not been established.

Two pituitary hormones, prolactin and ACTH, have also been implicated as endocrine determinants of salinity preferences. Considerable evidence suggests that prolactin and cortisol cycles can control seasonal behaviour in migratory birds and fish (Meier and Fivizzani 1980). The circadian rhythms of prolactin and cortisol concentrations in the blood of the gulf killifish, *Fundulus grandis*, and the sparrow, *Zonatrichia albicollis*, have different phase relationships in different seasons. Meier and Fivizzani (1980) suggest that the temporal synergism between the circadian rhythms of prolactin and cortisol acts as a central endocrine mechanism regulating the seasonal biology of migratory species. For the killifish, simultaneous injections of prolactin and cortisol led to preferences for higher salinities than when the two hormones were injected 12 h apart (Fivizzani and Meier 1978). These responses were explained in terms of prolactin's effect in promoting osmoregulation in hypotonic media and cortisol's equivalent role in hypertonic media. Simultaneous injections of prolactin and cortisol into killifish also led to fattening (Meier *et al.* 1971) and increases in gonadal weights (Fivizzani and Meier 1976), whereas injections 12 h apart had little effect on these parameters.

Since photoperiod modifies salinity preferences of stickleback, and since salinity tolerances of stickleback are modified by prolactin and ACTH injections, Audet *et al.* (1985*b*) examined whether the temporal synergism model holds for anadromous completely plated stickleback. Stickleback were held at a daylength of 24 h to avoid photoperiodic influences on the circadian cycles of these hormones. These fish showed a marked preference for high salinities. Cortisol and ovine prolactin were injected either simultaneously or 12 h apart to mimic presumed seasonal differences in the temporal synergism of the circadian rhythms of these hormones. Control fish were either uninjected, or were injected only with prolactin, only with cortisol, or with NaCl, 0.65 per cent. After the fish were maintained under these conditions for 8 d, the salinity preferences were determined. The prolactin injections markedly increased the preference for low salinities, and the simultaneous injection of cortisol and prolactin masked this effect. However, when prolactin and cortisol were injected 12 h apart, the prolactin effect was evident. These results support the temporal synergism model, even though the specific responses of the stickleback differ from those of the killifish.

To further test this model, Audet *et al.* (1986*b*) verified whether circadian cycles in serum cortisol levels occur in anadromous completely plated stickleback and whether such cycles are modified by photoperiod. No significant circadian rhythmicity could be demonstrated, perhaps owing to the considerable variability among fish. However, serum cortisol concentrations were higher in stickleback acclimated to a long daylength than in those acclimated to a short daylength. This difference is due primarily to increased cortisol levels in males acclimated to a long daylength, and may reflect the stress of increased territorial and aggressive interactions among these males. Alternatively, this correlation may reflect a role of cortisol in the control of reproduction in males. While these results do not support the temporal synergism model, seasonal differences in prolactin levels remain a means by which salinity preferences may be modulated in threespine stickleback. However, the difficulty of measuring prolactin with heterologous radioimmunoassays (Nicoll *et al.* 1981; Prunet *et al.* 1985) prevents determination of blood levels of prolactin in stickleback.

In several regards, it would be interesting to study the environmental and endocrine control of salinity preferences of the low-plated morph. This morph shows little seasonal change in prolactin cell histology and generally in pituitary histology, but its tolerance of seawater is affected by thyroid hormones. Furthermore, although its osmoregulatory mechanism seems less effective than that of the completely plated morph, the low-plated morph tolerates brackish or salt water during much of the year. The loss of efficient seawater adaptability presumably occurred after the colonization of freshwater habitats by anadromous stickleback. It would be exciting to determine the extent to which the environmental and hormonal control of salinity preferences have shifted during such a secondary loss.

In summary, the environmental control of salinity preferences in the completely plated stickleback is probably mediated by one or more of the following pituitary hormones: prolactin, ACTH through its stimulation of cortisol secretion, and thyrotropin through its stimulation of thyroxine and triiodothyronine production. Currently, photoperiod effects on plasma hormone concentrations have only been demonstrated for cortisol, but they cannot explain migratory movements by both male and female stickleback. Histological evidence suggests seasonal differences in the secretion of prolactin, ACTH, and thyrotropin. As with the effects of these hormones on salinity tolerance, it is possible that all three hormones are implicated in the *in vivo* regulation of salinity preferences. Arguing an adaptive utility for such multiplication of regulatory mechanisms is difficult, particularly when the selective importance of salinity preferences is not definitively established. However, similar examples of multiple hormonal controls are well documented in mammals (e.g. blood glucose levels, glomerular filtration rates), suggesting that functionality and not simplicity of regulatory control has been selected. Since a given hormone often has numerous effects, the fact that one of its effects duplicates that of another hormone is not disadvantageous for the organism, particularly if the temporal secretion of the hormones is well coordinated.

Temperature: preferences, tolerances, and acclimation

Stickleback migrating from oceanic waters to shallow inshore bodies of water encounter marked thermal variations. Clearly, environmental temperature has major effects on metabolism, growth, and fundamental biochemical processes. Anadromous stickleback from three populations have been found to prefer 4–8 °C (Røed 1979), 9–12 °C (Lachance *et al.* 1987), and 16–18 °C (Garside *et al.* 1977). These contrasting results may represent interpopulation differences or variation in the physiological condition of the fish owing to acclimation conditions. Overall, stickleback can be seen to prefer relatively cool temperatures (< 18 °C). In the field, anadromous stickleback nesting in tidepools will tolerate temperatures as high as 30 °C, whereas in the laboratory their upper lethal temperature is 28 °C. Acclimation salinity modifies this upper lethal temperature (Jordan and Garside 1972).

As stickleback will tolerate a wide range of temperatures but prefer cool temperatures, their eurythermality is distinct from that of many temperate zone fish for which the preferred temperatures are near the upper thermal limits. This distinction is also reflected by the cool thermal optima for egg development and growth in stickleback (Elliott 1981). Given that northern European and Canadian stickleback migrate at low temperatures and remain active during their breeding season, locomotion is maintained over a wide thermal range. By contrast to these stickleback, the southern California, USA, stickleback, *Gasterosteus aculeatus williamsoni*, has a critical thermal maximum (CTM) of 30.5 °C when acclimated to 8 °C and a CTM of 34.6 °C when acclimated to 22.7 °C (Feldmeth and Baskin 1976). Presumably the thermal optima of egg development, growth, and locomotion would be shifted to warmer temperatures in this stickleback.

The lower lethal temperatures of stickleback have not been systematically studied. Since northern European and Canadian anadromous stickleback overwinter in cold oceanic waters, they may be exposed to temperatures near or below the freezing point of their body fluids. Although this would suggest that stickleback possess the antifreeze proteins which protect many fish overwintering in cold oceanic waters, their possession of antifreeze has not been examined.

Thermal acclimation and compensation

The maintenance of locomotion over a wide thermal range suggests that stickleback could demonstrate thermal compensation of their aerobic capacity, much as do eurythermal fish from temperate zones. Generally, cold acclimation leads temperate-zone fish to increase the aerobic capacity of their muscles, both by increasing the proportion of aerobic fibres in their musculature and by increasing the abundance of mitochondria in the muscles

(Johnston and Dunn 1987; Guderley and Blier 1988; Sidell and Moerland 1989). This increased aerobic capacity enhances the sustained swimming capacity at low temperatures (Rome *et al.* 1985; Sisson and Sidell 1987). In contrast to temperate zone fish, stickleback have their thermal optimum in the centre of their thermal range (Elliott 1981). In stickleback, the pectoral muscles can sustain speeds up to 5–6 body lengths per second and the axial muscle is used primarily for burst swimming (Taylor and McPhail 1986; Whoriskey and Wootton 1987). The pectoral muscle has a higher aerobic capacity (as judged by mitochondrial enzyme activities) than the axial muscle (Vézina and Guderley 1992). The axial muscle is largely composed of fast glycolytic fibres (Kilarski and Kozlowska 1983; te Kronnie *et al.* 1983). In stickleback, an increase in the aerobic capacity of the musculature could be achieved by increasing the mass of the pectoral muscle, by increasing the proportion of aerobic fibres in the axial muscle, or by increasing the level of mitochondrial enzymes in either muscle.

To examine whether anadromous stickleback from the St Lawrence Estuary modify the aerobic capacity of their swimming musculature to compensate for the impact of temperature on locomotion, we acclimated males and females to 4 and 20 ° for 6 wk. The fish were fed daily with dry food at a ration of 1 per cent of their wet mass, because this ration leads to slight but equivalent growth at the two temperatures for Welsh, freshwater stickleback (Allen and Wootton 1982*a*). However, warm-acclimated fish had a markedly lower physical condition and a lower mass of axial and pectoral muscles than cold-acclimated fish. Furthermore, mitochondrial and glycolytic enzyme activities decreased with warm acclimation in both muscles. Over several years, the physical condition and enzyme activities were lower in males captured in late June (mean temperatures approximately 20 °C) than in those captured in early May (mean temperatures < 10 °C). The physical condition and enzyme activities in females change less than in males (Vézina and Guderley 1992).

To prevent the decrease in physical condition which could have been caused by an accelerated metabolic rate at high temperatures, fish were next fed at maximal rations during thermal acclimation (dry food at 2 and 8 per cent of their wet mass at 4 and 20 °C, respectively). After 6 weeks' thermal acclimation, the physical condition of the warm- and cold-acclimated fish did not differ. Since cold-acclimated fish had higher muscle activities of mitochondrial enzymes than warm-acclimated fish, we concluded that temperature modifies the enzymatic composition of muscle and that mitochondrial enzyme activities are enhanced in stickleback acclimated to low temperatures. The ninespine stickleback, *Pungitius pungitius*, from Isle Verte also increases the levels of mitochondrial enzymes with cold acclimation and decreases its physical condition as temperatures rise during the reproductive season (Guderley and Foley 1990).

As the zooplankton densities in the tidal pools at Isle Verte assure that stickleback are not food limited (Ward and FitzGerald 1983; Castonguay and FitzGerald 1990), it is intriguing that the physical condition of males decreases during the breeding season. The stickleback's budgeting of aerobic power may explain this tendency. As fish cannot exceed their VO_{2max} for significant periods, aerobic power must be carefully budgeted among the fish's different activities (Wieser *et al.* 1988; Wieser and Medgyesey 1990). The increase in metabolic rate caused by feeding and digestion decreases the aerobic power instantaneously available for energetically costly activities, such as reproductive aggression (Chellappa and Huntingford 1989) and other parental duties (FitzGerald *et al.* 1989). Although the postprandial increase in oxygen consumption has not been quantified for stickleback, maximum ration reduces the aerobic scope for activity by 50 per cent in largemouth bass, *Micropterus salmoides*, and blenny, *Blennius pholis* (Jobling 1981). In the fast-growing cod, *Gadus morhua*, the postprandial VO_2 is virtually equal to $VO_{2\,max}$ (Soofiani and Hawkins 1982). Fish hatchery managers have practical knowledge of power budgeting. Fish are never fed before being transported because the excitement of transport, coupled with the increase in metabolic rate due to feeding, quickly exhausts and often kills the fish (Priede 1985). Restriction of food intake may provide a compromise by which stickleback can improve territorial defence and parental behaviour, and thereby increase reproductive success. Selection might therefore favour males that decrease their levels of feeding during the reproductive period. Birds that nest in difficult habitats decrease their food intake to increase their reproductive success (Thompson and Raveling 1987; Parker and Holm 1990). By contrast, as female stickleback must produce energetically expensive eggs to assure their reproductive success, high rates of feeding would enhance reproductive success of females. In freshwater, low-plated stickleback, ovarian growth is sustained at the expense of somatic growth (Wootton *et al.* 1978; 1980*a*; Wootton page 133 this volume).

Stanley and Wootton (1986) have examined a related question in establishing the effect of ration on territoriality and nest building by low-plated stickleback from Afon Rheidol, a river in mid-Wales, UK. At a given density of males, the proportion that established territories and built nests increased with ration. Similarly, nest building and aggression increase at higher levels of food availability in brook stickleback (Smith 1970). These findings initially seem to contradict the above arguments concerning power budgeting and a possible selective value of reduced feeding. However, because, in Stanley and Wootton's (1986) study, the proportion of time males spent inactive increased with ration, the impact of power budgeting is evident. Probably, the increase in metabolic rate caused by food assimilation and processing reduced the stickleback's capacity for activity. Unfortunately, the actual food consumption and energy reserves in the males were not measured in either of the above studies. Such data would help evaluate Stanley and Wootton's suggestion that increases in short-term reserves are the means by which high rations increase the resource-holding power of male stickleback. Alternatively, the rise in nesting at higher rations could reflect the advantage of greater food availability for potential offspring. Accordingly, Smith (1970) suggested that brook stickleback may use increased food availability as a signal for the initiation of reproductive activities.

Another explanation of the decline in condition of the Isle Verte stickleback is that reproductive maturity may be directly linked to senescence. Such a relationship is certainly well documented for Pacific salmon (Mommsen *et al.* 1980; Ando *et al.* 1986). Only juveniles, 1+, and 2+ stickleback are observed in the Isle Verte population (Picard *et al.* 1990). Although it is not known whether these stickleback can reproduce in two successive years, our current thinking is that it is unlikely. As the proportion of 1+ individuals that reproduce is low, most fish in this population seem to postpone reproduction until age 2+. This situation suggests that, in the salt-marsh, once the stickleback reproduce, the decline in condition is inevitable. In an annual freshwater population of stickleback, the energy reserves in breeding males decline sharply at the end of the reproductive season. Non-breeding males survive beyond the reproductive season, but their energy reserves decline sharply in early winter when these fish disappear from the population (Chellappa *et al.* 1989). In combination with the data from the population at Isle Verte, these data are consistent with the concept that reproductive maturity and in particular reproduction *per se* initiate a senescence process.

Tolerance of hypoxia

Changes in availability of oxygen are a potential problem for fish reproducing in shallow tidal pools full of vegetation. For example, in the salt-marsh pools at Isle Verte, on the south shore of the St Lawrence Estuary, nocturnal oxygen levels can drop to 2.8 mg l^{-1} or lower. This drop could become a serious constraint for parental males which are faced with the conflicting demands of their own survival and that of their developing embryos. Nonetheless, parental males generally did not decrease the time spent fanning their eggs and shifted to aquatic surface respiration between fanning bouts (Reebs *et al.* 1984). The southern California stickleback, *G. aculeatus williamsoni*, which regularly encounters hypoxic conditions, shows its highest oxygen consumption at 2 ppm, and slightly lower but equivalent rates of oxygen consumption at 1, 5, and 7 ppm (Feldmuth and Baskin 1976).

Hypoxia can also become a serious problem for stickleback overwintering in still, ice-covered waters. Low temperatures will delay the gravity of the problems posed by hypoxia, by their effects both on oxygen solubility and on metabolic rate. Nonetheless, under such conditions, stickleback may need special mechanisms for finding and exploiting oxygen-rich waters. Although the response of *G. aculeatus* to such conditions has not been studied, the adaptations of the brook stickleback, *Culaea inconstans*, to this type of hypoxia are illustrated by Klinger *et al.* (1982). The gas bubbles that form beneath the ice significantly increase survival of brook stickleback. Its head shape seems particularly well adapted to the use of oxygen layers at the ice/water interface. Brook stickleback reduce their level of activity at low dissolved oxygen levels (below 0.5 mg l^{-1}) and show an acute capacity to detect waters with a higher oxygen content. This capacity leads them to orientate to water being expelled by a mudminnow pumping oxygenated water through a simulated crack in the ice (Klinger *et al.* 1982). This symbiotic relationship between brook stickleback and mudminnow seems a major determinant of the brook stickleback's presence in lakes where oxygen levels drop markedly during winter (Magnuson *et al.* 1988).

ENVIRONMENTAL AND HORMONAL CONTROL OF THE REPRODUCTIVE CYCLE

Anadromous stickleback experience changes in environmental conditions primarily because of their reproductive activities. One could thus expect that the environmental signals that coordinate physiological adaptation to changing environmental conditions also control the reproductive cycle. The most extensive studies on environmental control of the reproductive cycle in stickleback are those of Baggerman (1957). In her studies, a behavioural criterion was used to establish reproductive maturity, namely the ability of females to spawn (as indicated by stripping the females for their eggs) and that of males to build a nest. Using these non-invasive criteria, she studied the photoperiod and temperature conditions which control the reproductive cycle. Considerable information concerning the mechanisms of environmental and endocrine control of oocyte maturation in females, and of spermatogenesis and the development of secondary sexual characteristics in the males, has also been provided by Borg and co-workers. All in all, these studies leave the European stickleback in the enviable position in which the control of gametogenesis and of secondary sexual characteristics is better understood than for most fishes. However, the applicability of Baggerman's and Borg's findings to populations of stickleback with different life histories and phylogenetic histories must be determined.

Control of the female reproductive cycle

In her study of the female reproductive cycle, Baggerman (1957) used all three lateral plate morphs of stickleback obtained at different sites throughout Holland, as she found no differences in their responses to external conditions. This lack of difference is intriguing in view of the apparent differences in pituitary physiology between at least the Pacific anadromous, completely plated and the European freshwater, low-plated stickleback (Lam and Hoar 1967; Lam and Leatherland 1969; Benjamin 1980). However, the discrepancy in pituitary physiology may reflect the difficulties of comparing results among studies in which exact experimental conditions differ.

Environmental control of oogenesis

Baggerman (1957) found that during the first two months after hatching, female stickleback will not mature, even when kept under a normally stimulatory photoperiod (16L 8D) at 20 °C for almost a year (phase 0). Subsequently, the fish pass into phase 1, in which they stay sexually immature if maintained under short daylengths (8L 16D and 20 °C). For some phase 1 fish (1a), an acceleration of gonadal maturation is obtained by acclimation to 8L 16D at 4 °C. All fish in phase 1 will mature if acclimated to 16L 8D. Subsequently, females enter phase 2, in which maturation is inevitable at a regime of 8L 16D or 16L 8D at 20 °C. Maturation at long daylengths and high temperatures becomes increasingly rapid the closer the fish are to their natural breeding season (Baggerman 1989). Thus exposure to 16L 8D and 20 °C takes 30 days to bring wild fish to sexual maturity in November and 2 d in May. Finally, adult females are refractory to the effects of long daylengths for approximately 6 wk after the end of the breeding season. For Dutch stickleback, phase 0 lasts from hatching until late July, phase 1 lasts until late winter, and phase 2 lasts from early spring until the breeding season (Baggerman 1957). The timing of the reproductive cycle depends upon the general environmental conditions as well as the specific conditions at the breeding grounds. For example, anadromous stickleback breeding in salt-marsh pools at Isle Verte, Quebec, complete reproductive activities by mid- to late June, whereas individuals of the same population which breed in adjacent tidal rivers continue breeding until mid-July (Kedney *et al.* 1987).

Temperature modulates the effects of photoperiod on sexual maturity in a phase-specific fashion. In phase 1a, a combination of short daylength and low temperature stimulates sexual maturation. For phases 1b and 2, high temperatures stimulate maturation at long daylengths. In phase 2, high temperatures accelerate maturation at both short and long daylengths (Baggerman 1957).

In her early work, Baggerman (1957) used the capacity of females to spawn as her criterion of reproductive maturity. Consequently, the histological stages of the ovaries during the different phases are not clear. Subsequent research demonstrated that the gonadosomatic index (GSI; mass of the ovaries/mass of the somatic tissue) peaks during the breeding season (late April through late July) and that relatively low values are observed outside the breeding season (Borg and van Veen 1982; Baggerman 1989). During the early autumn, the oocytes contain little yolk, and primary yolk formation occurs in November through March.

In the population studied by Borg and van Veen (1982), the deposition of exogenous yolk begins in the first females in April, and is evident in increasing numbers of females through May and June. In the Dutch fish studied by Baggerman (1989), exogenous yolk deposition begins between late February and late March. The primary and secondary yolk both react with concanavalin

A, indicating that they are glycoproteins (Covens *et al.* 1988). The water content of mature eggs is significantly higher than that of ovarian tissue, although the difference is much less marked than that for fish with pelagic eggs (Craik and Harvey 1986). Although Borg and van Veen (1982) found large numbers of atretic eggs in most females in late summer and early autumn, and numerous overripe eggs in late summer, Baggerman (1989) indicates that atretic eggs were rare. During the breeding season, ovaries generally contain oocytes in several stages of development, including a developing clutch, a group of slightly enlarged pre-vitellogenic oocytes that will form the following clutch, and a pool of very small oocytes just beginning endogenous yolk accumulation. This variety suggests simultaneous development of batches of eggs (Wallace and Selman 1979; Borg and van Veen 1982) and accords with the stickleback's capacity for multiple spawning.

The effects of photoperiod and temperature upon ovarian histology vary with the season in which the stickleback are captured. Generally, oocyte maturation is stimulated by long daylengths at high temperatures (Schneider 1969; Borg and Ekström 1981; Borg and van Veen 1982). At high temperature, short daylengths only stimulate oocyte maturation in March and May, generally leading to considerable atresia at other times (Borg and van Veen 1982). In contrast to Baggerman's results with spawning readiness (see above), Schneider (1969) and Borg and van Veen (1982) found no consistent effect of photoperiod on ovarian histology at low temperatures. Borg and van Veen (1982) confirm Baggerman's finding that at high temperatures, the ovaries either mature or regress. It is unknown whether females in which the ovaries regress can subsequently produce mature oocytes.

Females held under constant short-day conditions (8L 16D at 20 °C) show little change in the GSI during the first 300 d, after which the GSI tends to decrease. Endogenous vitellogenesis begins in these stickleback, but oogenesis does not progress beyond this stage, and many atretic eggs are present. A small percentage of the females held either from a very early age or for their entire lives under constant short-day conditions matured. Changes in photoperiod and temperature conditions are therefore not essential for reproductive maturation, possibly because endogenous rhythms partly control the timing of reproductive maturation (Baggerman 1989). Stickleback show a circadian rhythm in photosensitivity and a circannual rhythm in sensitivity to photoperiodic stimulation (Baggerman 1980, 1985). Interindividual variance in the level of this photosensitivity would allow a small number of individuals to become reproductively mature even under constant short-day conditions.

Neuroendocrine control of oogenesis

The neuroendocrine mechanisms which mediate the effects of photoperiod and temperature have received considerable attention. Injections of various hormones into females maintained at inhibitory photoperiods revealed that high doses of mammalian luteinizing hormone were reasonably effective at stimulating gonadal maturation (Ahsan and Hoar 1963). These data are not particularly conclusive, however, because fish gonadotropins have been shown to stimulate spermatogenesis in goldfish at 0.01 $\mu g\ g^{-1}$ (Billard *et al.* 1982), a concentration approximately 20 000-fold lower than that used by Ahsan and Hoar (1963). Since melatonin levels are higher during the night than during the day, melatonin is a potential mediator of short-daylength control of physiological processes (Vivien-Roels 1983). High doses (4 $\mu g\ d^{-1}$) of melatonin inhibited the stimulatory action of long daylengths on stickleback oocyte maturation during late autumn and winter. Although these effects are consistent with the proposed inhibitory role of melatonin, low doses of melatonin (0.8 $\mu g\ d^{-1}$) enhanced ovarian weights and oocyte maturation under long daylengths. Progonadal effects only occur in females (Borg and Ekström 1981). Measurements of seasonal variations of melatonin levels are required to establish the physiological role of melatonin.

Although gonadotropins probably play a major role in controlling ovarian maturation, this has not been established definitively in stickleback. Generally, in vertebrates, gonadotropins are thought to stimulate ovarian oestrogen synthesis, which then stimulates hepatic vitellogenin synthesis (Jameson 1988). Incorporation of vitellogenin into the oocytes is stimulated by gonadotropins. Accordingly, in teleosts, gonadotropins are known to stimulate the production of vitellogenin and its transfer from the plasma into the oocytes, perhaps by a direct effect upon the transport process (Wahli *et al.* 1981; Wallace and Selman 1981, 1982). In stickleback, the available information is consistent with this pattern because oestrogen treatment of immature females increases plasma vitellogenin levels and incorporation of this vitellogenin into oocytes (Ollevier and Covens 1983). Vitellogenin is converted into two major yolk proteins which retain the antigenic determinants present in plasma vitellogenin (Ollevier and Covens 1983; Covens *et al.* 1987). Stickleback vitellogenin shares antigenic determinants with vitellogenins from rainbow trout, *Oncorhynchus mykiss*, and sea bass, *dicentrarchus labrax* (Covens *et al.* 1987).

Female stickleback can maintain viable ovulated eggs within the ovary for a certain period after which the eggs become overripe and the female's abdomen has a 'berried' appearance. This changed appearance is accompanied by a marked decrease in ovarian fluid production, a loss of egg viability, and an increase in the water content of the eggs (Lam *et al.* 1978). In 'berried' females, the corpora lutea and the ovarian epithelium regress, suggesting that decreased hormone secretion by the corpora lutea causes the diminished ovarian fluid production. Administration of progesterone caused marked fluid accumulation in the ovarian cavity, both in postspawning and in overripe females. In some postspawning females, administration of oestradiol leads to oocyte maturation, ovulation, and ovarian fluid accumulation (Lam *et al.* 1979). Progesterone production by the corpora lutea seems to be the primary means by which ovarian fluid production and hence egg viability is maintained. Lam *et al.* (1978) suggest that species such as stickleback, with elaborate courtship rituals, would benefit from maintaining such secretory activity in postovulatory corpora lutea for a certain period.

Control of the male reproductive cycle

In contrast to females, gametogenesis and reproductive readiness are not directly correlated in males. In adult males, spermatogenesis can start immediately after the reproductive season. If temperatures remain warm, spermatogenesis continues and fully mature sperm can be present in the testes in winter. However, reproductive readiness, as signalled by reproductive coloration, the transformation of the kidney into a glue-secreting organ, and the readiness of males to build nests, occurs much later in the annual cycle. See Bakker (page 361 this volume) for a consideration of the endocrine control of aggression during the male reproductive cycle. The environmental and endocrine control of the male reproductive cycle in European threespine stickleback has been extensively studied by Baggerman and by Borg and co-workers.

Environmental control of spermatogenesis and reproductive readiness

Baggerman's studies (1957, 1972, 1980) elucidated the environmental signals that bring males to sexual maturity by using the criterion of the male's capacity for nest building. Long daylengths in combination with warm temperatures are highly stimulatory in autumn and winter, whereas in spring, sexual maturity can also be attained at short daylengths. At low temperatures in winter and spring, long daylengths also stimulate maturation, but at a slower pace than at high temperatures. Baggerman's studies indicated that the factors stimulating reproductive readiness in males and females are generally similar. However, her studies did not indicate the relationship between sexual maturation and spermatogenesis. It has long been known that spermatogenesis in threespine stickleback is quiescent during the breeding season (Craig-Bennett 1931) owing to androgen inhibition (Borg 1981). This leads to an inverse relationship between spermatogenesis and androgen-dependent processes, such as the development of nuptial coloration and the transformation of the kidney into a glue-secreting organ (Wai and Hoar 1963; de Ruiter and Mein 1982). Early spermatogenetic stages occur under the short-day conditions that inhibit the expression of secondary sexual characteristics (Ahsan and Hoar 1963; Borg 1982*a*) and of reproductive readiness.

A complete view of the relationship between spermatogenesis and the development of secondary sex characteristics during the annual reproductive cycle was provided by Borg (1982*a*) for marine threespine stickleback. In the natural cycle, spermatogenesis occurs between the end of the breeding season and the beginning of winter. After January, the testes are largely composed of spermatozoa. The GSI shows a clear peak in October and declines to a plateau which is maintained during February, March, and April, and then declines further during the breeding season. The height of the kidney epithelium is low throughout the year and rises in May to peak values attained in June. Thus, on an annual basis an inverse relationship exists between spermatogenesis and the height of the kidney epithelium.

The impact of environmental conditions on reproductive maturation in male stickleback varies seasonally (Borg 1982*a*). The decline in the expression of secondary sex characteristics at the end of the breeding season is accelerated by high temperatures. Exposure to long daylengths and high temperatures during winter accelerates development of the kidney epithelium and inhibits spermatogenesis. By contrast, exposure to short daylengths and high temperatures leads to a drastic decline in the height of the kidney epithelium and an initiation of spermatogenesis. The negative correlation between the height of the kidney epithelium and spermatogenesis is as evident in experimental as in natural conditions (Borg 1982*a*). These studies indicate that the environmental control of the transformation of the kidney in males is similar to that of oocyte maturation, whereas the environmental control of spermatogenesis is distinct.

Influences of androgens on the male reproductive cycle

The endocrine control of the male reproductive cycle has received considerable study. As indicated earlier, the transformation of the kidney into a glue-secreting organ and the development of secondary sexual characteristics are androgen dependent, as administration of androgen to castrated males evokes these characteristics (Wai and Hoar 1963; Borg 1987). Furthermore, administration of an antiandrogen, cyproterone acetate, and castration lead to the same effects on the epithelium of the kidney of sexually mature male stickleback (Mourier 1976*b*). The interstitial cells in stickleback testes develop the ultrastructural characteristics typical of steroid-producing cells in early spring (Follenius 1968). Testicular androgen production is most probably responsible for the manifestation of secondary sexual characteristics in the spring. On the other hand, implantation of Silastic capsules filled with 11-ketoandrostenedione at the end of the breeding season, completely inhibited the initiation of spermatogenesis (Andersson *et al.* 1988). Administration of the artificial androgen, methyltestosterone, also inhibits spermatogenesis (Borg 1981; Borg *et al.* 1986).

Plasma levels of androgens vary seasonally. During the breeding season, 11-ketotestosterone dominates. The levels of 11-ketotestosterone peak during the breeding season and are directly correlated with the development of secondary sexual characteristics. By contrast, testosterone and 11-β-hydroxytestosterone levels were low throughout the year, and 11-β-hydroxyandrostenedione and 11-ketoandrostenedione levels were highest in early winter. During the period when active spermatogenesis occurs, plasma levels of these steroids were low. Castration reduced 11-ketotestosterone and testosterone levels without affecting the levels of the other steroids (Mayer *et al.* 1990*a*). Potential sites of extra-testicular androgen synthesis include the interrenal gland, which produces sex hormones as well as glucocorticoid and mineralocorticoid hormones. Implants of 11-ketoandrostenedione into castrated fish increased plasma 11-ketotestosterone levels (Mayer *et al.* 1990*a*), indicating extra-testicular conversion of androstenedione into 11-ketotestosterone. This conversion is catalysed by 17-β-hydroxysteroid dehydrogenase in blood cells of stickleback and several other teleosts (Mayer *et al.* 1990*b*). Although plasma levels of 11-ketotestosterone are high in stickleback, its biological effects are unclear as no binding to androgen target organs has been found in brown trout, *Salmo trutta* (Pottinger 1988) or goldfish, *Carassius auratus* (Pasmanik and Callard 1988).

Steroid metabolism

The testes of reproductively mature males convert pregnenolone into androstenedione, 11-β-hydroxyandrostenedione, and 11-ketoandrostenedione. The major steroid synthesized, 11-ketoandrostenedione, was made at lower rates at the end of the breeding season (Borg *et al.* 1989*a*). It is not enough to know which steroids are produced by the testes in order to predict the plasma steroid profile and the androgens which affect behaviour and secondary sexual characteristics, as the secreted steroids will be modified by enzymes in the blood cells (Mayer *et al.* 1990*b*). Nonetheless, 11-ketoandrostenedione was more effective than testosterone, 5-α-dihydroxytestosterone, and androstenedione in restoring reproductive behaviours of castrated stickleback (Borg 1987) and in modifying secondary sex characteristics (de Ruiter and Mein 1982; Andersson *et al.* 1988). As testosterone and androstenedione can be aromatized to oestrogens, whereas 11-ketoandrostenedione and 5-α-dihydroxytestosterone cannot, both aromatizable and non-aromatizable androgens modify reproductive behaviour in stickleback (Borg 1987). In contrast, aromatization of androgens to oestrogens is critical for androgen modification of behaviour in mammals and birds (Beyer *et al.* 1973).

An interesting paradox is evident here. On the one hand, structural differences between male and female sex hormones are relatively small, which suggests that the specificity of hormone–receptor interaction is high. Because administration of methyltestosterone to females evokes the transformation of the kidney into a glue-secreting organ (Mourier 1972) and elicits male reproductive behaviour patterns (Wai and Hoar 1963), gender differences are not caused by developmental differentiation of steroid receptors. On the other hand, the studies reviewed above indicate that several, structurally distinct androgens affect male reproductive characteristics. However, because dose–response curves are not available for these androgens, and as high doses have consistently been used, the structural specificity of steroid hormone receptors in stickleback remains to be established.

Potential neuroendocrine feedback mechanisms

In contrast to the testes, which do not aromatize androgens into oestrogens, the stickleback brain converts androstenedione to oestrogens and to testosterone (Borg *et al.* 1987*a*). This conversion was strongest in the pituitary and the nucleus preopticus (NPO), nucleus anterioris periventricularis (NAPv), nucleus posterioris periventricularis (NPPv), and nucleus lateralis tuberalis (NLT) of the diencephalon. Males and females show the same anatomical localization of brain aromatase activity. Reproductive females had a higher aromatase activity in the NPO than reproductive males, refractory males, males held under short and long daylengths, or males sampled in May or December. As experimental changes in daylength did not modify aromatase activity, photoperiod effects on reproductive maturation in males are not due to modifications in cerebral aromatase activity. However, because the enzymatic activity in the NPO of males is higher in May than in December, changes in aromatase activity may be involved in the seasonal differences in the effects of methyltestosterone on pituitary gonadotropic cells and testicular Leydig cells (Borg *et al.* 1985, 1986).

High aromatase activity has also been found in the NPO and NLT in other teleosts. In goldfish, circulating levels of gonadotropins are increased by lesions in the NPO and decreased by lesions in the NLT (Peter 1983). Furthermore, only aromatizable androgens exert feedback inhibition on gonadotropin release in mature catfish (de Leeuw *et al.* 1986) and stimulate gonadotropin release in immature trout (Crim *et al.* 1981). By analogy, Borg *et al.* (1987*a*) suggested that in stickleback, cerebral aromatase activity may be implicated in the control of gonadotropin secretion. Salmon gonadotropin-releasing hormone binds to stickleback pituitary, indicating that gonadotropin release is probably mediated by such a substance (Andersson *et al.* 1989).

Gonadal stimulation of cerebral aromatase activity is indicated by a series of experiments with stickleback in breeding and non-breeding condition. Stickleback were captured in winter and exposed to different photoperiods. Fish were then either sham-operated or castrated and implanted with Silastic capsules containing androstenedione or 11-ketoandrostenedione. Castration of stickleback that were in breeding condition markedly reduced aromatase activity in the NPO and the NAPv, and this reduction was reversed by treatment with the two androgens. Both aromatizable and non-aromatizable androgens reverse the effect of castration on aromatase activity. Furthermore, because castration of stickleback in non-breeding condition markedly decreased aromatase activity in the NPO and NAPv, Borg *et al.* (1987*a*,*b*, 1989*b*) suggest that the low levels of androgen in non-breeding stickleback are sufficient to activate aromatase activity in these areas.

Interactions between the pituitary and the testes

By analogy with other vertebrates, the endocrine activity of the testes in fish is most probably controlled by pituitary gonadotropins. Histological studies suggest that the secretory activity of gonadotropic cells in anadromous, completely plated stickleback is higher from spring to early autumn than in other seasons (Leatherland 1970*b*). The activity of gonadotropic cells is thus temporally correlated with the development of androgen-dependent characteristics. In the annual reproductive cycle of Swedish stickleback, ultrastructural indications of gonadotropic cell activity are also well correlated with the development of androgen-dependent characters, such as kidney hypertrophy, breeding coloration, and reproductive behaviour (Borg *et al.* 1987*c*, 1988). Environmental conditions that stimulate the development of secondary sexual characteristics led to the extension of the dilated granular endoplasmic reticulum in the gonadotropic cells. In contrast, spermatogenesis started when all indications of gonadotropic cell activity (i.e. nuclear size, cell size, cisternae of the granular endoplasmic reticulum, and the size of the Golgi complex) were decreasing. Similarly, short daylengths at high temperatures decreased evidence of secretory activity in the gonadotropic cells, increased spermatogenesis, and suppressed 3-β-hydroxysteroid dehydrogenase activity in testicular Leydig cells (Borg *et al.* 1987*c*). The 3-β-hydroxysteroid dehydrogenase is essential for the production of sex steroids. The negative correlation between gonadotropic cell activity and spermatogenesis may be caused by the stimulation of androgen synthesis by gonadotropins and a subsequent androgen inhibition of spermatogenesis (Borg *et al.* 1987*c*, 1988).

The above results suggest that spermatogenesis occurs only at low levels of gonadotropic cell activity, a response that differs from the trend observed for most vertebrates, in which gonadotropic hormones stimulate spermatogenesis. However, these results do not definitively rule out gonadotropic control of spermatogenesis. Spermatogenesis may be controlled by low levels of gonadotropins or by secretion of a specific gonadotropin. Generally, vertebrates are thought to have two gonadotropins, luteinizing hormone (LH) and follicule-stimulating hormone (FSH) (Jameson 1988), but considerable controversy has surrounded the existence of two gonadotropins in fish. Two gonadotropins with homologies to LH and FSH respectively have been isolated from chum salmon, *Oncorhynchus keta* (Suzuki *et al.* 1988*a*,*b*), suggesting that this pattern may apply to other fish, including the stickleback. Two types of pituitary gonadotropic cells have been identified in stickleback by Slijkhuis (1978), but Borg (pers. comm.) has not obtained similar results. Conceivably, spermatogenesis could be under the control of a gonadotropin that does not stimulate androgen synthesis, while the development of secondary sex characteristics and male reproductive behaviour could be controlled by the other gonadotropin, which stimulates testicular androgen synthesis.

The temporal separation of spermatogenesis and the development of secondary sex characteristics makes the stickleback an excellent system in which to examine the influence of distinct pituitary gonadotropins.

Generally in fish, androgens exert a negative feedback upon the pituitary gonadotropic cells, because castration enhances and steroid administration inhibits gonadotropic cell activity (Jameson 1988). In contrast, androgens exert a positive feedback upon pituitary gonadotropic cells in stickleback. Administration of methyltestosterone to winter fish maintained at short daylengths and high temperatures stimulates the development of secondary sexual characteristics, while decreasing the number of granules and increasing the dilated endoplasmic reticulum in pituitary gonadotropic cells. This pituitary ultrastructure resembles that of gonadotropic cells during the breeding season when they are most active. Furthermore, the testicular Leydig cells had larger nuclei after treatment with methyltestosterone (Borg *et al.* 1986). When either castrated or sham-operated winter stickleback were maintained at long daylengths and high temperatures, the gonadotropic cells of sham-operated fish contained a more dilated endoplasmic reticulum and fewer secretory granules than these cells in castrated fish (Borg *et al.* 1989*c*). These data indicate a positive feedback of testicular androgens on pituitary secretion of gonadotropins.

Potential photoreceptors

The mediation of the numerous effects of photoperiod upon the physiology of sticklebacks requires the involvement of photoreceptors. The lateral eyes, pineal tract, or other extraretinal photoreceptors are clear candidates. Winter stickleback, either blinded or unoperated, were submitted to long or short daylengths. Under long daylengths, both blinded and intact stickleback developed secondary sexual characteristics, while neither group showed such development under short daylengths. Spermatogenesis was well established in both groups at the short daylengths, but was arrested under the long daylength. Clearly, extraretinal photoreception occurs in stickleback (Borg 1982*b*).

The role of the pineal gland in photoperiodic control was examined by injections of melatonin into stickleback held under short or long daylengths at different seasons (Borg and Ekström 1981). High doses of melatonin inhibited the stimulatory effect of long daylengths on the transformation of the kidney epithelium in males during the winter. This result suggests that melatonin mediates the short-daylength inhibition of reproductive readiness (Borg and Ekström 1981). The pineal gland may serve as a transducer of photoperiodic information very early in stickleback life. During embryonic development, the stickleback pineal gland possesses well-developed photoreceptors (van Veen *et al.* 1980) well before the retina (Ekström *et al.* 1983). Opsin, the peptide component of photopigments, is present in the pineal photoreceptors early in embryonic development and only appears in retinal photoreceptors after hatching (van Veen *et al.* 1984). The close proximity of the pineal gland and the hypothalamus suggests that the pineal may be an important transducer of photoperiodic signals (Ekström and van Veen 1983). The nervous connections between the di- and mesencephalic areas and the eyes and pineal organ, respectively, show significant overlap and provide a possible means for information to be transferred between these two sensory systems (Ekström 1984). Furthermore, neural signals from two types of pineal photoreceptor cell as well as an integrative neural circuitry have been described (Ekström and Meissl 1989). This system, together with the peptidergic innervation of the pineal (Ekström *et al.* 1988), indicates that photic information from the pineal organ may be modulated by other systems.

In summary, Borg and co-workers' studies concerning the neuroendocrine control of the reproductive cycle in stickleback indicate an intricate web of interactions. Gonadotropic cells stimulate testicular production of androgen, which further stimulates gonadotropic cell activity. Androgen production by the testes inhibits spermatogenesis. Although male stickleback do not conform to the 'general' vertebrate pattern, the considerable variety of reproductive strategies in vertebrates (Jameson 1988) suggests that other, less well-studied species may share such deviations. The variety of developmental and reproductive patterns shown by stickleback (Baker Chapter 6 this volume) suggests that the control patterns established for European stickleback may not apply to all threespine stickleback populations. For example, a Japanese population which can breed virtually year-round (Mori 1985) clearly requires different environmental and possibly neuroendocrine controls on gametogenesis and on the development of secondary sexual characteristics from those in North American and European populations for which reproduction is highly seasonal. Similarly, the mechanisms controlling reproductive maturation in stickleback that require more than one year to attain reproductive maturity (Wootton 1976) must differ from those in the well-studied European stickleback.

The separation of spermatogenesis from the development of secondary sexual characteristics may provide an energetic benefit for the stickleback. Because spermatogenesis is completed before the transformation of the kidney and the development of breeding coloration occur, the energetic costs of these processes are uncoupled in time. Nonetheless, because in stickleback the testes do not exceed 1–2 per cent of body mass, whereas in other teleosts the testes can exceed 10 per cent of body mass, such energetic savings would be small. Alternatively, the uncoupling of spermatogenesis from the development of secondary sexual characters may reflect the disadvantages of a glue-secreting kidney. The decrease in osmoregulatory capacity caused by the transformation of the kidney into a mucus-producing structure would favour delaying this transformation until directly before the reproductive season.

Biochemical adaptation to changing environment conditions

Biochemical modifications provide the underpinnings of the physiological and behavioural changes during the reproductive cycle of stickleback. These modifications involve not only the metabolism of the messenger molecules that regulate tissue responses, but also the metabolic capacities of the tissues themselves. My laboratory's studies of the biochemical responses to thermal acclimation illustrate the marked impact of the nutritional, reproductive, and thermal status on the metabolic capacities of stickleback muscle (see above). Borg and co-workers' studies on cerebral aromatase activities also indicate the biochemical adjustments which occur on a seasonal basis. Although electrophoretic variation of enzyme polymorphism has been used to compare fresh-water and anadromous populations of stickleback (Withler and McPhail 1985), little attention has been paid to functional biochemical differences among the different populations of stickleback.

Some insight into biochemical differences between anadromous and resident freshwater stickleback populations comes from the studies of Jürss and colleagues who have examined biochemical and physiological attributes. Although certain properties such as the gill Na^+/K^+ ATPase activities did not differ, muscle activities of alanine aminotransferase and aspartate aminotransferase were considerably higher in the anadromous population (Jürss *et al.* 1982). Protein concentration was also higher in the muscle of the anadromous population. In the liver, the specific activities of alanine aminotransferase, aspartate aminotransferase, and glutamate dehydrogenase (expressed as international units per mg protein) are higher in the resident freshwater population. The greater relative mass of the liver in the anadromous stickleback gives this form a greater total activity of the three enzymes under study (Jürss *et al.* 1983). These differences between populations were maintained throughout the year, leading Jürss *et al.* (1985) to suggest that they have a genetic basis. As these enzymes can generate free amino acids, differential needs for cellular volume regulation could have established genetic differences between the freshwater and anadromous populations. These biochemical differences could be adaptive, because the freshwater population experiences relatively constant osmotic conditions, whereas the anadromous population is exposed to different salinities. In contrast, Gutz (1970) showed that low-plated stickleback rely upon volume regulation when exposed to different salinities, but completely plated stickleback increase oxygen consumption, presumably owing to osmoregulatory work by gills, kidney, and intestine.

FUTURE DIRECTIONS FOR RESEARCH ON THE PHYSIOLOGICAL ECOLOGY OF STICKLEBACK

Perhaps the primary conclusion that can be drawn from this overview of the physiological ecology and evolution of stickleback is that too few studies have examined and interpreted the physiological ecology of stickleback in an evolutionary perspective. In future studies, the physiological ecology of populations that differ in their reproductive strategies should be compared. Systematic comparison of the physiological ecology of the low, partial, and complete plate morphs would lend insight into the evolutionary processes favouring their divergence. Similarly, further elucidation of the physiological and biochemical differences between these morphs, as well as between populations of the same morph that have different life histories, would be highly useful. For example, the osmoregulatory strategies of low-plated stickleback from brackish water habitats should be compared with those of low-plated stickleback from purely freshwater habitats. The osmoregulatory capacities of the purely marine and anadromous completely plated stickleback would also be advantageous to examine. Given the polyphyletic origin of freshwater stickleback (Bell and Foster page 16 this volume; McPhail page 401 this volume), comparison of the osmoregulatory capacities of separate freshwater populations would indicate whether adaptation to fresh water consistently follows the same pathway.

Future experiments should examine the energetic cost of physiological mechanisms and of different behaviours, particularly in stickleback populations in which availability of energy affects fitness. However, even for stickleback populations for which food is abundant, the energetic cost of different activities is critical, since aerobic power must be allocated among the organism's different activities at any moment. The recent development of combined calorimetry and respirometry could be useful for such experiments. Thermal acclimation experiments with stickleback from the southern and northern extremes of its range would elucidate the extent to which its capacity for, and strategy of, thermal compensation has changed during its exploitation of different habitats. Similarly, populations from habitats that differ markedly in thermal stability (for example large lakes and small tidepools) would be interesting to examine in this respect. Such comparisons would help elucidate the adaptive modifications of which stickleback are capable during short-term acclimation, and whether the capacity for adaptive modification remains constant among different populations.

Combining the approaches of physiological ecology and population genetics would accelerate the interpretation of physiological ecology in an evolutionary perspective. One area that would clearly benefit from further study is the genetic basis for the correlation between plate morph and successful exploitation of low-salinity habitats. Although physiological ecology has generally been carried out within the conceptual framework of comparative physiology, in which evolution is certainly a central theme, too often the integration of the information in an evolutionary framework is missing.

ACKNOWLEDGEMENTS

My research is supported by grants from NSERC as well as by grants from les Fonds FCAC held conjointly with G. J. FitzGerald. I wish to express my gratitude for the encouragement from, and discussions with, G. J. FitzGerald as well as J. H. Himmelman.

5

Energy allocation in the threespine stickleback

R. J. Wootton

Energy is the basic currency of living organisms (Harold 1986). It gives the organism the potential to do the work required for maintenance, growth, and reproduction. Organisms are subject to the laws of thermodynamics (Brafield and Llewellyn 1982), making it possible to describe quantitatively the expenditure of energy. All forms of energy are interconvertible and so can be measured in common units of joules (J) (1 J = 4.18 cal).

A stickleback acquires its energy from the chemical bonds of the metabolizable components of food (or when an embryo, from the yolk provided by its mother), the proteins, lipids, and carbohydrates. Food also provides the basic chemical units for the synthesis of tissue. The total energy intake of an individual fish must be balanced by energy expended on maintenance plus any energy stored in the form of new tissue:

$$\text{Intake} = \text{Maintenance} + \text{Growth}. \tag{5.1}$$

Energy intake (C) is the energy content of the food consumed. Maintenance, used here in a wide sense, has several components. Some energy is lost in faeces (F), consisting of undigested or partly digested food, cells sloughed off from the alimentary canal, mucus, and other voided by-products of digestion. Energy is also lost in the nitrogenous excretory products (U) produced by the catabolism of protein. The major component of maintenance is the energy lost as heat generated by the metabolic processes as useful work is done. This loss can be partitioned into three components (Brett and Groves 1979). The first, resting or standard metabolism (R_s), is the work that has to be done to maintain the individual in a dynamic steady state. It is measured as the metabolic rate of a resting individual in a postabsorptive state. The second component represents the energy costs of swimming ('activity', R_a). The third is the energy cost of food processing ('digestion', R_d), often called specific dynamic action (SDA). This component may include the costs of synthesis of new tissue (Jobling 1985). Energy can be accumulated in new tissue through somatic growth and the deposition of storage products, (P_s), and as reproductive products (P_r). The energy budget of an individual can be written as:

$$C = F + U + R_s + R_a + R_d + P_s + P_r \quad (5.2)$$

in which all components are measured in energy units, usually joules (Wootton 1990). The power generated by metabolism can be expressed in watts (W), equivalent to 1 J s^{-1}.

Changes in the pattern of allocation of energy between components of the energy budget will often have important effects on the growth, reproductive output, and even the survival of an individual (Fig. 5.1). For example, an increased allocation to growth may reduce the time that an individual is vulnerable to a size-selective predator (Reimchen page 246 this volume). Because of these consequences for the components of fitness, energy allocation must be sensitive to natural selection (Alexander 1967; Calow 1985).

Selection will favour patterns of energy allocation that tend to maximize the lifetime production of young by individuals. Energy allocation provides the means by which reproductive success is achieved (see also Glebe and Leggett 1981), but that success is not defined in terms of energy. A successful energy allocation might be one that uses an energy income efficiently, but it might also be a pattern that uses a high rate of income relatively inefficiently but which generates a high rate of production of offspring. The laws of thermodynamics form a physical constraint within which natural selection operates. But the study of energetics taken on its own offers no guidance as to what patterns of allocation will maximize the lifetime production of offspring in specific environmental circumstances (see Baker Chapter 6 this volume).

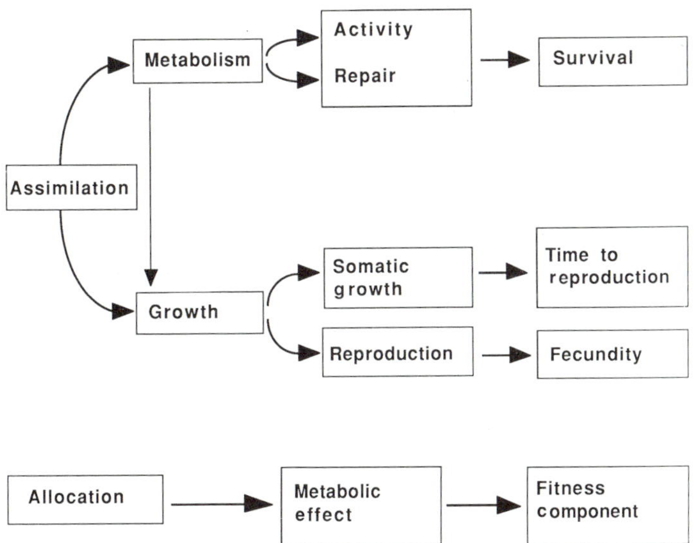

Fig. 5.1 Scheme of metabolic and possible fitness effects of energy allocation. (Modified after Calow 1985.)

The energy available to an individual is limited. Sometimes this limitation will result from a shortage of food. But even if food is plentiful, consumption rate is limited by the structures and mechanisms used to acquire and process food (Hart and Gill Chapter 8 this volume). Energy allocated to one component of the budget is at the expense of allocation to other components. Time is also a constraint. An individual may be able to increase its energy income by increasing the time it spends foraging, but this will be at the expense of the time spent in other activities such as sheltering from predators. Different activities have different energy costs, so that changes in the time spent in different activities will have implications for energy allocation.

The threespine stickleback, with its wide geographical range and its profusion of small variations on a common morphological and behavioural theme, offers great possibilities for understanding how patterns of energy and time budgeting relate to the reproductive success of individuals over a range of abiotic and biotic environmental conditions. Only a tentative start has been made on such a programme.

This essay reviews what is now known about energy allocation in the stickleback using data from both laboratory and field studies. It takes as its framework the energy balance equation, taking each term in that equation in turn and highlighting where important data are still required. An energy budget taken on its own is of little interest apart from demonstrating that the techniques used to measure each component are adequate. More important problems are these:

1. How do the patterns of energy allocation by individuals in a population change as environmental conditions change?
2. How do the patterns of allocation in different populations differ?
3. What are the consequences for growth, reproduction, and survival of differences in allocations?

CONSUMPTION

Prey selection and energy intake

The stickleback is catholic in its choice of prey (Wootton 1976, 1984*a*; Hart and Gill page 227 this volume), although its small body size restricts the size range of prey with which it can cope. Nevertheless, over a wide geographical range the diet tends to be dominated by two prey categories: zooplankton and the larvae and pupae of chironomids (Diptera) (Fig. 5.2). Interpopulation differences in the diet of the stickleback represent variations on a restricted theme. This restriction is emphasized even more by the similarity in the energy content of the typical prey (Table 5.1). Putting it crudely, a stickleback acquires approximately the same gross quantity of energy per unit dry weight of prey eaten, irrespective of which of the typical prey items are eaten. Prey will differ in their water content, ease of detection, and ease

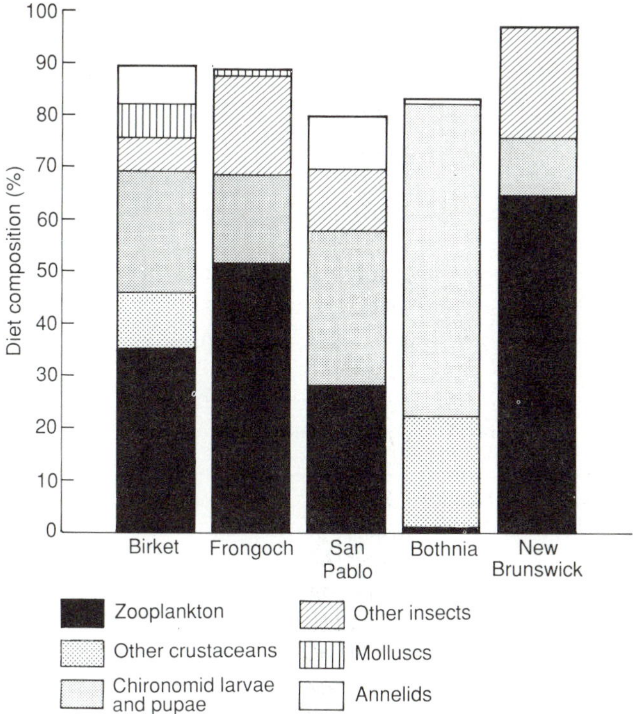

Fig. 5.2 Examples of diet composition of stickleback populations from a wide geographical range. Sources: River Birket, north-west England (Hynes 1950); Llyn Frongoch, mid-Wales (Allen and Wootton 1984); San Pablo Creek, California, USA (Snyder 1984); Bothnian Sea, Sweden (Thorman and Wiederholm 1986); New Brunswick, Canada (Delbeek and Williams 1987*a*). Prey types are indicated in the figure (other components, including unidentified debris, make composition up to 100 per cent).

of capture, and so will differ in profitability (energy gained per unit time) (Hart and Gill Chapter 8 this volume).

Interpopulation differences in morphology and behaviour related to feeding (e.g. Bentzen and McPhail 1984; Lavin and McPhail 1985, 1986; Hart and Gill page 210 this volume) have consequences for the acquisition of energy. An example is given by a study in Scotland, UK, by Ibrahim and Huntingford (1988, 1989*b*) on the stickleback from a large lake, Loch Lomond, and a small adjacent pond. The pond stickleback had a larger gape and inter-raker distance than the loch fish. When provided with either benthic prey or zooplankton in densities similar to those occurring naturally, experimental pond fish acquired energy at the rate of 0.034 J s^{-1} from zooplankton and at 0.026 J s^{-1} from benthic prey. Compared with the pond stickleback, the loch fish were significantly more effective at exploiting

Table 5.1 Energy contents of typical prey of the stickleback.

Food item	Range of energy contents		Sources[a]
	$kJ\,g^{-1}$ dry wt	$kJ\,g^{-1}$ wet wt	
Oligochaeta			
Tubificidae	21.63–23.34	3.18–3.60	1, 3, 4
Enchytraeidae	22.05–24.10	4.81	2, 5
Insecta			
Chironomidae	20.51–22.24	2.11–3.6	3
Crustacea			
Copepoda	23.45–24.62	2.30	3
Cladocera			
Daphnidae	17.13–24.48	1.07–1.55	3
Bosminidae	21.49–22.27		3
Chydoridae	22.62		3

[a] Sources: 1, Cole (1978); 2, Cui and Wootton (1988*a*); 3, Cummins and Wuycheck (1971); 4, Walkey and Meakins (1970); 5, Wootton *et al.* (1980*a*).

the zooplankton, 0.059 J s^{-1}, and not significantly less effective at exploiting the benthos, 0.011 J s^{-1}. A regression model developed by Wootton *et al.* (1980*b*) for sticklebacks feeding *ad libitum* on enchytraeid worms (Oligochaeta) predicts a daily maximum consumption of 450 J for a fish weighing 1 g at 15 °C. Although the calculations are crude, they suggest that loch fish feeding on zooplankton can consume their maximum daily intake in only 2 h, whereas the same fish feeding on benthos would take about 11 h. These values suggest the potential effects on time and energy allocation that differences in foraging effectiveness could have. However, Ibrahim and Huntingford's (1988) data were obtained from feeding sessions of only 5 min. Consequently, they provide no direct evidence that the differences translate into differences in the rate of energy acquisition over longer periods, which would have consequences for the components of fitness of growth, reproduction, and survival.

Maximum rate of consumption

The daily, maximum rate of food consumption by the stickleback is a function of body size, water temperature, and possibly other environmental factors including salinity and pH. Over the temperature range of 3–19 °C, maximum consumption of enchytraeid worms by sexually immature fish was predicted by the relationship:

$$\ln C_{max} = -4.28 + 0.93 \ln W + 0.91 \ln T \qquad (5.3)$$

where C_{max} is consumption in mg, W is fish weight in mg, and T is temperature in °C (Wootton *et al.* 1980*b*; Fig. 5.3). At higher temperatures, con-

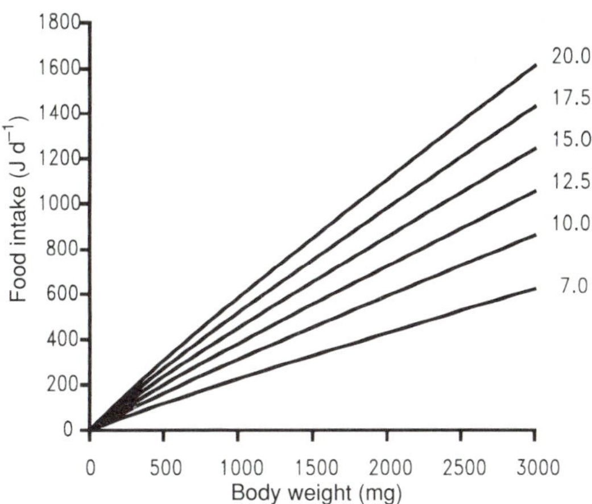

Fig. 5.3 Predicted effect of body weight and temperature over range 7.0 to 20 °C on maximum daily food consumption, C_{max}, based on regression model in Wootton *et al.* (1980*b*).

sumption probably declines sharply as in other teleost species (Elliott 1981).

Faris (1986) presented evidence that at pH lower than 6, appetite is depressed in freshwater residents. The effects of salinity are not known.

Under a defined set of environmental conditions, the maximum rate of consumption may change with the physiological condition of the fish. During a breeding season, sexually mature females readily consumed a daily ration of 16 per cent of their body weight of enchytraeid worms. At the end of the season, when egg production ceased, the same ration was only consumed over a period of several days, although neither temperature nor photoperiod changed. The females were still in a healthy condition two months later, so the changes in appetite were probably not due to senescence (Fletcher and Wootton unpubl. obs.). Such observations suggest that the fish do not always consume food at the maximum rate at which they can process it. Fish in a given physiological condition may have an optimal daily ration which is not necessarily the maximum possible ration. The regulation of appetite is an important but little-explored aspect of energy allocation in fishes.

Natural rates of consumption

There are several estimates of natural rates of food consumption. The use of different units to express consumption makes comparisons difficult, but the data suggest that a typical daily consumption is 2–10 per cent of body weight, depending on the temperature. Manzer (1976) estimated that planktivorous stickleback in Great Central Lake, Vancouver Island, Canada, were eating 6–8 per cent of their body weight daily. Stickleback living in brackish water in the Baltic Sea had daily rations estimated at 24.2 mg per g of fish at 10 °C, 48.3 mg g^{-1} at 14 °C, and 132 mg g^{-1} at 18 °C (Rajasilta 1980). Assuming that the average energy content of the prey was 4.81 J mg^{-1} wet wt (Cui pers. comm.) gives energy intake rates of 116, 232, and 634 J d^{-1} respectively.

The population in Llyn Frongoch, a small reservoir in mid-Wales, UK, had an annual consumption estimated at 2700–4200 mg wet wt (13–20 kJ a^{-1}) per average fish as it grew from 65.8 to 552 mg wet wt (Allen and Wootton 1984). For this Welsh population, in most months, the rate of consumption estimated from observed growth rates was less than the maximum possible consumption estimated from laboratory studies (Allen and Wootton 1982*b*). If Allen and Wootton's estimates are accurate, the growth rate of fish in the Frongoch population is energy limited, at least for portions of the year.

MAINTENANCE

Faecal and excretory losses

The proportion of ingested energy that is lost in faeces is largely determined by the quality of the food. Sticklebacks fed enchytraeid or tubificid worms, oligochaetes with soft cuticles, lost in faeces about 3–14 per cent of the energy ingested (Cole 1978; Allen and Wootton 1983). This proportion was lower in larger fish and at higher temperatures (Allen and Wootton 1983), but slightly higher when the intake rate was higher (Cole 1978). Similar research using prey with less digestible cuticles has yet to be done, although a field study suggests that the inclusion of such prey will increase the energy lost in faeces. The weight of faeces collected over 24 h from sticklebacks that had fed naturally in a Welsh lake was significantly higher than the weight produced by fish feeding at their maximum rate on an *ad libitum* supply of enchytraeid worms (Allen and Wootton 1983). For comparison, brown trout feeding on the isopod *Gammarus* lost 11–31 per cent of the energy in the food consumed, depending on temperature and ration size (Elliott 1979).

The major pathway of nitrogenous excretion is as ammonia/ammonium through the gills. The energy lost through nitrogenous excretion has yet to be estimated for the stickleback. A study of the European minnow, *Phoxinus phoxinus* (a small cyprinid), provides some comparable data. When fed on enchytraeid worms, the minnow lost 2–9 per cent of the ingested energy as nitrogenous wastes (Cui and Wootton 1988*a*). The loss tended to be proportionately greater at low rations and at higher temperatures. Brown trout fed *Gammarus* lost 3.6–15.1 per cent (Elliott 1979).

Metabolic expenditure

Energy lost neither in faeces nor in excretory products can be used to do useful work. The rate of energy expenditure can be directly measured by the rate of heat production (direct calorimetry). Although some attempts have been made to measure energy expenditures in this way (Lowe 1978), the high heat capacity of water means that changes in its temperature caused by the metabolic activities of the fish are small and difficult to measure accurately. Indirect calorimetry is usually used (Brafield 1985). If the fish is respiring aerobically, the rate of energy expenditure is related to the rate of oxygen consumption. This latter rate can be converted to a rate of energy expenditure by an appropriate oxycalorific coefficient, which depends on the substrate being respired. For a carnivore such as the stickleback, a value of 13.61 J mg O_2^{-1} respired is appropriate (Brafield 1985). If rates of carbon dioxide and ammonia production by the fish are also measured, a more accurate rate of energy expenditure can be estimated because these rates will reflect the substrate being catabolized (Brafield 1985). The respirometer should allow the fish sufficient space to make voluntary movements and should, preferably, allow the experimenter to impose known rates of swimming on the fish whose respiration rate is being measured. Studies on the rate of metabolism of the stickleback often fall short of these requirements. One of the better respirometers used is Lester's (1971). Unfortunately his presentation of the results makes it difficult to compare them with those from other studies that used less satisfactory respirometers.

Standard metabolism, R_s

Standard, or resting metabolism is difficult to measure accurately because the fish must be still and in a defined postabsorptive state. Lester (1971) and Meakins (1975) measured the decline in the oxygen concentration of water in a closed respirometer containing the fish. Meakins ensured that his fish were still by swimming them to exhaustion prior to measuring resting metabolism (although this procedure raises the possibility that the fish were repaying an oxygen debt incurred as they approached exhaustion). Meakins' (1975) data could be expressed as:

$$R(\mathrm{s}) = aW^b \tag{5.4}$$

where R (s) is the rate of oxygen consumption of the exhausted fish, W is body weight, and a and b are parameters with b taking values between 0.174 and 0.352 (Fig. 5.4). A weight exponent, b, for standard metabolism is typically less than 1.0 in teleosts (Brett and Groves 1979), showing that the weight-specific energy cost of maintenance declines as the weight of the fish increases.

The estimates of Lester (1971) and Meakins (1975) are similar. Meakins' (1975) equations predict that a 1 g stickleback has a minimum rate of energy expenditure of about 80 J d^{-1} at 15 °C (fish caught in August). Lester (1971) gives the standard metabolic rate for a fish 60 mm in length as 52 J d^{-1} at 10 °C. These values are about 17 per cent of the maximum rate of energy intake (311 J d^{-1} at 10 °C and 450 J d^{-1} *at* 15 °C) predicted for a 1 g fish by the model of Wootton *et al.* (1980*b*).

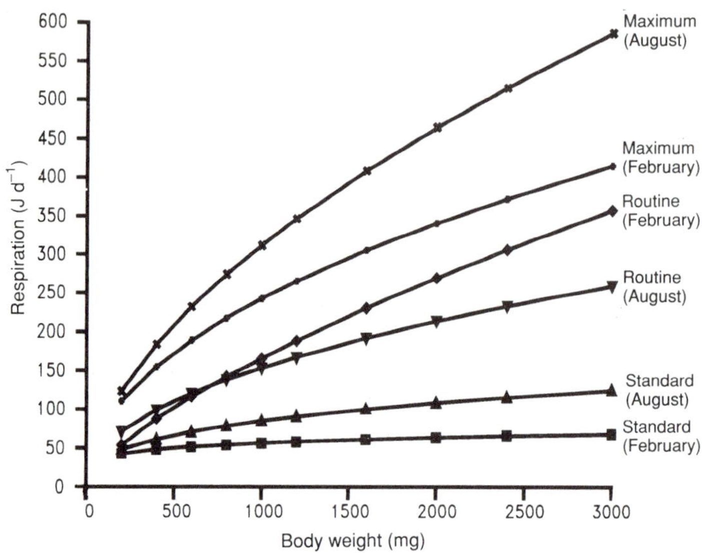

Fig. 5.4 Predicted effect of body weight on standard, routine, and maximum rates of metabolism, R_s, R_{rout}, and R_{max}, at 15 °C in fish collected in February and August, based on regressions in Meakins (1975).

Standard metabolism includes the energy costs of active osmotic and ionic regulation. For freshwater sticklebacks, such costs will remain almost constant, but for anadromous populations the costs will change as the fish move into waters of different salinities during migration. Gutz (1970) compared the respiratory rates of phenotypes that differed in the number of their lateral plates. When the low-plated morph was moved from fresh water into seawater, it adapted passively by increasing the pool of free amino acids and did not increase its energy expenditure. The completely and partially plated morphs adopted an active regulatory mechanism signalled by an increase in the rate of oxygen consumption. Calculations based on the respiration data of Gutz (1970) suggest that for a 1 g fish, the rate of energy expenditure in salt water was about 65–90 J d^{-1} higher in salt water than in fresh water at 20 °C. At 4 °C, the difference was about 45 J. However, these differences may also partly reflect differences in the level of spontaneous activity in fresh and salt water. McGibbon (1977) found that at 12 °C, a low-plated, freshwater stickleback showed no increase in the rate of oxygen consumption over a salinity range of 1.7–27.2‰ but there was a significant increase at 32.3‰.

The effects of other abiotic factors, including temperature, oxygen concentration, and pH, on the standard metabolism of the stickleback have yet to be quantified.

Activity metabolism, R_a

The energy costs of swimming can be estimated in two ways. In the first, the fish is provided with sufficient room in the respirometer to allow for spontaneous movement. Under these conditions, the routine rate of respiration is measured. This is the value commonly reported in studies of fish respiration rates (Brett and Groves 1979). The cost of swimming is the difference between the routine rate and the standard rate under the same conditions. Meakins (1975) and Wootton *et al.* (1980*a*) estimated the routine rate for fish held in a closed respirometer (Fig. 5.4, 5.5). At 15 °C, the routine rate predicted for 1 g fish was 153 J d^{-1} or about double the standard rate by Meakins, but 105 J d^{-1} by Wootton *et al.* (1980*a*). The effect of temperature on routine rate is also illustrated in Fig. 5.5. In a simultaneous comparison of routine metabolism by direct and indirect calorimetry, Lowe (1978) obtained estimates of 84 J g^{-1} d^{-1} (direct) and 80.9 J g^{-1} d^{-1} (indirect) (Brafield 1985). These studies did not define the amount of swimming shown by the spontaneously active fish, so any comparison of the results is difficult.

A second and better approach is that of Lester (1971), who forced fish to swim at a defined speed while the rate of respiration was measured. The range of speeds was from 0.9 to 1.8 body lengths per second (BL s^{-1}). Calculations based on Lester's (1971) data suggest that the rate of energy expenditure for a fish swimming at 0.9 BL s^{-1} ranges from 60 to 142 J d^{-1}. The range partly reflects differences in fish size. Assuming a fish of 55 mm,

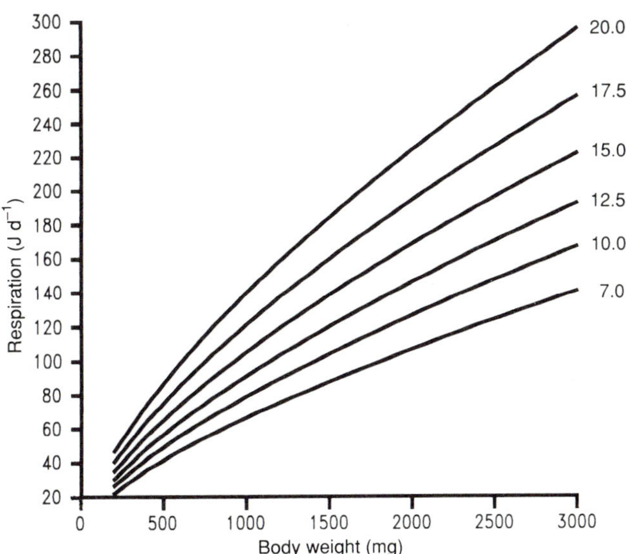

Fig. 5.5 Predicted effect of temperature over range 7–20 °C and body weight on routine metabolism, R_{rout}, based on regression in Wootton *et al.* (1980*a*).

this represents the energy cost (including standard metabolism) of swimming just over 4 km in a day.

Stickleback usually swim with a labriform mode of locomotion using the large pectoral fins. This mode is often associated with fish that live in complex environments such as rocky reefs or weed-beds and is correlated with manoeuvrability, rather than speed in cruising or fast starts and turns (Webb 1984). When sprinting, stickleback switch to a body and caudal fin mode of swimming. However, Taylor and McPhail (1986) and Whoriskey and Wootton (1987) found that stickleback using the labriform mode could swim for long periods at speeds as high as 5 BL s^{-1}. Taylor and McPhail (1986) describe morphological differences between the anadromous and freshwater stickleback which suggest that the former is better adapted for cruising. Anadromous forms may migrate upstream for many kilometres in spring. The energy cost of such prolonged swimming may be lower in stickleback with the morphology typical of anadromous (or limnetic) forms (Taylor and McPhail 1986). Well-designed experiments on sustained swimming of the anadromous, limnetic, and benthic forms are needed to relate the energy costs to the morphological adaptations, including differences in the number of lateral plates.

The small size of the stickleback means that swimming speeds of 5 BL s^{-1} represent absolute speeds of only 25–30 cm s^{-1}. This limitation essentially restricts sticklebacks to the slow-moving waters of lakes, ponds, lowland streams, or rivers (Wootton 1976). In coastal waters, tidal currents may aid onshore and offshore movements, with some of the energy required for transportation being provided by the motion of the water rather than the metabolism of the fish (Whoriskey *et al.* 1986). Adult and juvenile sticklebacks can occur many kilometres from the coast both in the North-west Pacific (Quinn and Light 1989) and in the West Atlantic (Cowen *et al.* 1991). The relative contributions of active and passive transport to these oceanic distributions are not known.

Apparent specific dynamic action, R_d

When a fish is fed, its rate of respiration increases to a peak, then declines as the meal is processed. This increase in energy expenditure associated with feeding has a small component attributable to the muscular work done during feeding and digestion, but the major portion is related to the biochemical processing of the food and the products of digestion (Jobling 1983). A portion of this expenditure may represent the energy costs of the synthetic processes involved in growth, particularly protein growth (Jobling 1985). R_d typically accounts for 9–20 per cent of the energy of the food consumed, the value partly depending on the chemical composition of the food (Jobling 1983). Although an increase in respiration associated with feeding has been observed in the stickleback (Wootton pers. obs.), quantitative analysis of the effect has not been completed.

Power budgeting

Priede (1985) has presented a stimulating if speculative analysis of energy expenditure in fishes. He notes that the total sustainable power output (i.e. energy per unit time) of a fish may be less than the maximum possible power demands of the three components, standard metabolism (R_s), activity metabolism (R_a), specific dynamic action (R_d) (see also Goolish 1991). Under these circumstances, the fish must regulate the allocation of the available power. Priede (1985) further argues that this allocation should, as far as possible, allow the fish to stay well within the limits of its scope for

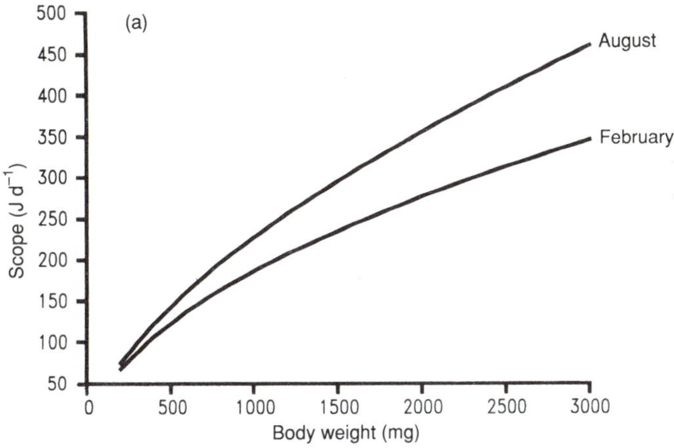

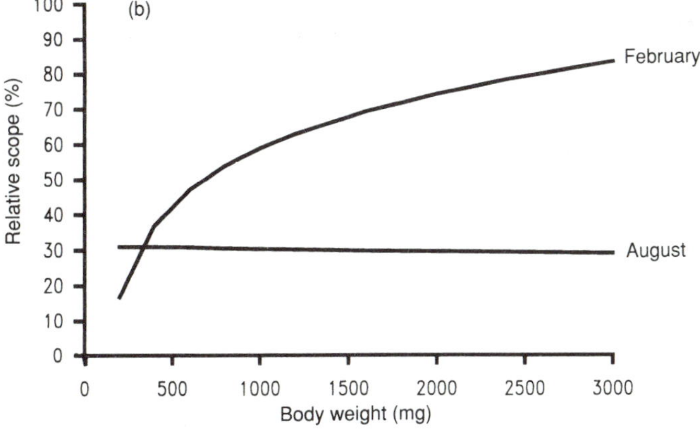

Fig. 5.6 Predicted scope (a) and normalized (relative) scope (b) for activity at 15 °C for fish collected in February and August. (Values based on regressions presented in Meakins 1975.)

activity. This scope is defined as the difference between the maximum and standard rates, i.e. $R_{max} - R_s$, where R_{max} may be measured as the rate of metabolism of a fish swimming at its maximum sustainable speed (Fry 1971). Meakins (1975) provides some data on scope (Fig. 5.6). Priede (1985) defines a normalized scope (S) as:

$$S = (R - R_s) / (R_{max} - R_s) \tag{5.5}$$

where R is the observed metabolic rate. In Meakins' experiments, fish that were collected in winter but acclimated and tested at 15 ° had S values ranging from 0.16 for the smallest fish (50 mg dry wt) up to 0.63 for the largest fish (300 mg dry wt). In contrast, fish collected in summer had an almost constant S value of about 0.3 irrespective of their size. Priede (1985) argues that the probability of dying increases greatly if a fish is forced to maintain very low or very high S values for long periods of time, and that evolutionary adaptations that tend to ensure that fish avoid such states are favoured.

GROWTH

In energy terms, growth ($P = P_s + P_r$) is the difference between the energy income and the energy expended in maintenance:

$$P = C - (F + U) - R. \tag{5.6}$$

Clearly, P, measured in units of energy, can be either positive (growth) or negative (degrowth). Growth is used in reference to individual fish, while the term 'production' refers to the new tissue generated by a population (Wootton 1990). Individual fish are not homogeneous masses. The allocation of growth energy to different components of the fish must be analysed, as well as the total growth energy. A simple classification of these components would be:

1. Structural components, forming the soma: these are organs and tissues essential for the immediate survival of the individual, including organs such as the heart and circulatory system, the central nervous system, skeleton, muscle, alimentary canal, etc.

2. Reproductive components: these are the organs that either produce the gametes or play some role in reproduction, and so include secondary sexual characteristics. These components are not essential for the survival of the individual, and could in times of energy shortage be exploited to maintain the soma. But, unless at some stage priority is given to investment in the reproductive components, the individual will leave no offspring.

3. Storage components: these act as reservoirs, which can be filled during times when the energy income greatly exceeds the maintenance costs and emptied when costs exceed income, thus buffering both the reproductive and

somatic components from the effects of temporary energy imbalances. The storage materials are usually lipids or glycogen.

The allocation of energy among these three components will depend on the phase of the life cycle and on the immediate environmental conditions. In practice, most studies of growth report only growth in length, a measure of axial size, and weight, a measure of bulk. These measures give only a crude picture of the growth dynamics.

Experimental studies on growth

Experimental studies can quantify the effects of environmental factors on growth rates. Ration size and temperature are the factors that have received most attention. The relationship between specific growth rate and ration is curvilinear, with growth rate tending towards an asymptotic value (Fig. 5.7) (Allen and Wootton 1982*c*). Although Pascoe and Mattey (1977) fitted a straight line to the relationship between growth rate and ration, visual inspection of their graphs shows clearly that the relationship was curvilinear. In both of these experiments growth was measured as change in weight.

When weight is used to measure growth, there is a complicating factor: the energy content per unit dry weight varies with the age and condition of the fish. Stickleback weighing 0.030 g dry wt had an energy content of 14.23 kJ

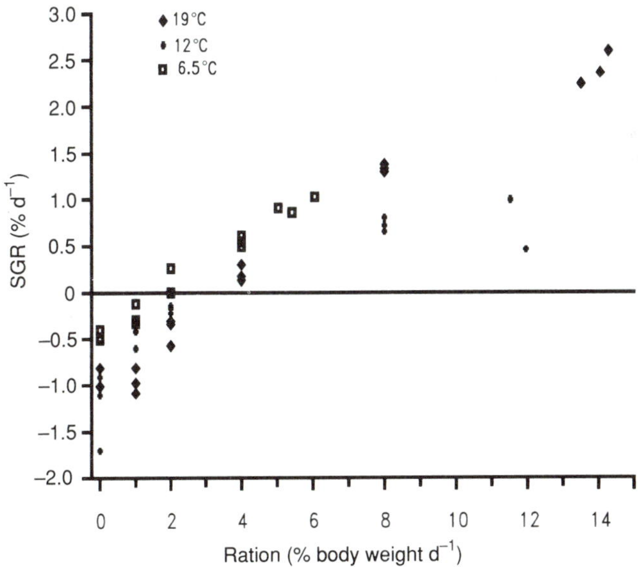

Fig. 5.7 Relationship between mean specific growth rate and ration, illustrated at three temperatures (N = 3) (Allen and Wootton 1982*c*).

g^{-1}, but this had increased to 19.87 kJ g^{-1} for fish weighing 0.270 g dry wt (Walkey and Meakins 1970). Sexually mature females had a higher energy content than mature males (24.27 v. 19.66 kJ g^{-1} dry wt, Meakins 1974). The difference represents the effect of the enlarged ovaries in the females. It is probable that fish receiving different rations also have different energy contents (Cui and Wootton 1988*b*), but growth measured as changes in weight will not reveal this difference.

In analysing the relationship between growth and rate of feeding, it is useful to define four ration levels: zero (C_0 = starvation), maintenance (C_{main}), optimum (C_{opt}), and maximum (C_{max}) (Brett 1979). The weight (or energy) loss at C_0 defines the resistance of the fish to starvation. C_{main} is that ration which allows the fish to maintain a stable weight (or total energy content). C_{opt} is that ration at which the fish maximizes its weight (or energy) increase per unit of food consumed. C_{max} defines the maximum voluntary rate of consumption, so that the difference, $C_{max} - C_{main}$, defines a scope for growth, which is analogous to the scope for activity as described above. Gross growth efficiency (GGE) is defined as:

$$\mathrm{GGE} = 100\ (P_s/C) \tag{5.7}$$

and is maximized at the optimum ration. Net efficiency (NGE) measures the growth per unit of food consumed after the requirements for maintenance have been subtracted, i.e.

$$\mathrm{NGE} = 100\ \{P_s/\ (C - C_{main})\}. \tag{5.8}$$

The curvilinear relationship between growth and ration means that C_{opt} is less than C_{max} (Brett 1979). The individual can maximize either growth rate (by consuming C_{max}) or growth efficiency (by consuming C_{opt}), but not both, except at low temperatures when the growth rate even at C_{max} is low.

At 15 °C, C_{main} for a stickleback weighing 250 mg wet wt was about 1.9 per cent of its body weight of enchytraeid worms per day (Allen and Wootton 1982*c*). This consumption is approximately 22 J d^{-1}, assuming that the energy content of enchytraeid worms is 23 J mg^{-1} dry wt, and their dry weight is 20 per cent of wet weight (Cui pers. comm.). This value is lower than R_s predicted by Meakins' (1975) equations or the routine metabolic rate predicted by Wootton *et al.* (1980*a*), and may indicate that the fish adapt physiologically to the low rations.

The effect of ration (expressed in J d^{-1}) and temperature on specific growth rate (expressed in terms of wet weight) is shown in Fig. 5.8. These relationships are based on an empirical regression model developed by Allen and Wootton (1982*c*) relating growth rate to ration, temperature, and body weight. This model can be used to predict the effect on growth rate of a change in foraging behaviour that leads to a measurable change in the daily ration. The same model also provided the basis for predictions of the effect of temperature on C_{main}, C_{opt}, and C_{max} (Fig. 5.9).

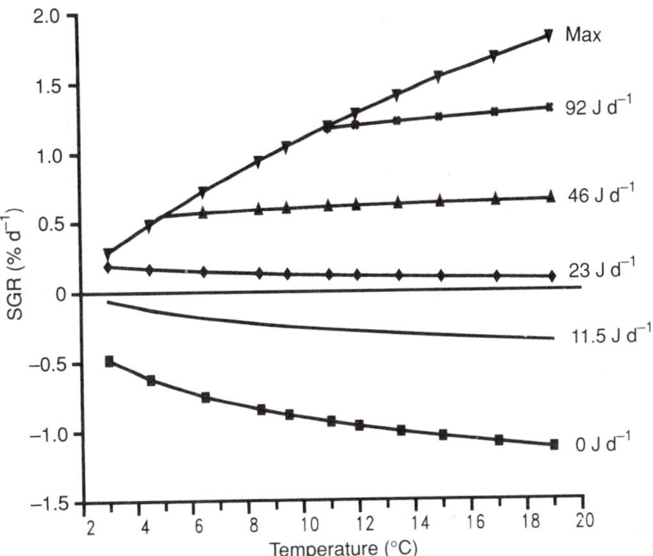

Fig. 5.8 Predicted effect of ration and temperature on specific growth rate, for a 250 mg fish. (Based on regression in Allen and Wootton 1982*c*.)

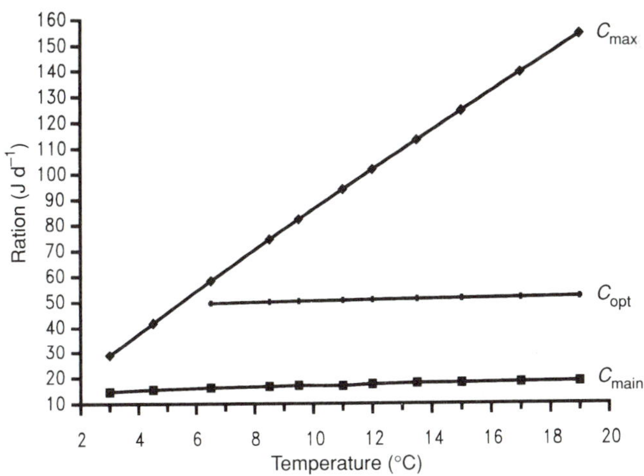

Fig. 5.9 Predicted effect of temperature on maintenance, C_{main}, optimum, C_{opt}, and maximum, C_{max}, rations. (Based on regressions in Allen and Wootton 1982*c*.)

At C_{main}, GGE is, by definition, zero. With a further increase in consumption, GGE increases up to a maximum at C_{opt} and then declines slightly between C_{opt} and C_{max}. Although slightly dependent on temperature, the maximum mean GGE for a stickleback fed on enchytraeid worms was 15–16 per cent, when both growth and food consumption were measured in wet weights (Allen and Wootton 1982*c*). Estimates of the growth and food consumption for the population in Llyn Frongoch suggested a GGE of about 15 per cent averaged over a year (Allen and Wootton 1982*b*). Lower values for GGE were found when *Tubifex* (Oligochaeta) was the food, and both growth and consumption were measured in energy units. Cole (1978) measured a GGE of 7.7 per cent at 7 °C and 12.5 per cent at 20 °C, whereas Walkey and Meakins (1970) estimated the GGE at 6.7 per cent at 15 °C. The effect of ration size on GGE makes a direct comparison of these results difficult.

Growth in natural populations

The threespine stickleback is typically a short-lived, seasonal breeder (Baker Chapter 6 this volume). In many populations, few or no fish survive much longer than a year (Wootton 1976, 1984*a*). This means that somatic and reproductive growth, plus any increase or depletion of stored reserves, are compressed into a single reproductive cycle. The species conforms to the typical teleostean growth pattern, with a high specific growth rate early in life, and the rate declining as the fish increases in size and approaches an asymptotic size (Allen and Wootton 1982*b*; Crivelli and Britton 1987).

The asymptotic size and size at sexual maturity reflect both genetic and environmental effects. McPhail (1977) reared the progeny of fish from populations taken from ten lakes on Vancouver Island, British Columbia, Canada. He demonstrated heritable differences among the populations for size at maturity of females. Similarly, Snyder and Dingle (1989) found differences in the size and age at maturity between the laboratory-raised offspring from an anadromous and a freshwater population from the Navarro River in California, USA. It is not clear whether these population differences are caused by differences in appetite, in the allocation of different proportions of the energy income to growth, or are the result of differences in the timing of the onset of reproduction.

An example that probably reflects environmental rather than genetic effects on growth is provided by a population living in a small backwater of the Afon Rheidol in mid-Wales, UK. In the period between 1972 and 1986, the mean size of fish aged about 15 months varied from 45 mm to 58 mm (Wootton pers. obs.).

Growth rates also reflect seasonal changes in the environment. In Llyn Frongoch in mid-Wales, the growth of the stickleback ceased during the late autumn and winter and resumed in the following spring (Allen and Wootton 1982*b*). In a population in southern England, the growth rate slowed in winter, but some growth was maintained (Mann 1971). For a population in the Camargue delta region in southern France, there was no check to growth in winter (Crivelli and Britton 1987). At the onset of maturation, the males stopped growing, although growth of females continued.

Growth of body components

Both field and laboratory studies of growth in total body weight may conceal differences in growth rates of the body components. Such differences can mean that some components maintain a positive growth rate while others are losing weight (or total energy content) (Wootton *et al.* 1978; Allen and Wootton 1982*a*). Prior to first spawning, almost all the investment in gonadal growth may be recouped and reallocated either to somatic growth or to storage if necessary. However, the structures required for the production and release of gametes have to be in place to take advantage of an opportunity to reproduce, so it may be advantageous to maintain some investment in the gonads during temporarily unfavourable conditions. The failure to invest adequately in gonadal maturation is illustrated in a population in Scotland, UK (Ukegbu 1986). A portion of the adult fish fail to spawn when they are a year old. Such fish survive up to their second winter of life, but die before the next breeding season (see also Chellappa *et al.* 1989).

Laboratory studies

Allen and Wootton (1982*a*) found that under experimental conditions, the soma (excluding the liver), liver, and ovaries of females from the Llyn Frongoch population differed in their sensitivity to the size of the food ration. Females were taken at month intervals directly from the field and exposed for a 21 d period to a range of rations from 0 per cent of body weight to *ad libitum*. For most of the year, ration size had no significant effect on the growth of the ovaries over the 21 d period, but the growth of both the soma and the liver was sensitive to ration. This sensitivity varied during the year. The liver was relatively insensitive to ration level during the breeding season in spring and summer, but highly sensitive in autumn and winter. The results suggest that the liver acts as a short-term buffer, ameliorating the effects of a food shortage on the soma and ovaries. The liver is also the site of synthesis of the precursors of the yolk deposited in the maturing oocyctes, and so in this role mediates between the soma and the ovaries.

Field studies of the growth of components

In two Welsh populations, ovarian growth was maintained over the first winter of life even when the hepatic and other somatic growth slowed or ceased (Wootton *et al.* 1978). Because of the small size of the ovaries, this growth did not represent a large absolute rate of increase in energy (Table 5.2). In the winter months, food consumption was low and ovarian growth represented a high proportion of the energy remaining after the losses in

Table 5.2 Estimates of the components of the overwintering energy allocation of an average female stickleback from Llyn Frongoch and the Afon Rheidol (mid-Wales, UK) (from Wootton *et al.* 1980*a*). *R*, respiration rate; *C*, consumption rate; *F*, faecal losses.

Site and month	Temp. (°C)	Mean length (mm)	Change in energy in		*R* ($J\,d^{-1}$)	*C* ($J\,d^{-1}$)	*F* ($J\,d^{-1}$)
			Soma ($J\,d^{-1}$)	Ovaries ($J\,d^{-1}$)			
Frongoch							
Sept.	13.0	31.8	5.8	0.2	32.8	47.8	1.5
Oct.	9.5	33.8	0.2	0.2	30.0	38.0	1.6
Nov.	6.5	35.0	−5.2	0.2	25.4	42.5	2.0
Dec.	5.8	35.6	−1.8	0.3	26.3	31.1	1.8
Jan.	6.5	35.7	0.3	0.4	26.3	38.0	1.9
Feb.	4.0	35.8	23.6	0.3	24.0	45.4	2.4
Mar.	5.0	36.4	3.6	2.1	26.5	61.0	2.7
Apr.	7.0	37.2	4.9	2.0	34.3	78.9	2.9
Total (kJ)			0.897	0.173	6.83	11.59	0.507
Per cent of *C*			7.8	1.5	59.1	100	4.4
Rheidol							
Sept.	12.5	36.8	12.3	0.2	48.4	64.4	1.9
Oct.	10.0	38.6	−0.2	0.0	46.3	55.8	1.9
Nov.	6.0	39.6	−31.6	0.2	37.0	44.1	2.1
Dec.	6.5	40.0	16.6	0.5	35.5	62.8	2.4
Jan.	6.5	40.0	1.2	0.9	39.7	46.7	2.1
Feb.	5.0	40.0	1.5	0.9	34.0	72.1	2.9
Mar.	4.5	40.8	0.2	4.6	35.0	84.8	3.2
Apr.	10.0	42.2	10.3	9.6	53.6	147	4.2
Total (kJ)			0.324	0.511	9.97	17.44	0.625
Per cent of *C*			1.9	2.9	57.2	100	3.6

faeces and routine maintenance were subtracted. In some months, the total energy content of the soma declined (Table 5.2), probably because of depletion of its lipid content. February marked the start of a period of rapid growth of the carcass, and March an acceleration in ovarian growth rate. Thus in spring, food consumption was sufficient to support investment in both the soma and the ovaries. After the start of the breeding season in May, somatic growth slowed or ceased, and the lipid content of the soma and liver declined.

A study of male stickleback in a Scottish population provides a comparable picture. During early winter, the lipid and glycogen contents of the liver and carcass were depleted, but then recovered in a period of growth in the spring. Over the breeding season, reproductively active males showed massive depletion of liver glycogen, liver lipid, and somatic glycogen, and to a lesser extent somatic lipid (Chellappa *et al.* 1989). The depletion of glycogen preceded lipid depletion. Males that did not reproduce entered their second autumn with higher lipid and glycogen contents than the few reproductively active males that survived. But the former then suffered rapid depletion of both lipid and glycogen during the autumn and had disappeared from the population by January.

Both the Welsh and Scottish populations were effectively annual. Gonadal growth was maintained over the first winter of life even at times when the soma and liver were being depleted. In the annual population in the Camargue (Crivelli and Britton 1987), the testes grew rapidly between September and December, and the ovaries between December and February. For both sexes, the soma grew rapidly between October and January, growth slowing or ceasing after the onset of breeding in February or March. No studies have been published on the pattern of growth of soma and gonads in those populations in which sexual maturity is reached at two or more years, nor on anadromous populations in which sexual maturity is reached at two or more years, nor on anadromous populations in which the energy costs of migration have to be met.

REPRODUCTION

The main pathways of energy expenditure on reproduction are the energy invested in the released gametes, plus any increased metabolism associated with the synthesis of the gametes or secondary reproductive characteristics and with the behavioural activities associated with reproduction. In the female stickleback, the expenditure is dominated by the eggs. In the males the main expenditure is probably associated with reproductive behaviour plus the energy content of the glue secreted during nest building. In terms of the energy budget (eqn 5.2), the expenditure of the female is measured principally by the P_r term, but that of the male by the R_a term.

Energy expenditure by females

In the laboratory, the female stickleback can spawn several times during the breeding season, at intervals of a few days. The energy represented by the eggs produced by a female is the product of the average energy content per egg, average clutch size, and the number of times the female spawns.

Energy costs of egg production

The energy costs of egg production can be met from two sources: from the food ingested, or by depletion of the soma, including the storage components. If insufficient food is consumed in the interval between successive spawnings to meet the cost of the next clutch, the cost of egg production is subsidized from the female's soma (Wootton 1977; Fletcher 1984).

Wootton (1977, 1979) estimated the daily energy intake that a female needed if she was to avoid losing weight during the breeding season (Table

Table 5.3 Predicted egg production per spawning in energy units, ration required for no net somatic weight loss, and efficiency of egg production of female stickleback during a breeding season (from Wootton 1979). (Ration as per cent body weight (% BW) is calculated assuming 1 J = 0.217 mg wet weight of food.)

Weight of female (mg)	Interval between spawnings (d)	Egg production per spawning (kJ)	Ration		Efficiency of egg production (%)
			(kJ d^{-1})	(% BW)	
800	4	0.501	0.683	18.6	18.3
800	5	0.470	0.597	16.2	15.8
800	6	0.447	0.512	13.9	14.5
1200	4	0.789	0.862	15.6	22.9
1200	5	0.741	0.777	14.1	19.1
1200	6	0.704	0.691	12.5	17.0
1600	4	1.089	1.041	14.1	26.6
1600	5	1.023	0.956	13.0	21.4
1600	6	0.972	0.870	11.8	18.6

5.3). Expressed in terms of the weight of enchytraeids consumed as a percentage of female body weight, the values in Table 5.3 range from 18.6 to 11.8 per cent. (In the experiment, the fish had been fed either minced beef or *Tubifex* worms.) In a subsequent experiment in which the females were fed enchytraeids alone, Fletcher (1984) estimated that females would have to consume about 8 per cent per day to avoid losing weight. These rations compare with a maintenance ration for a non-breeding female of about 2 per cent. The energy demands of egg production probably explain why somatic growth slows or stops during the breeding season (Wootton *et al.* 1978; Crivelli and Britton 1987), and explain the depletion of lipids and glycogen over the breeding season described above (Wootton *et al.* 1978).

Giles (1987*a*) provided indirect evidence of demands on metabolism that are imposed by the synthetic processes associated with multiple spawning. As fish were subjected to progressively decreasing oxygen concentrations, gravid females started using the oxygen-rich surface film (aquatic surface respiration) at significantly higher oxygen concentrations than did non-gravid females. The behaviour of the gravid females resembled that of females infested with plerocercoids of the cestode *Schistocephalus solidus*, a parasite that makes substantial energy demands on its host (Walkey and Meakins 1970; Lester 1971; Meakins 1974; see also Hart and Gill page 217, Huntingford *et al.* page 281 this volume).

The proportion of energy income invested in eggs over a breeding season, a measure of reproductive effort, can also be calculated from measurements of the total food consumption and total egg production (Wootton 1977;

Fletcher 1984). Fletcher (1984) found an inverse relationship between ration size and reproductive effort (see also Wootton 1985*a*). Although they produced fewer eggs over a breeding season than the females on high rations, the females on low rations were investing a higher proportion of their energy income in eggs. The high reproductive effort of females on low rations resulted from the subsidy from the soma and storage components for egg production, because rate of food consumption was insufficient to meet the costs of egg production. A more complete index of reproductive effort would be given by the ratio: (total energy content of eggs produced) (energy in food consumed + energy losses of soma and storage components). There was a weak but significant inverse correlation between reproductive effort and growth rate (Fletcher 1984; Wootton 1985*a*) over the breeding season. That is, at the phenotypic level, there was a trade-off between growth and breeding season fecundity. This inverse correlation was strongest for females receiving a 2 per cent daily ration, but not significant for females on daily rations of 8 and 16 per cent of body weight.

The relevance to natural populations of these laboratory studies will only become evident with studies of the fecundity and rates of food consumption in natural populations. Some circumstantial evidence—a decline in growth rate and a depletion of lipids and glycogen over the breeding season observed in natural populations (Wootton *et al.* 1980*a*; Crivelli and Britton 1987) and high postbreeding mortality (Hagen and Gilbertson 1973*b*)—suggests that the energy demands of reproduction exert a cost.

Breeding season fecundity

Food supply is an important determinant of the total number of eggs spawned in a breeding season by a female because of its effect on the number of spawnings (Wootton 1973*a*, 1977; Fletcher 1984). Under laboratory conditions, a well-fed female may spawn 15–20 times at intervals of 3–6 d, whereas a poorly fed female may not spawn. Figure 5.10 shows the predicted increase in breeding season fecundity with ration size (J d^{-1}) based on experiments with females fed enchytraeid worms.

Do females in natural populations express this physiological capacity to produce multiple clutches? Bolduc and FitzGerald (1989; see also Baker page 170, Whoriskey and FitzGerald page 199 this volume) showed that individually marked females in salt-marsh pools spawned between zero and three times before leaving a pool. In this habitat, where suitable food was judged to be abundant, egg production does not seem to be limited by energy but by the short time the females spend on the breeding grounds (Whoriskey *et al.* 1986). In contrast, Foster (pers. comm.) observed marked females spawning at approximately weekly intervals over a three month breeding season at Crystal Lake, British Columbia, Canada. Some females produced clutches throughout the season, although the majority did so for a shorter (≤ 2 month) period.

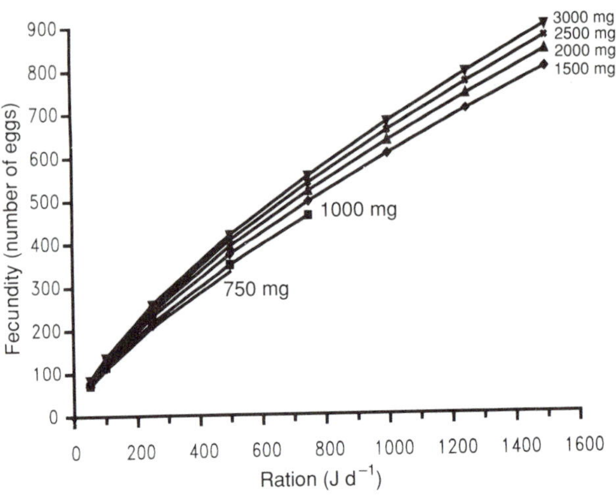

Fig. 5.10 Predicted effect of ration and body weight on breeding season fecundity. (Based on regressions in Fletcher 1984.)

Clutch size

The number of eggs in a clutch is a function of female size at the start of the breeding season (Wootton 1973*b*). The relationship between length (L) or weight and clutch size (F) takes the general form:

$$F = aL^b \text{ or } F = cW^d. \qquad (5.9)$$

The length exponent, b, is typically about 3.0 (as expected from the relationship between linear dimensions and volume), and the weight exponent, d, about 1.0 (Wootton 1973*a*; 1976). Both parameters a and b (or c and d) may show interpopulation variation (Hagen 1967; Wootton 1973*a*; Snyder and Dingle 1989; reviewed in Baker Chapter 6 this volume, Fig. 6.4). Such interpopulation differences suggest different levels of investment in egg production. Crivelli and Britton (1987) report the high length exponent (b) of 5.28 for the Camargue population, which lives close to the southern limit of the stickleback in Europe.

Unlike breeding season fecundity, clutch size is little affected by the amount of food eaten during the breeding season. Wootton (1977) found no significant relationship between the food eaten in the interval between successive spawnings and clutch size. In a larger series of experiments, Fletcher (1984) found that differences in daily ration accounted for only 4 per cent of the variance in clutch size, although the effect was statistically significant. In contrast, body size accounted for 50 per cent of the variance. This contrast suggests that food supply exerts its major effect on clutch size indirectly through its effect on growth and hence size at maturity (Wootton 1973*b*). Other studies have found that clutch size is little influenced by temperature fluctuations (Boulé and FitzGerald 1989), pH (Faris 1986), and a range of physicochemical variables including temperature, dissolved oxygen, and salinity (Bolduc and FitzGerald 1989). Both food and temperature have important effects on the components of the energy budget, so the insensitivity of batch fecundity to these factors is unexpected. Once a female is committed to producing a clutch, the size of that clutch is not adjusted to the level of energy income from the food consumed.

Clutch size varies among individuals within a population (even after correcting for differences in female size) and between clutches produced by the same female in the breeding season, but the causes of this variation are not known.

Energy content of eggs

The mean energy content of eggs from the Afon Rheidol population of mid-Wales is 22.6 J mg^{-1} dry wt (Wootton and Evans 1976), giving a mean energy content per egg of 7–9 J. Meakins (1974) gives a higher value of 26.2 J mg^{-1} dry wt for ovarian tissue. Interpopulation variation in egg size is discussed by Baker (page 160 this volume), and there may also be variation in the energy content.

Reproduction in the male

Under laboratory conditions, the energy accounting for reproductively active females is not technically demanding. But for males the situation is different, because the energy costs of reproduction are predominantly imposed by the behavioural activities associated with successful reproduction: territoriality, nest building, and parental care. In the absence of direct measurements of the energy costs of reproductive behaviour, two approaches have suggested that the costs are substantial. The first investigates the effect of different levels of energy income on reproductive behaviour (Stanley 1983; Stanley and Wootton 1986; Wootton 1985*a*). The second measures the depletion of the energy content of body components (Chellappa and Huntingford 1989; Chellappa *et al.* 1989; FitzGerald *et al.* 1989).

Males with a large body size may achieve greater reproductive success. Rowland (1989*a*) introduced pairs of sexually mature males into a tank simultaneously and found that the larger male usually dominated the smaller in the contest for a territory (see also Rowland page 302 this volume). A weight difference of 15 per cent was sufficient to give the heavier male a distinct advantage. In more complex experimental designs in which several males competed for territories, size has not always emerged as an important variable (van den Assem 1967; Sargent and Gebler 1980). Dufresne *et al.* (1990) examined the effect of size and age by keeping groups of small and large mature males in wading pools. Only when the mean size difference between large and small males was 16 mm or more were the larger males significantly more successful in obtaining territories.

Two aspects of size need to be considered: the first is absolute size; the second is the size of the energy reserves on which the fish can draw to meet the energy demands of the establishment and defence of a territory and of the subsequent courtship and parental activities.

Several effects of ration size on the reproductive biology of the male are recorded. As a male matures, its kidneys enlarge and start to synthesize the glue used in nest construction. When comparisons were made between sexually mature males of the same length, the kidneys of males on a low ration (2 per cent body wt d^{-1}) were significantly smaller than those of males on higher rations (6 per cent or 18 per cent d^{-1}) (Stanley 1983).

In large tanks, at a given density of males, proportionately fewer males on low rations held territories than in tanks holding males on higher rations. When males were transferred from the large tanks into smaller, individual tanks, a significantly smaller proportion of the males on low rations built nests (Stanley and Wootton 1986). After each isolated male with a nest was

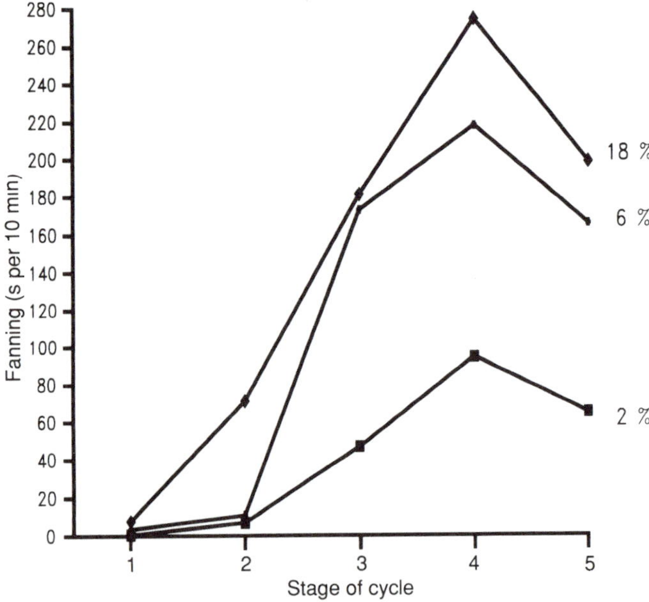

Fig. 5.11 Effect of ration on fanning by males fed rations of enchytraeid worms of 2 per cent ($N = 6$), 6 per cent ($N = 5$), and 18 per cent ($N = 10$) of body weight during a parental cycle from fertilization to hatching. Stage of cycle: 1, day after fertilization; 2, 25 per cent of cycle completed; 3, 50 per cent of cycle completed; 4, 75 per cent of cycle completed; and 5, 100 per cent of cycle completed (from Stanley 1983).

allowed to fertilize one clutch, the low-ration males spent less time fanning their eggs than males on higher rations (Fig. 5.11). This difference in fanning did not translate into a difference in the proportion of fertilized clutches that hatched (Stanley 1983).

The depletion over the breeding season of the lipid and glycogen in the carcass and liver of males in a Scottish population is also evidence of the energy costs incurred by breeding compared with non-breeding males (Chellappa *et al.* 1989). Even fights between pairs of males lasting between 11 and 204 s caused depletion of liver glycogen. The depletion increased the longer the fight lasted. At the end of the fight, the liver glycogen of the victorious males was significantly higher than in the losers (Chellappa and Huntingford 1989). Growth over a 12 d period was significantly lower in nest-holding males that in others (Stanley 1983). The energy costs of parental behaviour were indicated by the changes in the weight of males relative to their weight on the day they fertilized eggs (Fig. 5.12). FitzGerald *et al.* (1989) also noted a decline in the dry weight of males that completed a parental cycle compared with non-parental males over the same period. One-year-old breeding males lost more weight than non-breeding males over the same time period, although inexplicably for two-year-old fish, there was no significant difference between the two groups (Dufresne *et al.* 1990).

The consequences of the energy demands made on the reproductive male

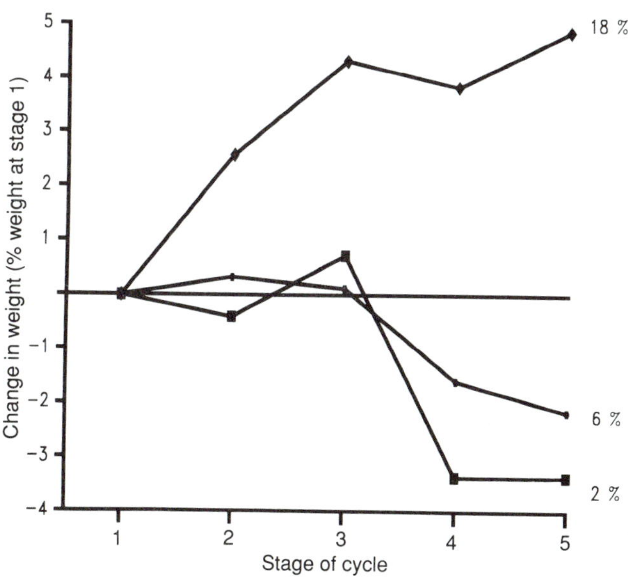

Fig. 5.12 Effect of ration on per cent weight change by parental males held at different rations (per cent body wt) during a parental cycle from fertilization to hatching. For further explanation see Fig. 5.11 (from Stanley 1983).

for other aspects of its biology may be illustrated by the study of Bentzen and McPhail (1984) on the foraging behaviour of the limnetic form of the stickleback in Enos Lake on Vancouver Island, British Columbia. When presented with a substrate containing benthic prey, only the males of the limnetic form fed. The females would strike at stray zooplankters in the water column, but did not exploit the abundant benthic prey. Bentzen and McPhail (1984) noted that the male limnetics must spend the summer months in a benthic environment to breed. This may impose a compromise on the males between pure planktivory and benthic feeding. In contrast, the females need only visit the benthic environment for the short time required for spawning and can then return to foraging on zooplankton in the water column.

Although a value can be calculated for the effect of an increase in the daily ration of a female on her breeding season fecundity (Fig. 5.10), it is not yet possible to relate the total number of eggs fertilized by a male to his energy income. Nor is it possible to compare the energy allocations of males and females until more is known about the rates of food consumption of the two sexes and the metabolic costs of reproductive behaviour.

DISCUSSION AND CONCLUSIONS

A central aim of studies of bioenergetics on species such as the threespine stickleback must be to relate the patterns of energy allocation described to the reproductive success of the fish. The study of energetics charts one pathway which causally relates both individual variation and the effects of environmental factors to differences in reproductive success. This is because these differences are likely to be related to differences in the patterns of energy allocation (Fig. 5.1). These differences may also relate to other patterns of allocation, such as those for nitrogen, calcium, or phosphorus, which are not considered here but could, in some circumstances, be important (Giles 1983*a*).

For stickleback above about 20 mm in length, there is now the clear possibility of describing the rates of energy income and expenditure and the pattern of allocation between maintenance, growth, and reproduction (Wootton *et al.* 1980*a*). Using the quantitative relationships between environmental factors, including food availability, temperature, and salinity, and the patterns of energy allocation, it will become possible to develop models that allow predictions of the likely effects of environmental change on reproductive success. The next step will be to describe, in quantitative terms, the ontogenetic changes in rates of income and expenditure and the pattern of allocation for a variety of populations, from a range of climatic zones (the stickleback is found from Arctic to Mediterranean zones) and for both resident and anadromous forms within the same watersheds. The patterns of energy allocation must then be related to the demographic characteristics of the populations. Such analyses would in turn be related to the predictions of life-history theory (see Wootton 1990; Baker Chapter 6 this volume) and those studies which explore the consequences of alterations in the life-table traits of age-specific survival and age-specific fecundity (e.g. Roff 1984; Stearns and Crandall 1984).

Such studies need to be carried out within protocols that allow comparisons to be made between groups of workers working in different parts of the world. The protocols need to agree on how the rate of food consumption, in energy units, can be accurately measured both in laboratory and field studies. The analyses of Wootton *et al.* (1980*b*) and Allen and Wootton (1984) suggest approaches to this problem. The measurement of the rates of energy expenditure at realistic levels of behavioural activity requires that particular attention be paid to the design of respirometers that allow simultaneous recording of oxygen consumption and activity. These recordings must be coupled with laboratory and field studies on the time spent by fish in different activities, both inside and outside the breeding season. A satisfactory estimate of the daily rates of energy expenditure by the free-living stickleback has yet to be obtained (Wootton *et al.* 1980*a*). Changes in the levels of its activity potentially provide the fish with an effective method of regulating its pattern of energy allocation (Priede 1985).

For biological traits, including those related to energy and time budgeting, there are two sources of variation: environmental and genetic (for reviews see Falconer 1989; Maynard Smith 1989). Some of the variation may be fortuitous, but the working hypothesis must be that much of the variation is adaptive. In the stickleback, there is good evidence for the adaptive significance of phenotypic variation in morphological and behavioural traits (Bell 1984; Wootton 1984*a*; Chapters by Bell and Foster, Reimchen, Huntingford *et al.*, Foster, McPhail this volume), and the same is likely to be true for variations in energy allocation.

At one level, the adaptive variation will reflect phenotypic plasticity. Fish with similar genotypes respond to changing environmental conditions with alterations in their time and energy allocations which minimize the consequences of the changes for the reproductive success of the fish (Slobodkin and Rapoport 1974). The range of phenotypic plasticity for such traits is not well known in the stickleback. Well-controlled breeding experiments are required to show that some of the variation in patterns of time and energy allocation has a genetic basis. The way ahead has already been pointed by the studies of McPhail (1977) on the age of first reproduction, and of Snyder and Dingle (1989) for several life-history traits.

The levels of parasitic infestation and predation experienced by populations are likely to be reflected in the patterns of energy (and time) allocation. Giles (1987*b*) found that exposure to a disturbing stimulus (being netted and placed in a strange tank) inhibited feeding of healthy sticklebacks for at least 96 h. In contrast, fish infested with *Schistocephalus* resumed feeding within

48 h of the disturbance. Calculations provide a crude estimate that the fright response in healthy fish had a total energy cost of 1.3 kJ (metabolic losses over 96 h plus income losses through failure to feed), whereas for infested fish this loss was only 0.24 kJ. The earlier resumption of feeding by the infested fish probably reflects the effects on the hunger of the fish of the energy demands made by the parasite (Walkey and Meakins 1970; Meakins 1974; Giles 1987*b*). But if the fright was caused by exposure to a predator, the earlier resumption of feeding could increase the risk of predation. The impact of size-selective predation on a stickleback population is also relevant (Reimchen 1980, 1988, 1990; Werner and Gilliam 1984). If a fish can grow rapidly through the sizes that make it vulnerable to predation, its chances of survival will increase. Large size also allows a faster absolute swimming speed, which may aid escape from predators. In other situations a slow growth rate may be advantageous if it ensures that the fish does not grow into a vulnerable size class (McPhail 1977). The nature of the predation to which a population is exposed will determine which size classes should be grown through quickly and which slowly (Reimchen Chapter 9 this volume). Predation by invertebrates such as dragonfly naiads is heaviest on small fish (Reimchen 1980, page 261 this volume). But under some circumstances, large sticklebacks could be at a disadvantage because of size-selective predation by fish, birds, and mammals (McPhail 1977).

An unexplored area is the effect on survival of maintaining investment in the gonads at the expense of the soma in times of energy shortage. The temporal patterns of energy availability will differ between populations (compare growth patterns noted by Mann 1971; Allen and Wootton 1982*b*; Crivelli and Britton 1987). Do the patterns of mortality within and between populations reflect the priority given to gonadal versus somatic growth as food supplies vary?

Two important unifying principles of biology are bioenergetics and the adaptive evolution through natural selection of life forms with Mendelian genetic systems. The studies of the past three decades on the biology of the threespine stickleback have encompassed physiological, behavioural, ecological, and evolutionary problems. These studies show that a programme of research can be mapped out which will give a coherent account of the way in which a complex organism uses energy and time to achieve reproductive success over a wide range of environmental conditions. The crux of such a programme will be the integration of studies on energetics with studies on the crucial demographic variables of size-specific mortality and fecundity.

ACKNOWLEDGEMENTS

I have to thank S. Cole, J. Allen, A. Faris, D. Fletcher, and B. Stanley, whose theses have provided much of the data presented in this essay. I also thank the three anonymous referees who did elegant hatchet jobs on the first version of this manuscript. I am grateful to The Royal Society and the British Council, whose travel grants have allowed me to keep in touch with other stickleback biologists in North America and Europe. I am especially grateful to the Natural Science and Engineering Research Council of Canada, whose Foreign Scientist Exchange Award allowed me to spend an important period in the incubation of this essay in Fred Whoriskey's laboratory and in association with Gerry FitzGerald and Helga Guderley. I note the two reseach studentships and one research grant with which the Natural Environment Research Council of the UK have supported my research since 1969.

6

Life history variation in female threespine stickleback

John A. Baker

A life history can be viewed as a suite of co-adapted traits fashioned by natural selection to solve a particular ecological problem (Stearns 1976; Tuomi *et al.* 1983). Selection may act on individual traits in different manners, producing an overall life history that is a compromise among varying selection pressures. Life histories are thought to vary among populations within species, and among species, in a manner that reflects adaptation to local environmental conditions (e.g. Leggett and Carscadden 1978). Because much of the variation in life history traits may have a genetic basis, conditions experienced by a taxon in the past may have moulded some aspects of life history that are not currently adaptive (Ballinger 1983; Dunham *et al.* 1988). Therefore, phylogenetic constraints and historical relationships among populations and species must be accounted for in assessing the evolution of life histories (Wanntorp *et al.* 1990). The evolution of life history traits may be further constrained by genetic correlations and covariances among characters (Trendall 1982; Tuomi *et al.* 1983; Gaillard *et al.* 1989) and by physiological and energetic limitations on variation.

Life histories are also notoriously labile. Immediate environmental conditions can modify life history characteristics within and between generations (see Caswell 1983 for a theoretical treatment, and Garrett 1982 and Trexler 1989 for specific examples in fish), although limits to the variation are imposed by genetic and physiological constraints. There may also be genotype–environment interactions (e.g. Berven and Gill 1983). It is perhaps because so many interacting factors affect the expression of life history traits that our understanding of how they evolve lags so far behind our understanding of the evolution of some other aspects of the phenotype.

The threespine stickleback species complex (Bell 1984*a*) possesses attributes that make it ideal for developing an empirical understanding of factors influencing vertebrate life history evolution. The taxon ranges across almost the entire Northern Hemisphere, and there is allozyme evidence for differentiation at this level (Buth and Haglund page 78 this volume). Moreover, within each of the major geographic regions, *G. aculeatus* displays at least three general lifestyles (strictly marine, anadromous, strictly freshwater;

Wootton 1984*a*), which suggests the presence of selection pressures unique to each lifestyle that cut across geographic areas. Additionally, freshwater populations inhabit lakes and ponds of varying characteristics, and streams ranging from small brooks to large rivers, and there is some evidence that lacustrine and stream populations may represent distinct, general lifestyles (Moodie 1972*b*; Lavin and McPhail 1985; Reimchen *et al.* 1985). Thus, comparisons can be made not only among marine, anadromous, and freshwater stickleback, but also among freshwater populations exposed to a range of selection pressures.

The freshwater lacustrine threespine stickleback promises to be a particularly fruitful lifestyle for investigating life history evolution. The morphologically relatively uniform marine and anadromous populations are thought to have given rise to the very diverse freshwater populations repeatedly over the evolutionary history of this complex (Bell 1984*a*, 1988; Bell and Foster page 16, McPhail page 401 this volume). In many parts of their range, present lacustrine populations must have been derived from marine or anadromous populations since the last glacial maximum (Bell 1984*a*, 1988; Bell and Foster page 16, McPhail page 401 this volume). In many instances, these populations may reside in the habitats in which they became land-locked following recession of the glacial ice. Because many of these habitats have no surface water connections to others containing stickleback, the populations are likely to be adapted to local conditions and relatively unaffected by gene flow. Thus, these populations provide a very promising opportunity for examining the selection agents responsible for modifying life history traits in this fish.

Additional, practical advantages include the ease with which the threespine stickleback can be studied and manipulated in many natural populations. This fish also can be reared easily in the laboratory, making it possible to evaluate the extent of genetic determination of traits, the correlations among life history traits and between these traits and other aspects of phenotype, and the nature of genotype–environment interactions (e.g. Snyder and Dingle 1989, 1990).

To date there has been no comprehensive review of life history variation in the threespine stickleback through interpopulation comparisons, although comparisons of two or more populations in limited geographic areas have been made for a few traits (e.g. Hagen 1967; McPhail 1977; Wootton *et al.* 1978; Mori 1987*b*; Snyder and Dingle 1989, 1990). The existence of extensive, often adaptive variation in morphology (e.g. Bell 1984*a*; Reimchen, McPhail Chapter 14 this volume) and behaviour (e.g. Chapter 9, Huntingford *et al.* Chapter 10; Foster Chapter 13 this volume) suggests that similar variation in life history will be discovered. Because life history traits are thought to be strongly affected by selection (Mousseau and Roff 1987), it is reasonable to expect that differences in selection regimes in the habitats occupied by threespine stickleback will have produced differentiation in life histories.

The primary goal of this chapter is to evaluate the available data on threespine stickleback life history, and to synthesize the information to form a preliminary overview of its life history evolution. Although there is relatively little information on the genetic bases of differences in life history or on genotype–environment interactions, substantial information does exist on phenotype variation within and among populations. Such great variation makes it likely that interesting, and perhaps unexpected, patterns will emerge from which testable hypotheses can be formed. A second goal is to highlight traits, geographic areas, and lifestyles for which more information is needed before a truly comprehensive picture of life history evolution can be drawn. Along with these deficiencies, I point out problems associated with assessing stickleback life history traits and indicate how these problems can affect interpretation of patterns. Finally, I argue that the threespine stickleback shows great promise for future investigations, not only because of the diversity of its life histories, but also because of the ease with which both laboratory and natural populations of this fish can be studied.

METHODS

Data for this review were compiled from the published literature, agency reports, and unpublished information. Life history traits for which data were sought included life span, mortality, growth, breeding season, size and age at first reproduction, egg size, clutch size, relative clutch mass, interclutch interval, and total number of clutches. Relative clutch mass was so rarely reported that it was not considered further. Data were compiled and analysed by major geographic region: Europe, eastern North America, western North America, and eastern Asia. Recent allozyme data (Buth and Haglund page 78 this volume) suggest that *G. aculeatus* populations form clades that correspond at least to some extent to these geographic areas.

Many factors can influence the expression of life history traits. In threespine stickleback these could include, but are not limited to: geographic area (= major clade), general lifestyle (marine, anadromous, lacustrine, stream-dwelling), latitude (= general environment?), morphology (including overall body size), and local environment. I have attempted to account for, or estimate the importance of, these factors in various ways. Techniques such as principal components analysis were not used because so few studies provided data on most life history traits, and these studies were distributed unevenly among geographic areas. Instead, simple correlation analysis was used to explore the relationships between life history traits and various factors. Generally, partial correlation coefficients were computed between the life history traits and latitude, with the effect of lifestyle held constant. Information on latitude and lifestyle was available for a large number of stickleback populations in most geographic areas, and these factors undoubtedly are important in shaping the observed life histories. Insufficient (e.g. almost no information for marine populations) or highly skewed data (e.g. information for anadromous populations mostly from a small range of latitudes within a geographic area) may have affected the accuracy of some correlations.

Local environment and morphology (with the exception of size) could not be adequately addressed. Local environmental conditions undoubtedly influence threespine stickleback life history greatly, perhaps more than any other single factor. However, the facets of local environment that are important (e.g. size of lake, predator abundance, productivity), and their interactions, are poorly understood. Also, comparable quantification of factors was impossible on the basis of available data. Similarly, although morphology and life history may be linked in threespine stickleback, there is at present no consensus as to the level at which effects occur. For example, the low-plated morph encompasses populations possessing up to about 18 lateral plates (total of both sides), and populations often consist of fish of a diversity of plate counts (e.g. Kynard 1972; Haglund 1981). In addition, asymmetry is often high. Accounting for such variability was beyond the scope of this review, and populations were simply categorized as completely, partially, or low-plated, or mixed, following Hagen and Moodie (1982).

As with some other taxa (e.g. Dunham and Miles 1985; Wilbur and Morin 1988; Elgar and Heaphy 1989), morphology in threespine stickleback (e.g. plate morph, body shape) appears to be at least partly confounded with general lifestyle. Marine and anadromous fish are almost always completely plated (Bell 1984*a*; Wootton 1984*a*), although populations sometimes consist largely of partially plated fish (e.g. van Mullem and van der Vlugt 1964; Aneer 1973). Freshwater populations may be low-, partially, or completely plated, or mixed, and in addition the distribution of plate morphs differs among major geographic areas (Hagen and Moodie 1982). In western North America, freshwater populations are usually low-plated, for example, while those in eastern North American are mostly completely plated.

Body size was one morphological trait that could be adequately investigated. Allometric relationships between body size and life history traits are common (Blueweiss *et al.* 1978; LaBarbera 1989). Therefore, analysis of covariance (ANCOVA) was used to explore the relationships between some life history traits and female size. Standard length (SL) was the criterion of size used in these analyses, since almost every study reported it, whereas few reported weight. Differences in mean trait values among various groupings of populations were tested using analysis of variance (ANOVA). Regression coefficients and trait means were tested following ANOVA or ANCOVA with unplanned pairwise comparisons using an experimentwise error rate. Prior to ANOVA or ANCOVA, data were inspected for approximate normality, and they were tested for homogeneity of variance using the log-anova test. If non-normality or heteroscedasticity were detected, appropriate transformations were employed. Differences in trait frequencies among lifestyles were examined using *G*-tests (Sokal and Rohlf 1981). Additional explanations of data handling for presentation or statistical analyses are given in the sections for each life history trait. Sample sizes vary for statistical tests owing to varying patterns of missing data among traits. Individual studies examined from one to 25 populations; when information was available for multiple populations from one area (e.g. McPhail 1977), each population was treated as an independent data point.

A number of problems are present in the overall data set, and those of importance to particular life history traits are discussed in the appropriate sections. In general, most studies of threespine stickleback have addressed only a few traits. Also, studies range from little more than general observations to very detailed investigations, and the data come from a mixture of field and laboratory work. The quality and quantity of information vary substantially among the four major geographic areas and among the four lifestyles. Additionally, the proportion of studies on each lifestyle varies considerably among the major geographic areas.

Categorizing populations as lacustrine or stream-dwelling seldom presents problems. However, differentiating among marine and anadromous populations is sometimes more difficult. Even stickleback that live in the sea most of the year show a preference for fresh water in spring (Baggerman 1957), typically moving into estuaries, bays, and tidepools to breed. Populations were considered anadromous if they migrated even a short distance to a distinct breeding habitat. Thus, the St Lawrence Estuary population at Isle Verte, Quebec (e.g. Craig and FitzGerald 1982; Whorisky and FitzGerald Chapter 7 this volume) was considered anadromous because fish migrate across a salt-marsh to breed in tidepools. The population studied by Coad and Power (1973*a*) was considered marine because it bred in tidepools immediately adjacent to the ocean.

As in any comparative study (Wanntorp *et al.* 1990), results can be distorted by non-independence of observations. In stickleback this could happen if freshwater populations in different geographic areas were derived from marine or anadromous ancestors having different life history characteristics. I have attempted to minimize this potential problem by analysing most traits within geographic areas. However, in some instances insufficient data were available, and analyses were performed by pooling information from all sources. The potential problems are acknowledged.

LIFE HISTORY TRAITS

Life span

The length of time that an organism can expect to live can have a strong effect on other life history traits (Michod 1979; Reznick 1983; Roff 1984). Mann *et al.* (1984), for example, have shown that in unproductive streams the life span of *Cottus gobio* is 7–8 yr, and females lay a single clutch of eggs each year. In contrast, in more productive streams life span is only 2–3 yr, but females produce several clutches of eggs annually. The importance of life span is not limited to populations that live to reproduce in two or more years. In annual populations, the shape of the survival function during the reproductive season is important. Both annual and longer-lived populations are affected by this latter function; however, in annual species it becomes pre-eminent.

Life spans of threespine stickleback have been reported in a large number of studies (Appendix 6.1), but the comparability of the estimated ages is unknown. Few studies examined age structure of populations over a series of years, and many used fish collected from within only a single reproductive season. Secondly, ages were sometimes determined using otoliths or spines, while in many other instances they were inferred from length-frequency plots. While length-frequency plots can sometimes accurately indicate age structure (e.g. Coad and Power 1973*b*), often there is ambiguity (e.g. Havens *et al.* 1984). In addition, inferring age-frequency from length-frequency requires relatively large, unbiased samples encompassing the entire length range of the species (Ricker 1975). Such samples are not always available. Finally, many data are from studies in which ages were presented only as supplementary material.

Threespine stickleback can live to age 5 yr in the laboratory, and there is one recent report of this species reaching age 8 yr in a very large-bodied, natural population (Reimchen 1992*a*); however, this potential does not appear to be generally realized in nature (Wootton 1976, 1984*a*). Beyond the conclusion that the maximum life span is probably 4 yr or less in almost all populations, few generalizations can be made. Nearly equal numbers of studies have concluded that this fish lives a maximum of 1, 2, or 3–4 yr (Appendix 6.1).

Wootton (1984*a*) suggested that populations at higher latitudes lived longer, a phenomenon that has been observed in other fish species (Leggett and Carscadden 1978). However, in threespine stickleback, life span (holding lifestyle constant) is not significantly correlated with latitude in Europe ($r_p = 0.26$, $N = 39$, $P > 0.10$), eastern North America ($r_p = 0.16$, $N = 6$, $P > 0.20$), or western North America ($r_p = 0.04$, $N = 24$, $P > 0.50$). Data for eastern Asia are insufficient for a statistical test. Most long-lived populations inhabit high latitudes, while low-latitude populations almost always have short life spans. The correlations appear to fail primarily because some high-latitude populations are also short-lived (e.g. Giles 1987*c*).

It was not possible to compare maximum life spans among the four geographic areas and lifestyles (e.g. two-way ANOVA) owing to the strongly unbalanced pattern of the data (Appendix 6.1). Pooling of data from all four geographic areas suggests that anadromous, lacustrine, and stream-dwelling threespine stickleback each have maximum life spans of about 3–4 yr (Fig. 6.1). Maximum life span is not equally distributed within these three life styles. Relatively few anadromous populations are annual, for example. In contrast, nearly 45 per cent of all stream-dwelling populations studied are

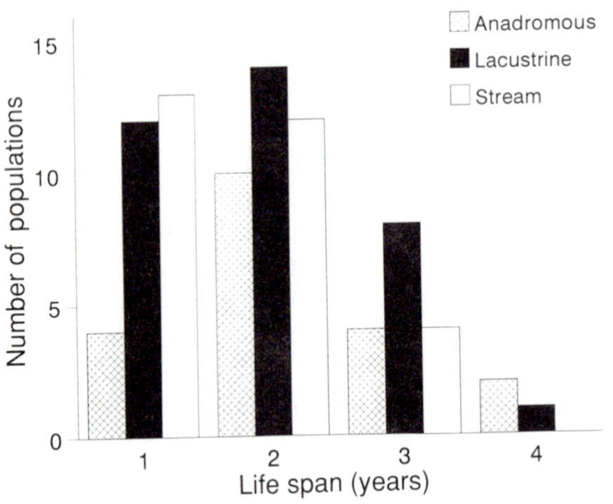

Fig. 6.1 Frequency distributions of reported maximum ages of three lifestyles of threespine stickleback. All four geographic areas (see text) are combined. Only a single report of maximum age for a marine population is available (Coad and Power 1973*a*). Distributions for all three lifestyles are significantly different from random: anadromous ($G_W = 6.362$, d.f. = 3, $0.10 > P > 0.05$); lacustrine ($G_W = 14.545$, d.f. = 3, $P < 0.005$); stream-dwelling ($G_W = 20.445$, d.f. = 3, $P < 0.001$).

reported to be annual (Fig. 6.1). The frequency distributions of maximum age for anadromous, lacustrine, and stream-dwelling populations pooled across all geographic areas differ significantly ($G_W = 27.659$, d.f. = 3, $P < 0.001$), with stream-dwelling fish differing from the other two lifestyles. In the only marine population studied, fish probably live 2 yr (Coad and Power 1973*a*).

Growth

The rate at which an organism grows can influence many other life history traits (e.g. Roff 1984; Stearns and Koella 1986). For example, slow growth may result in delayed maturity and/or smaller size at maturity (e.g. Reznick and Endler 1982), which may lead to reduced lifetime fecundity. Slow growth may also extend the time during which fish are vulnerable to certain predators (e.g. Giles 1987*c*; Foster and Ploch 1990).

Growth rates can be estimated by employing a mathematical model such as the von Bertalanffy equation and using size-at-age data determined from otoliths or other hard structures (Ricker 1975). This has been done infrequently for threespine stickleback. Fortunately, to compare growth rates of populations at a coarser level this may not be necessary. Growth in natural populations can be estimated by back-calculation of size at age from otoliths without proceeding to develop a formal mathematical statement. Growth can also be estimated by inspecting the change in size of known-age cohorts of fish from a relatively large number of samples taken at frequent intervals (e.g. Jones and Hynes 1950; Mann 1971; O'Hara and Penczak 1987), or even by inspecting length-frequency plots of fish taken at only a few times per year (e.g. Bertin 1925; Havens *et al.* 1984) if ages can be inferred reliably from the plots. Though estimates of growth derived from these methods may differ somewhat, even within the same population, and will estimate the population growth rate rather than the individual rate (Ricker 1975), they are probably sufficiently similar that almost all studies that report size and age (or that present size-frequency data) can be used.

Reliable data on growth of threespine stickleback beyond age 1 are too sparse and geographically scattered to be of comparative value. Reported sizes at age 1 appear to be more reliable and are certainly more abundant. In addition, size at age 1 will incorporate both differences in actual growth rate of juveniles and the length of time during the year in which they grow. As with life span, data on size are too unbalanced to permit direct tests among major geographic areas.

Pooled data suggest that anadromous threespine stickleback reach a significantly larger average length after one year than either lacustrine or stream-dwelling fish (Table 6.1), and stream fish are, in turn, significantly larger than those in lacustrine populations. The difference in mass, assuming a typical length–weight relationship for each lifestyle (stream-dwelling: Wootton 1976, Fig. 19; lacustrine: Pennycuick 1971; anadromous: Mori 1990*a*), would be even more substantial. Anadromous fish would weigh, on average, nearly 3.5 times as much as lacustrine fish at age 1, while stream-dwelling fish would weigh nearly twice as much. Anadromous populations

Table 6.1 Reported standard length at approximately one year of age for populations of four lifestyles of threespine stickleback.[a]

Lifestyle	No. of pops	Standard length (mm)				
		Mean	SE[b]	Median	CV[b]	Range
Marine	2	38.5	13.50	38.5	49.6	25–52
Anadromous	9	52.1	4.23	52	24.4	31–77
Lacustrine	16	36.5	1.23	35.5	13.5	30–46
Stream-dwelling	18	43.8	1.01	45	9.8	34–50

[a] Means of the anadromous, lacustrine, and stream-dwelling lifestyles differ significantly (one-way ANOVA using $\log_{10}$-transformed data, $F = 14.92$; d.f. = 2, 40; $P < 0.001$). The marine lifestyle was not included in the analysis because of the low sample size. Variances of transformed data are not significantly different for the three lifestyles tested (log-anova test, Sokal and Rohlf 1981: $F = 4.348$; d.f. = 2, 8; $P > 0.05$). The mean standard lengths of the three lifestyles differ significantly from each other ($P < 0.05$; GT2 test, Sokal and Rohlf 1981).

[b] SE, standard error of the mean; CV, coefficient of variation expressed as a per cent.

appear to be more variable in the size reached at age 1 than either freshwater lifestyle (Table 6.1).

Although there are few comparisons of growth of anadromous and freshwater fish from the same geographic area, those studies do support the general conclusion that anadromous fish grow more rapidly than freshwater ones. Narver (1969) indicated that anadromous fish in Alaska reached 65–90 mm TL (total length) at age 1, while lacustrine stickleback in the area reached only 50–65 mm at age 2. Although age was not determined, Hagen's (1967) data also suggested that anadromous stickleback were somewhat larger than resident stream fish. In an experimental test, Snyder (1988, 1991*a*) found that resident stream fish were smaller at all ages than migratory 'upstream' and 'estuary' fish, and the difference in growth rate was at least partly genetically based.

Growth rates of juvenile stickleback vary greatly. Mori (1987*b*), for example, found that stream-dwelling fish in Japan grew at a rate of about 0.5 mm d^{-1} early in life, and a similar growth rate was calculated for juveniles of the marine lifestyle by Cowen *et al.* (1991). In contrast, Picard *et al.* (1990) and Poulin and FitzGerald (1989*a*) determined that their St Lawrence Estuary, Canada, anadromous fish grew much more slowly, perhaps as slowly as only 0.19 mm d^{-1}. The Mediterranean anadromous population studied by Crivelli and Britton (1987) appears to be intermediate. Because many other factors (e.g. length of growing season) are also important, however, juvenile growth rates may bear little relationship overall to the actual size reached at age 1. Fish in the Japanese and Canadian populations discussed above reached a similar size at one year of age.

Growth can vary considerably among years within a population. Wootton (page 130 this volume) has shown that in a single stream-dwelling population, mean size at age 1 ranged from 45 to 58 mm SL over a 14 yr period. Similarly, Crivelli and Britton (1987) reported mean female SL of 43 and 52 mm (age 1 fish) in two consecutive years. In addition, marine populations may not be as homogeneous as has been thought, as Black (1977) found a significant positive correlation between body size and latitude along the coast of western North America.

Although the factors that control growth may be at least partly genetic, the seemingly higher growth of anadromous fish may also be due to the generally higher productivity of marine waters compared with fresh waters (Gross *et al.* 1988). The rather small mean size I estimated for the two marine populations (Table 6.1) might argue against this hypothesis; however, data for the marine lifestyle may be unreliable.

Mortality

Age-specific mortality can profoundly influence the evolution of almost all other life history traits (Michod 1979; Roff 1984; Ware 1984; Stearns and Koella 1986). However, despite the importance of age-specific mortality rate, no study of threespine stickleback adequately assesses this trait. Wootton (1984*a*: 233) recognized this deficiency in the available data when he remarked that it would be useful to make a '. . . comparison of the life-history traits found in populations in which juvenile and adult survivorships were measured. . . .'

Mortality can be envisioned as occurring in three distinct life stages in threespine stickleback: embryos and larvae in the nest, juveniles, and reproductive adults. Mortality of embryos is best documented, and considerable interpopulation variation has been found. Poulin and FitzGerald (1989*a*) found that 50–60 per cent of nests with eggs failed to produce fry in Quebec tidepools, and 42 per cent of such nests failed in a nearby river. A similar overall percentage of nest failures was reported by Kynard (1978*a*). Foster (1988) has reported nest failures as high as 68 per cent in one British Columbia, Canada, lake. Nesting success may vary throughout the reproductive season. Kynard (1972) found that only 23 per cent of nests that were started early in the reproductive season failed, whereas 98 per cent of late nests were unsuccessful.

Death of embryos in the nest may be caused by leeches, sculpins, and other heterospecific nest predators. Male stickleback are, however, relatively effective at recognizing and discouraging these predators (Moodie 1972*a*; Foster and Ploch 1990), and in many populations much of the embryo mortality probably results from the loss of the guarding male. For example, Whoriskey and FitzGerald (1985*a*) have shown experimentally that when the defending male is removed, nests in salt-marsh tidepools are virtually certain to be destroyed. In many other populations, however, stickleback themselves are the major cause of nest destruction (Foster 1988). Embryo losses in such populations may depend partly on the density of potential cannibalistic conspecifics. In one British Columbia, Canada, lake (Hyatt and Ringler 1989*a*), the probability of a male losing his nest to cannibals over a 7 d period ranged from about 20 per cent at low densities (1 fish m^{-2}) to 75 per cent at high densities (10–12 fish m^{-2}).

Hatchability differences have also been found. Poulin and FitzGerald (1989*a*) found that egg viability was only 94–98 per cent and 82.5 per cent in tidepool and river nests, respectively. Mori (1987*b*) found that the percentage of dead eggs in nests of three stream populations ranged from < 1 per cent to > 21 per cent, but was generally < 10 per cent. If males remove dead eggs these percentages may be substantial underestimates.

Direct measurements of juvenile mortality are scarce. However, studies in which large samples of both juvenile and adult threespine stickleback were collected (e.g. Bertin 1925; O'Hara and Penczak 1987) typically show several times as many juveniles in the autumn as reproductive adults in the spring, thus providing indirect evidence that mortality during the juvenile stage is high. A number of predators, including insects (e.g. Reimchen 1980; Foster *et al.* 1988) and fish (e.g. Giles 1987*c*), are known to feed on stickleback fry.

Giles (1987*c*) examined stomachs of perch, *Perca fluviatilis*, and found large numbers of juvenile stickleback from 15 to 25 mm SL. In an experimental test, Poulin and FitzGerald (1989*a*) showed that 15–20 mm total length fry experienced from 13 to 43 per cent mortality over a 2 wk period in tidepool enclosures. In a second experiment, larger fry (20–25 mm) experienced mortality rates up to 90 per cent over 2 wk. In neither instance were potential sources of the mortality identified. Foster *et al.* (1988) have shown that adult male threespine stickleback themselves are efficient predators on juveniles from about 5 to 15 mm SL. In accordance with this finding, Kynard (1972) found that when guarding males were removed from nests that had either eggs or very small fry, fry were quickly consumed by nearby males. However, Kynard also noted that few stickleback that he examined for evidence of cannibalism contained fry, whereas a fairly large percentage had eaten eggs.

Mortality rates of adults are largely unknown. Studies of populations in which the maximum life span is two years or more (e.g. Jones and Hynes 1950; Hyatt and Ringler 1989*b*) could assess adult mortality in a general way as the percentage of adults that survive to spawn in a second year. However, this life span is not typical of many threespine stickleback populations (Appendix 6.1), and it ignores the potentially important effect of within-season mortality pattern.

Mortality within a spawning season, both of the young produced and of the adults themselves, can obviously affect the way that females allocate reproductive effort. The probability that a predator will consume any particular stickleback probably varies throughout the reproductive season, and also with habitat (Reimchen page 250 this volume). Piscivorous birds, for example, undoubtedly fish more actively when they have young. Kedney *et al.* (1987) thought that predation on spawning stickleback was greater in tidepools, where a variety of piscivorous birds fed, than in a nearby river.

Parasites may also affect the seasonal partitioning of reproductive effort (e.g. Pennycuick 1971; Meakins 1974). However, McPhail and Peacock (1983) found that the greatest incidence of *Schistocephalus* infestation occurred at the end of the reproductive season, thus minimizing the effect of this parasite on the reproductive population.

Egg production is an energy-intensive process that may physiologically stress female threespine stickleback (Wootton 1979). The shape of the mortality–reproductive effort function is not known for any natural population. In an experimental situation, no significant correlation of mortality and reproductive effort was found, either during or after the reproductive season (Wootton 1984*a*). However, sample sizes were quite small, and differences in mortality would have to have been very large to be detectable. Reports of mass mortalities of adults at the end of the spawning season (e.g. Greenbank and Nelson 1959; Threlfall 1968; Kynard 1972; Crivelli and Britton 1987; Mori and Nagoshi 1987; Ziuganov *et al.* 1987; Hyatt and Ringler 1989*a*) suggest that in many populations, mortality rate may be relatively low during much of the spawning season, and then abruptly high at the end. However, the astonishing potential for some predators to consume adult sticklebacks (Reimchen page 264 this volume) argues that in some populations, mortality throughout the reproductive season may be substantial.

Breeding season

Within species, populations from high latitudes typically initiate reproduction later in the year than those from low latitudes, and a short breeding season is imposed by climate (Baggerman 1980; Hislop 1984). In part, the lower total reproductive potential due to the short breeding season may be compensated by changes in other life history traits, such as increased juvenile growth, increased clutch size, or reduced interclutch intervals. Also, though

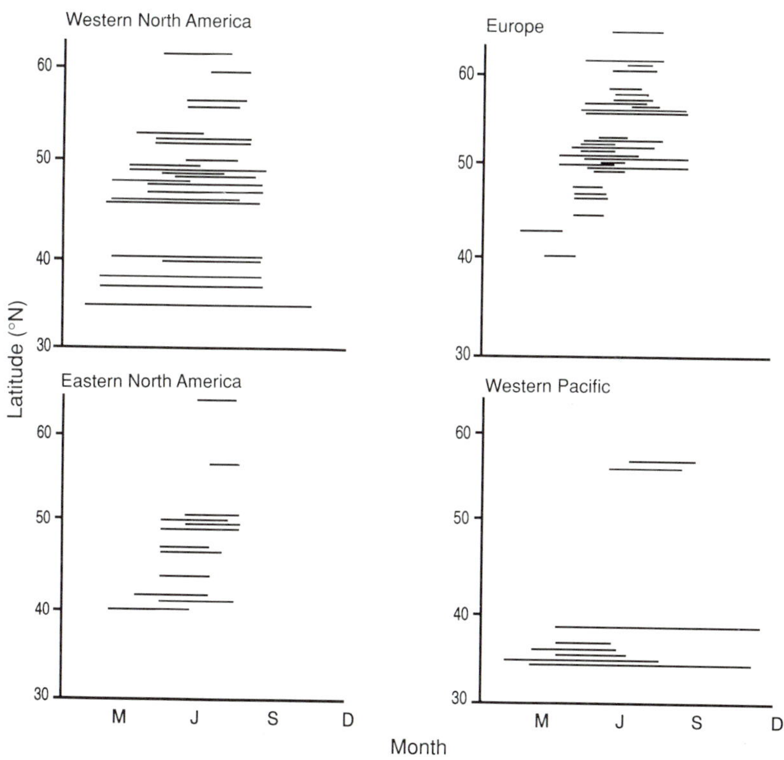

Fig. 6.2 Latitudinal distribution of date of initiation of spawning, and length of spawning season reported for populations of threespine stickleback from four geographic areas. Letters on the abscissa denote the months of March, June, September and December. Ordinate values are degrees north latitude. Initiation dates are likely to be relatively accurate; lengths of season may be more imprecise.

fish living at lower latitudes generally reproduce earlier than those at high latitudes, they do not always have longer breeding seasons.

Breeding seasons vary dramatically among threespine stickleback populations (Fig. 6.2). Within all four major geographic areas, the date of initiation of reproduction (all lifestyles pooled) becomes later in the year with increasing latitude (Europe: $r_p = 0.76$, $N = 28$, $P < 0.01$; eastern North America: $r_p = 0.77$, $N = 15$, $P < 0.01$; western North America: $r_p = 0.69$, $N = 28$, $P < 0.01$; eastern Asia: $r_p = 0.93$, $N = 8$, $P < 0.01$). In western North America, for example (Appendix 6.1), stickleback in southern California begin spawning in late February or early March (Vrat 1949; Baskin 1974), those in southern British Columbia generally in April or May (e.g. McPhail and Peacock 1983; Hyatt and Ringler 1989*a*), and those in Alaska in late May or June (e.g. Greenbank and Nelson 1959; Narver 1969).

Despite the strong overall trend, considerable variability exists in some areas. This local variation is particularly evident in western North America, where lacustrine populations near latitude 49 °N can begin reproducing as early as late March (Carl 1953) or as late as mid-May (Pressley 1981). There are even differences between stream and anadromous populations in the same stream (McPhail page 407 this volume). Although the data are distributed too unevenly to permit statistical tests, there are no obvious differences among geographic areas in the association of reproductive commencement and latitude (Fig. 6.2).

Although differences in sample sizes and latitudinal distribution of samples preclude separate correlation analyses for the four lifestyles, comparisons of natural anadromous and freshwater populations from western North America and Japan (Hagen 1967; Hay and McPhail 1975; Black 1977; Mori 1987*b*; Snyder and Dingle 1989) suggest that anadromous populations begin spawning somewhat later than nearby freshwater populations. Recent experimental evidence (Snyder and Dingle 1989) also supports this conclusion. However, in the St Lawrence River area of eastern Canada (Coad and Power 1973*a*,*b*; Worgan and FitzGerald 1981*b*; Kedney *et al.* 1987) and on the Kamchatka Peninsula of Russia (Ziuganov *et al.* 1987; Ziuganov and Bugayev 1988), freshwater populations are reported to begin reproduction later. This pattern may be related to the relative severity of marine and freshwater environments in these areas, an explanation that has been proposed to explain large-scale geographic differences in plate morph distribution (Hagen and Moodie 1982).

Length of spawning season shows a less consistent relationship with latitude (Fig. 6.2). A clear pattern is evident in western North America, where some southern California populations spawn from late February through at least October (Baskin 1974), Washington and southern British Columbia populations from about late April through July or early August (e.g. Kynard 1972; Hyatt and Ringler 1989*a*), and Alaska populations only from June through mid-August (e.g. Greenbank and Nelson 1959; Gilbertson 1980). Here, and in eastern North America, length of spawning season is significantly negatively correlated with latitude ($r_p = -0.75$, $N = 26$, $P < 0.01$; $r_p = -0.56$, $N = 14$, $P < 0.05$, respectively). In eastern Asia and Europe the relationships are not significant ($r_p = -0.34$, $N = 8$, $P > 0.45$; $r_p = 0.12$, $N = 17$, $P > 0.50$, respectively), though in eastern Asia the trend is in the expected direction. In Europe, however, spawning season and latitude appear to be truly uncorrelated. The southernmost European population studied, along the Mediterranean coast of France, spawns only about 50 d annually (Crivelli and Britton 1987), while some lowland populations at much higher latitudes reproduce for 100 d or more per year (e.g. Craig-Bennett 1931; Wootton *et al.* 1978; Ukegbu and Huntingford 1989). The Mediterranean population is interesting for yet another reason: it seems to cease spawning not in response to shortening day length, but because temperatures on the breeding grounds become too high very early in the year.

Anadromous populations appear to show less variation in length of spawning season than do freshwater populations. Anadromous fish seldom reproduce for more than 12–14 wk, and 8–10 wk is more typical (e.g. Hagen 1967; Mori 1987*b*; Ziuganov *et al.* 1987). Freshwater populations often spawn for much longer, and in some habitats nearly year-round reproduction has been documented (Baskin 1974; Mori 1985). However, at low latitudes where prolonged reproduction is most common, only the stream lifestyle has been extensively studied. In fact, lacustrine stickleback are relatively uncommon at low latitudes.

Size and age at first reproduction

Theoretical studies have suggested that selection can readily alter age or size at first maturity in fishes (e.g. Roff 1984; Stearns and Crandall 1984; Stearns and Koella 1986). In support of this prediction, wide variation in this trait has been demonstrated in many species of fishes, and the differences have been shown to have both genetic and phenotypic bases (Policansky 1983; Reznick 1990).

Female threespine stickleback most commonly reach sexual maturity at age 1 in anadromous, lacustrine, and stream-dwelling populations (Fig. 6.3). The distributions are not homogeneous ($G = 4.71$, d.f. $= 4$, $P < 0.05$), the stream lifestyle including a significantly smaller proportion of populations that first reproduce at age 2. The only marine population for which age-at-maturity data are available spawns at age 2 (Coad and Power 1973*a*).

Exceptions occur primarily at high latitudes. For example, among lacustrine populations, those in the Queen Charlotte Islands of British Columbia, Canada (Moodie 1972*b*; Reimchen 1980), that in the Chignik Lake system of Alaska, USA (Narver 1969), and one in northern Quebec, Canada (Power 1965) do not breed until age 2. Stickleback in Lake Aleknagik, Alaska (Gilbertson 1980) may not breed until age 3. Only two stream-dwelling

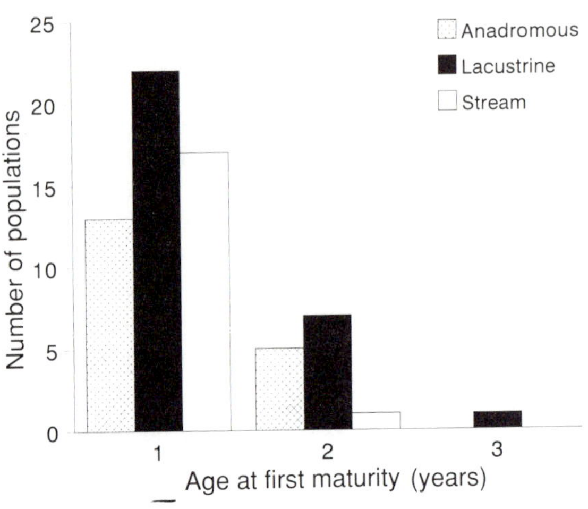

Fig. 6.3 Frequency distribution of reported ages at first reproduction for three lifestyles of threespine stickleback. All four geographic areas (see text) are combined. Distributions of age at first reproduction are significantly different from random for all lifestyles (anadromous: $G = 266$; lacustrine: $G = 2046$; stream-dwelling: $G = 3674$; all d.f. = 2, all $P < 0.001$).

populations are reported to delay reproduction to age 2, one in England (Craig-Bennett 1931) and one in the Queen Charlotte Islands (Moodie 1972b). Among anadromous stickleback, at least some populations in northern Europe (Wunder 1928, 1930; Münzing 1959; Aneer 1973), and those in the St Lawrence Estuary, Quebec (Craig and FitzGerald 1982) are reported to spawn at age 2. In spite of this tendency for populations at more northerly latitudes to mature at age 2, partial correlations of age at first reproduction with latitude (holding lifestyle constant) are not significant for Europe ($r_p = 0.23$, $N = 36$, $P > 0.15$), eastern North America ($r_p = 0.52$, $N = 6$, $P > 0.35$), or western North America ($r_p = 0.22$, $N = 31$, $P > 0.25$). Data for eastern Asia are inadequate for analysis (Appendix 6.1).

Populations living near the extremes of the threespine stickleback's range could potentially push the age at maturity to its physiological limits. Because of this species' generally short maximum life span, populations delaying reproduction past age 2 should be uncommon. Indeed, only a single population, from near the species' northern limit in western North America, is reported to mature at age 3 (Gilbertson 1980). At the opposite extreme, Snyder and Dingle (1989) found that laboratory-reared anadromous and freshwater stickleback from northern California matured at an average of 211 and 195 d (6.5–7 months), respectively, and Wootton (1984*a*) has noted that fish reared in the laboratory could be brought into breeding condition

Table 6.2 Reported standard length at first reproduction for populations of four lifestyles of threespine stickleback[a].

Lifestyle	No. of pops	Standard length (mm)				
		Mean	SE[b]	Median	CV[b]	Range
Marine	4	44.0	2.35	43.5	10.7	39–50
Anadromous	21	45.6	2.10	45	21.1	31–65
Lacustrine	62	40.2	1.22	38	23.9	24–68
Stream-dwelling	24	39.6	1.46	39	18.0	25–58

[a]Means of the four lifestyles differ significantly (one-way ANOVA using untransformed data, $F = 3.17$; d.f. = 3, 108; $P < 0.05$). The marine and anadromous lifestyles do not differ significantly, nor do the lacustrine and stream-dwelling lifestyles (specific linear contrasts, $P > 0.05$). Variances of the three lifestyles do not differ significantly ($F = 0.12$; d.f. = 3, 16; $P > 0.50$). The mean standard length for anadromous populations is significantly greater than that of the two freshwater populations ($P < 0.05$; GT2 test, Sokal and Rohlf 1981), which do not differ significantly.
[b]SE, standard error of the mean; CV, coefficient of variation expressed as a per cent.

in about 6 months. It is not known whether fish in any natural population reproduce at such ages, and the highly seasonal environments in which most stickleback live probably preclude this (Wootton 1984*a*). However, Baskin (1974) and Mori (1985) reported spawning in almost every month of the year in fairly constant environments. The possibility that some females in these populations mature at less than 1 yr of age remains to be tested.

Size at maturity varies enormously in threespine stickleback (Table 6.2; Appendix 6.1). Among anadromous populations, reported sizes range from possibly as small as 31 mm TL (Bertin 1925; but definitely as small as 35 mm TL, Crivelli and Britton 1987) to 65 mm SL (Narver 1969; Mori 1987*b*). Lacustrine populations vary over a wider range, with fish first maturing as small as 25 mm SL (McPhail 1977) and perhaps as large as 70 mm SL (Moodie 1972*b*). At least one stream-dwelling population may mature as small as 25 mm SL (O'Hara and Penczak 1987), while others may be as large as 58 mm TL (Giles 1987*c*). Mean length at first reproduction is significantly greater in anadromous and marine populations than in freshwater populations (Table 6.2); within the above pairs, mean lengths do not differ.

Minimum reproductive size of females is not correlated (effect of lifestyle held constant) with latitude in Europe ($r_p = 0.09$, $N = 38$; $P > 0.05$), eastern North America ($r_p = -0.45$, $N = 8$, $P > 0.05$), or western North America ($r_p = 0.21$, $N = 56$, $P > 0.05$). Data from eastern Asia do not permit a robust analysis (Appendix 6.1).

Some of the differences in size and age at first reproduction reported for threespine stickleback reflect genetic differentiation (McPhail 1977; Snyder and Dingle 1989). McPhail suggested that differences might be due to dissimilarity in predation regimes, but he was unable to demonstrate any association between the size or abundance of trout in the lakes (trout are major predators of stickleback; Reimchen page 260 this volume) and the size of female stickleback at first reproduction. McPhail (1977) also tested for an effect of 'lake productivity', suggesting that a smaller size at first reproduction might be favoured in lakes of low productivity. Again, no significant relationship emerged. In support of the 'productivity' hypothesis, Giles (1987*c*) found that fish from several unproductive 'peaty moorland' lakes in the Outer Hebrides matured at much smaller sizes than did fish from nearby, more productive mainland lakes. Additionally, Allen and Wootton (1982*c*) have shown experimentally that stickleback somatic growth is related to ration, and Wootton (1973*b*, 1977) has demonstrated that females on restricted rations mature later and at a smaller size than well-fed females.

Snyder and Dingle (1989) predicted that anadromous stickleback would be larger at first reproduction than resident (non-migratory) fish. Their experiments with California (USA) fish raised in the laboratory indicated that anadromous fish matured at a mean size nearly 4 mm larger. Broad-sense heritabilities suggested that the differences had a substantial genetic basis. The link between anadromy and life history in threespine stickleback is generally supported by the available data. Anadromous fish are on average significantly larger at first maturity than freshwater fish (Table 6.2). Mean SL for anadromous populations is 45.6 mm, while the comparable values for lacustrine and stream fish are 40.2 and 39.6 mm, respectively (Table 6.2). The exceptions to this generalization, however, suggest that age and size of reproduction are affected by a number of other factors that are not yet well understood.

There is little information on the potential effect of lateral plate phenotype on age or size at first reproduction of female threespine stickleback beyond that accounted for by partially correlated factors such as lifestyle. Haglund's (1981) data for a California stream population show no obvious differences in size among females of varying plate counts. In a Washington (USA) lake (Kynard 1972), the smallest females tended to be those having an 8/8 lateral plate phenotype (plates on the left and right sides, respectively), though the mean sizes of the 8/8 and 7/7 phenotypes were similar. Differences in reproductive characteristics of males within the low-plated morph have been documented (Kynard 1972; Haglund 1981). Haglund found that males with 11–14 lateral plates (total of both sides) accumulated more eggs per nest than did males with 10 or fewer total plates. Thus, it seems possible that lateral plate phenotype may also be associated with some female traits.

Egg size

Egg size is an important life history trait in fishes because it can affect clutch size and lifetime fecundity (Mann and Mills 1979; Wootton 1984*b*), embryo and juvenile size (Heins and Baker 1987; but see Lagomarsino *et al.* 1988), and juvenile growth (e.g. Wallace and Aasjord 1984) and survival (e.g. Blaxter and Hempel 1963; Marsh 1986). Unfortunately, published egg size data may be of limited comparability for many fish taxa owing to differing preservation techniques (e.g. Fleming and Ng 1987), differing or inaccurate interpretations of egg maturity stages (Heins and Baker 1988), and varying methods of measurement (Heins and Baker 1987).

After a clutch is recruited into vitellogenesis, the oocytes typically increase in size for a period of several days to several weeks (Wallace and Selman 1981). In one well-studied British population of *G. aculeatus*, oocyte diameter increases from about 0.6 mm to 1.50 mm, and weight from about 50 μg to over 340 μg over a 4–8 d interclutch period (Wootton 1973*b*; see also Wallace and Selman 1979). Thus, measurements of oocytes during the vitellogenic period can produce substantial underestimates of final egg size, and for accurate comparisons only females with clutches of ovulated eggs should be used (but see Heins and Baker 1988). However, even relatively early in the vitellogenic period, the oocytes that will constitute the next clutch are often obvious, and they may be quite large well before completion of growth (see Wallace and Selman 1979, Fig. 1). Oocytes gradually become more translucent and spherical during growth, and it is tempting to assume that nearly spherical, fairly translucent oocytes are finished growing. Nevertheless, unless ovulated eggs are present, the stage of maturity is difficult to determine with certainty. Unfortunately, in contrast to the relatively prolonged vitellogenic period, ripening, ovulation, and spawning occur quickly following completion of oocyte growth. Thus, females of most species, including stickleback, contain completely developed eggs for only a brief period during the production of a clutch, and most random field collections will contain few females appropriate for determination of ovum size (Heins and Baker 1988). Because ripe female stickleback are distinctive behaviourally and physically (Kynard 1972; Wootton 1974*a*; Baker and Foster unpubl. data) and can be relatively easily captured individually on the spawning grounds, studies of threespine stickleback could largely avoid this problem. An additional advantage is that threespine stickleback females appear to hold ovulated eggs for a comparatively long period (several hours or more) before spawning.

Ripe eggs themselves present several obstacles to determining accurately their size. As eggs ripen and are ovulated, the rate of yolk accumulation slows considerably and finally ceases; however, at this same time the eggs begin to increase in diameter through hydration (Wallace and Selman 1978). Again, these phenomena are characteristic of threespine stickleback (Wallace and Selman 1979). The increase in diameter is often considerable, and unless the eggs have hydrated to the maximum extent within the female, measurements of their diameter may be biased. Once spawned and fertilized, eggs may undergo further enlargement. Vrat (1949) found that diameters of freshly stripped threespine stickleback eggs increased from 1.30–1.36 mm to 1.71–1.73 mm when fertilized. Nevertheless, diameter is by far the most commonly used measure of egg size (Appendix 6.1).

Dry weight may be a more reliable estimator of egg size than diameter. However, determining dry weight accurately presents a number of problems itself, including considerations of drying conditions (temperature, duration) and method of preservation (if any). The effects of these factors are generally unknown, and comparative data on the diameter and mass of known-stage eggs would be of considerable value. Presently, I know of only a single data set for threespine stickleback that provides both measurements (Baker and Foster unpubl. data).

Finally, differences in fixation and storage can considerably affect egg size measurements (e.g. Fleming and Ng 1987). Threespine stickleback eggs have been treated in at least four ways: fixed and stored in Gilson's fluid only; fixed and stored in 10 per cent formalin only; fixed in 10 per cent formalin and stored in 40–50 per cent isopropyl alcohol; fixed and stored in 70 per cent ethanol only. The effects (if any) of these differing treatments remain unexplored.

These potential limitations notwithstanding, there appear to be large differences among threespine stickleback populations in the size of eggs produced (Appendix 6.1). Reported mean diameters of ripe or nearly ripe eggs range from about 1.25 mm (Coad and Power 1973*a*) to 1.99 mm (Baker and Foster unpubl. data). These extremes correspond to over a fourfold difference in volume, and also in mass if it is assumed that eggs of different diameters contain proportional amounts of yolk. Although this assumption has not been carefully tested, preliminary data of Baker and Foster (unpubl. data) suggest that there may be variation in the egg diameter–dry weight relationship among populations. Even so, the relatively few dry weight data available (Wootton 1973*a*; Baker and Foster unpubl. data) suggest differences of two- to threefold (about 300–850 μg), even among populations in close geographic proximity.

Egg diameters of the four threespine stickleback lifestyles (Appendix 6.1) differ significantly (Table 6.3). Mean egg sizes of lacustrine populations are significantly larger than those of the other three lifestyles, which do not differ. Comparisons of egg sizes of populations of two or more lifestyles from the same geographic area are generally consistent with this overall finding. Coad and Power (1973*a*,*b*) found that egg size in a Quebec, Canada, marine population was smaller than that in a nearby lake. Snyder and Dingle (1989) determined that laboratory-reared, completely plated fish from anadromous and near-estuary stream populations in a California river system did not differ, while the eggs of a well-inland, low-plated stream population in the same river were larger (Snyder 1990). Baker and Foster (unpubl. data) found that four of five Alaska lacustrine populations had much larger eggs than a nearby stream population, while the fifth had eggs no larger than those of the stream-dwelling fish. Mori (1987*b*) found that eggs of the anadromous lifestyle in Japan were significantly smaller than those of nearby stream populations, while Moodie (1972*b*) noted that a

Table 6.3 Reported egg diameter (mm) for populations of four lifestyles of threespine stickleback.[a]

Lifestyle	No. of pops	Egg diameter (mm)				
		Mean	SE[b]	Median	CV[b]	Range
Marine	4	1.43	0.094	1.40	13.1	1.25–1.66
Anadromous	10	1.53	0.051	1.56	10.5	1.22–1.70
Lacustrine	9	1.83	0.043	1.88	7.1	1.56–1.99
Stream-dwelling	12	1.59	0.051	1.60	11.1	1.30–1.81

[a] The means of the four lifestyles differ significantly (one-way ANOVA using $\log_{10}$-transformed data, $F = 7.30$; d.f. = 3.31; $P < 0.001$). Variances of transformed data are not significantly different (log-anova test, Sokal and Rohlf 1981: $F = 0.12$; d.f. = 2, 15; $P > 0.50$). The mean egg size for lacustrine populations is significantly larger than that of the other three lifestyles ($P < 0.05$; GT2 test, Sokal and Rohlf 1981), which do not differ significantly.
[b] SE, standard error of the mean; CV, coefficient of variation expressed as a per cent.

lacustrine and a stream-dwelling population from the Queen Charlotte Islands had almost identical egg diameters. Combined data for all four geographic areas suggest that egg diameters within the four distinct lifestyles of threespine stickleback have similar coefficients of variation (Table 6.3).

Some of the difference in egg size between freshwater and marine stickleback could simply be due to the greater ionic composition of marine waters, which would tend to cause eggs of the same mass to hydrate less than in fresh water. However, Marteinsdottir and Able (1988) found that fixation in high-salinity seawater resulted in egg diameters of *Fundulus heteroclitus* only 3 per cent smaller than fixation in fresh water. Thus, although this potential source of variation has not been tested in threespine stickleback, it is unlikely to contribute strongly to the observed differences.

Threespine stickleback egg diameter is significantly correlated with latitude in western North America after the effect of lifestyle is removed ($r_p = 0.64$, $N = 15$, $P < 0.05$), but the pattern appears to be more nearly a dichotomy than a gradual trend. North of Vancouver Island, Canada, egg diameters range from 1.71 to 1.99 mm, and there is clearly no trend within this area. From Vancouver Island south, eggs are much smaller, from 1.33 to 1.69 mm. This situation may be generated because most northern populations are lacustrine, and thus have relatively large eggs, while most southern populations are from streams. Egg diameter and latitude do not appear to be related in either Europe ($r_p = 0.03$, $N = 8$, $P > 0.95$) or eastern North America ($r_p = -0.30$, $N = 9$, $P > 0.45$).

Egg size in threespine stickleback is not related to fish size among populations within any of the lifestyles. Egg diameter of the giant, black Mayer Lake stickleback (Moodie 1972*b*) was not substantially larger than that of the much smaller, parapatric, low-plated stream population. Bertin (1925) noted no covariation of egg size and female size in a large number of European populations. Baker and Foster (unpubl. data) found that mean egg diameter was not correlated with mean female size in five Alaskan lakes in which mean female SL ranged from 44 to 76 mm. Egg size in individual populations is also unrelated to female size in freshwater (Coad and Power 1973*b*; Wootton 1973*b*; Mori 1987*b*; Snyder and Dingle 1989; Baker and Foster unpubl. data) and anadromous (Craig and FitzGerald 1982; Mori 1987*b*; Snyder and Dingle 1989) fish.

Egg size probably does not covary strongly with aspects of morphology such as plate morph in the threespine stickleback. Low-plated fish generally produce large eggs (see Table 6.3, lacustrine form). However, the anadromous population inhabiting the Mediterranean coast of France produces relatively small eggs (Crivelli and Britton 1987), even though it is low-plated. At the opposite end of the spectrum, preliminary data for a lacustrine, completely plated population in British Columbia (Baker and Foster unpubl. data) suggest that it produces fairly large eggs. Finally, the largest egg diameters reported for a stream-dwelling population are for one that includes partially plated fish (Mori 1987*b*).

In some populations, individual females vary substantially in egg size (Appendix 6.1). Although the evidence from dry weight data is much scarcer, the conclusion is similar. Wootton (1973*b*) found that egg weights for individual females from the same population ranged from 250 to 440 μg, and Baker and Foster (unpubl. data) found nearly a twofold difference in mean dry egg weights among females in some Alaskan populations. Obviously, before we can correctly interpret observed differences among stickleback populations, more information must be obtained on within-population variability.

Clutch size

Lifetime fecundity is often the quantity maximized in theoretical life history models (e.g. Roff 1984), and the size of individual clutches is an important determinant of this quantity. Unlike the case for many groups of squamate reptiles (Dunham *et al.* 1988), birds, and mammals (Blueweiss *et al.* 1978), clutch size in fishes is a strong function of female size, both across taxa (Blueweiss *et al.* 1978; Duarte and Alcaraz 1989) and within individual species, including threespine stickleback (Wootton 1979). Therefore, clutch size at any given spawning depends on two related factors: the rate of increase in clutch size with female size, and the actual size of that female. Females that defer reproduction, allocating the energy to growth instead, can produce larger clutches when they do spawn than can females that mature earlier at a smaller size. In populations having very low body size–clutch size regression slopes, females may be more likely to emphasize early reproduction at a relatively small size because increases in size will produce relatively small gains in fecundity. Conversely, in populations with steep slopes, delaying reproduction to grow even a few millimetres larger could result in much larger clutch sizes.

In comparison with egg size, reported clutch sizes for threespine stickleback are probably very accurate. Even early in vitellogenesis, the eggs that will form a clutch are easily distinguished. Thus, during the reproductive season a large proportion of adult females can contribute data on this trait in most populations. This is not true for many other North American fishes (Heins and Rabito 1986), in which the eggs forming the next clutch are not always easily distinguishable from the pool of non-vitellogenic eggs until relatively late in their growth. In addition, because female stickleback spawn an entire clutch at once, even females with ovulated eggs can be used for clutch-size determinations. Again, this is not the case for many other fish species, which may ripen and spawn a single clutch in many, small batches (e.g. Gale 1986; Burt *et al.* 1988).

Data on clutch sizes and on female size – clutch size relationships are available for a large number of threespine stickleback populations (Appendix 6.1). In nearly all studies, size was estimated simply as female length. Ideally, clutch size or its energy equivalent should be plotted against perhaps the 0.7 power of body mass in order to provide a measure of reproductive investment relative to metabolic activity. However, the data with which to do so (energy intake; body mass, composition, and energy content; clutch size and frequency; mortality curve) are not yet available for most threespine stickleback populations. Although plotting clutch size against length has potential problems and makes several assumptions, this approach does permit a first-order estimation of differences among populations. I do not imply that this approach necessarily provides comparable or accurate estimates of reproductive investment.

As an initial examination of the overall variation in regression slopes, I plotted all available data (Fig. 6.4). Regressions based on untransformed data were plotted as given by the authors; those based on log-transformed data were plotted after back-conversion to raw values. In all instances in which I could compare both raw and transformed data, there was no substantial change in the relationships of the regression lines among populations. For plotting, standard length (SL) was used, since most studies reported it; when total length (TL) was reported, SL was estimated as 0.88 TL (Havens *et al.* 1984; Crivelli and Britton 1987).

Although clutch size is a positive function of female size in all populations of threespine stickleback studied to date, an extremely wide range of regression slopes has been reported (Fig. 6.4; Appendix 6.1). As a preliminary evaluation, I tested data on slope with a one-way ANOVA to determine whether threespine stickleback lifestyles showed any consistent differences. Information was insufficient to simultaneously test among the individual geographic areas, so all data were pooled. The single marine population for which slope was reported was not included in the analysis. Anadromous,

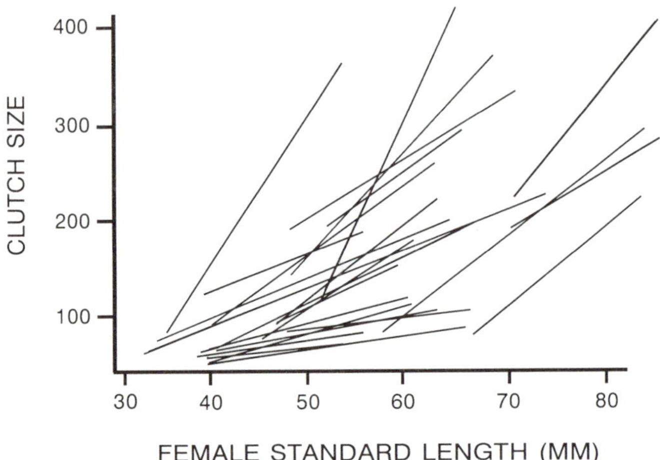

Fig. 6.4 Plots of least-squares regression lines relating clutch size (number of eggs) and standard length for populations of threespine stickleback. The plots are for raw values; data from studies presenting only transformed data were back-transformed to raw values prior to plotting.

lacustrine, and stream-dwelling populations did not differ with respect to mean slope ($F = 0.549$; d.f. $= 2, 22$; $P > 0.50$) and all were highly variable (Appendix 6.1). Regression slope was not correlated with latitude when the effect of lifestyle was partialled out (all populations combined: $r_p = 0.18$, $N = 31$, $P > 0.40$).

As a more detailed assessment, I used ANCOVA to compare ten populations for which samples sizes were adequate and for which the raw data were easily obtained (Table 6.4). For nine populations I used the data from published papers or provided by authors. In one case (Mori 1987*b*), data were digitized from a published figure. The ten populations differed significantly in slope ($F = 3.385$; d.f. $= 9, 402$; $P < 0.01$). Although the two populations with the highest body size—clutch size regression slopes are anadromous, and many of the populations with low slopes are freshwater, the mean slope of the group of marine and anadromous populations (Table 6.4) does not differ significantly ($F = 0.27$; d.f. $= 1, 8$; $P > 0.50$) from that of the five freshwater populations.

This overall result is supported by five comparisons of marine or anadromous populations with freshwater populations from the same area, and these comparisons span much of the range of the threespine stickleback. For southern British Columbia, Canada, analysis of data for anadromous and stream-dwelling populations presented in Hagen (1967) gives $F = 0.057$; d.f. $= 1, 76$; $P > 0.50$ (hybrids not included). The data presented for marine and lacustrine populations from eastern Canada by Coad and Power

(1973*a*,*b*) gives $F = 0.467$; d.f. $= 1, 33$; $P > 0.50$. Anadromous and stream-dwelling populations in California, USA (Snyder and Dingle 1989, Snyder 1990) were also found not to differ. Although Mori (1987*b*) reported significant differences between anadromous and freshwater populations (two stream-dwelling, one lacustrine) in Japan, his analysis used raw data. Females in the anadromous population were much larger than those in the freshwater populations, and a reanalysis of his data using log-transformed values to correct for the apparent curvilinear relationship of female length and clutch size indicated no significant differences among the populations ($F = 1.995$; d.f. $= 3, 95$; $P > 0.10$).

ANCOVA including only the anadromous and marine populations showed that there were significant differences in slope ($F = 4.688$; d.f. $= 4, 213$; $P < 0.01$) within this group. Interestingly, the marine population has the lowest slope and is the seeming 'cause' for the overall significant difference found for this group of populations. The regression slope for the marine population is significantly different ($P < 0.05$; Tukey–Kramer test) from that of the Camargue fish (Table 6.4), and may be ($0.10 > P > 0.05$) different from that of the Isle Verte population. None of the other populations differ statistically from each other. Strong conclusions cannot, obviously, be drawn for marine populations from this analysis, though the result is intriguing because it is different from that expected. Because the slopes differ significantly, adjusted mean clutch sizes cannot be compared with ANCOVA. However, predicted clutch sizes, at the overall unweighted mean SL of 53.1 mm, derived from the individual regression lines (Table 6.4), suggest that the clutch size of the Camargue population would greatly exceed that of the other four populations at this size.

The five freshwater populations do not differ in regression slope ($F = 0.453$; d.f. $= 4, 193$; $P > 0.05$). Adjusted mean clutch sizes for these populations are considerably different ($F = 82.1$; d.f. $= 4, 197$; $P < 0.001$), and their predicted clutch sizes at a common SL (Table 6.4) range from 77 to 158 eggs. Although this sample of populations is much too small to permit any generalizations, it is interesting to note that the two stream-dwelling populations (Hagen 1967; Mori 1987*b*) have significantly higher adjusted clutch sizes than do the three populations inhabiting lakes ($F = 183$; d.f. $= 1, 3$; $P < 0.001$).

The statistical demonstration of significant differences in length-adjusted clutch sizes among freshwater threespine stickleback populations substantiates what is obvious (Fig. 6.4): at any size one might select, populations differ enormously in the number of eggs produced per clutch. Overall, predicted clutch sizes at an arbitrarily chosen, but common, size of 55 mm SL range from about 10 eggs (Moodie 1972*b*) to 250 eggs (Wootton 1973*a*; Wootton *et al*. 1978). Even among populations in close geographic proximity, there can be substantial differences (Appendix 6.1). For example, on islands along the coast of British Columbia, predicted clutch sizes at 55 mm SL

Table 6.4 Regression and population mean clutch size statistics for ten populations of threespine stickleback.

Population	Life-style[a]	No. of fish	Clutch size–SL regression[b]			Population mean		Predicted clutch size[c]	Reference
			Slope	Intercept	r^2	SL (mm)	Clutch size		
Amory Cove, Quebec	M	22	1.2959	0.1100	0.496	58.7	253	222	Coad and Power (1973*a*)
Camargue, France	A	98	3.8234	−4.0248	0.538	40.7	134	372	Crivelli and Britton (1987)
Isle Verte, Quebec	A	35	3.3003	−3.4361	0.197	63.5	327	181	Craig and FitzGerald (1982)
Little Campbell River, British Columbia	A	40	2.1210	−1.3749	0.471	58.5	236	192	Hagen (1967)
Kahoku-gata Lagoon, Japan	A	24	1.6056	−0.4184	0.217	74.4	386	225	Mori (1987*b*)
Trouble Lake, Alaska	FL	30	2.5283	−2.4164	0.685	46.4	63	88	Baker and Foster (unpubl. data)
Bruce Lake, Alaska	FL	40	1.9344	−1.4496	0.381	49.6	67	77	Baker and Foster (unpubl. data)
2 lakes (pooled), Quebec	FL	15	1.7217	−1.0461	0.386	53.1	84	84	Coad and Power (1973*b*)
3 streams (pooled), Japan	FS	79	2.3709	−1.8919	0.674	52.2	151	158	Mori (1987*b*)
Little Campbell River, British Columbia	FS	39	2.2734	−1.7729	0.708	44.7	95	141	Hagen (1967)

[a] A, anadromous; FL, freshwater lacustrine; FS, freshwater stream; M, marine.
[b] Log–log regression of clutch size on female standard length (SL, mm).
[c] Predicted for an overall, unweighted mean female SL of 53.1 mm from individual population log–log regressions.

range from 10 to 130 eggs (Moodie 1972*b*; Hyatt and Ringler 1989*a*; Baker and Foster unpubl. data), and in a relatively small geographic area in Alaska, USA, they ranged from 45 to 160 eggs (Baker and Foster unpubl. data). In contrast, along the mainland coast of North America from southern British Columbia to southern California, clutch sizes of freshwater populations at 55 mm SL are quite similar, ranging from about 135 to 165 eggs (Hagen 1967; Kynard 1972; Baskin 1974; Haglund 1981; Snyder and Dingle 1989). The first two groups above comprise mostly lacustrine populations that vary greatly in size and morphology; the last group consists mainly of stream populations that are more similar morphologically.

Irrespective of the female size—clutch size relationship, threespine stickleback populations can differ substantially in the number of eggs produced per spawning by an average-sized female because populations differ considerably in the mean sizes of reproductive females (Appendix 6.1). Means for the ten populations I compared statistically (Table 6.4) show that although the Kahoku-gata Lagoon, Japan, population has a very low body size—clutch size regression slope, it produces large clutches, primarily because the females are large. One-way ANOVA using all available data on mean clutch size showed that anadromous populations have significantly larger clutches than populations of either freshwater lifestyle, which do not differ (Table 6.5). This overall result is consistent with the findings of Snyder (1990) for anadromous and freshwater fish from the same California, USA, drainage. The clutch size of the single marine population is almost identical to the mean for the anadromous populations. Clutch size for all data pooled is not correlated with latitude when the effect of lifestyle is held constant ($r_p = -0.23$, $N = 31$, $P > 0.20$).

Table 6.5 Reported mean clutch sizes for populations of four lifestyles of threespine stickleback[a].

Lifestyle	No. of pops	Clutch size (no. of eggs)				
		Mean	SE[b]	Median	CV[b]	Range
Marine	1	265	–	–	–	–
Anadromous	10	269	36	254.5	42.3	110–450
Lacustrine	11	144	27	106	63.3	37–317
Stream-dwelling	9	122	11	123	26.5	86–172

[a]The means of the anadromous, lacustrine, and stream-dwelling lifestyles differ significantly (one-way ANOVA using $\log_{10}$-transformed data, $F = 6.86$; d.f. = 2, 27; $P < 0.005$). The marine lifestyle was not included in the analysis. Log_{10}-transformed variances of the three lifestyles do not differ significantly ($F = 4.63$; d.f. = 2, 6; $0.10 > P > 0.05$). Mean clutch size for anadromous populations is significantly greater than that of the two freshwater lifestyles ($P < 0.05$; GT2 test, Sokal and Rohlf 1981), which do not differ significantly.

[b]SE, standard error of the mean; CV, coefficient of variation expressed as a per cent.

Seasonal aspects of clutch size have not been adequately investigated. In the only study of which I am aware, Bolduc and FitzGerald (1989) found that size-adjusted clutch size declined during the spawning season.

There are few data on the effect of female morphology (other than general female size) on clutch size. Kynard (1972) reported that females possessing a 7/7 lateral plate phenotype (i.e. seven plates per side) produced significantly larger clutches than those possessing an 8/8 phenotype, although the two phenotypes did not differ in slope. However, my reanalysis of his data (Kynard 1972: Fig. 20) using $\log_{10}$-transformed instead of raw data, indicates no difference in SL-adjusted clutch sizes ($F = 1.900$; d.f. $= 1, 72$; $0.25 > P > 0.10$). In agreement with Kynard, I detected no difference in slope between the forms ($F = 0.44$; d.f. $= 1, 71$; $P > 0.50$). Similarly, Haglund (1981) found no differences in mean clutch size for similarly sized females of a range of plate phenotypes.

Of course, the differences in clutch size observed among some lifestyles may partly be related to morphology (Snyder 1990), as marine and anadromous populations are generally completely plated whereas freshwater populations are typically low-plated. At present, however, the effect of lateral plate phenotype *per se* (if any) cannot be untangled from the effects of other factors.

Clutch interval and number of clutches

Lifetime fecundity is a function of both the size of individual clutches and the total number of clutches produced by a female during her lifetime. Despite the considerable importance of the number of clutches produced, so little information is available on this trait for natural fish populations that synthetic studies of reproductive tactics either limit themselves to semelparous species (e.g. Duarte and Alcaraz 1989) or assume that variability in the trait will not bias results (e.g. Winemiller 1989).

Data on interclutch intervals in threespine stickleback are mostly from laboratory studies. These suggest that an interval of 3–12 d may be typical of many populations. Baggerman (1957) thought that females in Europe could spawn from 15 to 20 times in a three-month breeding season, which suggests a 4–6 d interclutch interval. Snyder and Dingle (1989) found that laboratory-reared females from both migratory and non-migratory populations produced a second clutch 5–6 d after their first. Fish from a British river (Wootton 1973*a*, 1974*a*; Wootton and Evans 1976) spawned at 4–8 d intervals when fed varying ration levels and 3–4 d when fed a constant ration. Interclutch interval seems to have been estimated for only a single natural population, that at Isle Verte, Quebec, Canada. For this population, Bolduc and FitzGerald (1989) calculated interclutch intervals for tagged females ranging from 5.0 to 10.7 days, whereas Reiffers (1984) reported a mean interval of 17.5 days.

Environmental conditions may influence clutch frequency. Boulé and

FitzGerald (1989) recorded mean intervals of 8.2 and 11.5 d under constant and fluctuating laboratory temperatures, respectively, for the anadromous population from Isle Verte. When the mean temperature was held constant but the magnitude of the daily fluctuations was varied, this same population produced clutches only every 15.4–17.8 d. Bolduc and FitzGerald (1989) reported mean interclutch intervals ranging from 9.3 to 10.7 days for fish from this same population breeding in wading pools.

The total number of clutches spawned by individual females during an entire reproductive season has not been determined for any natural population. The range of opinions is substantial, however, and probably reflects a real difference among natural populations. Females in some populations undoubtedly spawn only a few clutches each year. Whoriskey *et al.* (1986), for example, suggested that fish at Isle Verte may spawn only once each season. Even though the studies of Bolduc and FitzGerald (1989) and Reiffers (1984) show that this may not always be true, the number of clutches spawned per female in this population does appear to be quite low. Similarly, Crivelli and Britton (1987) suggested that, while females in their Mediterranean population probably produced multiple clutches each year, the number of clutches was undoubtedly very low. This suggestion is substantiated by their observations that the spawning season was, at most, 50 d per year. At the opposite extreme, Baggerman (1957) thought that 15–20 clutches per year was possible, and Wootton (1973*a*) has demonstrated that laboratory-reared females can produce from 6 to 8 clutches over a 10 wk period.

Lifetime fecundity is unknown for any natural threespine stickleback population, primarily because the total number of spawnings produced by individual females is unknown. Data on the interclutch interval and on the overall length of the spawning season, in conjunction with age- and size-related clutch size information, could be used to estimate lifetime fecundity. For example, a well-studied population in the Afon Rheidol, Mid-Wales (Wootton 1973*a*,*b*; Wootton *et al.* 1978, 1980*a*), spawns over a 12 wk period annually. The population is annual, so that females reproduce only during a single summer. Fish from this population that were fed an intermediate ration level spawned about every 8 d, so that if a female survived for the entire season and remained healthy she would spawn about ten clutches. Mean female clutch size is 86 eggs (Wootton 1973*b*), and so a reasonable estimate of lifetime fecundity of female stickleback for this population is 860 eggs. This estimate is quite similar to that derived from laboratory studies of the same population (Wootton 1984*a*), in which females raised on a high ration through an entire reproductive season produced an average of 973 eggs. Unfortunately, similar estimates can at present be made for only a few populations.

DISCUSSION

Summary of female life history traits

This chapter presents an overview and synthesis of available data on intra- and interpopulation variation in life history traits of threespine stickleback. However, the suggestions made here concerning the causes and consequences of life history evolution must be considered preliminary. Not only are the data limited, they are often difficult or impossible to compare because of differences in methods of data acquisition. Nevertheless, the analyses presented here have provided some insights into the patterns and causes of life history evolution in threespine stickleback.

The interesting exception noted by Reimchen (1992*a*) notwithstanding, maximum life span appears generally to vary from only 1 yr to 4 yr in all four lifestyles of threespine stickleback. At least one-half of all populations appear to be annual, with most of the remainder living only 2 yr. Differences among the lifestyles are evident, with most anadromous populations living 2 yr or more, but most stream-dwelling populations being annual. Life spans of lacustrine populations are variable. The expected relationship of life span and latitude was not observed, seemingly because local environmental conditions have an overriding effect on this trait.

Size at age 1 yr appears to be greater in anadromous populations than in stream-dwelling populations, which in turn average larger than lacustrine populations. It is not clear whether this disparity is due to a faster growth rate in anadromous fish, or whether the length of the annual growing season is greater. For juveniles, the actual rate of growth of the lifestyles appears to be too variable to permit general conclusions.

Mortality rates are little known, especially within individual reproductive seasons. Information on size-specific mortality is needed for all life stages, but it would be particularly valuable for mature fish during the spawning season.

The reproductive season begins earlier at low latitudes than at high latitudes. However, the length of the season is not necessarily greater at low latitudes. Whether anadromous or freshwater populations begin spawning earlier appears to depend upon the geographic region, a relationship which may indicate a general climatic influence.

Sexual maturity is attained at age 1 in most populations. Although it might have been anticipated that age at first reproduction would be correlated with latitude in a species that ranges across nearly 40 degrees of latitude, this is evidently not so for threespine stickleback. Despite the limited range of age at maturity, size (at least as estimated by length) at maturity varies enormously overall among threespine stickleback populations. Marine and anadromous fish mature at significantly larger sizes than do fish of the two freshwater lifestyles. As with age, minimum reproductive size is not correlated with latitude.

Egg size is significantly larger in lacustrine populations than in marine, anadromous, or stream-dwelling populations. Although the latter three lifestyles do not differ significantly in egg size, freshwater stream fish do tend to have slightly larger eggs than either anadromous or marine fish. The variance introduced by pooling data from the entire geographic range of the threespine stickleback may obscure any distinctions. The differences are more readily apparent in direct comparisons of stream-dwelling populations with the other lifestyles within relatively limited geographic areas.

Clutch size is positively related to female length in all populations. The regression slope relating clutch size to length is quite variable overall, and the four lifestyles of threespine stickleback do not differ in mean slope. Size-adjusted clutch size is greatest in marine and anadromous threespine stickleback, and significantly smaller in freshwater stream fish; size-adjusted clutch size in stream-dwelling fish is greater than in lacustrine fish. The ranking, as would be expected, is the reverse of that for egg sizes. Unadjusted clutch sizes of anadromous and marine fish are significantly larger than those of the two freshwater lifestyles, owing to the larger mean body size in the former two lifestyles.

Clutch frequency is unknown for any natural population. However, laboratory studies, and some field work, suggest that the number of clutches produced per year ranges from perhaps just 1 to over 20 in different populations.

Implications of observed life history differences

The marine form of the threespine stickleback has remained morphologically consistent (Bell 1984*a*, 1988, page 446 this volume) for a period of at least 10 million years. In contrast, freshwater populations have repeatedly arisen, diversified rapidly, and become extinct during this same time (Bell 1988; Bell and Foster page 14 this volume). Present freshwater populations exhibit enormous morphological variation, often over very small spatial scales (McPhail 1984; Reimchen *et al.* 1985; Taylor and McPhail 1986; Baumgartner *et al.* 1988; reviewed in Bell and Foster page 8; Reimchen page 248 this volume). Variation in life history traits is similarly great, and it is comparable to that observed among some entire families of fishes (e.g. Reznick and Miles 1989). For example, egg mass varies over at least a four-fold range, size at first maturity from about 25 to 70 mm SL, and spawning season from only one month per year to nearly year-round.

Differentiation in some traits appears to be a general phenomenon associated with the transition from the ancestral marine environment to fresh water. For example, my analyses suggest that invasion of fresh water has generally coincided with an increase in egg size, a reduction in clutch size, a decline in growth rate, and possibly greater variation in average maximum size. Verification of these general patterns would be an important contribution to the more general studies of the evolution of marine–fresh water transitions and senescence. In addition, closer examination of populations in which differentiation has proceeded counter to the general trends would yield insights into the factors that determine the trends.

The egg size and clutch size patterns mirror those found in a comparison of freshwater and marine populations of semelparous fishes (Duarte and Alcaraz 1989), and it may be associated with an advantage to large eggs or fry in freshwater environments. Because male stickleback actively guard and care for their embryos and yolk-sac fry, egg size is unlikely to affect vulnerability to predation. In consequence, the principal effect of egg size may be on juvenile survival. Larger eggs probably hatch into larger larvae that grow faster, eat a wider size range of prey, and are better swimmers (Ware 1975; Heins and Baker 1987; Duarte and Alcaraz 1989; but see Lagomarsino *et al.* 1988). Thus, larger eggs may be selectively favoured if food for the larvae is less abundant or less consistently available in freshwater environments (Mori 1987*b*), or if avoiding predation, or growing rapidly through a relatively short growing season, are more important determinants of fitness in freshwater habitats. The presence of only a single significant correlation between latitude and egg size (in western North America), however, suggests that egg size is probably determined by a number of interacting factors. The evolution of large eggs would reduce clutch sizes unless compensation was achieved by evolution in other traits (e.g. body size, age at first reproduction, interclutch interval). Short growing seasons and lower food availability could also explain the apparent lower growth rates and generally smaller sizes shown by populations of many freshwater organisms in general (Gross *et al.* 1988).

Environmental factors may influence the expression of life history traits through phenotypic plasticity, possibly at several levels. In this review, significant correlations between date of commencement and length of breeding season with latitude, and differences in these traits among geographically close populations, suggest both general climatic and micro-geographic effects. Additionally, most threespine stickleback populations mature at age 1 yr, a strategy that is undoubtedly advantageous for this relatively short-lived species. The finding that most of the exceptions occur at high latitudes suggests that environmental factors (i.e. short growing seasons, cool temperatures) that limit growth are the controlling factors (see Moodie 1972*a*; Reimchen 1992*a*).

It is important to stress that we have little evidence concerning the extent to which interpopulation differences in life history traits are determined by genetic factors. When genetic differences have been sought, studies have often demonstrated at least a partial heritable basis for differences in trait values (e.g. McPhail 1977; Snyder and Dingle 1989; Snyder 1991*a*). Such studies are uncommon, however. It is likely that at least some of the observed differences are due to genetic changes caused by local selection. Genotype–environment interactions are also likely (Snyder and Dingle 1990), and the evolution of particular traits may be constrained, or augmented, by their covariation with other traits (Snyder 1991*b*). Clearly, genetic studies of threespine stickleback are demanded if our understanding of the evolution of this taxon is to advance much further.

The potential value of the threespine stickleback for studying life history evolution is substantial. The recent, explosive freshwater radiation in areas glaciated during the Pleistocene, and the very poor dispersal ability of freshwater stickleback, allow us to assume independent evolution of derived character states (Bell and Foster page 20 this volume). Because we also know the ancestral form, and because the approximate ages of many populations can be inferred from studies of glacial geology, we can study both the pace and the direction of change in relation to a wide variety of factors.

Interpopulation comparisons would provide data suitable for first-order analyses (= correlations) of the relationships between life history traits and environmental factors (e.g. food availability, predation). Hypotheses developed through such analyses could then be tested directly in the laboratory or field (Endler 1986). This approach would also help us to resolve the interactions among factors that influence life history trade-offs, such as those between egg size and clutch size.

Interpopulation comparisons also permit relatively direct tests of life history models that predict the evolution of particular character states under specific conditions. Roff (1988), for example, hypothesized that migratory populations should mature later and be larger-bodied than non-migratory populations. Snyder and Dingle (1989) tested this hypothesis using threespine stickleback, and their results agreed with these predictions. Other theoretical studies have explored the conditions under which selection might alter age or size at first maturity in fishes (e.g. Roff 1984; Stearns and Crandall 1984; Stearns and Koella 1986). One of these (Roff 1984) found that when lifetime fecundity was the trait being maximized, the optimal age at maturity of iteroparous species depended on the instantaneous rate of natural mortality and the constant (k) in the von Bertalanffy growth equation, but not on the cost of reproduction. Because these variables could be measured in natural populations of threespine stickleback, it should also be possible to test this model in the future.

Although the primary value of the threespine stickleback for life history research may lie in the existence of so many populations exposed to differing selection regimes, the species should also be extremely useful for studies of evolutionary processes within populations. Individual females can be captured easily, marked, and observed regularly in many populations. Thus, such features as size, body shape, and even foraging efficiency could be determined and correlated with reproductive output.

The value of future studies, especially for interpopulation comparisons, will depend on the use of uniform methodologies for data acquisition. Better data are needed on almost all aspects of life history, and they should be accumulated for multiple populations of each lifestyle from all major geographic areas. Particular attention should be given to obtaining data representing all combinations of lifestyles, lateral plate phenotypes, and geographic areas. This need not be accomplished in a single, enormous study, nor could it be. However, this need for an ultimate synthesis of numerous studies makes it imperative that researchers obtain comparable data. Particular areas of concern involve accurate determination of egg size and age, consideration of seasonal effects, and determination of clutch frequencies. For example, if clutch size varies during the reproductive season (e.g. Bolduc and FitzGerald 1989), and if analyses do not take this into account, spurious differences could be suggested. It is also imperative that studies investigate as many aspects of life history as possible in each population, even if some are not addressed in detail. Selection agents should also be suggested whenever possible.

In conclusion, although present data available for threespine stickleback may be insufficient to test life history models, the substantial interpopulation variation in life history traits, the high probability that selection mechanisms can be identified and quantified, and the relative ease of working with this fish in the laboratory and in many natural populations suggest that future tests will not only be possible, but most fruitful.

ACKNOWLEDGEMENTS

I thank Steve Ross, David Soltz, and George Williams for providing comments that greatly improved the manuscript. Alain Crivelli, Gerard FitzGerald, and Brian Coad generously provided raw data. Much of this review was accomplished while I was a visiting scientist at the Department of Ecology and Evolution of the State University of New York at Stony Brook. The support of the department, and in particular that of Mike Bell, Jeff Levinton, and Dennis Slice, is greatly appreciated.

APPENDIX 6.1

Reported life history data for threespine stickleback [a].

Location	Lat.	Plate morph	Life-style	First repro.		Maximum		Mean female size	Mean clutch size	Clutch–SL slope	Egg size		Repro. season	References
				Age	Size	Age	Size				Range	Mean		
Europe														
Scandinavia	60–70	C?	M?										6–7	Bertin (1925)
Norway	60	L	FL	1	31?	2?	61?	36?					5–8	Jakobsen *et al.* (1988)
	59.5	M?	A				72	45					6/7–?	Bertin (1925)
Denmark	55–58	M?	A				67	55			1.60–1.70	1.65?	6–?	Bertin (1925)
	56	C?	M				90?						L5–E7	Rasmussen (1973)
Sweden	59	M	A	2	40	4	67						6?–M7	Aneer (1973)
	55–57	M	M										5–6	Borg (1985)
	56	M?	A					54					6/7–?	Bertin (1925)
	56	C	M										L4–E8	Borg and van Veen (1982)
Scotland	57.5	L	FL(3)	1	24–39	1	36–40	31						Giles (1987*c*)
	57.5	L	FL(2)	1	30–53	1	51–55	43						Giles (1987*c*)
	57.5	L	FS(2)	1	37–58	1	47–59	44						Giles (1987*c*)
	57.5	L	FL	1?	35?	2?	65						L5?–?	Ukegbu and Huntingford (1989); Ukegbu (1986)

Appendix 6.1 *cont.*

Location	Lat.	Plate morph	Life-style	First repro.		Maximum		Mean female size	Mean clutch size	Clutch-SL slope	Egg size		Repro. season	References
				Age	Size	Age	Size				Range	Mean		
	55.5	L	FS(3)	1	34–39	1	59–68	37–43					L4–L8	Ukegbu and Huntingford (1989); Ukegbu (1986)
Netherlands	52	M	A	1	38	2?	76	56?					M4–E6	van Mullem and van der Vlugt (1964)
	52	P?	A										4–7	Baggerman (1957, 1980)
	52	L	FS										M4–M7	Goldschmidt (pers. comm.)
England	54.5	L	FL	1?	35?	3	48							Jones and Hynes (1950)
	54	L	FL	1	30?	1?	59	40?						Chappell (1969)
	53.5	C	A?			3	86							Jones and Hynes (1950)
	53.5	L	FS	1	28	3	66							Jones and Hynes (1950)
	53.5	L	FS	1	25?	2	59							O'Hara and Penczak (1987)

Location	Lat.	Plate morph	Life-style	First repro.		Maximum		Mean female size	Mean clutch size	Clutch–SL slope	Egg size		Repro. season	References
				Age	Size	Age	Size				Range	Mean		
	51	L	FL			4	65						L4–L8	Pennycuick (1971)
	51	L	FS(2)	1	30?	1	54	40?					4–5	Mann (1971)
	52	L	FS	2	45	3	55						L4–M8	Craig-Bennett (1931)
	51–52	L	FS								1.20–1.70	1.45?	5/6–?	Bertin (1925); Swarup (1958)
	52	L	FL	1	30	1	51	38?					5	Wootton *et al*. (1978); Allen and Wootton (1982*a*)
	52	L	FS	1	40?	1	55?					1.50	5–8	Wootton *et al*. (1978, 1980*a*); Wootton (1985*b*)
	.				36?		67?	43	86	3.323	250–440[b]	340[b]		Wootton (1973*b*)
Poland	52	L?	FS?	1?		1?		35?						Penczak (1981)
Germany	53	M?	A?			3–5	60						5/6–?	Bertin (1925); Bock (1928)
	53	C	A	2	50?	3	70+		70?					Wunder (1928, 1930)
	53	M?	A	2	48?	3	56?							Münzing (1959)
	53	C	A	1	45?	3?	78		90?					Leiner (1929, 1930, 1931)

Appendix *cont.*

Location	Lat.	Plate morph	Life-style	First repro.		Maximum		Mean female size	Mean clutch size	Clutch–SL slope	Egg size		Repro. season	References
				Age	Size	Age	Size				Range	Mean		
	53	C?	A		51?		72?	62	320		1.49–1.76	1.65	L4–7?	Fries (1965); Ehrenbaum (1904); Schneider (1969)
	49	L	FL?										4/5–?	Bertin (1925)
Belgium	51	L	FS?		42		60	53					4/5–?	Bertin (1925)
	51	M?	A	1	42	1	58?							Heuts (1947*a*, *b*)
Austria	48	L	FS?										4/5–?	Bertin (1925)
Switzerland	46	L	FS				70	53				1.50	4/5–?	Bertin (1925); Fatio (1882)
Italy	41	L	FS?				74	67			1.6–1.7	1.65?	3/4–?	Bertin (1925)
Spain	35–43	L?											3/4–?	Bertin (1925)
France	47–49	L?	A(4)	1	39–?	1–2?	59–?	51						Bertin (1925)
	48.5	L?	FS?	1?	37	1?	54				1.5–1.6	1.55?		Bertin (1925)
	48.5	C?	M		42?		59?						M4–?	Bertin (1925)

Location	Lat.	Plate morph	Life-style	First repro.		Maximum		Mean female size	Mean clutch size	Clutch–SL slope	Egg size		Repro. season	References
				Age	Size	Age	Size				Range	Mean		
	46	L	FS		36									Bertin (1925)
	44.5	L?	A	1	45	1	59	48						Bertin (1925)
	43.5	L	A	1	31	1	52	45						Bertin (1925)
	43	L	A	1	35	1	63	41	134	3.8234		1.49	3–4	Crivelli and Britton (1987)
	43–50	L,M?	A,FS								1.5–1.9		4/5–?	Bertin (1925)
Eastern North America														
Greenland	60–70	M?											6/7–?	Bertin (1925)
Newfoundland	60?	M?									?–2.0			van Vliet (1970)
Quebec	57	L?	FL	2	40?	3?	61	47?					L6–7?	Power (1965)
	50	L?	FS	1	40?	2?	65?	46?					E6–7?	Power (1965)
	50	P?	FS	1	42	3?	63?	50?					6–7?	Power (1965)
	50	P	FL	2	50	2	67	56	88	1.7217	1.26–1.86	1.56	5–7	Coad and Power (1973*b*)
	50	C	M	2	39	2	71	61	265	1.2959	1.01–1.60	1.25	E5–M7	Coad and Power (1973*a*)
	48	C	A	2	51	2	70	64	366	3.3303	1.30–1.52	1.39		Craig and FitzGerald (1982)
								3.81[b]	187		1.60–1.64	1.62	5–6	Boulé and FitzGerald (1989)

Appendix 6.1 *cont.*

Location	Lat.	Plate morph	Life-style	First repro.		Maximum		Mean female size	Mean clutch size	Clutch–SL slope	Egg size		Repro. season	References
				Age	Size	Age	Size				Range	Mean		
	.					2?	70?	63	273		1.02–1.29	1.22	E5–M7	Kedney *et al.* (1987)
											1.65–1.75	1.70		Baker (unpubl. data)
	.								214				5–6	Worgan and FitzGerald (1981*b*); Bolduc and FitzGerald (1989)
Maine	44	C	M?									1.66	5–7?	Bigelow and Schroeder (1953)
Massachusetts	41.5	C	M									1.31	L4–L6	Wallace and Selman (1979)
	41.5	C	M?									1.50		Newman (1915)
	41.5	C	M										5–7	Nichols and Breder (1927); Kuntz and Radcliffe (1917)
Connecticut	41.5	C	M?										5–7	Pearcy and Richards (1962)

Location	Lat.	Plate morph	Life-style	First repro.		Maximum		Mean female size	Mean clutch size	Clutch–SL slope	Egg size		Repro. season	References
				Age	Size	Age	Size				Range	Mean		
New York	41	C	M		45		64?						?–E6	Perlmutter (1963)
	41	C	M		50		70?						3–5	Rowland (1970)
Western North America														
Alaska	61.5	L	FL(5)	1–2?	36–68	1–2?	51–84	44–76	37–317	1.93–3.06	1.71–1.99			Baker and Foster (unpubl. data)
	61.5	L	FS	1?	42		63	54	123	3.6894	1.68–1.78	1.72		Baker and Foster (unpubl. data)
	61	L	FL(3)	1?	35–41	2–3?	65–82						5–7/8?	Engel (1971)
	61	L	FL(7)	1–2?	32–57	2–3?	46–73	43–61					5–M7?	Havens *et al.* (1984); Wenderoff (1982)
	59	L?	FL	3	50?	3	75						7–M8	Gilbertson (1980)
	57	L	FL(2)	1–2	38–39	2	64–79						L5–E8	Greenbank and Nelson (1959)
	57	L	FL	2	50?		65						6–7?	Narver (1969)
	57	P?	A	1	65?		90							Narver (1969)
British Columbia	53.5	L	FL	2	40?	3?	60?						4/5–?	Reimchen (1980)
	53.5	L	FL	2	70?	3	116	90	257	3.9617		1.90	5–M8	Moodie (1972*b*)

Appendix 6.1 *cont.*

Location	Lat.	Plate morph	Life-style	First repro.		Maximum		Mean female size	Mean clutch size	Clutch–SL slope	Egg size		Repro. season	References
				Age	Size	Age	Size				Range	Mean		
	53.5	L	FS	2	45?	3	60?	51	105			1.87	5–M8	Moodie (1972*b*)
	49.5	C	FL				84?						6?–7?	Manzer (1976)
	49.5	L	FL										4–M6	Larson (1976)
	49.5	L	FL(2)								1.88–1.91		E4–E9	Baker and Foster (unpubl. data); Carl (1953)
	49.5	L	FL										5–M7	MacLean (1980)
	49.5	L	FL										M5–M7	Pressley (1981)
	49.5	L	FL										M5–E8	Pressley (1981); Foster (pers. comm.)
	49.5	L	FL										L4–M7	Foster (pers. comm.)
	49	L	FL	1	35?	3	69	60?	76	2.6535			L4–E9	Hyatt and Ringler (1989*a*,*b*)
	49	L	FL(25)	26–53										McPhail (1977)

Location	Lat.	Plate morph	Life-style	First repro.		Maximum		Mean female size	Mean clutch size	Clutch–SL slope	Egg size		Repro. season	References
				Age	Size	Age	Size				Range	Mean		
	49	L	FS	1?	30		55	45	95	2.2734			3–E7	Hagen (1967); Hay and McPhail (1975)
	49	C	A	1?	33		65	58	236	2.1210	1.20–1.60	1.40?	M5–M9	Hagen (1967); Hay and McPhail (1975); Lam *et al.* (1978)
	48.5	L	FL	1		1			106				L4–8	McPhail and Peacock (1983)
	48.5	L	FL										M5–E8	Foster (pers. comm.)
Washington	48	L	FL	1	35	1	65	55	164	2.1694			M4–E9	Kynard (1972); Hagen and Gilbertson (1972)
	47	L [c]	FS										3–7	McPhail (1969)
	47	L [c]	FS										L2–E9	McPhail (1969)
California	39	C	FS/A	< 1[d]	38			48?	87?	3.59		1.68	3–8	Snyder and Dingle (1989)
	39	C	A	< 1[d]	47			52?	110?	2.83		1.70	5–8	Snyder and Dingle (1989)
	39	L	FS	< 1[d]	44							1.76		Snyder (1990)
	38	L	FS								1.20–1.60	1.30	2/3–8	Vrat (1949)
	36.5	L	FS								1.10–1.50	1.36	2/3–8	Vrat (1949)
	34	L	FS	< 1?	42		68	50?	165?	1.4788			2–10[e]	Baskin (1974, 1975)

Appendix 6.1 *cont.*

Location	Lat.	Plate morph	Life-style	First repro.		Maximum		Mean female size	Mean clutch size	Clutch–SL slope	Egg size		Repro. season	References
				Age	Size	Age	Size				Range	Mean		
	34	L	FS		39		54	45	123	1.6253				Haglund (1981)
Eastern Asia														
Kamchatka, Russia	56	L	FL										L6–E9	Ziuganov *et al.* (1987)
	56	C	A										M6–M8	Ziuganov *et al.* (1987)
Japan	43	C	A[f]				88	83	450	3.8545				Mori (1990*a*)
	43	C	FL[f]				70	62	226	3.5452				Mori (1990*a*)
	38	C	FL										4–12	Hirai *et al.* (1973); Yamanaka (1971)
	36	C	A		65		90						3–E6	Mori (1987*b*)

Location	Lat.	Plate morph	Life-style	First repro.		Maximum		Mean female size	Mean clutch size	Clutch–SL slope	Egg size		Repro. season	References
				Age	Size	Age	Size				Range	Mean		
	35	L	FS[g]	1	43–48	1	65						4–M6	Mori and Nagoshi (1987); Mori (1987*b*)
	36.5	C	A		64		80	75	401	1.6056	1.20–1.60	1.46	L3–5	Mori (1987*b*)
	35	L/P	FS(2)		47	1?	85	57	172	2.3708	1.60–2.05	1.81	2–8	Mori (1987*a*,*b*)
	35	L	FS(2)	1?	32–39		58–61	51	145	2.2805	1.40–1.86	1.71	3–11[e]	Mori (1987*b*, 1985)

[a] Lat. approx. °N latitude. [Note: '.' in Lat. column indicates that the population studied was the same as that above]. Plate morph: L, low; P, partial; C, complete; M, mixed. Lifestyle: A, anadromous; FL, freshwater lakes; FS, freshwater streams; M, marine. First repro. age (yr) and size (SL, mm) at first reproduction. Maximum, maximum age (yr) lived and size (SL, mm) attained. Mean female size, mean SL of reproductive females. Mean clutch size, mean number of eggs in mature females. Clutch–SL slope, regression slope of $\log_{10}$ clutch size on $\log_{10}$ female SL. Egg size, diameter (mm). Repro. season, duration of reproductive activity: months are numbered from January (1) to December (12); E, early; M, mid; L, late. '?' indicates uncertainty in the data. [Note: not all tabulated data were specifically presented by all authors; in many instances I estimated trait values from other information in the referenced sources.]

[b] Somatic wet weight (g) for females, dry weight (μg) for eggs.

[c] 'Red' (upper) and 'black' (lower) fish of McPhail (1969).

[d] Laboratory-reared populations. Mean female and clutch sizes after producing two clutches, therefore these values may under-estimate the true population means.

[e] Some spawning in almost every month.

[f] Mori (1990*a*) believes that this population consists of two 'forms', larger anadromous fish and smaller resident fish. The data are separated along these lines.

[g] Reared in semi-natural outdoor concrete troughs.

7

Ecology of the threespine stickleback on the breeding grounds

Frederick G. Whoriskey and Gerard J. FitzGerald

A central theme of the behavioural ecology of reproduction of natural populations is the study of the factors causing variation in individual reproductive success (genetic contribution to future generations) (Clutton-Brock 1988). A knowledge of their relative importance is necessary to formulate predictions about the evolution of an animal's morphology or behaviour. Pertinent questions include: how widely does breeding success vary among individuals of each sex? How much variance in reproductive success is contributed by different components of reproductive success (survival to breeding age, reproductive life span, fecundity, mating success, and offspring survival)? To what extent does survival change with age? How do environmental, phenotypic, developmental, or genetic factors affect reproductive success in each sex?

Although only some of these questions have been adequately answered for the threespine stickleback, *Gasterosteus aculeatus*, this small fish is an excellent subject with which to address many important issues in behavioural ecology. Not only is this stickleback amenable to laboratory observations and manipulations, it can be readily observed *in situ* during the reproductive season. Males arrive on the breeding grounds in the spring, establish territories, look for materials with which to build nests, and court females to obtain eggs. During the spawning act, guardian males must prevent rival males from stealing fertilizations or eggs. Paternal care consists mainly of oxygenating the eggs during incubation by fanning currents of water over them with the pectoral fins, and removing dead or diseased eggs (reviewed by Wootton 1976; Rowland page 337 this volume). Males guard their eggs and fry against a variety of predators including conspecifics (see also Rowland page 339, Foster page 394 this volume). After the eggs are laid, the female has no additional role in rearing of the young.

The purpose of this review is to summarize what is known about the factors affecting some of the components of reproductive success, and, by indicating gaps in our knowledge, to provide suggestions for future work. Emphasis is placed on an anadromous population that breeds in salt-marsh tidepools along the St Lawrence Estuary near the village of Isle Verte,

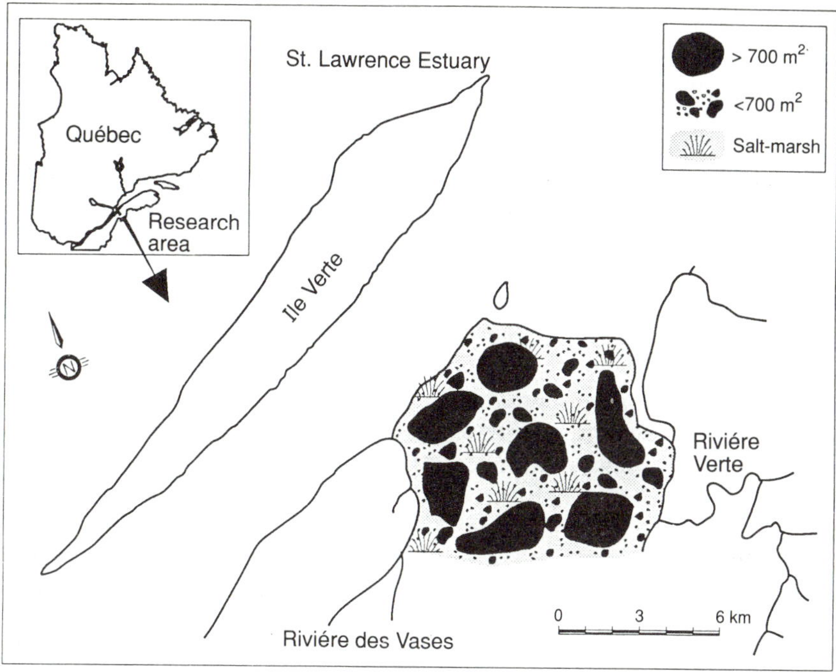

Fig. 7.1 Map of Isle Verte study site, showing series of tidepools where work on an anadromous population of threespine stickleback has been conducted since 1977. As there are about 887 tidepools at the site, this map is schematic and cannot depict all the smaller pools. For additional details see FitzGerald *et al.* (1992).

Quebec, Canada, (Fig. 7.1) because more ecological data have been collected at this site than at others. The numerous pools high in the intertidal zone of the Isle Verte salt-marsh offer abundant opportunities for field observation and experiment. This breeding habitat differs considerably from that used by freshwater populations of the threespine stickleback in that it experiences rapid and extreme fluctuations in the physical environment. Whenever possible, data are also presented from populations breeding in other habitats.

MEASURES OF REPRODUCTIVE SUCCESS

Field biologists are obliged for practical reasons to use measures that estimate reproductive success (indices). The indices used for male threespine stickleback have been the number of eggs in nests at the time of collection (e.g. Kynard 1978*a*; FitzGerald 1983) or the number of males hatching fry (Ward and FitzGerald 1988). These indices may under-estimate reproductive success because (1) it is possible that some of the males whose eggs were collected would have completed additional breeding cycles within the same or in another breeding season (Kynard 1978*a*; FitzGerald *et al.* 1986), (2) the male might have fertilized eggs in the nests of other males (van den Assem 1967), and (3) some of the eggs fertilized by the male could have been stolen by a rival and reared successfully in his nest (van den Assem 1967; Whoriskey and FitzGerald 1985*a*). In other situations, egg or fry counts may over-estimate reproductive success because (1) the eggs collected might have failed to hatch because of predation or unfavourable environmental conditions, (2) some of the eggs in the nest of the male could have been fertilized by a rival male, and (3) some of the eggs in the nest of a male might have been stolen from the nest of another male after the other male had fertilized them.

Most examinations of reproductive success of threespine stickleback have been done in the laboratory (reviewed by Wootton 1976, 1984*a*; Rowland Chapter 11 this volume). More recently, field investigations of some of the factors affecting individual reproductive success have been conducted (e.g. Kynard 1978*a*, 1979*a*; FitzGerald 1983; Whoriskey and FitzGerald 1987; Ward and FitzGerald 1988). This chapter concentrates on summarizing these studies. Additional information on the breeding ecology of the threespine stickleback is given in Wootton (1976, 1984*a*), FitzGerald and Wootton (1986), and in the chapters of this volume by Baker, Foster, and Rowland.

Variation in male reproductive success is more thoroughly studied and is treated first in some detail. Then, because of the relative paucity of data, factors influencing variation in female reproductive success and the survival of independent juveniles are treated sequentially.

DURATION AND TIMING OF THE BREEDING SEASON

The length of the breeding season ranges from year-long in one Japanese population (Mori 1985) to about 50 d in southern Quebec (FitzGerald 1983) and southern France (Crivelli and Britton 1987). It is not known what factors determine its onset and length. Possible factors include changes in food, in water temperature, and in photoperiod. However, both the onset and the length of the breeding season can vary over short geographic distances. In tidepools on the southern shore of the St Lawrence Estuary, fish finish breeding at the end of June, whereas in Lac Gauthier, only 40 km to the north, fish breed well into August (FitzGerald, unpubl. data). The only pattern to emerge from a comparison of different populations is that the breeding season generally begins later in the spring in northern areas (Wootton 1984*a*; Baker page 155 this volume).

Differences in the time of breeding can occur for different forms of the threespine stickleback that breed in the same lake. In Lake Azabachije, Kamachatka (Russia), an anadromous completely plated form breeds much earlier than a resident low-plated form (Ziuganov and Bugayev 1988). It seems that temperature and photoperiod are involved in determining the onset and termination of breeding, but the relative importance of these cues may vary among populations. In Lake Azabachije the anadromous populations bred at 5–6 °C whereas the resident population spawned at much higher temperatures. Mori (1985) suggested that the constant 15 °C temperature at his site in Japan allowed the fish to breed throughout the year. However, in southern France, breeding ceases with no noticeable environmental changes (Crivelli and Britton 1987; FitzGerald, pers. obs.). In this population it appears that endogenous factors (e.g. acclerated senescence associated with reproduction) cause fish to cease breeding.

Although laboratory studies indicate that photoperiod may influence the timing of breeding and the length of the breeding season (Wootton 1976, 1984*a*; Baker page 155 this volume), the situation is more complex in the field, as illustrated by the above example comparing the two Quebec populations.

BREEDING HABITAT

The threespine stickleback breeds in sloughs, ponds, rivers, lakes, drainage canals, freshwater and saltwater marshes, tidal creeks, and sublittoral zones of the sea. It is not known why certain areas within these habitats are chosen but other seemingly suitable areas are ignored. Warner (1988, 1990) has shown that both the physical characteristics of a site and a tradition to use the same spawning area determine where the bluehead wrasse, *Thalassoma bifasciatum*, spawns. The annual breeding cycle of most populations of the threespine stickleback makes it unlikely that cultural transmission plays a role in choosing a breeding site, although it is possible in biannual populations. It is not known whether this stickleback returns to its natal site to breed. The choice of a breeding area is probably determined principally by physical factors such as the presence of suitable nesting material and cover.

An anadromous population of threespine stickleback migrates from its overwintering areas in the middle St Lawrence Estuary to a coastal saltmarsh near Isle Verte, Quebec, where it breeds in tidepools in May and June (Picard *et al.* 1990). The fish enter the marsh with high spring tides. The pools are separated from one another and from the estuary during the two-week interflooding intervals. The choice of pool is critical for a fish because some pools will dry out, killing the adults and fry before they can return to the estuary (Whoriskey and FitzGerald 1989). Fish avoid those pools that will dry out, although it is not known how they do so. Unlike lizards (Stamps 1988), they evidently do not use conspecifics as a cue for habitat selection, because newly arriving fish were unaffected by the presence of fish already in the pools (Whoriskey and FitzGerald 1989).

Male threespine stickleback may form breeding 'aggregations' or colonies, as evidenced by the fact that nests are often clumped when seemingly suitable habitat nearby is unused (e.g. Moodie 1972*a*,*b*; Kedney *et al.* 1987). There are advantages and disadvantages to coloniality. Whereas territorial males can combine efforts to defend their nests against raiding shoals of females and non-territorial males, the same males may have increased risks of egg stealing, courtship interruptions, and stolen fertilizations from rival territorial fish (e.g. Loiselle and Barlow 1978). The costs and benefits of coloniality in the threespine stickleback need to be investigated.

Nest site characteristics

Males build nests at sites that differ in water depth, substrate type, amount and type of cover, and distance from the shore. The type of available materials used in nest construction may help conceal nests. Some nests are more difficult to locate than others because they blend well with the substrate (pers. obs.). Perhaps non-human predators also have difficulty in locating well-camouflaged nests. Nests are found at depths ranging from a few centimetres in tidepools (e.g. FitzGerald 1983) to 6 metres in lakes (Kynard 1978*a*). In shallow tidepools, nests are especially vulnerable when water levels recede because of evaporation (Poulin and FitzGerald 1989*a*). In lakes, nests in shallow water are vulnerable to waves during storms (Kynard 1978*a*), and in rivers they are vulnerable to fluctuations in water level (Wootton 1972*a*).

Male threespine stickleback build nests on different types of substrates, but the effects of this variation on reproductive success have not been studied. Nests are found on sand, mud, rocks, and detritus (reviewed in Wootton 1984*a*). Anadromous and stream-resident species of threespine stickleback in the Little Campbell River, British Columbia, Canada, nest on sand and mud surfaces, respectively (Hagen 1967, McPhail page 408 this volume). Although the threespine stickleback typically nests on a firm substrate, the 'white stickleback' (an undescribed threespine stickleback species characterized by white male nuptial coloration and divergent reproductive behaviour) nests above the substrate in plumes of algae. In the same location, the threespine stickleback nests on the bottom of the substrate (Blouw and Hagen 1990; Jamieson *et al.* 1992*a*).

Nest sites of the threespine stickleback also differ in the amount of concealment they afford the nest. Cover can be provided by rocks, fallen logs, vegetation, and virtually any other object. Within populations there can be considerable variation in nest concealment (Moodie 1972*a*; Kynard 1978*a*, 1979*a*), and in Enos Lake the benthic species nests mostly in vegetation whereas the limnetic species uses open areas (Ridgway and McPhail 1984, 1987; McPhail page 421 this volume). The nest of the white stickleback is partially concealed by the algae in which it nests, whereas the sympatric threespine stickleback nests in the open (Blouw and Hagen 1990; Jamieson *et al.* 1992*a*).

VARIATION IN MALE REPRODUCTIVE SUCCESS

There is great variation in male reproductive success within and among populations. In some years at Isle Verte, only about 30 per cent of the male threespine stickleback obtain territories and nest sites, and of these only about 30 per cent obtain any eggs (FitzGerald 1983). In a sample of 150 nests collected at the site, egg numbers ranged from 0 to 2107 (Lachance 1990). Haglund (1981) also reported that only about 30 per cent of the territorial males in Sespe Creek, California, USA, had eggs in their nests at collection. Nests with eggs contained an average of 385 eggs (range 23–877).

Males can also differ in the number of reproductive cycles they complete during one or more breeding seasons. In the laboratory, males from a freshwater population in Wales can complete up to five reproductive cycles within a breeding season and live up to 5 yr (Wootton 1976). Individuals in other populations can complete more than one cycle in a breeding season (e.g. Kynard 1978*a*; Whoriskey *et al.* 1986). It is not known whether those individuals that breed in one season survive to breed in a subsequent one. In an anadromous population from southern France, individuals almost certainly breed only once (e.g. Crivelli and Britton 1987).

A high parental effort in an initial reproductive cycle is energetically costly, and this high cost may reduce a male's chances of surviving to a second breeding season (Chellappa *et al.* 1989; FitzGerald *et al.* 1989). Moreover, reproductive males may be more vulnerable to predation than non-reproductive ones (Whoriskey and FitzGerald 1985*b*). Increased vulnerability could occur because nesting males are brighter and hence more visible, and (or) activities associated with reproduction make them less wary and more visible. At sites where year-long breeding occurs, it is likely that different fish breed at different times, and not that the same fish breed continuously (Mori 1985).

Factors potentially causing variation in male reproductive success include intra- and interspecific competition for territories, nesting materials, nest sites, matings, and fertilizations.

Intraspecific competition for territories

Competition occurs when animals of the same or different species interfere directly or indirectly with each other's use of resources. Territories are essential resources and may be in short supply in some habitats (Haglund 1981). One way to determine whether space limits the number of nesting males is to remove resident males and observe whether the vacated territories are refilled by other previously non-breeding males. Sometimes newly vacated territories are quickly filled (e.g. Black and Wootton 1970; Haglund 1981), but in Isle Verte tidepools, only 4 per cent of the emptied sites were reoccupied within 24 h (Whoriskey and FitzGerald 1985*a*). Furthermore, there were many seemingly suitable nest sites available in the pools.

A second test for competition for territories is to manipulate densities of adults and evaluate the effect upon male reproductive success. Manipulations showed that the percentage of males nesting, and the average territory size, decreased as fish densities increased in tidepools (Whoriskey and FitzGerald 1987; see also van den Assem 1967; Stanley and Wootton 1986). Many males failed to nest at low densities when seemingly suitable space was available. Furthermore, the median number of eggs obtained per male was unrelated to density, indicating that other unmeasured factors are more important determinants of reproductive success than competition for territories at this site.

Competition for nest sites was demonstrated in the laboratory by Sargent and Gebler (1980), who allowed males to compete for a limited number of flowerpots in wading pools. Males nesting in pots spawned earlier and more often, had a better hatching success, and suffered fewer stolen fertilizations, nest raids, and territorial encounters than males nesting outside the pots. It is still not known how important such competition is in nature.

When males compete for territories, the outcomes of interactions may be influenced by body size, age, male nuptial coloration, and aggressiveness (Rowland page 298; Bakker page 348 this volume). Because these characters are highly intercorrelated they are often difficult to discriminate, particularly in the field.

Interspecific competition for nest sites

Male threespine stickleback may compete for nest sites with other species, although this phenomenon is little studied. Kynard (1979*b*) reported that a decline in the stickleback population of Wapato Lake (Washington, US) followed the introduction of the pumpkinseed sunfish, *Lepomis gibbosus*. He speculated that the threespine stickleback may have lost nesting space to the sunfish. The species may also compete with the mudminnow, *Novumbra hubbsi*, where the two coexist (Hagen *et al.* 1980). In tidepools, male threespine stickleback outcompete blackspotted stickleback, *G. wheatlandi*, for territories and nest sites (FitzGerald and Whoriskey 1985; Gaudreault and FitzGerald 1985). In laboratory studies of competition for territories and nest sites, the threespine stickleback outcompetes blackspotted (Rowland 1983*a*), fourspine, *Apeltes quadracus*, (Rowland 1983*b*), and ninespine stickleback, *Pungitius pungitius* (Ketele and Verheyen 1985).

Male threespine stickleback may also compete with salmonids for nesting sites, although this has not been observed directly. Gaudreault *et al.* (1986) found that breeding ninespine stickleback chased juvenile brook charr, *Salvelinus fontinalis*, from their feeding territories. The charr held feeding territories adjacent to the ninespine territories. Because the threespine stickleback is sympatric with several species of salmonids throughout its range, such interference competition may be common and should be investigated.

Mating success

Mating success in threespine stickleback can be measured by the number of eggs a male obtains. This number depends upon the number and size of the females he can attract to spawn in his nest. Females may prefer to spawn with some males rather than others because of differences in male phenotype, degree of parasitism, and nest contents (Rowland page 313 this volume). For example, females responded more strongly to the larger of two dummy males (Rowland 1989*c*), indicating a possible preference for larger males in the wild. McLennan and McPhail (1989*a,b*) showed that those males that courted most intensively had the brightest nuptial colours, and Milinski and Bakker (1990) suggested that females choose the brightest males because these males are more likely to be free of parasites than duller males. However, breeding males of the Isle Verte population of threespine stickleback harbour few parasites, and there is no association between male nuptial colouration and parasites (FitzGerald *et al.* unpubl. data). Ward and FitzGerald (1987) studied female choice of Isle Verte threespine stickleback. In the laboratory, they found that females 'chose' males based upon their level of aggressiveness rather than their degree of nuptial coloration when territory size and quality were similiar. Both highly aggressive males and very timid males were unsuccessful in getting mates. The most aggressive males broke off courtship prematurely to attack rivals, whereas the most timid males courted less often than other males. This finding supported FitzGerald's (1983) earlier field result that the most aggressive males obtained the fewest eggs.

Female choice may also be affected by nest and nest site characteristics. Although this idea has not been tested directly in the field, in the laboratory, males with concealed nests spawned earlier and more often than those with more exposed nests. They also suffered fewer stolen fertilizations, nest raids, and encounters with rivals (Sargent and Gebler 1980).

In the field, more eggs have been found in nests in deeper water (e.g. Moodie 1972*b*; Wootton 1972*a*; Kynard 1978*a*) and nests in or near cover (e.g. Moodie 1972*b*; Kynard 1978*a*; FitzGerald 1983), suggesting that these nest site characteristics affect female choice. Deep sites may be more attractive because they are less vulnerable to drying and wave action (Wootton 1972*a*, Kynard 1978*a*; Poulin and FitzGerald 1989*a*), and possibly because they are less vulnerable to predators. Concealed sites may be preferred because the young are less vulnerable to predators (e.g. Moodie 1972*b*).

Nest size may also be a factor in determining mating success. In the Isle Verte tidepools, the small one-year-old (1+) males build smaller nests than larger two-year-old (2+) males. Some large 2+ females were unable to enter the nests of the 1+ males (Lachance pers. comm.).

Nest contents evidently influence mating success because once a male obtains an initial clutch, he may become increasingly successful in obtaining additional clutches. Ridley and Rechten (1981) concluded that female threespine stickleback prefer to spawn with males guarding eggs in preference to those with empty nests. Since their study, this finding has been confirmed with other species of fish (e.g. Marconato and Bisazza 1986; Unger and Sargent 1988; Peterson 1989; Sikkel 1989). However, Jamieson and Colgan (1989) criticized the experimental methodology and statistics of Ridley and Rechten, and concluded that it was not the presence of eggs but a more vigorous courtship by males with eggs that caused their greater mating success. In the above studies, females were presented with a choice between an empty nest and one with about one clutch of eggs. However, it is possible that females avoid nests having large numbers of eggs because of egg crowding. Belles-Isles *et al.* (1990) showed that female threespine stickleback prefer to spawn with males having one clutch rather than no eggs, but when females were given a choice between one and either two, three, or four clutches, they did not spawn in the nests with the most eggs, suggesting that egg crowding may occur. Another possibility is that males court less when they have more than one clutch. In nature, it is likely that all or some of these factors interact to determine male mating success.

Stolen fertilizations

Male *G. aculeatus* may steal fertilizations from other males, thereby enhancing their own reproductive success and decreasing that of their rivals. This behaviour involves sneaking into a rival's territory while he is courting, and then following the female through the nest, fertilizing the eggs before the resident. The behaviour occurs both in the laboratory (van den Assem 1967) and in the field (Kynard 1978*a*). At Isle Verte, Rico *et al.* (1992) used DNA fingerprinting to assign paternity of embryos collected from nests in the wild. They found that two of the 17 nests examined contained eggs that had been fertilized by sneaker males. It would be interesting to know whether some males specialize in sneaking, and whether some individuals are more likely to be victimized than others. At Isle Verte the 1+ fish produce a greater volume of sperm, and have more viable sperm, than 2+ fish (de Fraipont pers. comm.). As the 1+ fish are less successful than 2+ fish in obtaining nest sites (Dufresne *et al.* 1990), it is possible that the 1+ fish specialize as sneakers in 'parasitizing' the mating success of the older fish.

Egg stealing

Guardian males can lose eggs to rival males who take them back to their own nests (e.g. Wootton 1971*a*). Egg stealing occurs regularly in the laboratory after sneak spawnings (Jamieson pers. comm.), so perhaps the thief is simply recovering eggs that he had previously fertilized. However it is unclear why he should not let the cuckolded male raise his young. A useful study would be to determine under what circumstances egg stealing occurs in the field. Are stolen eggs raised to fry or are they eaten by the guardian male? Perhaps a male steals eggs as a means of attracting females. Another possibility is that by adding foreign eggs to his own eggs, he decreases the risk that his own eggs will be stolen or eaten (dilution effect). Eggs and fry are vulnerable to a number of invertebrate and vertebrate predators (see below).

Paternal care

After the eggs have been fertilized, the male drives the female from the territory and begins to care for the eggs. The principal activities during the egg stage are oxygenation ('fanning') and aggressive defence. Following hatching the male may guard the fry for up to 2 wk. He attempts to keep the fry together by retrieving strays and spitting them back into the centre of the brood. These activities are time-consuming, energetically costly (e.g. Chellappa *et al.* 1989; FitzGerald *et al.* 1989), and expose the male to risks of predation. As a consequence, the time a male spends with his current brood is probably inversely related to the time and energy he has for subsequent broods. The length of the parental phase depends upon how long it takes the eggs to hatch and how long before the fry are large enough to avoid the male's attempts at retrieval. From the time the male has fertilized eggs until he is ready to begin another cycle, three or more weeks may have elapsed.

Parental investment theory

Predictions generated by parental investment theory (e.g. Trivers 1972; Pianka and Parker 1975; Carlisle 1982) can be effectively tested with paternal male threespine stickleback (e.g. Pressley 1981; Sargent and Gross 1986; Ukegbu and Huntingford 1988; Lachance and FitzGerald 1992). Parental investment (PI) is any material contribution or behaviour performed by a parent that increases the fitness of offspring while decreasing the parent's ability to produce future offspring. The pattern of investment should reflect a trade-off between the relative value of current broods versus future broods. The shape of the trade-off curve should be determined by the reproductive value of the parent (its chances of producing additional offspring) as determined by the age of the parent (Sargent and Gross 1986). Three key predictions of PI theory are that PI should increase with (1) the age of the parent, (2) the number of young in the current brood, and (3) the age of the brood. Each of these will be discussed below.

Two studies (Ukegbu and Huntingford 1988; Lachance and FitzGerald 1992) that have investigated the relationships between parental investment and male age in threespine stickleback failed to demonstrate the expected positive association. Ukegbu and Huntingford (1988) observed a seasonal decline in risk taking (a measure of PI) of parental males from an annual population, rather than the predicted increase, as the males aged over the course of the breeding season. They suggested that this behaviour was a relic from an ancestral condition in which the population was biannual or that the behaviour was a laboratory artefact. However, Lachance and FitzGerald (1992) also failed to detect a positive association between male age and PI. He found no differences in risk taking over the breeding season or between the two age classes of males in the biannual salt-marsh population at Isle Verte, Quebec.

Pressley (1981) provided support for a positive relationship between PI and both brood size and brood age. He demonstrated that males from two lacustrine populations in which males live only 1 yr did accept increased risk in defence of the young as brood size and age increased. In contrast, males in the biannual population in the Isle Verte salt-marsh did not display increased levels of parental care when tending older and larger broods (FitzGerald and van Havre 1985; Lachance and FitzGerald 1992).

Lachance and FitzGerald (1992) argued that the differences in results between Pressley's (1981) study and his could be explained by habitat differences, because environmental conditions are more unpredictable in tidepools than in lakes. In the tidepools, extreme temperatures, low dissolved oxygen levels, and pool desiccation can kill parents or cause them to leave the salt-marsh before the eggs can hatch (FitzGerald *et al.* 1986; Whoriskey *et al.* 1986; Poulin and FitzGerald 1989*a*). In such extreme habitats, the best strategy may be to invest maximally in the current brood regardless of its size or age, because the probability of obtaining a second brood may be low, and high levels of investment may maximize the probability of hatching prior to the onset of adverse conditions. In lacustrine habitats, where risks to the parent and offspring are relatively low (e.g. Pressley 1981), changes in parental investment with changes in the reproductive value of the young may be greater and more readily detectable.

Although high levels of environmental uncertainty and individual variation in behaviour reduce the probability that differences in PI associated with differences in the reproductive value of parents or offspring will be detected (e.g. Pressley 1981), the absence of any detectable associations in the salt-marsh population suggests that PI is not adjusted to these factors in this population. The apparent differences in PI relative to brood value between the salt-marsh population (Lachance and FitzGerald 1992) and two lake populations (Pressley 1981) suggest that different selection pressures have produced adaptive interpopulation variation in this aspect of stickleback reproductive behaviour.

FEMALE REPRODUCTIVE SUCCESS

Little is known about the factors determining individual variation in female reproductive success in nature. Probably the best estimate of female reproductive success that field biologists can achieve is the total number of eggs produced in her lifetime. In the laboratory, the threespine stickleback can spawn a clutch of several hundred eggs every 3–5 d over a two to three month breeding season (Wootton 1976, page 135 this volume). If the breeding season is relatively short, females may spawn only once or twice in their lives (Crivelli and Britton 1987; Bolduc and FitzGerald 1989). When marked females in the Isle Verte tidepools were observed daily, they stayed on the breeding grounds for only about 2 wk (Whoriskey *et al.* 1986; Bolduc and FitzGerald 1989), about enough time to produce a single clutch at this site. In other habitats, females remain on the breeding grounds much longer, potentially producing many more clutches in a season (reviewed in Baker page 171 this volume).

Bolduc and FitzGerald (1989) compared the fecundity of females breeding in tidepools at Isle Verte with that of Isle Verte females breeding in the laboratory. They found that total seasonal egg production was higher in the laboratory, but that there was no significant difference in clutch size between the two groups. In the field, physicochemical factors (water temperature, levels of dissolved oxygen, pH, and salinity) explained a statistically significant but small amount of variation in the various measures of reproduction (i.e. daily egg production, total egg production, clutch size, interspawning interval).

Even temperature, which varies dramatically in the Isle Verte tidepools on a daily basis (McQuinn *et al.* 1983), had little effect on egg production when manipulated in the laboratory to mimic the natural temperature regime (Boulé and FitzGerald 1989). Females kept under fluctuating temperatures produced more eggs per clutch, but had longer interspawning intervals than fish kept at a constant 20 °C. Total seasonal egg production and egg size did not differ between the two groups, although fish kept in fluctuating conditions survived longer and were in better condition than those kept at 20 °C.

Bolduc and FitzGerald (1989) argued that the time spent on the breeding ground *per se* was the principal determinant of the number of spawnings an individual completed. We used some of Bolduc and FitzGerald's unpublished data to compare the patterns of individual variation in reproductive success in the laboratory and field during a single breeding season (Fig. 7.2). The field data were from individuals that survived on the breeding grounds at Isle Verte for periods ranging from 25 to 57 d. In the tidepools 34 per cent of the females never spawned and another 32 per cent produced only one clutch, whereas in the laboratory more than 77 per cent of the fish spawned twice or more over a comparable time period. These differences between laboratory and wild fish are highly significant ($\chi^2 = 67.3$, d.f. = 1, $P < 0.001$). Hence, contrary to the conclusion of Bolduc and FitzGerald (1989), our analysis suggests that field conditions may limit reproductive output in some circumstances, but we have yet to determine the cause.

There are a number of reasons why females in nature are unlikely to achieve the same reproductive output as in the laboratory. Lower average temperatures in the field may reduce egg production if time for breeding

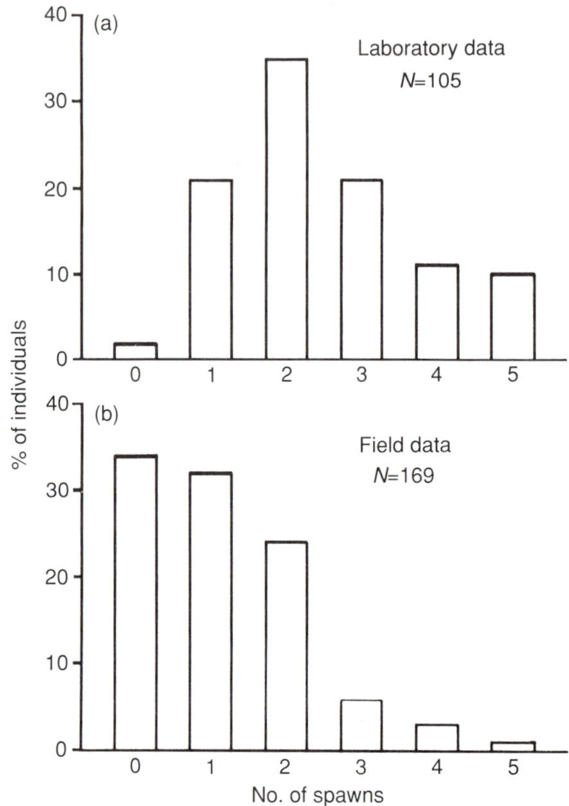

Fig. 7.2 Multiple spawning by *Gasterosteus aculeatus* during a single breeding season in salt-marsh tidepools near Isle Verte, Quebec, Canada. The proportion (per cent) of females that spawned, one, two, three, four, or five times (a) in laboratory pools and (b) in tidepools is shown. See Bolduc and FitzGerald (1989) for details.

is limiting (e.g. Boulé and FitzGerald 1989). Females may be killed by predators, or the presence of predators and (or) parasites may reduce their food intake (Lima and Dill 1990) and hence egg production. Harsh physical conditions may also lead to a suppression of feeding activity (Whoriskey *et al*. 1985) with similar consequences for egg production (Wootton page 133 this volume). Furthermore, the availability of mates may limit spawning opportunities (Ward and FitzGerald 1988).

In contrast to that of males, the reproductive ecology of female threespine stickleback has received little attention, and this is clearly an area for additional research.

ECOLOGY OF EGGS, FRY, AND JUVENILES

Mortality is usually great during the early life history stages of fish, and a small change in the daily or weekly rate of mortality can cause a severe effect so that an entire year class will be small or even non-existent (Wootton 1990). For a complete understanding of the factors affecting the population dynamics and ecology of threespine stickleback, it is essential to know the relative importance of biotic and abiotic factors in determining the survival of the early life history stages.

Although paternal care presumably evolved to minimize the effects of predation upon the eggs and fry and to provide adequate oxygenation to the eggs (Potts 1984), such care may not always be sufficient to ensure a high success. If the male dies before the eggs hatch (e.g. killed by a predator; Whoriskey and FitzGerald 1985*b*), they may all die, either because of predation or owing to suffocation.

Survival of embryos and fry

Abiotic factors

High water temperatures, low levels of dissolved oxygen, and siltation may affect egg survival. In extreme cases, the water bodies may dry out before the eggs hatch or the fry can escape to deeper water (Wootton 1972*a*; Poulin and FitzGerald 1989*a*). Abiotic conditions may directly and indirectly affect the survival and growth of fry. Poulin and FitzGerald (1989*a*) compared the growth rates of threespine stickleback fry in a physically harsh tidepool habitat with those in a more benign river. Growth rates were higher for river than for tidepool fish. This difference may have been caused by differences in food supplies between the sites, although the tidepools are a productive habitat (e.g. Castonguay and FitzGerald 1990). A more likely possibility is that the higher pool temperatures combined with low levels of dissolved oxygen may have added higher physiological cost(s) for the rapidly growing tidepool fry compared with cost(s) encountered in the river.

In tidepools, threespine stickleback first nested near the pool banks and used the middle areas only when near-shore sites were taken (Gaudreault and FitzGerald 1985). FitzGerald *et al.* (1992) showed that nests farthest from the banks were most vulnerable to raids by conspecifics trying to attack the eggs.

It is likely that nest site characteristics are closely coupled with male care in ensuring successful rearing of the brood. Whoriskey and FitzGerald (1985*a*) removed males from their nests and compared the fate of the nest contents during the next 24 h. There was no relationship between the likelihood that a nest would be raided for nests differing in distance from the shore, per cent cover, depth, nearest neighbour distance, species of nearest (stickleback) neighbours, and material used in nest construction. The impact of environmental factors may be more pronounced in some years than in others in highly unstable environments such as tidepools, but less variable in milder and more stable habitats such as rivers and lakes.

Predators and parasites

Paternal care is probably essential for the survival of the offspring. When 58 males were removed from their nests before the young had hatched, only 11 nests survived the first 24 h (Whoriskey and FitzGerald 1985*a*). Frequent causes of adult mortality include predators (Riemchen Chapter 9 this volume), parasites (Wootton 1976), and, in some habitats, suffocation (e.g. Poulin and FitzGerald 1989*a*). However, male defence of young is not perfect, and several predators, including the prickly sculpin, *Cottus asper* (Foster and Ploch 1990), the leech, *Haemopsis marmorata* (Moodie 1972*a*), and possibly large predatory aquatic insects (Benzie 1965), prey on embryos and fry in the nests of males.

It is not known for any population what proportion of the eggs and fry are lost to predators, but intraspecific predation (cannibalism) appears to be a major source of mortality in many populations (Whoriskey and FitzGerald 1985*c*; Hyatt and Ringler 1989*a*,*b*, Foster page 395 this volume). Rival territorial males, non-territorial males, and females are all potential cannibals.

Filial cannibalism (eating of kin) is of particular interest to evolutionary biologists. Rohwer (1978) suggested that guardian males eat some of the eggs in their nests in order to maintain themselves in sufficient condition to raise present and future broods. This might occur because males might not be able to obtain enough other food while caring for their young. However, filial cannibalism in the threespine stickleback occurs in habitats with abundant food supplies (e.g. Whoriskey and FitzGerald 1985*c*) and when males are well fed (Belles-Isles and FitzGerald 1991). The adaptive significance of male filial cannibalism in fish remains to be elucidated (FitzGerald 1992*a*; FitzGerald and Whoriskey 1992).

Females are also frequent egg cannibals in some populations of threespine stickleback (Whoriskey and FitzGerald 1985*c*, 1987; FitzGerald and van Havre 1987; Foster 1988; Foster *et al.* 1988; Ridgway and McPhail 1988; Hyatt and Ringler 1989*a*,*b*). In these populations, females form shoals of a few to several hundred individuals to attack nests guarded by males. Although males can defend nests effectively against solitary males, they cannot do so against large groups (e.g. Foster 1988). Thus, group foraging enables cannibalistic females to overwhelm the defensive abilities of males, a phenomenon similar to that described for several species of coral reef fishes (e.g. Barlow 1974; Robertson *et al.* 1976; Foster 1985*a*,*b*, 1987). As shown for two species of group-foraging coral reef fish (Foster 1985*a*,*b*, 1987), large shoals of threespine sticklebacks may be more effective than smaller shoals. At Isle Verte, large pools contain more adult threespine stickleback than do smaller pools, and the shoals formed in large pools are larger. As expected, males in large pools are less successful at defending their nests from shoals of cannibalistic fish than are males in small pools (FitzGerald *et al.* 1992).

Eggs may be superior to alternative prey because they are more easily digestible or contain some essential nutrient not found in other foods (Belles-Isles and FitzGerald 1991). In support of this possibility, female threespine stickleback prefer conspecific eggs to blackspotted stickleback eggs, even though the same numbers of eggs were offered (FitzGerald 1992*b*).

It would be extremely useful to know the proportion of eggs lost to the different types of cannibalism and to different predators, in order to predict what antipredator tactics the parents and young would be expected to evolve (see Huntingford *et al.* Chapter 10 this volume). It is likely that the relative importance of cannibalism and interspecific predation on the young stages varies among populations and perhaps even within populations.

Competition

Although not systematically investigated, there is a possibility that competition can occur at the egg and (or) early fry stages. Eggs could compete for oxygen in nests having high egg density (Reebs *et al.* 1984). In tidepools at high temperatures, oxygen levels can be below 1 ppm for several hours, and the males are forced to abandon oxygenation of their eggs to seek shelter in cooler waters. In these cases, eggs may suffer sublethal effects of oxygen deprivation. Eggs that have been laid first and covered by subsequent clutches may suffer greater oxygen deprivation than eggs on the perimeter of the egg mass, with consequent effects on egg survival, time to hatching, and morphology of the developing embryo (Ali and Lindsey 1974).

In the Isle Verte threespine stickleback population, bigger females produce larger eggs (Wootton and Whoriskey unpubl. data; but see Baker page 164 this volume). As males may mate with multiple females, their nests can contain eggs of different sizes. If competition for oxygen occurs, there may be longer-term effects on the behaviour, growth, and survival of fry. For example, large eggs may hatch sooner and produce faster-growing fry, which are better able to resist stresses, such as high temperatures and low oxygen conditions (Ware 1975).

Survival of independent juveniles

The juvenile stage is the period in which the immature young are independent of their father. Little is known about the ecology of this stage for threespine stickleback, although it is likely that mortality is due to both biotic and abiotic factors. Juveniles are probably less vulnerable to deteriorating environmental conditions than eggs and fry because of their greater mobility. Juvenile stickleback can probably avoid unfavourable temperatures and salinity conditions as can adults (FitzGerald and Wootton 1986), although this has not been studied in juveniles.

Juveniles are vulnerable to predators and parasites. The predators that feed on juveniles are diverse, including several species of fish, birds, and insects (Reimchen Table 9.1 this volume). Crivelli and Britton (1987) reported that egrets, *Egretta garzetta*, and grey herons, *Ardea cinera*, fed large numbers of juvenile threespine stickleback to their nestlings. Adult conspecifics are among the piscivorous fish that feed on juvenile threespine stickleback, although their impact is difficult to measure (Foster *et al.* 1988). As is the case for juveniles of other prey fishes (e.g. Werner and Gilliam 1984), predators can affect habitat use by juveniles. In particular, small juveniles vulnerable to cannibalism are confined to vegetation until large enough to be invulnerable to this form of predation. In contrast, insects that perch on weeds and ambush juveniles can cause them to leave the vegetation (Foster *et al.* 1988).

Parasites may also affect the growth and survival of juveniles. Although little is known about the effects of endoparasites, some recent experiments have shown that juveniles with a heavy infestation of the blood-sucking fish louse, *Argulus canadensis*, were less likely to survive than uninfected fish (Poulin and FitzGerald 1987, 1988, 1989*b*,*c*). The parasites rest on the pool substrate and attach themselves to fish that pass near by. In pools with parasites, fish swam near the pool surface, whereas in pools without parasites, fish swam near the bottom and stayed near vegetation. As there were no other 'predators' of juveniles in the pools, the observed differences in microhabitat use were probably caused by the parasites. It is unlikely that differences in microhabitat used were caused by differences in the kinds and distributions of food, as all juveniles were feeding upon zooplankton in the open waters of the pools (Poulin and FitzGerald 1989*a*; Castonguay and FitzGerald 1990).

There is little evidence that competition is a major factor affecting the growth or survival of juvenile threespine stickleback. Delbeek and Williams (1987*a*) and Poulin and FitzGerald (1989*a*) examined the food habits of sympatric sticklebacks (threespine, blackspotted, ninespine, and fourspine stickleback) in eastern North America. At both sites, food supplies were abundant and there was strong overlap in diet among juveniles of the species present. Poulin and FitzGerald (1989*a*) manipulated densities of sympatric and allopatric populations of juvenile sticklebacks in tidepools and found no consistent effects of either intraspecific or interspecific competition in these experiments. Although their results must be considered preliminary, as only a limited range of densities was used, the evidence to date indicates that competition for food is unlikely to be important for juvenile sticklebacks in highly productive coastal environments. Similar data have not been collected in other habitats.

CONCLUSIONS

Our goal was to review some of the factors important in determining reproductive success of threespine stickleback upon the breeding grounds.

Many basic questions remain to be answered with respect to those factors. Much work on stickleback biology has been done in the laboratory, and while such studies will continue to be useful in helping to explain variation in reproductive success of the threespine stickleback, it is clear to us that more field studies are needed. The field studies that do exist are relatively few in number, of limited duration, and are usually descriptive. More field experimentation would be valuable. This should be possible in habitats such as tidepools, backwaters of small streams, and in some small lakes. In order to make generalizations about the selective pressures acting to shape morphology and behaviour, researchers must study more than one population of threespine stickleback, as different populations are influenced by different selection pressures. Only by addressing the same questions with the same methodology in different populations will progress be made in determing how a suspected selection pressure has acted to determine morphology and behaviour.

Some basic questions concerning the reproductive ecology of male threespine stickleback that remain to be answered include: How many times does an individual breed over his lifetime? Does an individual that reproduces in his first year of life reduce his chances of reproduction in future years? What proportion of the males actually fertilize eggs? Any attempt to predict individual reproductive success based on phenotype and (or) nest site characteristics will fail if the eggs have been fertilized by a rival male.

For females, field work is needed to determine what proportion of the females contribute to the gene pool. How often do females spawn and how many eggs do they produce over their lifetime? It is likely that the reproductive potential of females differs among populations as a consequence of variation in fish size and longevity, the intensity of competition for food, numbers of predators and parasites, and the severity of the abiotic environment.

More research is needed on early life history stages as very little is known about events affecting the survival of young. The breeding season is only a short part of a threespine stickleback's life, yet we know little about factors affecting the survivorship of adults and their progeny outside the breeding season. Is there any correlation between the number of eggs an individual produces or fertilizes and the number of its progeny surviving to reproduction? If not, the usefulness of much of the current work on stickleback ethology and behavioural ecology for an understanding of the ecology of this species is problematic. Much of the work reviewed in this chapter has attempted to link phenotypic characteristics (e.g. body size, nuptial coloration, territory size) to variation in measures (indices) of reproductive success (e.g. number of spawnings, number of eggs in nests, whether or not fry hatch). Even in these short-term studies, however, our ability to predict individual reproductive success is poor. Why some males are obvious 'winners' and others obvious 'losers' still eludes us in most cases. The importance of nest raiding and group behaviours have received little attention in past studies, and these factors clearly affect individual fitness (Foster page 394 this volume).

ACKNOWLEDGEMENTS

Our research on sticklebacks is supported by the Natural Sciences and Engineering Research Council of Canada (NSERC) and les Fonds pour la Formation de Chercheurs et l'Aide à la Recherche (FCAR, Québec). We thank V. Boulé, H. Guderley, R. J. Wootton and two anonymous referees for comments on various versions of the manuscript.

8

Evolution of foraging behaviour in the threespine stickleback

Paul J.B. Hart and Andrew B. Gill

'. . . the idea that animals are designed is dead, killed by Hume, buried, . . ., by Darwin, but however comprehensively it is disposed of, like the walking dead it haunts us still.'

Ollason (1987)

Through foraging behaviour the stickleback matches its needs to the demands of the habitat. The flexibility of the fish is constrained by its morphology and physiology, which reflect the history of selection through which generations of ancestors have passed. In this chapter we consider the threespine stickleback as a decision-making system buffeted by the features of past and present environments. We first describe the structure of the habitats in which the fish live, as this will determine the context in which selection has taken place. Next we consider the morphological attributes of the fish, as these determine the constraints within which decisions must be taken. The principal part of our discussion reviews what is known about behaviours such as deciding when to feed, searching for prey, exploiting patchy resources, and selecting a diet. Development of behaviour is considered where appropriate.

The chapter is written within the framework of the 'adaptationist programme' (Gould and Lewontin 1979; Mayr 1983), which assumes that in some sense natural selection has led to the 'design' of organisms that solve the problems of their life in an optimal way. We realize that the 'design' paradigm is not universally accepted, hence the quotation from Ollason (1987) at the beginning, and we are fully aware of its shortcomings (Hart 1989). The paradigm has had a powerful influence on the organization of diverse data on foraging and we hope that our review may point towards a reassessment of the status of foraging theory.

Throughout the chapter we regard the fish as a problem-solving system that has been moulded by natural selection to function in the appropriate environment. First we consider the structure of that environment; next we discuss the apparatus the fish has with which it can behave, as this constrains what it can do. The bulk of the chapter then discusses what is known about the way the fish behaves, together with results which help us understand why the fish behave as they do from a functional viewpoint. Whenever the word 'stickleback' is used alone it means *Gasterosteus aculeatus*.

THE FORAGING ENVIRONMENT

The structure of a fish's habitat sets the context for its feeding decisions, determining features such as encounter rates, ease of detection, and search tactics. An analogous situation is the way in which the properties of water determine the body form of fish (Webb 1982), although the feedback between behaviour and environment leads to a dynamic relationship in ecological time.

Threespine stickleback are found in both running and still freshwater habitats and in the sea (see Bell and Foster page 4 this volume). The most important division for the fish's feeding habits is between the pelagic (or limnetic) and littoral (or benthic) habitats: each of these presents the fish with particular feeding problems. Even within a particular habitat, micro-habitat use is dynamic. For example, *G. aculeatus* in salt-marsh pools are found nearer the bottom when feeding but nearer the surface when inactive (Walsh and FitzGerald 1984).

The differences in demands made by the littoral and limnetic habitats are so great that the threespine stickleback has adapted specially to them (Larson 1976; Bentzen and McPhail 1984; Lavin and McPhail 1986; Ibrahim and Huntingford 1988; McPhail page 418 this volume). In the Cowichan Lake drainage system, British Columbia, Canada, Lavin and McPhail (1986) found three allopatric morphotypes, benthic, limnetic, and intermediate. Benthic forms are morphologically specialized to feed at the lake bottom. This form is typical of small lakes characterized by an extensive, structurally complex, shallow littoral zone in which they feed. In contrast, the limnetic form is specialized for feeding on plankton and is typically found in large, deep lakes. Benthic stickleback can ingest larger prey than can limnetics and can handle them more rapidly. The limnetic and intermediate types are more adept at taking prey from the water column, a skill that is a function of their greater gill raker density and the structure of the mouth (see below and McPhail, page 418 this volume).

The two habitats differ greatly in structural complexity. Within the littoral, threespine stickleback still use the water column to search in (Ibrahim and Huntingford 1989*a*), but they must search for food through a maze of plants and over the uneven bottom. It is likely that encounter rate between stickleback and their prey is strongly influenced by the density and structural complexity of the vegetation through which they must search. Equally, when feeding on the bottom, they must search a complex surface, and encounter rates with prey are probably affected both by the substratum complexity and by the need to sort prey from detritus. For example, bluegill sunfish, *Lepomis macrochirus*, searching for prey at similar densities in open water, on bare sediment, and in vegetation, had encounter rates of 0.77 prey s^{-1}, 0.01 prey s^{-1} and 0.02 prey s^{-1}, respectively (Mittelbach 1981*a*).

The patchy distribution of invertebrates makes the littoral habitat even more complex (Downing 1986). As an example of this, gastropods in an English pond showed preferences for particular plant species which themselves were patchily distributed (Lodge 1985). For example, *Planorbis vortex* was associated with graminoid emergent macrophytes, especially *Glyceria maxima* (reed sweet-grass), and *Lymnea pereger* was associated with submerged macrophytes, in particular *Elodea canadensis* (Canadian pond weed). Even within a plant, the distribution of invertebrates is not uniform. Using artificial plants, Macan and Kitching (1972) found that *Leptophlebia*, *Cloeon*, and *Gammarus*, all eaten by stickleback, were much more abundant on long 'plants', tending to prefer the upper parts of long leaves. Highly branched plants have more invertebrates, mainly as a result of the increased surface area (Voights 1976; Gerrish and Bristow 1979).

Threespine stickleback feed only in the light (Wootton 1984*a*). For many fish, prey movement is the main means of prey detection (Bone and Marshall 1982). When *Coenagrion puella* (damselfly) larvae had perches to rest on, they were attacked less by threespine stickleback (Convey 1988). With no perches, the damselfly larvae moved more, which made them more visible to the fish. The response of fish to increased habitat complexity varies with prey type. For example, increasing structural complexity of the littoral caused largemouth bass, *Micropterus salmoides*, to switch from a search to an ambush hunting strategy (Savino and Stein 1989). Savino and Stein (1989) also found that fathead minnows, *Pimephales promelas*, were more vulnerable than bluegill sunfish to both pike, *Esox lucius*, and largemouth bass at high stem densities.

The open, three-dimensional nature of the limnetic habitat means that the relation between predator and prey is less constrained by physical structure than in the littoral, and is therefore more changeable on a short time scale (George 1981; Jakobsen and Johnsen 1987). The copepod *Acartia hudsonica* changed its vertical distribution and migration behaviour in response to predation from threespine stickleback (Bollens and Frost 1989). The turbidity of water is also an important factor in determining the encounter rate with prey (Werner and Hall 1974; Eggers 1982). In the limnetic region, encounter rates will be a function of the reactive field of the fish, prey size, shape and colour, and water clarity (Eggers 1982). Many of these features will change with season.

The distribution of limnetic zooplankton is patchy at several different scales varying from centimetres to hundreds of kilometres (George 1981). George divides the causes of patchiness into three groups: behavioural, advective, and reproductive. Behavioural processes operate at the level of the individual animal and produce small-scale distributional changes (< 1cm–1 m). Advective processes are responsible for medium-scale patchiness (1 m–1 km), while large-scale patchiness (1 km–100 km) can be caused by reproductive differences between populations subject to different food or temperature regimes. From the point of view of a foraging stickleback in a freshwater lake, small-scale aggregations are likely to be most important. In a sheltered bay in Windermere, UK, Colebrook (1960) found that swarms of potential stickleback prey, *Daphnia hyalina* and *Diaptomus gracilis*, were about 1 m in diameter and confined to the top 10 cm. Prey animals were found between patches, but the swarm was denser by about two orders of magnitude. The swarms contained a high proportion of juvenile *Daphnia*, which suggested some form of behavioural cause for aggregation. The location of these small-scale patches in a lake is governed by medium-scale advective processes (George 1981). For example, George and Edwards (1976) showed that wind-induced surface transport caused a 20-fold difference in the density of *Bosmina longirostris* (which are prey to stickleback) between the windward and leeward ends of Eglwys Nynydd, a small lake in Wales, UK.

THE FORAGING APPARATUS

The outward appearance of a threespine stickleback is well known (Wootton 1976; Bell and Foster page 5 this volume). Here we concentrate on those parts of the fish which may determine its feeding performance, namely mouth and gill raker morphology, body form in relation to foraging movements, and eye position and size.

Threespine stickleback have a terminal mouth with protrusible, toothed jaws (Anker 1978; Bowne page 31 this volume). The exact position of the mouth can vary considerably, often in association with differences in the shape of the body. Individuals that live in running water and that are more streamlined can have the mouth nearer the ventral surface than those living in still water (Taylor and McPhail 1986; Balbi 1990). This difference of the threespine stickleback may be related to the feeding modes adopted in still and moving water (Balbi 1990). Fish from a pond up-ended when feeding on benthic prey so that the long axis of the body made a right angle to the bottom (Tugendhat 1960). Fish from a river never took up the 90 ° angle, keeping below 45 ° (Balbi 1990). A vertical stance in running water would be hard to maintain, and having the mouth lower on the head makes it easier for the fish to take prey when tilted at a low angle. Fish from the river also had a marked keel on the lower jaw. In tanks these fish were seen to move along the bottom, body held at 45 °, with the lower jaw resting on the bottom as if the fish were attempting to disturb and dig up benthic organisms.

The mouth and gill raker morphology of the threespine stickleback causes them to feed more effectively on plankton than do the other species of stickleback (*G. wheatlandi*, *Pungitius pungitius*, and *Apeltes quadracus*)

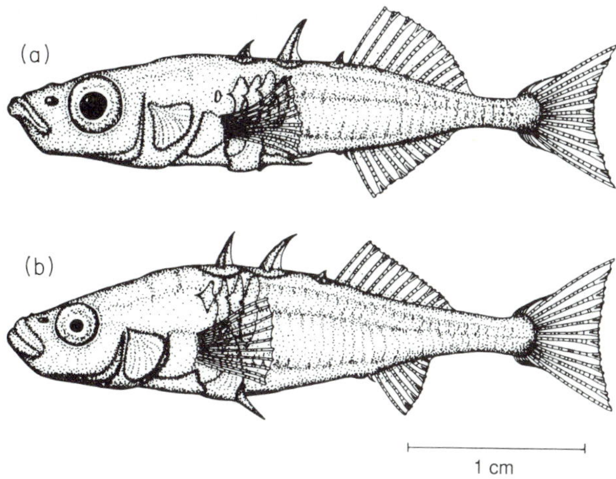

Fig. 8.1 Differences in body form of threespine stickleback with consequences for feeding. The Enos Lake species pair, (a) limnetic and (b) benthic species (see McPhail page 418 this volume).

with which they are sympatric in salt-marsh tidepools (Delbeek and Williams 1987*b*). They tend to have longer and more numerous gill rakers, enabling them to retain small prey more efficiently than do the other species. Although these trophic differences could be the product of competition for food in tidepools, they may well reflect ecological interactions outside the breeding season when the threespine stickleback is offshore.

The trophic morphology of the lacustrine benthic and limnetic forms also reflects differences in prey type (McPhail page 418 this volume) (Fig. 8.1). Compared with benthics, which prey on relatively large benthic invertebrates, the planktivorous limnetic form has a long snout, a narrow mouth, and more numerous gill rakers. This holds both for allopatric benthic and limnetic populations and for the benthic and limnetic species pairs. McPhail (page 418 this volume) discusses the morphological correlates of trophic differentiation of lacustrine stickleback in detail.

The style of stickleback swimming is usually described as labriform (Lindsey 1978; Taylor and McPhail 1986; Delbeek and Williams 1987*b*). They use their pectoral fins for most of their locomotion, only using the whole body when making a fast start (Webb 1978). The threespine stickleback in tidal salt-marsh pools had larger pectoral and caudal fin areas and a shorter and wider caudal peduncle than other species of stickleback (*G. wheatlandi*, *Pungitius pungitius*, and *Apeltes quadracus*) in the same habitat. These differences were interpreted as adapting the species to the more pelagic lifestyle it assumes in the salt-marsh habitat (Delbeek and Williams 1987*b*).

The diverse habitats in which threespine stickleback are found are associated with intraspecific differences in swimming performance and body form (Fig. 8.1). Anadromous individuals from the Salmon River, British Columbia, Canada, became fatigued less easily when swimming at 5 body lengths (BL) s^{-1} than did fish resident in the same stream (Taylor and McPhail 1986). They also had larger pectoral fins and less robust bodies than freshwater residents, which were better at burst swimming, being able to cover 2.67 cm as opposed to 1.98 cm during the first two stages of a fast start (Webb 1978). The differences discovered were correlated with the different demands made on the two life history forms. The anadromous individuals make long migrations and are mainly pelagic when overwintering in the sea. Also, in the pelagic zone, foraging is likely to involve searching behaviour with more swimming (O'Brien *et al.* 1990). Stickleback permanently resident in the river are mostly found in weedy zones with low flow-rates where they must avoid predators and catch prey by stalking and darting (Taylor and McPhail 1986).

The eyes of *G. aculeatus* are large, being about 25 per cent of head length (Scott and Crossman 1973), with the retinal area forming 3.5 per cent of the body surface (Beukema 1968). Limnetic stickleback tend to have larger eyes than benthic fish (Bentzen and McPhail 1984; McPhail 1984) and larger fish eyes have been shown to have greater powers of discrimination (Hairston *et al.* 1982). In contrast, the surface area of the olfactory organ is 0.5 per cent of the body surface in threespine stickleback, indicating that olfaction does not play a large part in the fish's life. The rods and cones in fish eyes are arranged in a regular mosaic. Each unit of the mosaic contains a combination of cones with maximal absorption at wavelengths of 452, 604 and 529 nm (Lythgoe 1979). This combination gives the fish a good response to red, the colour which best retains its distinctness in the green water of freshwater habitats (Lythgoe 1979). During feeding, *G. aculeatus* has a preference for red prey (Ohguchi 1981), the colour being more important than movement when there is a choice between the two (Ibrahim and Huntingford 1989*b*). Some evidence indicates that this difference in response may be because red is more conspicuous rather than preferred (Ohguchi 1981). Several prey types, such as chironomids and copepods, often have red pigment which must make them particularly attractive to the fish. It would be interesting to know whether the retinae of the benthic morphotype as defined by Bentzen and McPhail (1984) has a different balance of pigments from the limnetic type. Presumably foraging for benthic prey requires competence in a light regime that is different from that in the pelagic realm.

Predation can also impose constraints on feeding behaviour. Unlike many fish, *G. aculeatus* has armour which reduces the risk of predation by many vertebrates (Hoogland *et al.* 1957; Reimchen 1983, Chapter 9 this volume). As the armour is not completely effective, antipredator behaviour is also an important first line of defence (Huntingford *et al.* Chapter 10 this volume). Antipredator behaviour, especially avoidance of predators, can affect foraging opportunities for prey species. For example, small bluegill sunfish are often forced by the threat of predation to stay in the weedy margins of lakes where foraging opportunities are not as good as in the pelagic habitat (Werner *et al.* 1983). Similarly juvenile threespine stickleback also tend to stay near weed to avoid being eaten by bigger conspecifics (Foster *et al.* 1988). Pike take *G. aculeatus* only when other, spineless prey are not available (Hoogland *et al.* 1957). Possibly because of their armour, threespine stickleback forage longer in good habitats than do ninespine stickleback, *Pungitius pungitius*, and the banded killifish, *Fundulus diaphanus* (McLean and Godin 1989). In an experimental aquarium, threespine stickleback visited a patch of weed significantly less often than did ninespine stickleback, irrespective of the availability of food in the weed or in the open water (Hart and Hamrin unpubl. data). This difference supports the notion that threespine are bolder because of their heavier armour.

FORAGING BEHAVIOUR

We have described aspects of the threespine stickleback's foraging environment and have given details of the apparatus that the animal has available to gather food. That feeding is important to fitness is shown by the way in which morphology has responded to different selection pressures found in different habitats. Such change is only possible between generations, and it is within a generation that behavioural responses become important. Through behaviour the animal can modify its response to changing conditions. It may do this by bringing into play a series of simple preprogrammed rules which allow adaptation to a well-defined set of different conditions, or the fish may learn about changing events and adjust. We have followed Stephens and Krebs (1986) and split the behavioural problems that a fish faces into four groups: deciding whether or not to feed, searching for prey, dealing with patchily distributed food, and selecting a diet from the range of prey types available. For each topic we first describe what is known about the phenomenon for stickleback in the wild, then discuss how the behaviour has been explained in an evolutionary context. We also discuss how hunger, breeding state, and the threat of predation influence the behaviour. Our main aim is to identify the general principles, such as energy maximization, that might underlie the stickleback's behaviour, and the special features of its lifestyle, such as nest building by males, that may have led to unique adaptations.

A word at this point about what must be the *magnum opus* of stickleback feeding studies, the long paper of Beukema (1968), which covers topics that we deal with in the four different sections. Beukema's paper describes experiments which examined daily food intake, the role of hunger in determining prey capture rate, stickleback searching behaviour and how it is affected by learning, visual acuity, the influence of fish size on capture rate, and aspects of prey preference. Beukema was mainly interested in the ecological role of predators and describes his results in terms of a 'risk index', R, which is defined as a proportionality factor relating numbers of prey eaten (N) to their density (D) and the time spent feeding (t), $N = RDt$. Beukema states that '. . . it was precisely the aim of this investigation to define quantitatively the influences that govern the value of the risk index for the combination of predator and prey studied.' The paper was written before the metaphor of animals as cost–benefit analysers had taken firm hold. The largest part of the 126-page paper is based on data gathered from only eight fish, five males and three females, which ranged in size from 1.6 to 3.0 g. The study cannot be ignored because it often contains the only data there are on some aspect of stickleback foraging, but its limitations must be recognized.

The decision to feed

Annual and diurnal variation in food intake

The temperature cycle of stickleback habitats causes an annual cycle of food intake (Wootton page 116 this volume). As with all temperate fish, stickleback eat less in the winter than in the summer; a typical pattern is shown in Fig. 8.2 for a population of stickleback from Llyn Frongoch, Wales, UK (Allen and Wootton 1983). Food consumption was estimated from faeces produced by fish caught 4 h after sunrise and left for 24 h in tanks set in the lake margin. The highest consumption occurs between May and August with a peak in June. Beukema (1968) found that at 11–12 °C stickleback ate between 33 and 60 per cent of what they ate at 18–20 °C. At the lower temperature the fish's stomach did not empty overnight, and as we shall see, stomach fullness is probably a controlling factor for the switching on of feeding behaviour. These patterns are no different from those in the majority of temperate fish species, in which temperature is the main influence on intake (Wootton 1990).

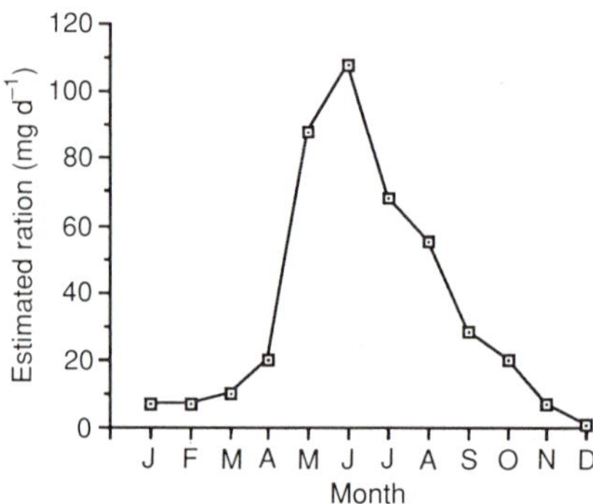

Fig. 8.2 The variation in food intake by threespine stickleback from Llyn Frongoch, Wales, UK (redrawn from Allen and Wootton 1983).

The daily pattern of intake has been studied in detail in the laboratory (Tugendhat 1960; Beukema 1964, 1968) but less often in the field. In the laboratory, Beukema (1968) found that threespine stickleback fed at a more uniform rate if a period of deprivation was ended with a morning feed: stickleback which ended a period of abstinence with an afternoon feeding session fed rapidly in the first hour but the rate then tailed off. Beukema also found, as did Tugendhat (1960), that stickleback that had not fed for 24 h fed at their highest rate at the beginning of the day. Fish in Great Central Lake, Vancouver Island, British Columbia, Canada, had full stomachs just after dawn and just before dusk (Manzer 1976), which led to a twice-daily alternation between feeding and non-feeding. Female *G. aculeatus* in a Canadian salt-marsh fed almost exclusively in the morning in June, whilst nesting males fed at any time of day (Worgan and FitzGerald 1981*a*). The males also fed infrequently. A later study in the same habitat (Walsh and FitzGerald 1984) found no clear-cut periodicity in feeding activity as indicated by gut fullness. Samples for the second study were taken on wet, overcast days with low air temperatures (2 °C) and a strong wind. There was some evidence that degree of cloudiness and wind strength influenced feeding behaviour. Snyder (1984) found that breeding males in a stream fed less frequently than females, but non-breeding males continued to feed through the breeding season.

The influence of hunger

Thirty-year-old studies of stickleback feeding treat the animal as an isolated individual detached not only from conspecifics but also from any other task. Both Tugendhat (1960) and Beukema (1964, 1968) went to great lengths to study the influence of deprivation and satiation on food intake rate. As might be expected, stickleback that had been starved for 72 h fed faster when given access to food than did those starved for 24 h, although at the end of a 40 min or 1 h feeding session all fish had consumed about the same number of prey. Tugendhat (1960) found that the number of completed prey captures declined very rapidly in the first few minutes of a feeding session but then rose again to a lower peak before declining slowly towards the end of a 1 h period. She proposed that the shape of the curve was indicative of the way in which stretch receptors in the stomach changed their response as the stomach filled. The controlling role of stretch receptors is also discussed by Beukema (1964), although he suggests in addition that longer-term deficits for energy and nutrients, as expressed by the term ‘systemic need’, have an influence, as illustrated by the higher feeding rate seen in fish that have been deprived of food for a long time (Colgan 1986). The way in which systemic need and stomach fullness interact is shown in Fig 8.3.

With no other distractions, a feeding stickleback does not suddenly stop catching prey. The rate of prey capture falls off exponentially, but the fish continues to initiate prey capture at a steady rate, even when nearly full

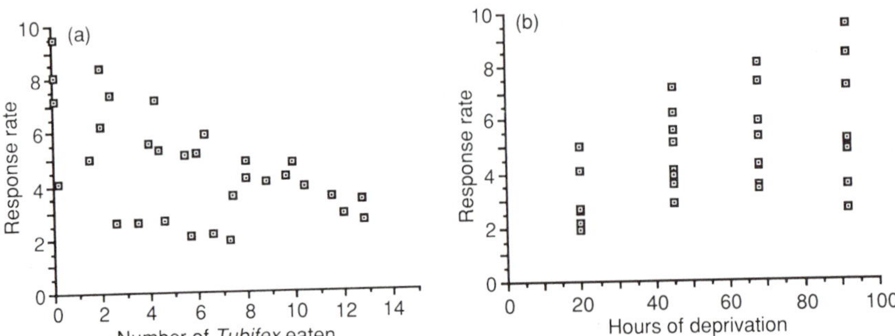

Fig. 8.3 The rate at which stickleback attack inedible objects in the experimental arena as a function of (a) the number of *Tubifex* already eaten, and (b) the time in hours of food deprivation. A stepwise regression shows that *Tubifex* eaten (T) and deprivation (D) explain significant (F-test, $P = 69.10$, d.f. = 2, 29 $P < 0.001$) amounts of the variation in response rate (R). The multiple regression is $R = 4.04 - 0.39(\pm 0.04)T + 0.05(\pm 0.01)D$. The figure is drawn and the analysis carried out from data extracted from Fig. 6 of Beukema (1968).

(Tugendhat 1960). The difference between hungry and replete fish is that the latter complete fewer feeding attempts and spend more time moving away from the area where food is available. Also each move away is accompanied by so-called 'comfort movements', such as raising the spines and S-bend, which occur as the fish stops and turns back to the feeding area. These would imply that the fish is in a state of indecision, and the comfort movements are a type of displacement activity (Tugendhat 1960).

The influence of predation risk on a hungry stickleback

In nature a stickleback is unlikely to be able to feed uninterrupted for 40 min. It will have to watch for predators, and during the breeding season, substantial time will have to be devoted by males to nest building, courting, and parental care. During some of these activities, fish will give feeding a low priority, and the interesting question, which can be tackled best from an evolutionary perspective, is how does the animal schedule its behaviours so as to maximize fitness? Heller and Milinski (1979), Milinski (1984*c*), and Godin and Sproul (1988) have studied the way in which a fish balances the risk of starvation against the risk of predation.

Although fishes suffer from food deprivation less rapidly than do small mammals and birds, small planktivores, such as the bleak, *Alburnus alburnus*, were found by lvlev (1961) to be less resistant to starvation than were piscivores such as pike. A group of 12 threespine stickleback subjected to starvation suffered a 42 per cent mortality after 35 d (Walkey and Meakins 1970). Heller and Milinski (1979) observed that a hungry stickleback preferred to attack the densest region of a swarm of *Daphnia*, but as feeding motivation waned, the fish increasingly favoured the less dense regions. Heller and Milinski (1979) argued that when very hungry, the stickleback chose a region of the prey swarm that would give the highest feeding rate, so allowing it to rapidly reduce its hunger, and therefore its risk of starvation. In the light of Walkey and Meakins' (1970) study, the risk of starvation may not have been as appreciable as originally thought. Feeding at a high rate in the dense part of the swarm carries with it the cost of not being able to watch for predators so effectively because the stickleback's attention is almost totally engaged in fixating and grasping prey (Ohguchi 1981). As the fish's hunger falls, the benefit of a high intake rate is outweighed by the cost of poor vigilance, and the fish changes to the lower-density regions where it is easier to watch for predators.

Conventional optimal diet models (Stephens and Krebs 1986, Chapter 2) do not account for this change in costs with increasing satiation. Heller and Milinski (1979) devised a model, following Sibly and McFarland (1976), which predicted the optimal feeding rate for the encountered food density and the hunger state of the fish. A principal prediction of the model was that capture rate should decline exponentially with time and increased satiation. The data of Tugendhat (1960), Beukema (1964, 1968), and Heller and Milinski (1979) were in agreement with the prediction. Other more detailed predictions were also confirmed. The general rule, expressed simply, is that the stickleback should take the most expensive prey only when very hungry, and then become more selective for the most profitable prey when hunger is low. This rule will only hold so long as the availability of the most profitable prey is too low to allow the fish to achieve the desired high feeding rate on them alone. This result was confirmed by Hart and Ison (1991) for stickleback feeding on a sequence of the isopod *Asellus aquaticus* ranging in length from 3 to 9 mm. As fish became satiated, they began to reject 8 and 9 mm prey that had long handling times, and concentrated instead on prey between 3 and 6 mm, which yielded greater energy per unit handling time.

In a later study Milinski (1984*c*) showed that when feeding on a dense prey aggregation, hungry stickleback more often failed to detect a model kingfisher, *Alcedo atthis*, than when they were feeding on a low density of prey. They also failed more often to see the model bird when feeding at a high rate on a dense swarm than when feeding at a low rate. This work shows that confusion has a detectable cost to the fish. Milinski (1984*c*) proposed that the limitations on the nervous system of the fish prevented it from processing more than a certain amount of information. This idea has not subsequently been examined further.

Stickleback are often infected with the tapeworm (Cestoda) *Schistocephalus solidus*, which fills the fish's body cavity and makes huge demands on its physiology. As evidence of this demand, Walkey and Meakins (1970) found that under laboratory conditions all 12 threespine stickleback infected with *S. solidus* were dead after 31 d of starvation. Fifty per cent had died after only 15 d. In contrast, only 42 per cent of unparasitized stickleback were dead at the end of 35 d of starvation. The parasitized condition was used by Milinski (1985) to understand better the interaction between hunger and threat of predation. *Tubifex* worms were placed at varying distances from a live fish predator in an aquarium, and parasitized and non-parasitized threespine stickleback were allowed to forage separately. Uninfected fish preferred to take *Tubifex* worms from stations as far away from the predator as possible, and they fed at a lower rate than they did in the predator's absence. Infected fish did not react to the predator, feeding as often at stations close to the predator as when it was absent, and they fed at the same rate under both treatments. Giles (1983*b*) and Godin and Sproul (1988) used the same principle but had a model heron as predator. They found that, after being frightened by the 'heron', heavily parasitized stickleback fled shorter distances, remained motionless and cryptic for shorter times, returned sooner to forage in a patch near the predator, and remained active longer in the patch. All these times were positively correlated with parasite load. A less convincing result was that parasitized fish stayed longer on a food patch and consumed more food under the threat of predation when food density was higher. Godin and Sproule (1988) argued that this result showed that fish were prepared to take a bigger risk when food resources were abundant.

All the analyses we have discussed have assumed that an increase in hunger is the main cause of decreases in risk sensitivity. It is of course possible that there is also direct manipulation of the stickleback's behaviour through other physiological pathways (reviewed in Milinski 1990 and Huntingford *et al.* page 281 this volume).

Feeding or reproduction?

During the breeding season, males defend a territory, build nests, court, and participate in parental care. At the same time, they must find time to feed and to be vigilant against predators. We now look at an experiment by Noakes (1986) which studied how male threespine stickleback allocate time to feeding during the reproductive phase.

Mature male *G. aculeatus* were kept in a long aquarium divided into three sections (Noakes 1986). The fish was then encouraged to build a nest in one end section. At the same time the animal had to move through the central compartment to the opposite end section to find food. Once the fish were trained they were allowed to choose, during a limited time each day, between carrying out nest-directed activities or feeding. As the fish moved from one end compartment to the other, passing through the central area, they were trapped for varying lengths of time. Their choice of food or nest after trapping was recorded as being indicative of their dominant motivational state. In some treatments, fish had been deprived of food, whilst in others the male was shown another male or a gravid female in the vicinity of the nest. Males were more likely to be nest dominant if they were satiated or had just been shown a male in the nest compartment. Hungry fish tended to be food dominant. In general, Noakes (1986) judged that nest-related activities were dominant and feeding was subdominant.

The male's choice of activity was also influenced by how long it was trapped in the central compartment. A long sojourn usually meant that the fish returned to nest-related activities. This gave a clue to the proximate mechanism that allows the animal to schedule its activities in an optimal way. Two ways in which an animal could do this have been proposed (McFarland 1974; Colgan 1986): 'competition', in which changes in the causal factors for a second activity lead to the ongoing behaviour being displaced, and 'time-sharing', in which a current dominant activity terminates itself, so disinhibiting a second subdominant behaviour. Noakes (1986) argued that if time-sharing is the mechanism, then trapping the fish for a short time whilst it is carrying out the subdominant behaviour will lead to it returning to the same activity when released from the trap. Trapping the fish carrying out the subdominant behaviour for a longer period will mean that the causal factors for the dominant behaviour will rise above the critical level during the period of trapping, so that when the fish is released it will return to the dominant behaviour. Noakes (1986) concluded that time-sharing is the proximate mechanism that stickleback use to schedule their behaviour. As with all proximate mechanisms, the assumption is that time-sharing is a simple rule that allows the fish to approach the optimal scheduling of behaviour.

There is still much room for future work on how threespine stickleback organize their time. Although Noakes (1986) concluded that time-sharing is the mechanism being used, his evidence is not convincing. He determined the dominance status of feeding and nest-related activities by counting the number of returns to a particular compartment resulting from six interruptions. In only 29 out of 76 cases could dominance be ascribed, a result leading to doubts about the generality of the final conclusion.

Searching for prey

Stickleback searching

The feeding movements of threespine stickleback searching for benthic *Tubifex* are described by Tugendhat (1960). She wrote, 'Sticklebacks feeding on ground-living *Tubifex* worms approach the ground and tilt their bodies from the horizontal swimming position to fixate the prey that is partly submerged in the sand. This fixation may be followed by a rapid sideways twist and snapping movement with which the fish grasp their prey. However, the fish may remain in the tilted position and turn or move slightly to poise over a new prey. Also, they may return to the horizontal position and after swimming a short distance again tilt and fixate on a new prey.' The enclosure used by Tugendhat (1960) was small so that the fish did not have to move far to find prey, which themselves were at high density. Beukema (1968) had his fish moving through a 187 dm^2 aquarium divided into 18 interconnected cells. He wrote, 'Swimming in the stickleback generally does not take the form of a single smooth movement covering a long distance, but consists of a series of consecutive small displacements, carried out in a leisurely fashion. . . . Swimming is called 'searching' when the fish swims in a peculiar jerking way, as if ever again turning its attention to new objects, which, however, are imperceptible to the observer. . . .' We can find no description in the literature of how *G. aculeatus* moves in nature when feeding on benthic prey, but personal observation confirms the patterns described by Tugendhat (1960) and Beukema (1968).

The diversity of feeding habits in stickleback makes it unlikely that one set of rules will have evolved for prey-searching. Each situation will make different demands on the fish. In any case, searching behaviour has not been modelled to the same degree as has prey choice or patch choice (Hart 1986; Stephens and Krebs 1986; but see O'Brien *et al.* 1990). This is not the place to describe the generalities of optimal searching, but in what follows we will account for what is known about the search behaviour of *G. aculeatus*.

Experience and searching

There are two elements to searching: choosing the path that maximizes some currency and choosing the speed that achieves the same goal (Hart 1986). Stickleback have been used to study the first of these, but not the second.

The eight stickleback used by Beukema (1968) became very familiar with the hexagonal world in which they were housed. Within the aquarium, the 18 interconnecting hexagons were arranged as 12 outer and 6 inner compartments. The main prey used, *Tubifex* again, were always placed in one of the outer compartments. The encounter rate per cell was calculated for the fish and compared with the rate of 0.04 encounters per cell expected if cells were visited at random. Also calculated was the path that would give the fish the highest encounter rate, which was 0.20 encounters per cell. The observed encounter rate was between the two, ranging from 0.079 to 0.090, showing that the stickleback were not moving at random, yet they did not reach the optimum. For all fish, encounter rate increased with the number of trials performed, showing that they were learning. The principal variable changing with experience was the rate of turning, although a fish that had just discovered and eaten a prey increased its rate of turning (see below). Encounter rate was independent of hunger.

Stickleback searching for buried *Tubifex* worms distributed sparsely over a 9 × 27 cm feeding area changed their behaviour in response to discovering a prey (Thomas 1974). If the detected prey was eaten, the fish increased its rate of turning immediately afterwards, showing what has been called 'area-restricted searching'. If the fish decided to reject the prey, then as it left the discovery area it decreased its rate of turning, so travelling quickly away from the point of rejection. This was called 'area-avoiding searching' by Thomas (1974). In functional terms, Thomas (1974) proposed that increased turning after prey ingestion would lead to increased prey discovery, assuming that prey are clumped in nature. That this is so has been shown earlier, but there is no experimental evidence to show that a fish that turns more after prey ingestion gains more net energy than one that does not.

Little work has been done on how fast a stickleback moves through its habitat when searching for prey. Ware (1978) proposed that a pelagic fish should search for prey at a speed that maximized its growth rate. He calculated that a 40 cm fish at 15 °C should move at about 2.9 body lengths (BL) s^{-1}. More generally, optimal speed was predicted to be a function of fish length and prey density. The stickleback in Beukema's (1968) system averaged about 1 BL s^{-1}, which is roughly as predicted by Ware's model. The prey in Beukema's system were at a very low density (2.68×10^{-6}) cal cm^{-3}, and Ware's model predicts that swimming speeds should decrease with prey density from a peak at about 7.5×10^{-4} cal cm^{-3}. Of course, threespine stickleback were not catching pelagic prey in Beukema's experiments, and the fish may have to spend more time over each individual benthic item than when catching pelagic prey. This rather crude analysis emphasizes that foraging speeds in *G. aculeatus* are unknown.

Exploiting patches of food

Feeding in patches

The section on stickleback habitats provided evidence for the clumping of potential food organisms, in both the littoral and limnetic zones. Exactly how fish exploit these patches has not been studied in the field. Nearly all published work examining how stickleback exploit patchy resources is based on laboratory experiments that start from the assumption that *G. aculeatus* feed on prey concentrated into patches. Some field evidence indirectly supports the assumption. For example, Manzer (1976) found that in October the guts of threespine stickleback from Great Central Lake, British Columbia, Canada, were packed with *Bosmina* and *Holopedium*. Sampling through the day showed that *Holopedium* was a particularly abundant dietary item in fish collected between 07.00 and 10.00 h. Such concentrations of one prey type in fish gathered over a short interval of time are consistent with the idea that fish feed on large aggregations. Similarly, Jakobsen and Johnsen (1987) found that in the summer, female threespine stickleback in Lake Kvernavann, Norway, were mostly full of *Daphnia longispina* and *Bosmina longispina*. One of the few observational studies reporting what fish do in the natural habitat (Foster *et al.* 1988) showed that small threespine stickleback spend much time near weed, snapping at pelagic plankton. In salt-marsh pools, active *G. aculeatus* spend most time on the bottom whereas they tend to be higher in the water column when inactive (Walsh and FitzGerald 1984).

Milinski (1979, 1986) mentions observing stream-dwelling *G. aculeatus* feeding in streams on drift, a patchy resource.

The experimental work on patch use by stickleback would be easier to interpret if there were some quantitative data on how fish exploit patchy resources. Although it would be difficult to do, it would be good to know how long fish stay in a swarm of prey, whether they take long runs of the same species, or whether they spend long periods moving systematically over the bottom of a pond.

How patches are exploited

Optimal foraging theory has concentrated mostly on how long a predator should stay in a patch and predicts the magnitude of variables such as residence time and giving-up time (Hart 1986; Stephens and Krebs 1986). Experimental work with stickleback has had different aims. Most of the early work by Milinski (1977*a*,*b*, 1984*c*), Milinski and Heller (1978), Heller and Milinski (1979), and Ohguchi (1981) examined how stickleback react to the characteristics of swarming prey, such as density and movement, whilst later researchers have studied how stickleback compete with each other for patchy resources (Milinski 1979, 1984*b*, 1986, 1988; Regelmann 1984; Milinski and Regelmann 1985).

Prey swarms protect individuals against predation (e.g. Williams 1964; Pitcher 1986). Predators have had to evolve strategies for overcoming the confusion effect of prey aggregations. Aspects of this problem were discussed in the section on the control of feeding. As with so much of feeding behaviour, hunger is a key factor in influencing what the fish does. Hungry stickleback make more early attacks on the central region of a *Daphnia* swarm (Milinski 1977*b*) and increase their attacks on the swarm's margin as they become satiated. Further experiments showed that it was the densest region, rather than just the centre of the swarm, that was attacked first.

Spatial oddity seems to increase the likelihood of attack. For example, Milinski (1977*b*) demonstrated that two *Daphnia* separated by a gap from the main swarm were more often attacked first than were two that were merely on the edge of the swarm. In a detailed study, Ohguchi (1981) showed that threespine stickleback also tended to attack odd-coloured prey disproportionately often, and that *G. aculeatus* were confused by the velocity with which prey moved. Ohguchi (1981) also showed the way in which prey movement and colour interact to reduce attack frequency, and how increasing prey density causes a decrease in the number of attacks on the swarm.

To summarize Ohguchi's (1981) 79 page paper in a sentence seems negligent. We acknowledge the importance of the paper as a contribution towards understanding how swarming reduces predation, but do not include a more detailed discussion because we are mainly interested in functional explanations for the way in which fish attack swarms. For this purpose Ohguchi's (1981) most important finding is that a threespine stickleback is at the limits of its information-processing capacity when attacking high-density regions of a swarm (see also Milinski 1984*c*). In this state the fish cannot carry out other tasks effectively (Milinski and Heller 1978; Heller and Milinski 1979; Milinski 1984*c*).

A hungry fish risks starvation. In this state reducing hunger becomes important, and as a result feeding becomes a priority. At the same time, the fish will be under a certain risk of predation which in some habitats can be considerable (Reimchen Chapter 9 this volume). For example, Whoriskey and FitzGerald (1985*b*) estimated that bird predation removed about 30 per cent of the stickleback population in a Canadian salt-marsh during May and June. In a large-bodied population of *G. aculeatus* in Drizzle Lake, Queen Charlotte Islands, British Columbia, Canada, 13.4 per cent had structural damage and skin lacerations ascribed to bird and fish predation (Reimchen 1988, page 269 this volume). At high hunger levels the risk of predation must be traded against the risk of death by starvation, and Heller and Milinski (1979) develop and test the model described in the section on when to feed.

It was shown by Milinski and Heller (1978) that a stickleback frightened by a model kingfisher attacked low-density regions of a *Daphnia* swarm, where the feeding rate was lower but the chance of detecting the predator was presumably higher. The change in feeding as a result of seeing the kingfisher model was predicted by a model based on Pontryagin's maximization principle. Subsequently, Milinski (1984*c*) showed that stickleback feeding on a high-density swarm of prey more often missed seeing a model kingfisher than when they were attacking low-density swarms. This result demonstrated that high risk caused by poor vigilance could be a potential cost of feeding at a high density.

Competing for food in patches

In the breeding season, hungry female *G. aculeatus* in Canadian tidepools choose to join shoals of conspecifics with about 15 members (van Havre and FitzGerald 1988). Satiated females join shoals with about 45 members. The fish in the tidepools are subject to a high risk of predation (Whoriskey and FitzGerald 1985*b*). These data are compatible with the hypothesis that satiated fish put safety first and join large shoals where defence against predation is greatest (e.g. Pitcher 1986). Hungry fish are more interested in food and are prepared to trade off the lower competition in a small school against the higher risk of being eaten themselves. The mechanics of competition for food in groups of fish have been studied by Milinski in the context of the ideal free distribution (IFD) first proposed by Fretwell and Lucas (1970).

The IFD is an evolutionarily stable strategy (or ESS for short) which, if adopted by a population as a pure or mixed strategy, cannot be invaded by any alternative strategy (Parker 1984). The IFD describes the way in which individuals should be distributed when they are competing for a patchy resource. Six equal-sized stickleback were placed in an aquarium which had a supply of *Daphnia* at either end (Milinski 1979). Two experiments were done: in the first, prey were delivered every 2 s at one end of the tank and every 10 s at the other, producing a ratio of inputs of 5:1. In the second experiment, the time intervals were 2 s and 4 s, giving a ratio of 2:1. After 3–4 min feeding, the six stickleback had divided themselves into two groups which had the same ratio to each other as did the input rates, 2:1 and 5:1. In the second experiment, the prey inputs were swapped from one end to the other so that the low-input end became the high-input end and vice versa. As a result the fish also changed ends, so a new distribution was established but still in proportion to the prey abundance. The match between fish group ratios and input ratios, however, was not as exact as before the switch.

The IFD is contingent upon the following assumptions: (1) patches have different profitabilities, (2) increasing competition decreases the patch profitability, (3) there is no resource guarding, so predators are free to leave or enter as they wish (hence the name of the distribution), (4) all individuals choose the patch where their expected gains will be the highest, and (5) all individuals have the same ability to gather food. This last assumption was shown to be untrue by further work.

By tagging individuals, Milinski (1984*b*) was able to show that, during the course of a trial, the number of *Daphnia* caught by six fish varied between individuals from over 30 to fewer than 10. This difference between individuals was consistent over time. The better competitors sampled at the start of the trial and then settled down to exploit the chosen patch. The less skilful fish continued throughout the trial to switch from one end of the tank to the other. Despite these differences in competitive ability, the overall distribution of individuals mimicked an IFD. A theoretical analysis by Sutherland and Parker (1985) showed that several distributions of good and bad competitors could produce the same sums of competitive abilities in the two patches. In only one would the densities of the competitors be in proportion to the input rates, so mimicking an IFD. It is this distribution that lay behind Milinski's (1986) results.

The idea that some individuals have a lower gain from patch exploitation is not compatible with the idea of an IFD, where all individuals are supposed to gain the same. The sorts of distributions predicted by Sutherland and Parker (1985) and shown by threespine stickleback indicate that when competitors differ in abilities, then each will behave so as to maximize its rewards, within the constraints determined by its capacities. In this sense, the strategy that evolves is dependent on the phenotypic limitations of the individuals. This type of strategy was called a phenotype-limited ESS by Sutherland and Parker (1985). Although pay-offs to individuals differ, each is doing the best it can, given its particular abilities.

Of four possible ways in which six good and six poor competitors can divide themselves between two patches with an input ratio of 2:1, only one

Table 8.1 Four ways in which 12 fish can distribute themselves over two patches varying in quality with a 2:1 ratio, given that six are twice as efficient at catching prey as the other six. Also shown is the ideal free distribution (IFD) where all 12 fish have equal competitive abilities. (From Sutherland and Parker 1985.)

Solution	Number in good patch		Number in poor patch	
	Good competitors	Bad competitors	Good competitors	Bad competitors
A	6	0	0	6
B	5	2	1	4
C (IFD mimic)	4	4	2	2
D	3	6	3	0
E (IFD)	8	–	4	–

mimics an ideal free distribution (Milinski 1988) (Table 8.1). The resultant division has four good and four poor competitors feeding on the good patch, and two good and two poor competitors feeding on the poor patch. This assumes that good competitors are twice as effective as poor competitors. Why is it that this one distribution that mimics the IFD was the one observed in Milinski's (1979, 1984*b*) experiments? In the stickleback example it is likely that the behaviour of the good competitors determines the final outcome (Milinski 1988). They decide early on how they will distribute themselves and do so in proportion to patch input rates (i.e. a 2:1 ratio). The poor competitors then have to fit in where they can. Houston and McNamara (1988), using an approach based on statistical mechanics, show that the observed distribution is the most likely one of the four possible.

How stickleback exploit patches when others are present has demonstrated the complex decisions that these small fish can make. The distributions observed in experiments are likely to be modified by risk of predation, reproductive state, and hunger. As yet there are no data showing how these three factors might interact. As all three factors are likely to be present simultaneously in nature, work is needed to examine how the factors interact to produce the changing pattern of aggregations observed in the field (van Havre and FitzGerald 1988).

Deciding which habitat to use

When confronted with two points of prey input, the stickleback first sampled the patches (Milinski 1979, 1984*b*). How does the fish sample the habitat and what rules is it using? Where there are two patches to choose from, the problem of how to sample so as to maximize the pay-off has been called the two-armed bandit problem (Krebs *et al.* 1978) and has been applied to stickleback by Thomas *et al.* (1985). The term 'two-armed bandit' derives from an analogy with a slot machine, normally called a 'one-armed bandit'. In the problem tackled here, there are two sources of reward rather than just one.

Threespine stickleback were trained to choose between one of two adjacent compartments in an aquarium. Each compartment had an associated probability of a pay-off, which in the experiments was the ubiquitous *Tubifex* worm. One worm was given as a reward with a probability that depended on the schedule being used. One compartment always had a higher probability (p) of reward than did the other (q). It was argued that the fish might use three methods to choose a compartment, given that its goal was to maximize pay-off over a fixed number of trials. It could choose exclusively the side with the highest probability of a pay-off and this would yield the maximum pay-off. Alternatively it could use a probability-matching rule in which it would devote time to compartments in proportion to the probability of getting a pay-off. Finally, the fish could choose compartments at random. Predicted pay-offs from these three alternatives were calculated for varying values of q; p was held constant at 0.9. By the end of the 11 d experiment, all fish were choosing exclusively the compartment which gave the highest pay-off. Over the first 6 d, the fish could have been either choosing at random or using the probability-matching rule. After day 6, fish steadily increased the proportion of visits to the most rewarding compartment until they visited it all the time.

Further analysis of the results showed that the greater the difference between p and q, the less time it took the fish to learn which was the best side. The improvement in pay-off mainly occurred from day to day rather than within days of the trials. The relevance of this result to animals in the wild needs checking, as fish might not be confronted with the same patches from one day to the next.

A second experiment gave fish the same sort of choice, but this time one group of fish was rewarded with just one *Tubifex* and fish in the second group were given three. Fish in the first group did not grow during the course of the trials whereas those in the second did, and this difference was taken as evidence that the first group of fish were hungrier. The hungry fish mostly maximized their pay-off over the last part of the trial period, but the less hungry individuals tended to behave as if they were probability matching. It was suggested by Thomas *et al.* (1985) that, as with most animal behaviour, the fish has things other than just food gain to consider as it forages. They proposed that probability matching in semi-satiated fish is a compromise between the needs of optimally exploiting the environment and gaining new information about it.

The study by Thomas *et al.* (1985) shows that *G. aculeatus* can sample and learn about patch quality, but their results are not specifically applicable to groups of fish exploiting prey patches. The important difference in group foraging is that the learning rule used by solitary individuals may not be the best one when there are competitors about. An ESS learning rule, devised by Harley (1981) and modified by Regelmann (1984), called the relative pay-off sum (RPS) learning rule, is described in detail by Milinski (1984*b*) and Hart (1986). Using the rule, Milinski (1984*b*) was able to predict the behaviour of six stickleback as they exploited two patches which had input rates with a ratio of 2:1. The RPS rule predicts that good competitors should decide first which patch to exploit and then should change patch or switch less often than poor competitors. The rule also predicted that longer travel times between patches should decrease the amount of switching.

The stickleback studied by Thomas *et al.* (1985) improved their performance from day to day, suggesting they remembered the rewards gained over previous days. The RPS learning rule includes a memory factor that devalues past pay-offs. Milinski (1984*b*) and Regelmann (1984) assumed that each successive pay-off during a foraging bout was devalued by a factor of 0.97 per time interval, which was close to the value originally proposed by Harley (1981). Others have argued that animals have a memory window over which they average the reward rate from patches (Krebs and Cowie 1976). Milinski and Regelmann (1985) offered threespine stickleback two prey patches in a training period during which 20 or 30 *Daphnia* were delivered at one item per 6 s. The fish were mostly fed in patch one. After the training session, fish were fed 10 prey in patch one, and then kept away from food for an individual-specific period of time after which they were offered five more prey in patch two. After the prey had been delivered in patch two, the way the fish used the two patches was monitored for 90 s. Fish that experienced a pause greater than 90 s between the delivery of food in patches one and two preferred patch two after feeding. Fish that experienced a pause less than 90 s remained with patch one. The results were best explained by proposing that the memory of the quality of patch one faded with increasing time since it last yielded food, the critical time being around 90 s.

Throughout this section we have concentrated on functional explanations of how threespine stickleback should exploit patchy resources. The picture that is emerging is that the fish's decision is influenced by resource density, the presence of conspecifics, the threat of predation, hunger, and experience. All of the work reported describes how the fish would behave in the laboratory under experimental conditions. To understand the relevance of these results to stickleback in the wild, information is needed on the frequency with which the juxtaposition of variables used in laboratory experiments occurs in the natural habitat. We also need to know how the variables fluctuate through time.

Diet selection

The diet of threespine stickleback and its variation with time

It is easy to catch fish, open their stomachs, and look at what they have been eating. In recent times, starting with Hynes (1950), many studies have been made of the gut contents of *G. aculeatus* (e.g. Maitland 1965; Manzer 1976;

Table 8.2 Gut contents data from three separate studies. The data for the Great Central Lake, Vancouver Island, British Columbia, Canada show seasonal change of diet during 1970 (Manzer 1976). Lavin and MacPhail (1986) sampled different habitats of the Cowichan Lake system, Vancouver Island, during May 1983 for the three stickleback morphotypes (intermediate, open water and littoral). Both sets of data are mean number of items per stomach. The final three columns show the percentage stomach volume of the main food items in the 1978 seasonal diet of sticklebacks in the Broälven Estuary, Sweden (Thorman 1983). T denotes a trace, i.e. less than 1 organism.

Food item	Great Central Lake					Cowichan Lake system			Broälven Estuary		
	Month					Phenotype			Month		
	Apr.	Jun.	Jul.	Aug.	Oct.	Int.	Open	Litt.	Jul.	Sept.	Oct.
Limnetic food											
Rotifera	–	15	91	52	56	–	–	–	–	–	–
Cladocera:	–	–	–	–	–	–	1.3	–	–	–	–
Holopedium	–	39	37	115	129	–	–	–	–	–	–
Bosmina	2419	235	37	18	116	–	–	–	–	–	–
Daphnia	–	1	T	T	T	–	–	–	–	–	–
Alona	–	T	2	2	1	–	–	–	–	–	–
Copepoda:											
Epischura	2	46	1	17	4	T (Epischura + Diaptomus)	3.4 (Epischura + Diaptomus)	– (Epischura + Diaptomus)	–	–	–
Diaptomus	–	–	–	T	98				–	–	–
Cyclops	88	T	–	T	38	T	–	–	–	–	–
Copepodids	–	–	–	6	20	–	–	–	–	–	–
Nauplii	–	–	–	10	9	–	–	–	–	–	–
Harpacticoid	–	T	–	T	1	–	–	–	–	–	–

	Apr.	Jun.	Jul.	Aug.	Oct.	Int.	Open	Litt.	Jul.	Sept.	Oct.
Insecta:											
Simulidae (adult)	–	–	–	–	–	T	–	T	–	–	–
Tipulidae (adult)	–	–	–	–	–	T	–	–	–	–	–
Unidentified	–	–	–	–	–	T	T	T	–	–	–
Zooplankton eggs	18	4	10	1	8	–	–	–	–	–	–
Estuarine zooplankton	–	–	–	–	–	–	–	–	84	–	–
Benthic food											
Chironomid larvae	–	5	T	1	T						
Chironomid pupae	–	2	T	T	T	3.4	T	4.3	–	82	–
Sediment	–	–	–	–	–	–	–	–	–	–	81
Ostracoda	–	–	T	T	T	10.4	1	14.4	–	–	–
Pelecypoda	–	1	–	–	–	–	–	–	–	–	–
Acari	–	–	–	T	T	–	–	–	–	–	–
Planaria	–	–	–	–	T	–	–	–	–	–	–
Odonata	–	–	–	T	–	–	–	–	–	–	–
Fish	–	–	–	T	–	–	–	–	–	–	–
Amphipoda	–	–	–	–	–	T	–	–	–	–	–
Chaoborus	–	–	–	–	–	–	–	T	–	–	–
Megaloptera (larvae)	–	–	–	–	–	–	–	T	–	–	–
Megaloptera (adult)	–	–	–	–	–	–	–	T	–	–	–
Ephemeroptera	–	–	–	–	–	T	–	T	–	–	–
Gasterosteus eggs	–	–	–	–	–	T	1.4	3.3	–	–	–
Hydracarina	–	–	–	–	–	–	–	T	–	–	–
Nematodes	–	–	–	–	–	T	–	T	–	–	–

Reimchen 1982; Thorman 1983; Thorman and Wiederholm 1983, 1984, 1986; Snyder 1984; Walsh and FitzGerald 1984; Zander *et al.* 1984; Delbeek and Williams 1987*a*, 1988; O'Hara and Penczak 1987; Hangelin and Vuorinen 1988; see also Fig. 5.2 this volume). The direction of these studies has changed over the years, so that very early studies were more likely to concentrate on the phenomenon of gut content in a descriptive way, while later studies have usually included quantitative estimates of niche breadth and given an analysis of competition.

The food of threespine stickleback was studied in detail over a 2 yr period in Great Central Lake, Vancouver Island, British Columbia, Canada, by Manzer (1976); the diet for the 1970 summer period is shown in Table 8.2. Diet was established from stomach contents of seine-caught fish. The table also shows data from Lake Cowichan on Vancouver Island from Lavin and McPhail (1986). These data show how diet is influenced by where in the habitat the fish forage. Table 8.2 also shows diet data for stickleback in the Broälven Estuary, west coast of Sweden, from Thorman (1983), illustrating differences attributable to habitat.

In tidal salt-marshes on the eastern coast of Canada, Delbeek and Williams (1988) found that differences in diets were mainly related to the range of prey types available. Shifts in diet with time were stated to be due only to changing prey abundances at different sites, rather than to changing preferences. To study prey selection, Delbeek and Williams (1988) used Strauss' (1979) linear selection index $L = R_i - P_i$, where R_i is the proportion of prey type i in the diet and P_i is its proportion in the environment. Delbeek and Williams (1988) concluded that the values of L mostly show that fish were selecting prey in proportion to their abundance (i.e. L was close to 0). We contend that Table I of their paper shows that threespine stickleback preferred a small subset of prey, namely the limnetic Calanoida, Cyclopoida, Rotifera (mostly taken by juvenile *G. aculeatus*), and the benthic *Gammarus*, fish eggs, Ostracoda (by juveniles), and invertebrate eggs. For all these prey, L was greater than $+0.5$, indicating selection. The fish were also actively rejecting certain very abundant items such as harpacticoids, diatoms, and nematodes, for which L was less than -0.5. These items may not be available ordinarily to stickleback because of their distribution.

The influence of habitat, fish size, sex, and parasites on natural diets

As discussed, availability plays a part in determining the diet of *G. aculeatus*, so that differences in diet would be expected between marine, estuarine, and freshwater habitats. In the sea off the mouths of the rivers Elbe and Eider in northern Germany, stickleback ate polychaetes (*Nereis*) and amphipods (*Corophium* and *Gammarus*), all from the benthos (Zander *et al.* 1984). Limnetic copepods were also very abundant in stomachs. Within the copepod group, predominance changed from a majority of harpacticoids in the sea with some calanoids, to a predominance of cyclopoids in the river habitat.

The diet of stickleback from Great Central Lake, Vancouver Island,

Canada, varied with fish size, sex, and female reproductive condition (Manzer 1976; Table 8.3). In salt-marsh pools of the St Lawrence Estuary, Canada, male stickleback ate more fish eggs and fewer chironomid larvae than did females at the start of the breeding season (Walsh and FitzGerald 1984). Males also took a larger mean prey size although they were on average 5 mm shorter than females. In Lake Kvernavann, Norway, in 1981, male stickleback fed mainly on benthic organisms throughout the year, whilst the female diet was dominated by zooplankton in summer but resembled that of males for the rest of the year (Jakobsen *et al.* 1988).

Table 8.3 Gut contents data from Great Central Lake sticklebacks. The first four columns show the percentage occurrence of organisms in the diet of different size groups of fish during July and August 1970. In the last three columns the stomach contents of non-gravid and gravid females and sexually mature males are expressed as a percentage of stomachs with the food item, sampled from Great Central Lake, Vancouver Island, British Columbia, during May and July 1971 (Manzer 1976).

Food item	Size group (mm)				Female		Male
	<30	30–49	50–69	70+	Non-gravid	Gravid	
Rotifera	65	85	49	9	9	–	10
Cladocera:							
Holopedium	68	88	81	71	23	32	17
Bosmina	75	50	36	28	–	4	6
Alona	46	19	19	–	5	–	2
Copepoda:							
Epischura	50	79	83	57	18	54	16
Diaptomus	3	6	2	–	–	–	1
Cyclops	4	–	–	–	14	11	22
Copepodids	7	10	8	–	5	–	9
Nauplii	18	–	–	–	–	–	–
Harpacticoid	7	10	8	–	5	–	4
Insecta:							
Chironomid larvae	42	22	21	14	27	4	17
Chironomid pupae	16	35	21	–	18	11	22
Coleoptera	–	–	–	–	5	–	1
Ceratopogonidae	–	–	–	–	14	–	11
Other	7	5	7	–	18	14	15
Eggs:							
Zooplankton	13	29	23	28	9	14	19
Stickleback	–	–	–	–	–	4	9
Other:							
Pelecypoda	–	–	2	–	–	–	15
Ostracoda	6	6	3	–	5	–	5
Acari	2	–	–	–	–	–	9
Araneida	–	3	–	–	–	–	1
Fish	–	–	10	–	–	–	–
Isopoda	–	–	3	14	–	–	1
Amphipoda	–	–	–	–	5	4	7

During the breeding season in 1982, stickleback eggs formed 27 per cent by weight of the diet of stickleback in Kennedy Lake, Vancouver Island, Canada (Hyatt and Ringler 1989*b*). In Myvatn, Iceland, stickleback eggs were an important part of the diet of stickleback from the north part of the lake, which had lower food availability than the south end, where eggs were eaten less frequently (Adalsteinsson 1979). Hyatt and Ringler (1989*b*) concluded that stickleback eat eggs as a strategy to maximize time available for reproductive activities that will lead to the successful completion of one or more breeding attempts, as had been suggested by Rohwer (1978). This interpretation is disputed by Foster (1990; see also Foster page 389 this volume), who pointed out that it is mainly benthic-feeding stickleback that eat eggs and that it is the mode of feeding that increases the encounter rate with egg, rather than a deliberate strategy.

Healthy female stickleback and those infected with the tapeworm *Schistocephalus solidus* ate different food types in Lake Kvernavann (Jakobsen *et al.* 1988), although the difference varied with time. In the summers of 1980 and 1981, infected females fed mostly on benthic organisms. In the autumn of 1981, chironomid larvae formed 83 per cent of the stomach content of infested fish but only 17 per cent of the stomach contents of healthy fish. Both groups ate similar amounts of fish eggs. The difference in diet was thought to be largely a result of different habitat use by the infested fish, which were most likely to be found in open areas. *Gasterosteus aculeatus* from Boulton Lake, Queen Charlotte Islands, Canada, that were infected with the cestode *Cyathocephalus truncatus* ate more amphipods than did healthy individuals (Reimchen 1982). The proportion of amphipods (the intermediate hosts of the cestode) eaten increased with the number of cestodes contained in the fish. These data illustrate how parasites can change primary host behaviour to achieve better transmission to secondary hosts (see also Huntingford *et al.* page 281 this volume).

Explanations of diet choice

Evidence from natural populations and laboratory experiments shows that prey choice is influenced by numerous factors (Fig. 8.4), although the number of factors recognized is dependent on the degree to which they are dissected. A tendency to imagine that diet choice can be understood in terms of just one or two factors has led to a huge body of theory based on the optimality principle (Stephens and Krebs 1986). The simplest optimality model of diet choice, the basic prey model (BPM for short), has diet being determined by the interplay between the energy content of prey types, their handling times, and their abundance, as reflected through encounter rate. This basic model was used to explain how threespine stickleback responded to changing densities of two sizes of *Daphnia* (Gibson 1980). The proportions of the two prey sizes were not satisfactorily accounted for by the apparent size hypothesis (O'Brien *et al.* 1976), which states that an animal always chooses the prey

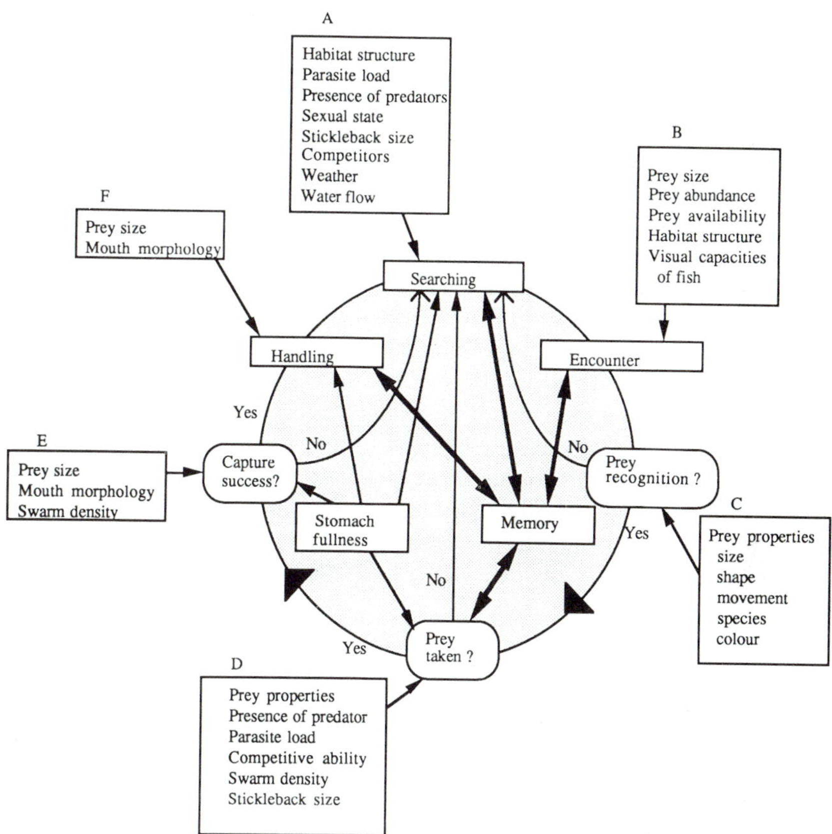

Fig. 8.4 A schematic representation of the internal and external factors that influence prey selection by threespine stickleback. The circle represents the cycle of behaviours employed (in rectangles) and decisions that must be taken (in round-cornered rectangles). 'Yes' and 'No' after each decision point show how the behavioural sequence would be altered by the decision. The arrows indicate the flow of influences. Boxes A–F list the factors that have been shown by experiment or field work to influence the behaviour in question. Internal factors are inside the circle.

type that appears to be the largest. At high prey densities, large prey will most often be perceived as the largest; as a result they should be taken more often. In Gibson's (1980) experiment the stickleback ate more small prey than expected from the apparent size hypothesis. Gibson's (1980) test of the BPM was weak for two reasons: the experimental design violated the assumptions of the model (see Stephens and Krebs 1986, p. 19), and the design was such that the theory predicted that both sizes of *Daphnia* should be taken under all experimental treatments. A more convincing test would have varied the rate of encounter with the most profitable prey, so that the optimal diet changed from including both prey to taking large prey only.

We have already described the experiments by Heller and Milinski (1979) which showed that simple optimality models failed to account for the way stickleback fed on swarming prey. Hunger was the critical factor controlling the fish's behaviour. It plays a similar role in prey choice (Hart and Ison 1991). In an attempt to overcome the problems of Gibson's (1980) experiment, Hart and Ison (1991) built a feeding arena that allowed the delivery of a sequence of different sizes of *Asellus* in a way that satisfied the BPM's assumptions. The fish were offered two sequences which differed in the rate at which three size groups were encountered. The largest prey size should have been excluded from the diet under the first treatment and included under the second, but the fish did not behave as predicted. A closer analysis of the way the fish reacted to the sequence of prey showed that fish were only prepared to take expensive items (large prey with a long handling time) when they were hungry. As they filled up, they became more selective and concentrated on prey that yielded the greatest energy per unit time. This is a similar outcome to the one obtained by Heller and Milinski (1979) for fish feeding on swarms of prey.

The BPM and its later developments (Stephens and Krebs 1986) are applicable only to solitary predators. Because stickleback often feed in groups, the competitive ability of an individual can influence prey choice (Milinski 1982). When offered a mixed size range of *Daphnia* in the company of another stickleback, good competitors took a larger mean prey size than did poor competitors. On their own, good competitors retained their interest in large prey as did poor competitors in small prey. The poor competitors fed as generalists, and Milinski (1982) suggested that fish learn a 'sure attack' distance which increases with competitive ability. Each fish is performing to the best of its ability, or is adopting a phenotypically limited strategy.

Fish infested with parasites have lower swimming speeds and cannot match the feeding performance of healthy fish (Milinski 1984*a*). As a result they should behave like the poor competitors of the earlier experiment and take less profitable but less contested prey types. When offered a choice between small and large *Daphnia*, stickleback infected with the sporozoan *Glugea anomala* and larvae of the cestode *S. solidus* chose small prey. The frequency with which they chose small prey increased with parasite load. Seeing that the strategy adopted by parasitized fish would depend on the number of competitors and the abundance of profitable prey, Milinski (1984*a*) suggested that infested fish learn the best strategy, perhaps in the same way that the poor competitors without parasites learn a 'sure attack' distance.

One assumption of the BPM is that animals do not learn as they gather food. A further example showing that they do comes from a study by Ibrahim (1988). Stickleback were reared with non-moving prey, and their feeding performance was then compared with that of fish taken from the wild. Laboratory-reared fish took longer than wild-caught fish to strike at chironomid larvae and *Daphnia*. Fixation times and handling times were longer for naive laboratory-reared fish when taking all prey types offered. The conclusion was that naive fish were less efficient foragers but it only took a week for them to improve their performance to the level of wild fish.

When feeding on two colour morphs of *Asellus aquaticus*, threespine stickleback chose the commoner form irrespective of its colour (Maskell *et al.* 1977). Stickleback feeding on differing densities of *Daphnia* and *Cloeon* always preferred the former to the latter, but their preference for *Daphnia* decreased as the total prey density increased (Visser 1982). A similar result was obtained with *G. aculeatus* feeding on a mixture of mackerel larvae, *Scomber scombrus*, and mixed zooplankton (Kean-Howie *et al.* 1988); as the proportion of fish larvae fell, the preference for them increased. Predator confusion may be the mechanism behind the change observed. As the density of prey increases, it becomes easier for the fish to track and capture the rarer prey type, as observed for stickleback attacking *Daphnia* in dense and sparse parts of a swarm (Heller and Milinski 1979; Ohguchi 1981). In effect, the profitability of the less preferred prey type is changed by the increasing cost of confusion generated by the high density of the preferred prey (Visser 1982). The main cost of confusion is suggested to be a higher risk of predation. As with good and poor competitors, this is another example of how prey profitability is a function of the conditions in which the prey is found and of the fish's internal state.

The presence of predators may make some habitats too risky to forage in, so excluding prey types from the fish's diet (Werner *et al.* 1983). Juvenile stickleback in Crystal Lake, British Columbia, Canada, were prevented from foraging in open water by the threat of predation from adult threespine stickleback (Foster *et al.* 1988). In this instance, being confined to the weeds did not change the diet of the juveniles, because they still ate zooplankton, but they may have been prevented from feeding in more profitable habitats. The differences in diet found in parasitized and healthy stickleback in Lake Kvernavann, Norway, were in some measure due to different habitat use determined by the threat of predation (Jakobsen *et al.* 1988). As shown by Milinski (1985), infected fish take bigger risks, and in Kvernavann, they spent more time foraging in open water. They were preyed upon much more in this habitat than were healthy fish, which kept in the weeds.

The size and appearance of prey also have an important influence on whether they are selected. When stickleback were offered *Tubifex* together with larvae of *Drosophila*, the fish still ate *Tubifex* at the same rate but gradually increased the number of *Drosophila* eaten (Beukema 1968). The fish learned to recognize the larvae as prey but never came to prefer them to *Tubifex*. As the fish became satiated, they rejected *Drosophila* before *Tubifex*. In contrast, the numbers of *Tubifex* eaten decreased when they were offered together with enchytraeid worms, and after an initial period of learning, the fish took more enchytraeids than *Tubifex*.

Although these results show that stickleback have preferences for particular prey types, they do not show what features of the prey are favoured. Binary choice experiments showed that small stickleback preferred large *Tubifex* and medium chironomid larvae to large *Daphnia* and small chironomid larvae (Ibrahim and Huntingford 1989*b*). These choices led the fish to the more profitable prey items, although such simple preferences did not always do so. Further work (Ibrahim and Huntingford 1989*c*) showed that with one stimulus dimension varying at a time, stickleback preferred red and pale over dark colour, fast over slow movement, straight and rectangular over globular shape, and larger over smaller size. When two stimulus features were varied together, fish preferred red to movement, which in turn was preferred to shape, which was preferred to size. If red was replaced by a pale colour, then movement became more important than colour.

Diet choice as a dynamic process

It may well be that the parameters of the basic prey model underlie the choice the fish makes, but the profitability of each prey type is not a static characteristic, nor is the internal state of the fish. Profitability varies with the prey type, but also with the state of the fish and the context in which the fish encounters the prey. A large *Daphnia* encountered by a solitary stickleback changes its profitability when encountered in company with a better competitor. Having a parasite can alter the fish's profitability assessment yet again. Experience with a prey type can alter the handling time, and large numbers of the most profitable type (calculated as energy gained per unit of pursuit and handling time) can increase the confusion effect and, as a consequence, the chance of death by predation. Nearly all variables are influenced by hunger. We have tried to summarize the complex network of factors influencing diet choice in Fig. 8.4, which depicts prey choice as a cyclic process using the metaphor of time's cycle (Gould 1987). We are convinced that this way of looking at the process conveys how diet choice is a repeating cycle but with each turn characterized by a unique combination of factors. It looks increasingly likely that standard static optimization models are inadequate and that a better approach might be the discrete dynamic models of Mangel and Clark (1988).

This chapter is about the evolution of foraging behaviour. The studies of threespine stickleback in Canadian lakes have shown how morphology and behaviour have responded to different feeding environments (Foster 1990; McPhail Chapter 14 this volume). It would be valuable if similar studies could be made of other characters that have been identified as determining diet. For example, it would be instructive to repeat the stickleback experiments done by Kean-Howie *et al.* (1988), which showed a changing preference for a prey type with its abundance, but this time using stickleback from habitats with and without predators. Guppies, *Poecilia reticulata*, and Hart's rivulus, *Rivulus harti*, foraged more persistently under predation threat when they came from a population subject to high predation hazard (Fraser and Gilliam 1987). Juvenile creek chub, *Semotilus atromaculatus* (Gilliam and Fraser 1987) and bluegills (Gotceitas 1990) foraging under the threat of predation minimize mortality per unit of energy gained rather than maximize long-term net energy. Assessment of predation risk can be moderated by the characteristics of the species, as shown by McLean and Godin (1989). They found that the tendency to flee from an approaching predator was inversely related to the amount of body armour in four populations of three fish species. *G. aculeatus* were bolder than were *Pungitius* spp.

FUTURE DIRECTIONS AND THOUGHTS ON FORAGING THEORY

Filling the gaps

To make proposals about future research is like suggesting that someone count the points on a line; however much we find out about stickleback there will always be more gaps to fill. Despite this it is possible to pick gaps that appear more important than others. What we know about stickleback foraging suffers from the convenience of the fish as an experimental animal. Much of the experimental work has studied problems that may or may not exist in nature. For example, we now know a great deal about how stickleback behave in relation to swarms of *Daphnia*, but what we do not know is how often fish in nature are faced with the same problems. So we can say that if 12 stickleback are faced with two patches of *Daphnia* that differ in quality with a ratio of 2:1 and the fish vary in competitive ability, then the good competitors will choose their patch in relation to each other, after which the poor competitors will distribute themselves. The resulting distribution will be like the ideal free distribution. What we do not know is how often this combination of events occurs in nature. The result is that our knowledge of threespine stickleback behaviour is modular. We can say that given that the set of states $\{1, 2, \ldots, n\}$ occurs, then the consequence will be X. The big question for understanding how fish behave in nature is how often $\{1, 2, \ldots, n\}$ occurs in the life of the fish. In the context of Fig. 8.4, we would want to know how much the factors in boxes A–F vary from one turn of the cycle to the next, and over what time scale. We would also want to know whether the variation is itself cyclical or random.

We propose that in the immediate future it would be interesting to know more about the time budget and pattern of movements of a foraging *G. aculeatus* in its natural habitat, and how the two features are influenced by the factors we have discussed, such as threat of predation, distribution of food types, hunger, sex, and breeding state. Experiments are needed on the way in which the various influences on patch and diet choice are integrated into a suitable sequence of activities. It would also be valuable to have more experimental data on how fish behave towards prey types other than *Tubifex* and *Daphnia*. These suggestions presuppose that investigators of stickleback behaviour are intent on understanding how stickleback function in nature. It could be argued that many are more interested in examining general behavioural principles using the stickleback as a convenient vehicle for investigation. If that is the case, filling these gaps becomes less urgent.

The future of foraging theory

'The arrow of homology and the cycle of analogy are not warring concepts, fighting for hegemony within an organism. They interact in tension to build the distinctions and likenesses of each creature.'

Gould (1987).

This is not the place for an extended discussion of foraging theory, but the detailed knowledge available on stickleback gives us a unique perspective that leads to some interesting points. Through the ingenious studies of Milinski we begin to see that a foraging fish is constantly evaluating its own state and the condition of the environment, and adjusting its behaviour appropriately. As we have suggested in Fig. 8.4, it is possible that many of the factors affecting foraging act through variables such as encounter rate, prey energy content, and handling time, which have been identified in foraging models as determining prey choice. What static models do not account for is the dynamic nature of the key variables and of the internal state of the fish (Hart 1989). Continuous-time dynamic models (Stephens and Krebs 1986, Chapter 7) attempt to deal with the problem but are likely to be too intractable mathematically to handle the observed complexity. The discrete-time dynamic models developed by Mangel and Clark (1988) offer greater flexibility and can cope with a wide range of complexity. They make it possible to replace maximization of energy gain with maximization of fitness, which seems more in accord with what animals do. What this approach does not handle is the unique nature of the life of each fish. Discrete-time dynamic modelling assumes repeatability of influences even though repetition is only statistical. Life histories, like all histories, are contingent (Gould 1989), and what models cannot account for are unique events the influence of which bears little relation to the animals' abilities. What we need is more information on how often the cycle in Fig. 8.4 is influenced by unique events with a significant impact on the individual's chance of survival.

The contrapuntal metaphors of time's cycle and time's arrow, so eloquently discussed by Gould (1987), are also implicit in Fig. 8.4. The quotation from Gould (1987) at the head of this section makes clear how the metaphors help to disentangle the two aspects of behaviour that must be considered for a full understanding. Regularities of the feeding environment, signalled by the cycling nature of the processes, can lead to adaptations through natural selection. It is on the truth of this aspect that optimality models have been built. Our understanding cannot be complete without also considering the uniqueness of each turn of the cycle, leading to time's arrow. All the factors in boxes A–F of Fig. 8.4 will change from one turn of the cycle to the next, so that each time the fish must assess anew its response. The significance of these unique combinations of events has yet to be explored.

We think that the idea of optimization has stimulated a large amount of interesting research but has also sometimes closed minds to the dynamism and complexity that is evident if only one looks at what animals really do. Our review illustrates this diversity, and we feel that the data and ideas produced by research on stickleback foraging can form the basis for new developments based on the idea that the fish are constantly updating their behaviour in order to survive.

ACKNOWLEDGEMENTS

This chapter was written in the shadow of Manfred Milinski. He was to have written the chapter himself but was prevented from doing so by unforeseen events (contingency again). As a result we were offered the task and hope that we have given a fair review of a large volume of research. We thank Manfred for reading a draft. We also thank Susan Hart, Larry Greenberg, Gary Mittelbach, and Dolf Schluter for carrying out the same task. David Stephens also made detailed comments on the first draft which helped greatly to improve it. Most of the first draft was written while Paul Hart was on sabbatical in the Limnology Department at Lund University, Sweden. He gratefully acknowledges help and facilities offered by Stellan Hamrin and funding provided by The Royal Society, London and Kungliga Vetenskapsakademien, Stockholm. Andrew Gill was supported by a grant from the UK Natural Environment Research Council (GR3/7293).

9

Predators and morphological evolution in threespine stickleback

Thomas E. Reimchen

The wealth of investigations on threespine stickleback during the last half-century has established the species as a model organism for evaluating population differentiation. Stickleback are particularly attractive for these studies because they exhibit extensive variability within and among populations and occurs in a diversity of habitats, from open oceanic waters to isolated bog ponds. Differences in predation levels, which are an underlying theme in diverse studies of prey defences (Curio 1976; Endler 1986; Vermeij 1987; Greene 1988), are also suspected of occurring among stickleback populations and are associated with increased expression of spines, bony armour, and escape responses (Hagen and Gilbertson 1972; Moodie and Reimchen 1976*a*; Gross 1977; Giles and Huntingford 1984; reviews in Wootton 1976 and Bell 1984*a*).

In this chapter, I focus primarily on predation as an ecological component in the biology of stickleback, and secondarily on some of the evolutionary implications of predation to morphology (behavioural aspects of defences against predators are treated in detail by Huntingford *et al.*, Chapter 10 this volume). Despite the diversity of studies on stickleback, there is not yet a detailed assessment of the age-specific causes and amounts of mortality within the life history of any stickleback population. Such an assessment would seem fundamental to evaluating morphology, behaviour and variability. To this end, I present results from a long-term investigation on sources of mortality among insular populations of stickleback from the Queen Charlotte Islands, western North America. Some of the data provide insight into the broader issues of predation levels, predator foraging efficiencies, and selection intensities in natural populations.

PREDATOR DIVERSITY OVER THE GEOGRAPHICAL RANGE OF STICKLEBACK

Although the large dorsal and pelvic spines of threespine stickleback are a substantive deterrent to gape-limited piscivores (Hoogland *et al.* 1957), this prey has been found in the diet of a remarkable array of species.

Common predatory fish such as perch, *Perca* spp., pike, *Esox* spp., and salmonids, *Salmo* spp. and *Oncorhynchus* spp., regularly consume stickleback (Hartley 1948; Frost 1954; Greenbank and Nelson 1959) and were the focus for initial experimental and field investigations of stickleback functional morphology (Hoogland *et al.* 1957; Moodie 1972*a*; Moodie *et al.* 1973; Hagen and Gilbertson 1973*b*; Moodie and Reimchen 1976*a*). Avian piscivores, including loon, grebe, merganser, heron, and kingfisher are widely distributed and prey on stickleback (Munro and Clemens 1937; Penczak 1968; Bengtson 1971; Huntingford 1976*a*; Gross 1978; Reimchen 1980; FitzGerald and Dutil 1981; Giles 1981; Giles and Huntingford 1984; Whoriskey and FitzGerald 1985*b*). As well, some of the macroinvertebrates found in stickleback habitats, such as leeches, dragonfly naiads, bugs, and beetles are piscivorous (Moodie 1972*a*; Reimchen 1980; Foster *et al.* 1988). Conspecific predation on eggs and fry may be prevalent (Greenbank and Nelson 1959; Foster 1988; Foster *et al.* 1988; Hyatt and Ringler 1989*b*).

Predators of stickleback vary in size from 0.3 g backswimmers (Hemiptera) to 300 kg fur seals (Pinnipedia) and comprise at least 68 species (Table 9.1) from seven major taxa: Cnidaria (1 sp.), Hirudinea (1 sp.), Insecta (4 spp.), Pisces (22 spp.), Reptilia (1 sp.), Aves (34 spp.), and Mammalia (5 spp.). The diets of marine species are least well known and one presumes that the list of stickleback predators will expand as the diets of other species are examined. Even the extinct great auk, *Pinguinus impennis*, is known to have preyed on threespine stickleback (Olson *et al.* 1979), so it seems probable that many pelagic avian piscivores could exploit this prey. The regular utilization of tidepools by juvenile stickleback (Weeks 1985*b*; Whoriskey and FitzGerald Chapter 7 this volume) offers unexplored associations.

Such a broadly based predation regime for a single species is the consequence of several factors. The small body size of stickleback (7–110 mm), their slow swimming speed, and their abundance would make them potentially suitable for a variety of predators, but more importantly, stickleback occur in a greater diversity of habitats than most fish. They are found in muskeg ponds, littoral, limnetic, and benthic lake habitats, streams, rivers, marine estuaries, tidepool habitats, subtidal pelagic habitats, and recently, stickleback have also been observed in open oceanic waters 100 km from the continental coastline (Williams and Delbeek 1989; Cowen *et al.* 1991). Clearly, there would be a broad range of vertebrate and invertebrate piscivores over these habitats and geographical distances.

The relative importance of each group of predators in different habitats and in different geographical areas is poorly known. Predatory fishes are dominant sources of mortality in several European and North American populations where detailed diet analyses have been made (Frost 1954; Moodie 1972*a*; Hagen and Gilbertson 1973*b*; Reimchen 1990). These fishes may be less frequent at the southern edges of the freshwater distribution of stickleback in Europe (Gross 1978), and they are commonly absent from northern ponds in Europe and North America where limnological conditions are unsuitable or where access is restricted. Avian piscivores, because of their mobility, probably occur in more stickleback localities than any other single group of predators. In some localities, stickleback are the primary prey for mergansers, terns, and kingfishers (Sjoberg 1985, 1989; Raven 1986). Rad (1980) noted that the nesting areas of red-breasted merganser in Norway closely track the distribution of threespine stickleback on which adult and pre-fledged birds feed. Stickleback harbour a diversity of parasites that require birds as definitive hosts (see Wootton 1976 for review), attesting to the general utilization of this prey by avian piscivores. Although birds are seldom as numerically abundant as predatory fish, they have much higher metabolic rates and eat approximately 18 per cent of their body weight in fish per day (Nilsson and Nilsson 1976), as compared with 1–3 per cent for predatory fish (Elliott 1976). Their contribution to total mortality in any single population could be substantial (Reimchen 1980; Whoriskey and FitzGerald 1985*b*).

Predation on stickleback by mammalian piscivores has not been extensively investigated. River otter consume stickleback (Erlinge and Jensen 1981), as do mink (Gerell 1968, cited in Wootton 1976). Water shrews, *Sorex palustris*, regularly take brook stickleback, *Culaea inconstans*, in central North America (Roberts pers. comm.), but have not been reported as predators on threespine stickleback in Europe where the ranges overlap. Fur seals captured 60 km off the coast of British Columbia had stomachs filled with stickleback (Biggs pers. comm.). Human exploitation of anadromous stickleback has been reported in northern Europe (Berg 1965, cited in Gross 1978).

Freshwater macroinvertebrates including odonates, hemipterans, and coelopterans are found in most stickleback habitats and occasionally consume stickleback fry (Reimchen 1980; Reist 1980*b*; Foster *et al.* 1988). Their importance might be greater in localities where predatory fish are absent or where macrophytes and submerged debris provide the appropriate foraging substrates for these predators (Reimchen 1980). Leeches are also widely distributed and prey on stickleback eggs (Moodie 1972*a*). Leeches capture and consume adult stickleback confined in fish traps (Reimchen unpubl. obs.), but the importance of this in nature is unknown. Marine invertebrates have not been evaluated for their contribution to stickleback mortality. There is evidence for passive consumption by jellyfish (Rasmussen 1973).

PREDATOR DIVERSITY WITHIN LOCALITIES

Multiple predator species can occur in a single locality. In one of the few systematic studies of predation in a stickleback population, Moodie (1972*a*) observed that leeches ate eggs from nests, prickly sculpin ate eggs and small stickleback, whereas cutthroat trout ate subadult and adult stickleback.

Table 9.1 Taxonomic diversity of predators on threespine stickleback. Citations are limited to studies where direct evidence of predation was observed (stomach contents or visual observations).

Group	Species	Reference[a]
Cnidaria	Jellyfish *Aurelia* sp.	1
Oligochaeta	Leech *Haemopis marmorata*	2
Insecta	Dragonfly *Aeshna palmata*	3
	Water scorpion *Ranatra* sp.	4
	Backswimmer *Notonecta* spp.	4
	Giant water bug *Lethoceros americanus*	4
Pisces	Dolly Varden *Salvelinus malma*	2
	Arctic charr *Salvelinus alpinus*	5
	Atlantic salmon *Salmo salar*	6
	Rainbow trout *Oncorhynchus mykiss*	7
	Cutthroat trout *Oncorhynchus clarki*	2,8–10
	Coho salmon *Oncorhynchus kisutch*	11,45
	Herring *Clupea harengus*	12
	Sea scorpion *Taurulus bubalis*	12
	Eel *Anguilla anguilla*	13
	Cod *Gadus morhua*	12
	Coalfish *Pollachius virens*	12
	Garfish *Belone belone*	12
	Pike *Esox lucius*	13,14
	Perch *Perca fluviatilis*	12
	Pikeperch *Stizostedion lucioperca*	12
	Western mudminnow *Novumbra hubbsi*	15
	Bass *Morone labrax*	12
	Chub *Leuciscus cephalus*	12
	Northern squawfish *Ptychocheilus oregonensis*	16
	Prickly sculpin *Cottus asper*	2,44
	Aleutian sculpin *Cottus aleuticus*	17
	Threespine stickleback *Gasterosteus aculeatus*	4,5,18–22
Reptilia	Two-striped garter snake *Thamnophis couchi*	23
Aves	Red-throated loon *Gavia stellata*	24,25
	Arctic loon *G. arctica*	24
	Pacific loon *G. pacifica*	26
	Common loon *G. immer*	25
	Pied-billed grebe *Podilymbus podiceps*	11
	Horned grebe Podiceps auritus	11
	Red-necked grebe *Podiceps grisegena*	11
	Great-crested grebe *Podiceps cristatus*	27
	Western grebe *Aechmophorus occidentalis*	26
	Double-crested cormorant *Phalacrocorax auritus*	11
	Grey heron *Ardea cinerea*	12,28
	Great blue heron *A. herodias*	11,26,29
	Black-crowned night heron *Nycticorax nycticorax*	30
	Scaup *Aythya marila*	31
	Scaup *Aythya* sp.	11,26
	Tufted duck *A. fuligula*	31

Table 9.1 *(Cont)*

Group	Species	Reference[a]
	Oldsquaw *Clangula hyemalis*	11,26,31
	Bufflehead *Bucephala albeola*	11,26
	Barrow's goldeneye *B. islandica*	31
	Hooded merganser *Lophodytes cucullatus*	11,26
	Common merganser *Merganser merganser*	11,26,32
	Merganser *M. serrator*	27,31,33,34
	Gull *Larus ridibundus*	35
	Gull *L. canus*	35
	Herring gull *L. argentatus*	29
	Ring-billed gull *L. delawarensis*	29
	Common tern *Sterna hirundo*	36,37
	Arctic tern *S. paradisaea*	36
	Greater yellow legs *Totanus melanoleucus*	29
	Lesser yellow legs *T. flavipes*	29
	American crow *Corvus branchyrhynchus*	29
	Bronzed grackle *Quiscalus quiscula*	29
	Kingfisher *Alcedo atthis*	38,39
	Belted kingfisher *Ceryle alcyon*	11,26
Mammalia	North American river otter *Lutra canadensis*	11
	European river otter *L. lutra*	40
	Mink *Mustela vison*	41
	Fur seal *Callorhinus ursinus*	42
	Human *Homo sapiens*	43

[a] References: 1, Rasmussen (1973), cited in 12; 2, Moodie (1972*a*); 3, Reimchen (1980); 4, Foster *et al.* (1988); 5, Greenbank and Nelson (1959); 6, Jakobsen *et al.* (1988); 7, Hagen and Gilbertson (1973*b*); 8, Armstrong (1971); 9, Nilsson and Northcote (1981); 10, Reimchen (1990); 11, present study; 12, Gross (1978); 13, Hartley (1948); 14, Frost (1954); 15, McPhail (1969); 16, Hagen and Gilbertson (1972); 17, Baxter (1956), cited in 5; 18, Semler (1971); 19, Wootton (1979*a*); 20, Kynard (1978*a*); 21, Whoriskey and FitzGerald (1985*c*); 22, Hyatt and Ringler (1989*b*); 23, Bell and Haglund (1978); 24, Madsen (1957); 25, Reimchen and Douglas (1980); 26, Reimchen and Douglas (1984*a*); 27, Giles (1984*b*); 28, Giles (1981); 29, Whoriskey and FitzGerald (1985*b*); 30, FitzGerald and Dutil (1981); 31, Bengtson (1971); 32, Munro and Clemens (1937); 33, Sjoberg (1985); 34, Sjoberg (1989); 35, Giles and Huntingford (1984); 36, Lemmetyinen (1973); 37, Becker *et al.* (1987); 38, Eastman (1969); 39, Raven (1986); 40, Jenkins *et al.* (1979); 41, Gerell (1968), cited in Wootton (1976); 42, Biggs pers. comm.; 43, Berg (1965); 44, Pressley (1981); 45, Zorbidi (1977).

Consequently, stickleback are exposed to different predatory regimes during their ontogeny. In a spine-deficient stickleback population without sympatric predatory fish, odonate naiads took juvenile stickleback while seven species of avian piscivores (common loon, red-necked grebe, horned grebe, common merganser, red-breasted merganser, hooded merganser, and belted kingfisher) took subadults and adults (Reimchen 1980). Mortality in an estuarine population of stickleback in Quebec was due to at least eight species (great blue heron, black-crowned night heron, herring gull, ring-billed gull, greater yellow legs, lesser yellow legs, American crow, and bronzed grackle) (Whoriskey and FitzGerald 1985*b*). Up to six species of avian piscivores have been observed in lakes in the Outer Hebrides

(Giles 1987*c*). Therefore, any differences in spatial or temporal components to foraging activity by different predators might be expected to influence the phenotypic distribution within the prey population (Reimchen 1980).

QUANTIFYING MORTALITY AND PREDATION INTENSITY

Several methods have been employed to estimate predation intensity on stickleback populations. During some of the first such studies in western North America, presence or absence of predatory salmonids was equated with presence or absence of predation (Hagen and Gilbertson 1972; Moodie and Reimchen 1976*a*). Although crude, this method yielded predictive differences in morphological traits including body size, lateral plate number, and relative spine length of the stickleback. This method was improved by examining stomach contents of trout and measuring the proportion of trout containing stickleback among different populations (McPhail 1977).

Gross (1977, 1978), working on European populations of stickleback, undertook a more rigorous assessment of predation intensity. He compiled a list of known or probable predators on stickleback that included 15 species of fish and five bird species. Using distributional records for each species and some site-specific information on importance of stickleback in the diet of the piscivores, he classified localities into one of three predation levels: none, low, or high. There were significant associations between predation level and a diversity of morphological traits, including relative spine length and number of lateral plates. Although diet was not directly examined for most of the predatory species in his studies, and macro-invertebrates were not considered as a source of mortality, the study by Gross clearly emphasized the predictive value of such predation indices and demonstrated the complexity of predator associations operating on stickleback populations.

Seasonal differences in predation intensity have been addressed. The highest proportion of trout stomachs containing stickleback occurred in winter and the lowest proportion occurred in spring and summer (Moodie 1972*a*; Hagen and Gilbertson 1973*b*). Trout collected during winter also had more stickleback per stomach than those collected during spring, further suggesting increased predation levels during winter (Hagen and Gilbertson 1973*b*). Using estimates on metabolic rates and daily food consumption, Reimchen (1990) concluded that predation levels were highest in summer and that the increased proportion of trout stomachs containing stickleback during winter was the result of reduced stomach evacuation rates during cold temperatures, rather than increased predation. Avian piscivores such as red-breasted merganser and common loon typically prey most intensively on stickleback in spring and summer (Rad 1980; Reimchen and Douglas 1984*a*; Sjoberg 1989), and within these periods can exhibit a narrow pulse of activity when most predation occurs (Reimchen and Douglas 1980).

EVIDENCE FOR PREDATORS AS SELECTION AGENTS ON STICKLEBACK

Differential predation on phenotypes appears to be a common theme in studies of stickleback morphology, yet direct evidence for this remains limited. Loss of the typical, red nuptial coloration of threespine stickleback has occurred in a number of populations. In two cases, this loss has been interpreted as a consequence of strong selection against conspicuous red males by predatory salmonids (Semler 1971; Moodie 1972*a*). This interpretation has recently been questioned, however, because a geographical survey of populations in the Queen Charlotte Islands, British Columbia, Canada, detected no association between predation regime and extent of nuptial colour (Reimchen 1989). In Washington State, USA, males with black, rather than red, nuptial colour, occur in association with the western mudminnow *Novumbra hubbsi*. Originally, predation by the mudminnow on fry was thought to have favoured the black nuptial phenotype (McPhail 1969), but subsequent work (Hagen and Moodie 1979) failed to find field evidence of fry predation by the mudminnow. Thus, evidence that predation has favoured loss of red nuptial colour is equivocal at best.

Highly divergent adult body sizes of stickleback occur in some populations and are associated with extensive trout predation. Large size may provide a size refuge against these gape-limited predators (Moodie 1972*a*; Reimchen 1988, 1990, 1991*a*) while small adult size may represent selection for early reproduction if opportunity for escape from predators is small (McPhail 1977). However, there is no empirical evidence for predator-mediated selection on adult size that has partitioned out the numerous additional factors such as gravidity and longevity that could also influence selection on body size.

There is stronger evidence that predation has influenced selection of spine lengths. In Mayer Lake in the Queen Charlotte Islands, the stomachs of cutthroat trout contained stickleback which had proportionately shorter spines than did those in the population, suggesting that long-spined stickleback had an advantage against predatory fish (Moodie 1972*a*). Geographical surveys in north-western North America (Hagen and Gilbertson 1972) and Europe (Gross 1978) have documented positive associations between predation by vertebrates and spine length, as would be expected if long spines provided defence against gape-limited piscivores.

In contrast, predation by macroinvertebrate piscivores such as odonate naiads could favour the loss of spines if spines facilitate the capture and manipulation of stickleback by these predators (Reimchen 1980). Although this is an intriguing possibility, preliminary experiments in this study detected no difference in escape probabilities among spined and non-spined phenotypes. Reist (1980*b*, 1983) detected differential predation by aquatic insects on spine phenotypes of brook stickleback, *Culaea inconstans*, but this was a consequence of differential avoidance of capture, rather than escape after capture. Postcapture escapes from predatory invertebrates were infrequent in each of these experiments, and as a consequence, the specific importance of spine loss during manipulation has not yet been effectively addressed.

Analyses of stomach contents of piscivorous fish also provide evidence of differential predation on lateral plate phenotypes. Hagen and Gilbertson (1973*b*) found that in rainbow trout, stickleback with seven lateral plates were eaten less frequently than expected in each of three years. This phenotype had a 7 per cent to 40 per cent reduction in predation levels relative to other plate phenotypes, and the greatest increase in frequency in the population occurred during the season when predation appeared to be most intense. In Mayer Lake, stickleback with eight lateral plates were marginally but significantly more common in the stomachs of cutthroat trout than were other phenotypes (Moodie 1972*a*). In contrast with these results from predatory fish, stickleback with five lateral plates appear to be less vulnerable to predation by two-stripe garter snake than are other plate phenotypes (Bell and Haglund 1978), demonstrating that predators can differ in their effect on lateral plate phenotypes.

Geographical surveys indicate additional evidence for selection on lateral plate phenotypes. In north-western North America, a mode of seven lateral plates occurred where predatory fish were present and a lower mode where these were absent (Hagen and Gilbertson 1972; Moodie and Reimchen 1976*a*). In Europe, modes of five to seven plates were found where pike and perch were abundant, whereas modes were lower outside the distribution of these predators (Gross 1978). Apart from number of lateral plates, there is evidence for decreased variance and decreased asymmetry in lateral plates where predatory fish were present, suggesting increased normalizing selection on these traits (Moodie and Reimchen 1976*a*).

The causes of the differential vulnerability of lateral plate phenotypes are unknown. Because lateral plate phenotypes display different evasive responses to predatory fish, this differential vulnerability may arise from genetic linkage between loci influencing behavioural and lateral plate expression (Moodie *et al.* 1973). Aggression levels and lateral plate phenotype are correlated, and this correlation could alter susceptibility to predators (Moodie 1972*b*; Huntingford 1981). It is also possible that the characteristic plate modes and plate positions observed in populations is directly functional if phenotypes differ in escape rate during manipulation by predators (Reimchen 1983, 1992*b*). The relative importance of either behavioural or morphological attributes to predator defence awaits more detailed study.

Predatory fish can have a selective influence on vertebral expression in stickleback. Stickleback fry which were variable in both total vertebral number and in the ratios of abdominal to caudal vertebrae had non-random survival when exposed to predation by pumpkinseed sunfish, *Lepomis gibbosus* (Swain and Lindsey 1984; Swain 1986). Differences in burst swimming velocity of fry occur among vertebral phenotypes, and may account for the differential survival observed in experiments and the temporal changes in frequencies of phenotypes in wild-captured fry (Swain 1992*a*,*b*). Such differences during early periods of the life history might explain the spatial differences in vertebral phenotypes within lakes, but would not readily account for seasonal changes in vertebral number of adult stickleback (Reimchen and Nelson 1987).

PREDATORS AND MORPHOLOGY IN QUEEN CHARLOTTE ISLAND STICKLEBACK

Morphological diversity

Threespine stickleback of the Queen Charlotte Islands exhibit a remarkable degree of morphological diversity among populations (Moodie and Reimchen 1976*a*; Reimchen 1983; Reimchen *et al.* 1985). The size of gravid females ranges from 27 mm SL to 110 mm in different lakes. The pelvis is absent in two populations, while dorsal and anal spines are deficient in these and four additional populations. Lateral plates vary concomitantly with spine expression. Where spines are long, the lateral plates which buttress the spines are generally well developed (Fig. 9.1(a)), but where spines are reduced in length or number, lateral plates are weakly expressed and frequently absent (Fig. 9.1(c),(d)) (Reimchen 1983, 1984).

Substantial morphological differentiation also exists over small geographical distances. Populations with robust spines and plates occur within several kilometres of those in which spines and plates are absent. Within a watershed, mean vertebral number is as variable as that found throughout the circumboreal distribution of the species (Reimchen *et al.* 1985). Between parapatric populations at stream–lake boundaries, where there is major opportunity for gene flow, morphologically discrete forms occur with adult body size of the lake form being twice that of the stream species (Moodie 1972*a*; Reimchen *et al.* 1985). These and other data on overlapping populations indicate speciation (McPhail page 411 this volume).

Evolutionary causes of this variation within and between populations may be very diverse. None of the divergent traits such as gigantism or spine reduction is unique to the Queen Charlotte Island populations; they may also occur in scattered localities throughout the distribution of stickleback (Larson 1976; Moodie and Reimchen 1976*b*; Campbell 1979; McPhail Chapter 14 this volume). The situation observed on the Queen Charlotte Islands is exceptional primarily in that the variation in this small geographic area encompasses the complete range of variation found throughout the circumboreal distribution of the species. Previous investigations (Moodie and Reimchen 1976*a*) concluded that differences in predation intensities, defined as the presence or absence of predatory fish, were a major factor in the morphological variability among populations. More recent research

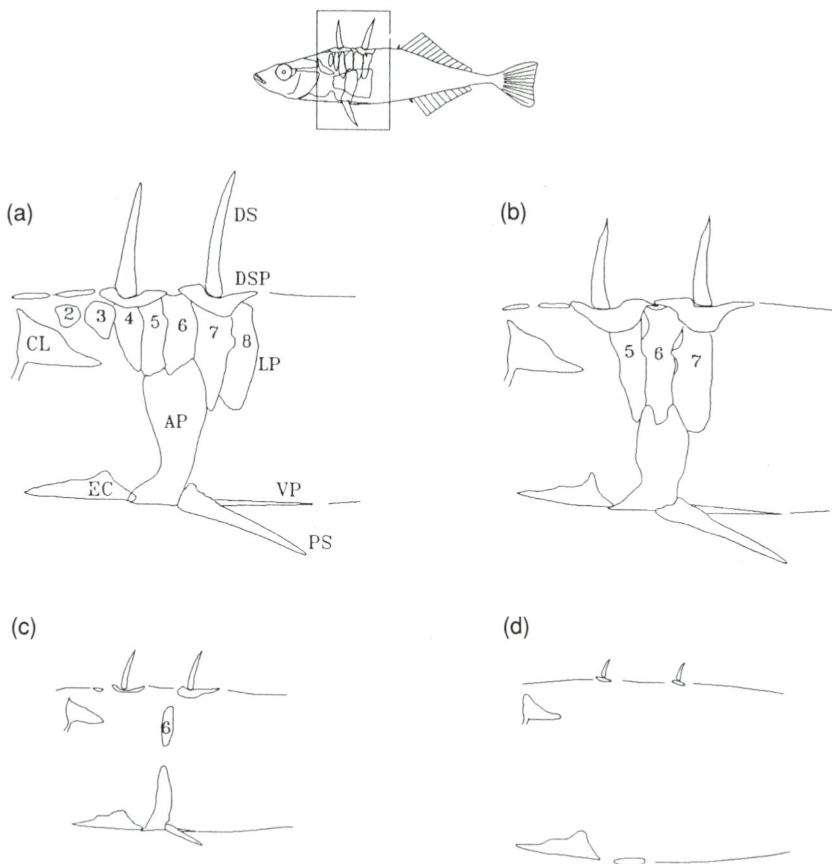

Fig. 9.1 Variation of lateral plate and spine expression among Queen Charlotte Island stickleback populations: (a) Eden Lake, (b) Hickey Lake, (c) Solstice Lake, (d) Serendipity Lake. Abbreviations: AP, ascending process; CL, cleithrum; DS, dorsal spine; DSP, dorsal spine plate; EC, ectocoracoid; LP, lateral plate (position shown by number); PS, pelvic spine, VP, ventral plate.

has demonstrated that avian piscivores and macroinvertebrates could also be significant predators in some of these localities (Reimchen 1980). These initial studies involved collections from 25 localities, less than 10 per cent of potentially habitable lakes in the archipelago. From 1975 to 1988, I made collections at all lakes in the archipelago to expand the analysis of the relationship between stickleback morphology and predation regime.

Evaluating predator occurrence

Distributional data on piscivores was obtained during the geographical surveys of stickleback. The presence of predatory fish was determined by gill netting, trapping, or rod-and-line fishing. Absence of large fish from some lakes was usually associated with poor drainage or extremes in water chemistry such as pH values near 4.0. However, even these localities may have occasional occupancy by anadromous salmonids that move into the upper reaches of most watersheds during periods of high run-off. Habitat occupation by avian piscivores was difficult to classify. These are often solitary or occur in pairs, and may readily be overlooked in sheltered habitats and on large lakes. Some species such as common merganser are absent from lakes during the day but occur near dusk and dawn (Reimchen and Douglas 1984*a*). Others such as grebes are only present during autumn and winter (Reimchen 1980). Consequently, some lakes were revisited at different times of the year and in different years. Macroinvertebrate abundance was also logistically difficult to ascertain because they were not regularly seen in visual surveys and were unpredictable in traps. Their occurrence was often detected by examining submerged debris. Several representative lakes were monitored weekly throughout the year for predators (Reimchen 1980; Reimchen and Douglas 1984*a*, see section below on Drizzle Lake).

Predator assemblages and habitat

Each habitat on the Queen Charlotte Islands has a characteristic assemblage of species. The occurrence and relative abundance of different taxa are associated with diverse factors such as shoreline cover, surface area, water depth, water colour, and productivity (Reimchen unpubl. data). For example, rainbow trout are common in clear water whereas cutthroat trout predominate in stained water. Flocks of avian piscivores are often abundant on dystrophic lakes but are rarely seen on oligotrophic lakes of comparable size. Large diving birds such as common loon did not occur on ponds less than 2 ha (Douglas and Reimchen 1988). Macroinvertebrates were common where cover such as submerged branches or vegetation was prevalent.

The major habitat types and the assemblage of species can be summarized as follows:

(1) Dystrophic ponds (< 1 ha) all with intermittent drainage: macroinvertebrates common (including odonate naiads and beetle larvae), small avian piscivores occasionally present (horned grebe, hooded merganser, red-throated loon, oldsquaw duck, belted kingfisher), trout absent, Dolly Varden occasionally present;

(2) large dystrophic lakes (1–200 ha) with intermittent drainage: small and large-bodied avian piscivores (common loon, red-throated loon, red-necked grebe, horned grebe, double-crested cormorant, common merganser, other diving ducks, belted kingfisher), macroinvertebrates present, predatory fish absent, otter occasional;

(3) large dystrophic lakes (1–200 ha) with open drainage: small and large-

bodied avian piscivores common (common loon, red-throated loon, grebes, double-crested cormorant, mergansers, other diving ducks, belted kingfisher), predatory fish resident (cutthroat trout, Dolly Varden, sculpin), otter usually present, macroinvertebrates present;

(4) Oligotrophic lakes (1–1800 ha) with open drainage: predatory fish common (rainbow trout, cutthroat trout, Dolly Varden, sculpin), avian piscivores present but probably less abundant than in dystrophic lakes (common loon, belted kingfisher, mergansers), otter present, macroinvertebrates present;

(5) small forested creeks: predatory fish usually common (rainbow trout, cutthroat trout, Dolly Varden, salmon, sculpin), avian piscivores present (hooded merganser, belted kingfisher, great blue heron), macroinvertebrates common, otter usually present;

(6) large unforested creeks and rivers: predatory fish common (cutthroat trout, rainbow trout, sculpin), avian piscivores present (double-crested cormorant, oldsquaw, common merganser, great blue heron, belted kingfisher), macroinvertebrates less evident, otter usually present.

Among these habitat types, which are part of a continuum, I recognize two basic predator assemblages: (A) invertebrate/bird (habitat categories 1, 2) and (B) fish/bird (habitat categories 3–6).

Given the taxonomic complexity of these predator regimes among each habitat type, it is not clear how predation intensity can be easily evaluated in a broad geographic survey. Stickleback in larger lakes appear to experience greater risks of predation because there is a greater diversity of predatory fish and birds in such lakes. However, these habitats also have a greater diversity of alternate prey (i.e. juvenile salmonids), a factor that could reduce the intensity of predation on a particular prey type. Stickleback in small dystrophic lakes would seem to experience low predation intensity because predators are seldom common, yet as demonstrated at one of these localities (Reimchen 1980), even a small number of avian piscivores combined with a paucity of alternate prey can produce substantial predation intensity. Ranking populations with respect to predation intensity, while a laudable goal, is in practice not possible without data on amount of predator-induced mortality in the survivorship curve for each population.

Lateral plates, predator assemblage, and lake area

Lateral plates have been the focus for many studies of stickleback. I extracted data on number of plates (on left side of the stickleback) from 58 lake populations (listed in Moodie and Reimchen 1976*a*; Reimchen 1983, 1988; Reimchen *et al.* 1985) and examined these in relation to predation regime. One of these populations was monomorphic for the complete morph ($\bar{X} = 33$ plates per side), while the remainder had only low morphs (grand

$\bar{X} = 4.1$, range 0–10.8). Partitioning for predator type yields a mean of 1.8 plates (range 0–4.6) for the bird/invertebrate assemblage ($N = 31$) and a mean of 6.6 (range 1.2–10.8, $N = 27$) for the fish/bird assemblage (F-ratio = 131; d.f. 1, 57; $P < 0.001$).

Lake area (ha) and lateral plate number are positively correlated in the bird/invertebrate, but not in the fish/bird assemblage (Fig. 9.2). Populations in small lakes (< 3 ha) average only a single lateral plate (Fig. 9.1(c)) whereas those in the largest lakes with the bird/invertebrate assemblage (200 ha) have three or four plates (Fig. 9.1(b)). None of the six populations with reduction in number of dorsal and pelvic spines occurs in habitats with the fish/bird assemblage.

Is any of this variation in lateral plate number adaptive? The average values near seven lateral plates in populations exposed to predatory fish and birds are consistent with those found in the small preliminary survey of this region (Moodie and Reimchen 1976*a*) and are similar to those found elsewhere in the geographical distribution of this stickleback (Hagen and Gilbertson 1972; Gross 1978). The seven lateral plates include the four structural plates that buttress the dorsal and pelvic spines and the three plates in the immediate postcranial and supracleithral region (Reimchen 1983). These anterior plates (Fig. 9.1(a)) also provide physical protection against a toothed predator (Reimchen 1992). That the populations with five

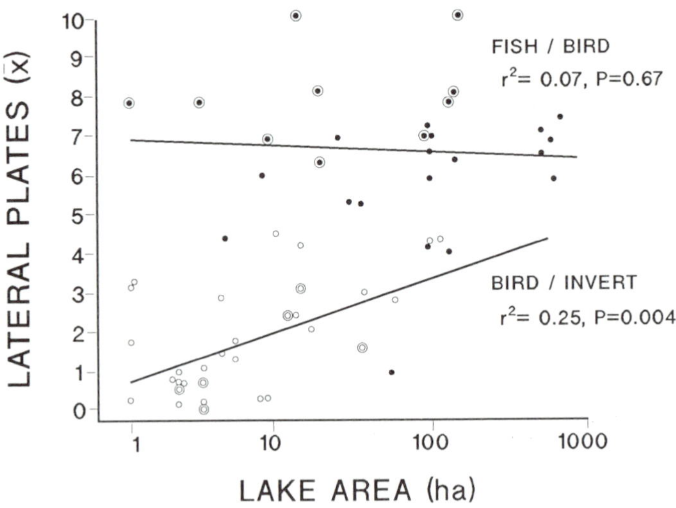

Fig. 9.2 Variation of mean number of lateral plates in relation to lake surface area and predator assemblage. Each point represents a separate population. Fish/bird communities are shown as solid points, and bird/invertebrate communities as open points. ● Cutthroat trout major fish predator; ⊙ rainbow trout major predator; ○ populations with normal spine complement; ◎ populations with spine deficiencies. All samples have $N > 50$.

to seven plates had cutthroat trout as a dominant predator while those with more than seven lateral plates had rainbow trout as the dominant predator (Fig. 9.2) suggests fine-scale tuning of the population, perhaps owing to differences in pursuit or manipulation behaviour of these predators.

Populations in larger lakes without predatory fish but with a diversity of avian piscivores have an average of three or four lateral plates. The difference in number of lateral plates from those observed in localities with predatory fish involves loss of plates at the anterior three positions (Fig. 9.1(a),(b)) and does not affect the major structural plates that buttress the dorsal and pelvic spines. Unlike toothed predators which can puncture the epidermis of the prey, avian piscivores hold the fish by compression, and only superficial scarring results (Reimchen 1988). The anterior lateral plates found where toothed predators are prevalent may not provide any advantage when birds are the dominant predators. These could conceivably be a disadvantage during pursuit by predators if burst swimming performance is compromised. Freshwater stickleback of the low-plated morph have a higher body flexure and higher burst swimming speed than the complete morph in anadromous stickleback (Taylor and McPhail 1986). If such differences also occur among stickleback with and without the anterior three plates, loss of these plates could be favoured when pursuit escapes were a major component of the total foraging failures by the predators (see below). Even small differences in burst velocity might be important against diving birds such as loons or grebes which submerge for only brief periods.

It is unclear why populations in small ponds and lakes exhibit a reduction in the number and size of the structural plates, as this would weaken the effectiveness of the spines (Fig. 9.1(c),(d)). As well, reduction of these structural plates is associated with reduction in size of spines (Reimchen 1983, 1984) and loss of spines (Fig. 9.2). Nelson (1969) suggested that reduced spines might be useful during escape into dense vegetation. Although the bog ponds have few macrophytes, they often have soft organic substratum. If stickleback with reduced spines and lateral plates are able to avoid capture by gaining refuge in the ooze more quickly than spined individuals, this trait could be favoured. Giles and Huntingford (1984) considered but rejected this hypothesis in analyses of spine deficiencies from the Outer Hebrides because there were no detectable differences in escape behaviour among individuals from populations with normal and reduced spine expression. However, differences between phenotypes within populations were not addressed, and therefore Nelson's suggestion remains untested. Alternatively, reduction in spines and plates (Fig. 9.1(c)) in progressively smaller lakes could result from a gradual shift in the predation regime. As lake size decreases, species diversity of avian piscivores is reduced, and it is possible that macroinvertebrates account for a greater proportion of stickleback mortality. If so, this might produce the observed reduction in spines and plates, given the grappling manipulation behaviour of these predators (Reimchen 1980; Reist 1980*b*). However, in the Outer Hebrides, predatory fish are also present in populations with spine and lateral plate reduction; Giles (1983*a*) suggested that reduced calcium levels in the water could be the selective agent. An association between low calcium levels and loss of armour is also present in Alaska (Bell *et al.* 1985*a*, in press) and in the Queen Charlotte Islands (Reimchen unpubl. data), yet this is coupled with characteristic predation regimes. These competing hypotheses will not be satisfactorily resolved until causes of mortality and predator-foraging behaviour are evaluated in some of these populations.

Current deficiencies in data

Although considerable progress has been made towards determining the role of predators as selection agents in stickleback populations, more attention is required on specific predator–prey interactions in each locality. Previous efforts at estimating predation intensity, either by categorizing localities according to the presence or absence of predatory fish or by compiling lists of potential predators, are not sufficient to yield realistic estimates. Furthermore, these data may provide little, if any, useful information on selection intensities because selection will be dependent on foraging efficiencies of the predators.

PARTITIONING CAUSES OF MORTALITY: DRIZZLE LAKE

Despite the wealth of studies on predator–prey interactions in numerous taxa (see Vermeij 1987 for review), the causes and amount of age-specific mortality throughout the life history of the prey have not been determined for any population or species, including stickleback. This information is fundamental to evolutionary studies of predation. During ontogeny, spines on the stickleback are expressed shortly after hatching, but lateral plates which buttress the spines are not fully expressed until the stickleback reach about 25–40 mm SL (Hagen 1973; Bell 1981; unpubl. data). Predation attempts that occur prior to full development of lateral plates, such as are observed for macroinvertebrate predators (Reimchen 1980; Foster *et al.* 1988), would have profoundly different evolutionary consequences from attacks on adults. Seasonal differences in mortality could alter the nature of selection. For example, the relative advantage of lateral plate phenotypes of stickleback against predatory fish depends on whether the stickleback were acclimated to summer or winter temperatures (Moodie *et al.* 1973). Habitat of stickleback changes with season, with fish moving offshore during autumn and onshore during spring. This alteration can be expected to impose different predator regimes (Reimchen 1980; Werner *et al.* 1983). Habitat preferences of fish generally shift during ontogeny, and there is evidence for corresponding changes in ecological interactions (Werner and Gilliam 1984). Yearly differences in predation regime could alter strength and direction of selection. The occasional extremes, rather than average conditions, may generate the important evolutionary effects on the population (Boag and Grant 1981). The relevance of each of these ontogenetic, spatial, or temporal factors could be compounded if there were multiple causes of mortality with distinctive selective effects on phenotype.

For this purpose, a 10 yr investigation of the interactions between stickleback and their predators was undertaken at Drizzle Lake on the Queen Charlotte Islands in western Canada. Several aspects of this study have been completed, including the analyses of seasonal and spatial patterns of avian piscivore activity (Reimchen and Douglas 1980, 1984*a*), examination of functional relationships between spines and lateral plates (Reimchen 1983), spatial and temporal variation in vertebral phenotypes within the lake (Reimchen and Nelson 1987), description of predator-induced injuries, their frequencies over time and among size classes (Reimchen 1988), and the analysis of yearly size-specific mortality caused by cutthroat trout (Reimchen 1990). Other aspects, such as evaluating predator foraging efficiencies, are continuing. Here I synthesize available data on the predator complex at Drizzle Lake and consider age-specific causes and levels of mortality through the life history of the stickleback.

Description of study area

Drizzle Lake is located on the north-eastern corner of the Queen Charlotte Islands, western Canada (Fig. 14.1), on a low-elevation plain (< 100 m) covered with *Sphagnum* bogs and coniferous forests. The lake (112 ha) is dystrophic with simple bathymetry, reaching a maximum depth of 30 m. Its water is deeply stained (transmission at 400 nm = 67 per cent; Reimchen 1989) and aquatic macrophytes are rare. The locality, apart from being representative of the broader region, was chosen for several reasons. Stickleback reach a large body size (110 mm), are highly melanic, and are endemic to the locality. Although exposed to trout predation, stickleback have fewer lateral plates than are observed in virtually all other equivalent populations (Moodie and Reimchen 1976*a*). Due to its remoteness, the lake has received negligible human-induced or other known ecological disturbance in recorded history. This is important for evaluating life history and phenotypic variability because the genetic and phenotypic structure of the population should be intact. The watershed has been given protected status as an ecological reserve and should remain undisturbed in the future.

Methods for estimating prey and predator abundance

Mark–recapture methods were employed to estimate population sizes of adult stickleback and cutthroat trout (Reimchen 1990). The size distribution and abundance of stickleback were determined from horizontal and vertical transects over the lake. These represented 96 sites in the lake which were generally sampled every two months from 1980 to 1983. Densities of littoral fish (salmonids and stickleback) were also surveyed using beach seines. Numbers and species of avian piscivores were determined daily throughout much of the year over a 5 yr period, and records were kept of foraging positions on the lake (Reimchen and Douglas 1984*a*).

Evaluating mortality

Causes of mortality in the life history of the stickleback were determined using a combination of techniques, including stomach-content analyses of gill-netted predatory fish (Reimchen 1990), shoreline surveys for regurgitated pellets from kingfishers and spraints from otters, and littoral collections of macroinvertebrates. Observations were made through spotting scopes on the foraging behaviour of avian piscivores and otters (Reimchen and Douglas 1980, 1984*a*). Prey were identified to the lowest taxon possible (species in most cases).

The size of stickleback consumed by predators was determined by several methods. Those in trout stomachs were measured directly if there was only limited digestion. For disarticulated stickleback in trout stomachs and in kingfisher pellets, I measured pelvic spine length, which is a good estimator of standard length ($r^2 = 0.92$, $N = 679$, $P < 0.001$). In otter spraints, dorsal and pelvic spines were fractured but the hypural plate was generally intact, and this was used to estimate stickleback size. Maximum width of this plate measured on radiographs of reference specimens provided a good predictor of stickleback body length ($r^2 = 0.93$, $N = 95$, $P < 0.001$). Odonates captured in the lake were placed in small aquaria and within 24 h generally produced intact faecal pellets which were preserved and later examined for fish bones.

For all of the avian piscivores that brought stickleback to the surface, I made rough estimates of stickleback size relative to bill length. A number of manipulation events were videotaped, allowing more detailed measurements (methods in Reimchen and Douglas 1984*b*; Reimchen 1988). Where stickleback were large relative to the bill, manipulation time by the bird was much greater than with smaller stickleback, as might be expected for gape-limited piscivores (Werner 1974). Accordingly, I was able to use relative manipulation time to recognize three general size classes (30–50, 50–70, 70–90 mm) in foraging events that were too distant to permit direct measurement. Several of the diving ducks, such as scaup and white-winged scoter, swallowed most of their prey beneath the surface. For these infrequent foragers, I estimated an average prey size based on combinations of bill size, dominant foraging habitat in the lake, and size availability of stickleback. Such data deficiencies are identified in the results.

The number of stickleback eaten yearly was estimated for each predator, based on daily caloric requirements, proportion of stickleback in the diet at monthly intervals, length distributions of stickleback in the diet, and average number of predator foraging days per month. The last was based on data over a 5 yr period. Daily caloric requirements (*D*cal) for predatory fish were calculated as

$$D\text{cal} = 15.116 * W^{0.767} * e^{0.138} * T \tag{9.1}$$

(Elliott 1976), where W is weight of the predatory fish (g) and T is water temperature (°C). This was converted to g assuming a caloric equivalence of 5.0 J kg^{-1} (Cummins and Wuycheck 1971). These calculations provide a close approximation to the actual daily consumption by trout (Reimchen 1990).

The daily weight of fish consumed (*D*g) by avian piscivores was estimated as

$$\log_{10} D\text{g} = -0.293 + 0.85 * \log_{10} W \tag{9.2}$$

(Nilsson and Nilsson 1976), where W is weight of bird (g). I was able to compare actual daily fish consumption with *D*g for a common loon which brought the majority of the prey to the surface prior to swallowing. On two days, separated by two weeks, an individual bird was observed continuously from dawn to dusk and all prey captures were recorded. The loon ate 135 fish (estimated total weights 300–405 g) and 198 fish (estimated weight 380–495 g) on the two days. Predicted daily consumption (*D*g) for a 3000 g loon is 460 g, which is comparable to the actual value. The average numbers of foraging days for avian piscivores are shown in Table 9.2.

Daily fish consumption by the river otter ranges from 10 to 23 per cent of its body weight (Erlinge 1968; Chanin 1985). I assume a value of 10 per cent and an average otter weight of 13 kg (Chanin 1985).

Piscivores: diet, abundance, and yearly occurrence

Piscivory was observed in 21 species over the 10 yr study period, and in all of these, stickleback were present in the diet. This assemblage of predators comprises birds (15 spp.), fish (4 spp.), mammals (1 sp.), and odonates (1 sp.) (Table 9.2). Fish and odonates were resident, while the remaining species were seasonal itinerants, occupying the lake for variable periods. Population estimates and number of foraging days varied considerably among species and among seasons (Reimchen and Douglas 1984*a*). All size classes of stickleback were eaten, ranging from 10 to 100 mm SL (Fig. 9.3).

Table 9.2 Species that preyed on stickleback observed at Drizzle Lake, Queen Charlotte Islands, between 1976 and 1985. Non-foraging species are excluded. Mean foraging days yr^{-1} is the total number of individuals × number of foraging days. Stickleback in diet is proportion of predator's total diet.

Species	Years observed										Population estimate	Stickleback in diet (%)
	76	77	78	79	80	81	82	83	84	85		
Resident												
Cutthroat trout *Oncorhynchus clarki*	+	+	+	+	+	+	+	+	+	+	220	0.80[a]
Dolly Varden *Salvelinus malma*	+	+	+	+	+	+	+	+	+	+	100	0.15[a]
Coho salmon *Oncorhynchus kisutch*	+	+	+	+	+	+	+	+	+	+	4000	0.05[a]
Stickleback *Gasterosteus aculeatus*	+	+	+	+	+	+	+	+	+	+	75 000 adults	0.015[a]
Dragonfly *Aeshna palmata*	+	+	+	+	+	+	+	+	+	+	?	0.17?[a]
											Average foraging days yr^{-1}	
Seasonal visitors												
Red-throated loon *Gavia stellata*	+	+	+	+	+	+	+	+	+	+	118	0.10[b]
Pacific loon *Gavia pacifica*				+							4	1.00[c]
Common loon *Gavia immer*, summer	+	+	+	+	+	+	+	+	+	+	710	0.50[b]
winter	+			+	+	+		+	+		99	1.00[c]
Pied-billed grebe *Podilymbus podiceps*					+						1	1.00[c]

	76	77	78	79	80	81	82	83	84	85		
Horned grebe *Podiceps auritus*	+	+	+	+	+	+	+	+	+	+	24	1.00[c]
Red-necked grebe *Podiceps grisegena*	+	+	+	+	+	+	+	+	+	+	196	1.00[c]
Western grebe *Aechmophorus occidentalis*							+				<1	1.00[c]
Double-nested cormorant *Phalacrocorax auritus*	+	+	+	+	+	+	+	+	+		42	1.00[b]
Great blue heron *Ardea herodias*	+	+		+	+	+	+		+	+	<1	0.20[b]
Scaup *Aythya* spp.	+	+	+	+	+	+	+	+	+	+	12	0.20[b]
Oldsquaw *Clangula hyemalis*	+	+	+	+	+	+	+	+	+	+	51	1.00[b]
White-winged scoter *Melanita deglandi*	+	+	+	+	+	+	+	+	+	+	11	1.00[b]
Bufflehead *Bucephala albeola*	+	+	+	+	+	+	+	+	+	+	253	0.05[c]
Hooded merganser *Lopodytes cucullatus*	+		+	+	+	+	+	+	+	+	95	1.00[c]
Common merganser *Merganser merganser*	+	+	+	+	+	+	+	+	+	+	53	1.00[c]
Belted kingfisher *Ceryle alcyon*	+	+	+	+	+	+	+	+	+	+	258	0.91[a]
River otter *Lutra canadensis*					+	+	+	+		+	15	0.95[a]

[a] Dietary data derived from stomach or pellet analysis.
[b] Dietary data imferred from foraging habitat, prey availability, and gape.
[c] Dietary data derived from surface manipulation behaviour.

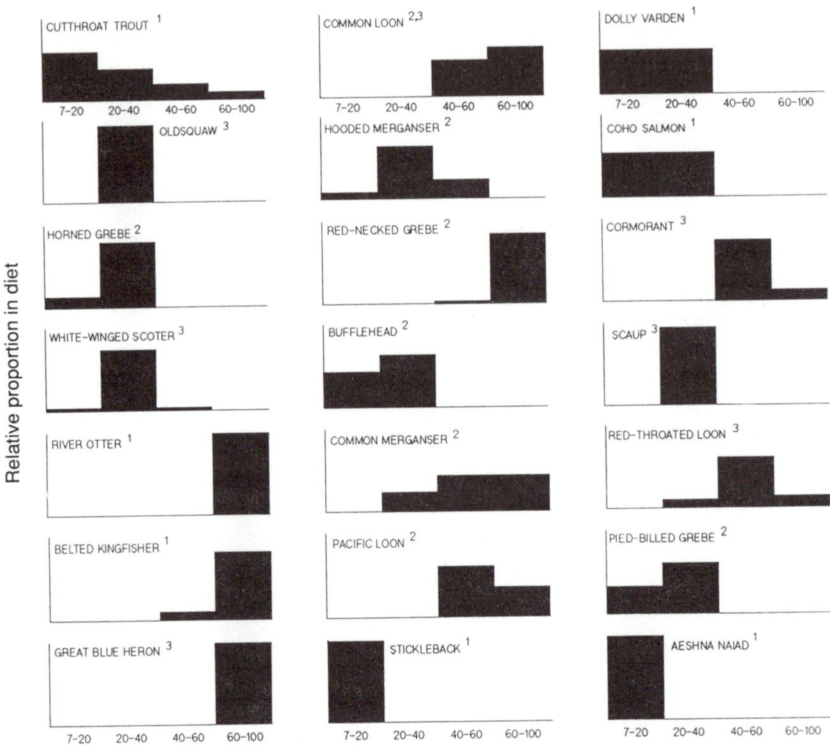

Fig. 9.3 Estimated size classes of stickleback eaten by predators. Data on size distributions are (1) derived from stomach or pellet analysis, (2) observation of surface manipulation behaviour, and (3) inferences from foraging habitat, prey availability, and gape.

Cutthroat trout

Stickleback constituted more than 80 per cent of the diet (Reimchen 1990). Of 1900 fish extracted from trout (range in SL 120–410 mm) over an 8 yr period, all but 8 were stickleback. There was a mean of 5.8 stickleback (range 0–55) per trout stomach, with the prey size ranging from 10 to 95 mm SL. Ratios of stickleback SL to trout SL were lowest in summer, increasing during autumn and winter and reaching a maximum in late spring. Mark–recapture methods provided population estimates of about 90 trout in winter and 250 in summer.

Dolly Varden

Benthos (primarily trichopteran larvae) was the dominant prey in 84 per cent of all stomachs ($N = 67$), while fish eggs were dominant in 16 per cent. Seven of the stomachs contained masses of trout eggs, and three contained an entire stickleback nest including eggs, fry, vegetation, and adhering sand. Low levels of piscivory (16 per cent of all stomachs) were detected throughout the year, and in all cases, the prey were small stickleback (< 40 mm). The number of Dolly Varden that were large enough (120–250 mm SL) to consume stickleback was estimated at 150 during winter and 25 during summer. During winter, many anadromous Dolly Varden (250–500 mm SL) enter the lake, but stomach content analyses ($N = 160$) from November to April showed no evidence of foraging.

Coho salmon

Non-foraging adult coho salmon enter the lake in November en route to stream spawning gravels. Juveniles, which appear in the lake the following spring and remain for 1 yr prior to migration to the ocean, primarily eat insects and benthos throughout the year. In 109 coho collected during the peak of the stickleback breeding season (April–July), I found no evidence of predation on fry, and only one stomach contained stickleback eggs. However, of 151 stomachs from August to November, two coho had consumed juvenile stickleback (1.3 per cent). From seining, I estimated the coho population at about 4000 individuals (40–130 mm SL).

Stickleback

Stickleback primarily took plankton, and secondarily benthos. During summer months, all size classes of stickleback engaged in low levels of conspecific predation. Of 67 adults (> 70 mm SL) examined, 11 (16.4 per cent) contained eggs and 2 (3 per cent) had fry. Of 134 subadults (50–70 mm), 7 (5.2 per cent) contained eggs while 2 (1.5 per cent) had fry. Of 543 large juveniles (30–50 mm), none had eggs in the stomach but 7 (1.9 per cent) had fry. Of 509 small juveniles (15–30 mm), none had eggs and a single individual (0.2 per cent) had fry. There were an estimated 75 000 (range 30 000–120 000) adult stickleback in the lake. Based on recaptures, the adult cohort is composed of at least five year classes (3–8; Reimchen 1992*a*). From nest densities and the number of eggs per nest, yearly production of fry was estimated at 12 million (range 4 million to 24 million).

Macroinvertebrates

Over the 10 yr data-collection period, which included numerous visual observations, beach seining, and minnow trapping, I rarely encountered any macroinvertebrate piscivores. Odonate naiads were present but rare. Leeches (Hirudinea), which are egg predators (Moodie 1972*a*), were rarely seen.

Eight odonate naiads were collected from the littoral zone when stickleback fry were abundant. Only one of these contained the bony remains of a stickleback fry (about 12 mm). Naiads placed in aquaria quickly stalked, attacked, and consumed fry, suggesting familiarity with this prey. The population of naiads was small and I did not estimate the size.

River otter

Analyses of spraints collected in 1979 ($N = 14$) and 1988 ($N = 8$) yielded the remains of 325 fish (323 stickleback and 2 salmonids). In both years, more than 90 per cent of the stickleback taken by otter were of adult size. Stickleback consumed were larger in 1988 than those in 1979 (83 mm v. 78 mm, $P < 0.01$, *t*-test). With a spotting scope, I observed 65 prey captures, all of which were adult stickleback.

Belted kingfisher

Three hundred and twelve prey were identified in 106 pellets of the belted kingfisher collected over 3 yr. These include stickleback ($N = 283$, 90.7 per cent), salmonids ($N = 16$, 5.1 per cent), and odonate naiads ($N = 10$, 3.2 per cent). The diet was similar between months and between years. The majority of stickleback captured were adults (75–85 mm SL).

Common loon

Loons that inhabited the lake in spring and autumn predominantly consumed stickleback (Reimchen and Douglas 1980). In autumn 1979 and spring 1981, I observed 674 prey brought to the surface and swallowed; all were stickleback. Most of the stickleback captured were 50–70 mm SL (range 40–90 mm).

During summer, large numbers of common loons (up to 89 individuals per day) foraged regularly, but they rarely brought prey to the surface (6 adult stickleback in approximately 2000 dives). The loons were taking either small stickleback or small salmonids, either of which can be swallowed underwater. Minnow traps set in limnetic regions where the loons foraged yielded exclusively stickleback. Large-mesh gill nets set for salmonids in this region were usually empty, but occasionally had coho salmon fry and rarely cutthroat trout. Traps set about 30 m from shore, which is generally as close to shore as common loons will forage, yielded predominantly stickleback (98 per cent) and rarely coho salmon fry and Dolly Varden. I searched the shoreline during periods of intensive loon foraging activity and frequently found injured fish with 'aviscars' (Reimchen 1988) at the drift line. Among 580 fish collected with injuries, there were 576 stickleback, 2 trout, 1 Dolly Varden and 1 coho fry. The evidence, although circumstantial, indicates that in summer common loons forage primarily on subadult stickleback (40–60 mm SL) and secondarily on larger size classes of fish (60–80 mm). For estimating total prey consumption, I will assume conservatively that stickleback represent 50 per cent of the diet during summer.

Red-throated loon

These loons foraged intermittently on the lake at dusk (Reimchen and Douglas 1980). During this period, I saw 23 fish captured by the loons, all of which were subadult stickleback (50–70 mm SL). On three instances, we observed pre-fledged loons capture stickleback in the nesting territory. During parental feeding of the young loons, all the prey brought to the young were marine fish and none were stickleback (Reimchen and Douglas 1984*b*).

Red-necked grebe

Grebes were regular seasonal residents on the lake, usually occurring in low numbers (1–5) between October and May. Stickleback were the only prey brought to the surface ($N = 63$). Based on number of stickleback consumed at hourly intervals over the day ($\bar{X} = 6.7$ fish h^{-1}, range 0–12), total daily consumption would exceed the estimated daily caloric requirements by 30 per cent. As a consequence, I assume that the grebe is not swallowing other species of prey such as salmonids beneath the surface. The average size of captured stickleback was estimated at 65 mm (range 50–90 mm SL).

Common merganser

Large flocks of mergansers regularly stayed overnight on the lake in spring and autumn (Reimchen and Douglas 1984*a*). In autumn, these flocks rarely foraged, but during April and May, I observed intensive surface foraging on stickleback at twilight. Stickleback are abundant at dusk just beneath the surface, and the mergansers captured the fish without underwater pursuit. Most of the captured stickleback appeared to be about 25–40 mm SL, which is the most abundant size class in the lake during spring.

Other predatory birds

Piscivores such as horned grebe and oldsquaw duck were seasonal residents (usually winter) and were uncommon, with rarely more than two individuals occurring on the lake at any time. Young-of-the-year stickleback (10–40 mm SL) appeared to be the predominant food item of these species. Double-crested cormorant, present during January and February, did not return prey to the surface. I sampled fish in the area of the lake where the cormorant foraged, and as I caught only stickleback I infer that this is the principal item in the diet. A single pair of hooded merganser nested adjacent to the lake in several years, and during the pre-fledging period, the adults and young were irregularly observed within 20 m of shore foraging on young-of-the-year stickleback. Other birds such as bufflehead were common winter residents that ate primarily trichopteran larvae and only rarely were observed capturing stickleback. Based on their foraging positions and prey capture, I estimated that 5 per cent of the bufflehead diet was composed of stickleback.

Numbers of stickleback eaten yearly by predators

An estimated total of 562 000 stickleback are consumed per year by all predators combined. Cutthroat trout and common loon were the two major predators (> 100 000 fish yr^{-1}). The remaining species each consumed < 30 000 fish yr^{-1} and can be considered minor predators (Fig. 9.4). I have been unable to make meaningful estimates of prey consumption by odonate naiads and stickleback. Conspecific predation could be very high, since juveniles, subadults, and adult stickleback ate fry.

Partitioning of total consumption between size classes of stickleback yields different rankings of predator importance. The estimated mortality of stickleback greater than 40 mm SL (i.e. > 12 months of age, full expression of lateral plates) was 228 000 individuals yr^{-1}. Avian piscivores accounted for 69 per cent of this total, while trout and otter consumed 30 per cent and 1 per cent respectively. Among adult stickleback (> 70 mm), 26 082 individuals were taken yearly. Common loon took 51.8 per cent, red-necked grebe 16.4 per cent, trout 11.9 per cent, and otter 10.8 per cent. Therefore, predators that were minor with respect to total mortality for all size classes combined (Fig. 9.4), and were uncommon on the lake,

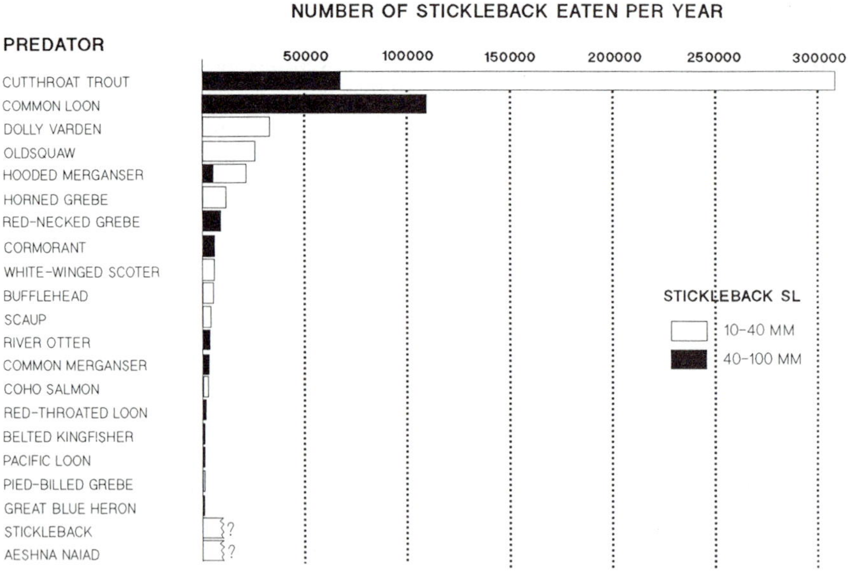

Fig. 9.4 Number of stickleback consumed annually for each predator at Drizzle Lake (averaged over five or more years). Rare species such as western grebe (one sighting in 10 yr) are not included.

become very important when considering mortality of subadult and adult size classes. Conversely, cutthroat trout, the most important predator overall, is only a minor predator on adult stickleback.

Seasonal differences in stickleback mortality

Total numbers of stickleback eaten by predators differed seasonally (Fig. 9.5(a)). The two major predators had their greatest yearly consumption during summer months. This was due to the very regular influx of large numbers of common loon onto the lake (Reimchen and Douglas 1980) and to the increased metabolic requirements of resident trout during the higher summer temperatures. Most minor predators were limited to a single season occupancy, but the cumulative effect produced a relatively uniform mortality level throughout the year. When only subadults and adults are included in this mortality (Fig. 9.5(b)), the seasonal trends are similar, although there is a more distinctive peak in late summer and autumn during the residency of hooded merganser and red-necked grebe.

Yearly differences in stickleback mortality

Considerable yearly heterogeneity occurs in predation by some of the foragers. Total yearly consumption by common loon ranged from 55 000 (1984) to 157 000 (1987). This species normally foraged during autumn and winter, yet none were present in the autumn of four of ten years. Red-necked grebe, which is one of the major sources of mortality in winter and spring, taking up to 9000 stickleback from January to May, were absent from the lake during this period in 1978, 1982, and 1985, although they were present

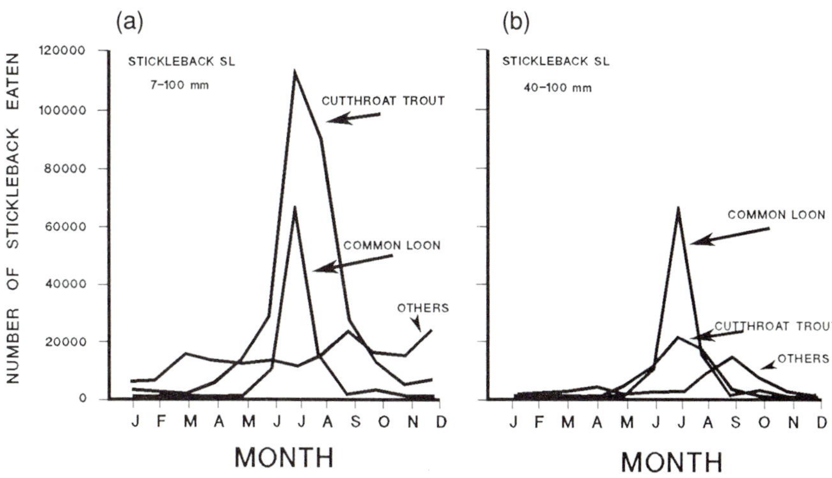

Fig. 9.5 Monthly consumption of stickleback for major and minor predators at Drizzle Lake: (a) total stickleback eaten, (b) subadult and adult stickleback eaten.

during the autumn of each year. Common merganser, which was abundant each year, rarely foraged on the lake (Reimchen and Douglas 1984*a*), yet in winter 1981, and early spring 1985, extensive foraging occurred, with some 11 000 stickleback taken during each period. River otter occurred in only five of ten years; in 1979, a single individual was resident for 3 wk and took about 10 000 adult stickleback. Oldsquaw duck, rare or absent during most years, was resident during the winters of 1980 and 1981, and consumed an estimated 15 000 stickleback each year. Hooded merganser, which takes up to 50 000 individuals yearly, took fewer than 1000 individuals during four years when the pair did not breed.

Lake foraging positions

There were consistent differences among species in the use of foraging localities on the lake, the major distinction being the relative proportion of littoral and limnetic activity (Fig. 9.6). Belted kingfisher only foraged in littoral regions within 10 m of shore. Some species such as hooded merganser, horned grebe, bufflehead, and river otter were largely within 100 m of shore. Others such as the common loon, red-necked grebe, oldsquaw duck, and common merganser exploited both littoral and limnetic habitats, although common loon rarely foraged in water less than 1 m deep. From fyke and gill netting throughout the year, cutthroat trout appeared to be most common in littoral regions during spring and autumn, and were often netted in water less than 1 m deep (Reimchen 1990). Individuals were not typically captured in open-water regions at any depth. Some limnetic predation may occur because I observed offshore movement of trout in July and August (Reimchen 1990; see also Andrusak and Northcote 1971).

The depths at which different piscivores foraged are poorly known. Avian piscivores would have been restricted to near the surface as the black waters of this lake are aphotic below 2–3 m depth. Horned grebe, common merganser, and otter pursued and often captured stickleback immediately beneath the surface, behaviour evident from the hydrodynamic wake on the surface during their dives. Among the predatory fish captured in gill nets which had been set at various depths and distances from shore (Reimchen 1990), trout and salmon were usually within 3 m of the surface, whereas Dolly Varden were benthic. Surface pursuit by trout was commonly observed in littoral but not in limnetic regions.

Predator foraging efficiencies

The development of defensive adaptations of prey to pursuit and manipulation by the predator should be associated not only with predation intensity, but also with the amount of unsuccessful predation (Vermeij 1982). If a predator is highly efficient, then few prey will escape and there is little

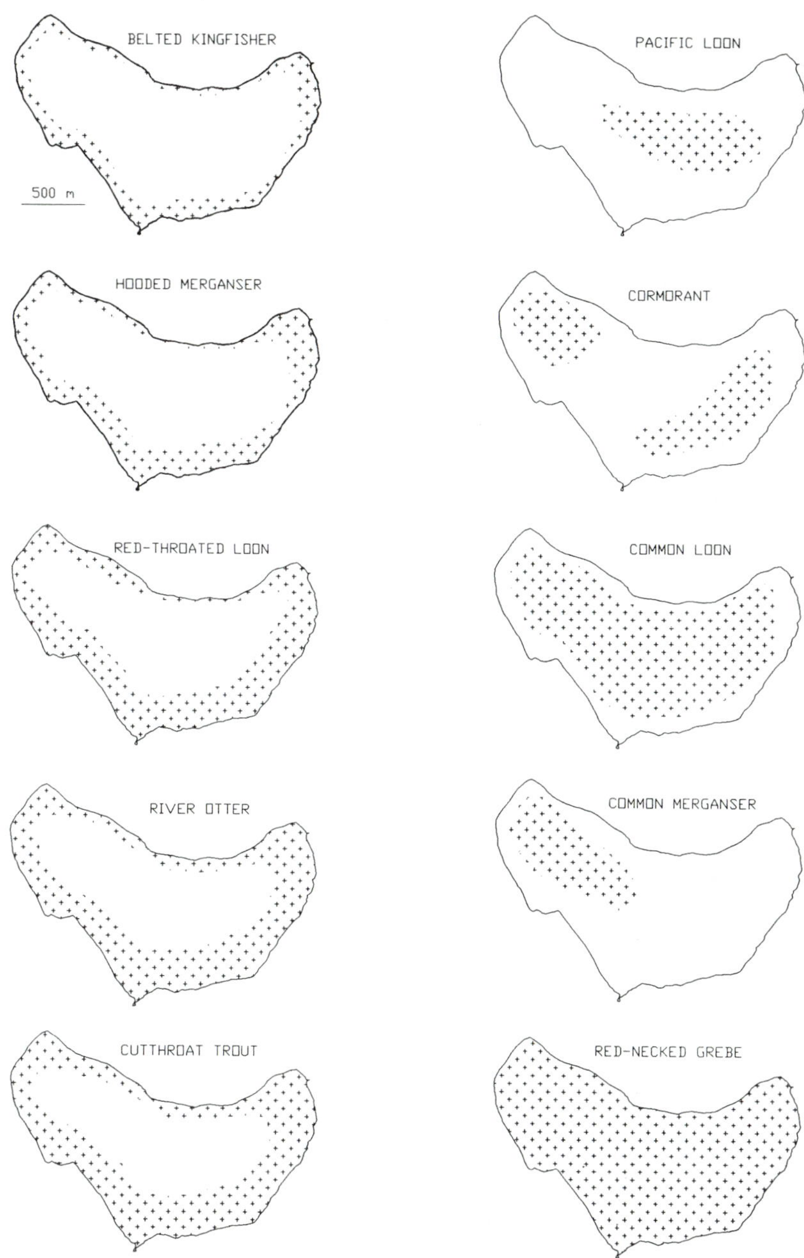

Fig. 9.6 Dominant foraging localities at Drizzle Lake for selected piscivores. Species are ranked from most littoral (*top left*); mainly littoral, some limnetic (bottom left); littoral and limnetic (bottom right); to most limnetic (top right).

opportunity for differential survival. In contrast, if the predator is inefficient, and consumes the same number of individuals as the efficient predator, then many more prey will escape and there is a far greater opportunity for selection. Thus, the substantive predator-induced mortality of stickleback observed at Drizzle Lake, which comes from a diversity of predator species and involves all size classes of stickleback, will be of no selective consequence unless there are foraging inefficiencies by one or more of these predators and body alignment (Reimchen 1991*b*).

The major stages to a foraging event are search, pursuit, and manipulation. Evaluating natural foraging efficiencies of these stages is difficult because such events are seldom observed with sufficient frequency to allow quantification. Substantial experimental data exist, and there is major variation between and within species dependent on attributes of the predator, such as experience, and on attributes of the prey, such as body size (Krebs 1978; see Vermeij 1982 for review).

Three fundamental questions must be addressed to evaluate the evolutionary importance of the assemblage of stickleback predators observed at Drizzle Lake. First, is there evidence to suggest predator inefficiency in one or more of the foraging phases? This will determine the potential for adaptation of the prey. Second, do predator species differ in their efficiencies? If so, this may result in some predators being much more important for the evolution of the prey population than would be predicted from their contribution to total mortality. Third, which of the variable traits of the prey maximize the probability of escape during each of the search, pursuit, and manipulation phases for each of the predators? This would provide insight into whether there are opposing selective forces among the three phases or among the different species of predator. Currently, I have an incomplete assessment of these questions but can provide data on efficiencies for several predator species.

Search and pursuit efficiencies

Among dives recorded in winter months ($N = 1025$), common loon returned fish to the surface on 679 instances ($\bar{X} = 65.3$ per cent over 41 different time blocks, 95 per cent confidence limits, 57–73 per cent). During similar periods of the day and in the same lake regions, and so where encounter rate should be comparable to that of the common loon, red-necked grebe captured 63 fish in 257 dives ($\bar{X} = 29.2$ per cent over 15 time blocks, 95 per cent confidence limits = 16–42 per cent). Efficiency was significantly different for these two species (F-ratio = 23.7, d.f. = 1,54, $P < 0.001$). As noted previously, it is unlikely that the grebe is swallowing additional fish underwater. Assuming that these efficiencies are realistic, it is possible to estimate number of pursuit failures. Common loon from autumn to spring consume a predicted 13 440 stickleback; since this represents about 65 per cent of the prey they initially dove for, 7195 search and pursuit failures should have occurred. Red-necked grebe consume a predicted 8553 stickleback during the same period, which should result in 29 316 additional failures. Consequently, although the red-necked grebe accounts for only 39 per cent of the winter consumption by the two species, it accounts for 80.3 per cent of their failures. If there are any attributes of the stickleback associated with their ability to avoid detection or evade capture during pursuit (for example, behavioural responses, body size, swimming speed), the red-necked grebe could theoretically have much greater selective influence than the common loon. During summer, common loon swallowed most of their prey beneath the surface, and I am unable to estimate efficiency. If it is comparable to that in winter, then there would be a total of 55 774 search and pursuit failures yearly by this predator, about twice as much as the yearly failures by red-necked grebe. These represent minimum values for both species since there may be multiple pursuits during each dive.

Manipulation efficiencies

Following capture, stickleback are subject to varying amounts of manipulation prior to swallowing. Some of these manipulations are unsuccessful, the major evidence for this being the occurrence of predator-induced injuries on stickleback collected from the natural population. Analyses of 8718 stickleback (predominantly adults) sampled over a 3 yr period at Drizzle

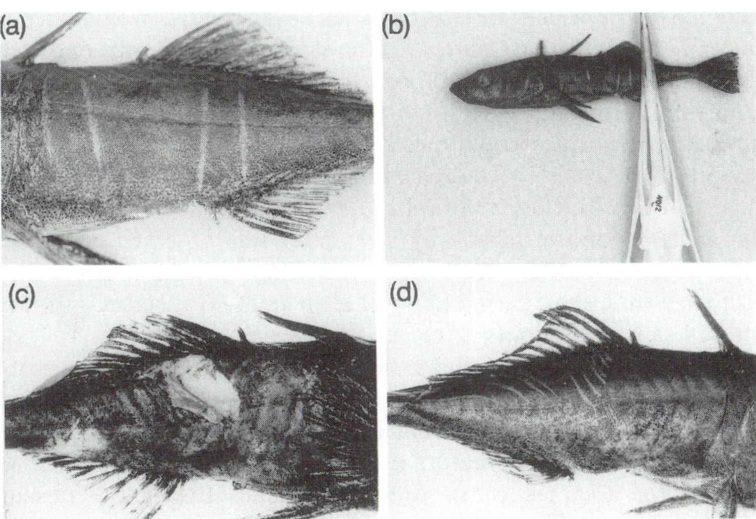

Fig. 9.7 Predator-induced injuries on stickleback at Drizzle Lake: (a) characteristic 'aviscar' on adult stickleback (SL 7.5 cm); (b) bill of red-necked grebe positioned on trunk for perspective; (c) skin laceration incurred from cutthroat trout attack; (d) tooth scars from cutthroat trout attack.

Lake (Reimchen 1988) showed that 13.4 per cent of the population have predator-induced injuries such as fractured dorsal and pelvic spines and skin lacerations. About one-third of the injured fish have 'aviscars', characteristic imprints of the bill profile from bird attacks (Fig. 9.7(a),(b)), about one-third of the fish have skin punctures and scratches from trout attacks (Fig. 9.7(c),(d)) (unpubl. data), while the remainder have fractured spines without skin injuries, probably from bird manipulation. Injuries were rare or absent on small stickleback (< 50 mm) but became progressively more frequent on larger and older individuals, reaching up to 35 per cent in the largest fish (80–90 mm).

Manipulation efficiencies were evaluated for three piscivores. Of 679 stickleback captured and brought to the surface by the common loon, I observed only 5 escapes (0.74 per cent). Of 63 stickleback captured by red-necked grebe, 10 escaped (15.9 per cent). There may, of course, have been additional escapes underwater. The higher frequency of failures by the grebe is probably a consequence of their smaller body size (1000 g) than that of the common loon (3000 g). This disparity is also reflected in their prolonged manipulation periods for fish of similar size. Common loon took an average of 13.5 s (range 1–58 s) prior to swallowing the stickleback, while red-necked grebe required 129.9 s (range 27–900) (ANOVA, $P < 0.001$). Although stickleback size could not be estimated with any precision, the extended manipulation periods and escapes all occurred when stickleback were large (> 70 mm).

Common loon, which ate 104 187 stickleback yearly, would therefore produce 777 manipulation failures. Of the 8553 stickleback consumed yearly by red-necked grebe, there will be an additional 1617 escapes during manipulation. However, even a marginal increase in failures by common loon beneath the surface would substantially increase manipulation failures. I suspect that this occurs because I regularly found injured adult stickleback at the drift line within several days of loon foraging bouts in summer.

Manipulation of stickleback by river otter was observed during their periodic occupancy of the lake. Upon surfacing with the stickleback, the otter chewed for a short period ($\bar{X} = 17$ s, range 11–37 s) before swallowing the prey. I saw no instances of escape in 65 separate captures.

It is possible roughly to evaluate the relative importance of some of the major predators at this stage. Overall, cutthroat trout consume the greatest number of stickleback (55 per cent of total), followed by common loon (18 per cent) and minor predators such as red-necked grebe (1.6 per cent) and river otter (0.6 per cent). Among all adult stickleback consumed, common loon take 52 per cent, red-necked grebe 16 per cent, trout 12 per cent, and otter 11 per cent. Examination of injuries in the natural population indicates that about one-half of the injuries are attributable to bird attack and one-third to trout attack (Reimchen 1988, unpubl. data). From direct observation of predator manipulation efficiencies on the lake surface, it is possible to account for 908 manipulation failures involving adult stickleback, of which red-necked grebe contributes 89 per cent, common loon 11 per cent, and otter none. Two conclusions derived from this analysis are that the major predator in the life history of the stickleback (i.e. cutthroat trout) produces fewer manipulation failures than avian piscivores, and among the latter, a relatively minor predator, the red-necked grebe, may contribute as many or more manipulation failures as the common loon, which is a major predator.

Pursuit, manipulation, and body size

From observations on the foraging behaviour of some of the piscivores and data on frequencies of injured stickleback in the population (Reimchen 1988), it seemed likely that probability of escape was directly related to body size of stickleback. The large size of adult stickleback at Drizzle Lake (70–110 mm), combined with the robust dorsal and pelvic spines, which increase effective diameter by 130 per cent, could represent an adaptation to gape-limited piscivores. I have tested this hypothesis using trout (Reimchen 1991*a*).

Six cutthroat trout (range in SL 19–34 cm) were collected from Drizzle Lake and placed in 400 l aquaria or in netted circular enclosures (3.14 m^2) in the lake. After their length was recorded, stickleback were dropped individually (N=1581) into the centre of the enclosure; they immediately accelerated towards the edge, during which time the trout gave chase.

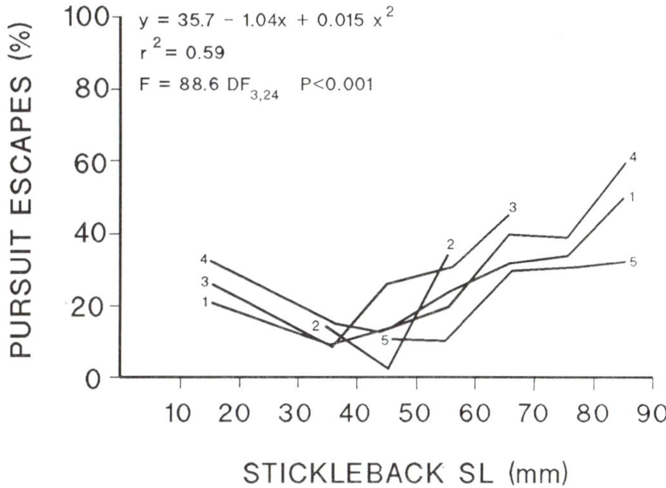

Fig. 9.8 Stickleback escape frequencies in relation to standard length (SL) during pursuit by cutthroat trout. Each line represents data from a different trout: 1, 19.0 cm SL; 2, 19.5; 3, 21.0; 4, 25.5; 5, 31.5 and 34.0 (data combined). Number of feedings per trout: 1, 464; 2, 97; 3, 357; 4, 514; 5, 149 (from Reimchen 1991*a*).

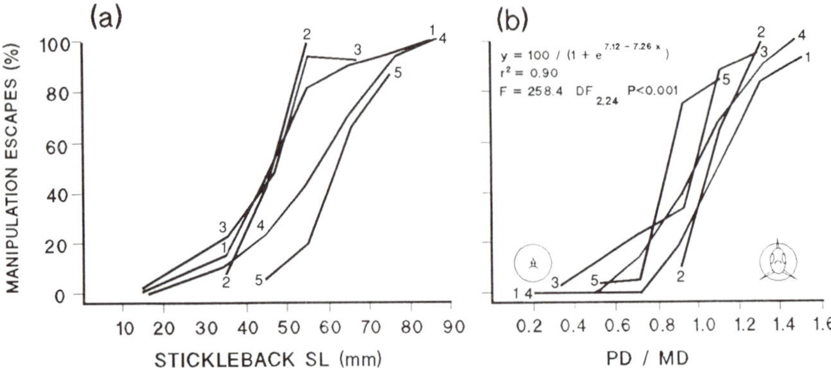

Fig. 9.9 Stickleback escape frequencies during manipulation by cutthroat trout versus (a) standard length (SL) of stickleback, and (b) relative body size (stickleback diameter with dorsal and pelvic spines erected divided by mouth diameter of trout, PD/MD). Insets show a schematic view of a stickleback in the mouth of trout when PD/MD is 0.2 and 1.4. Each line on the graphs represents data from a different-sized trout: 1, 19.0 cm SL; 2, 19.5; 3, 21.0; 4, 25.5; 5, 31.5 and 34.0 (data combined). Number of separate manipulation events per trout: 1, 316; 2, 71; 3, 267; 4, 389; 5, 96 (from Reimchen 1991*a*).

In the pursuit phase, there is a curvilinear relationship with escape frequency, small and large individuals having an increased probability of escape (Fig. 9.8). Stickleback between 30 and 50 mm SL were at the greatest risk to trout. A second-order polynomial described 60 per cent of the variance ($P < 0.001$).

In the manipulation phase, small stickleback (<30 mm SL) rarely escaped after capture, but as stickleback size increased, there was a sharp increase in escape, reaching more than 90 per cent for the largest stickleback (>70 mm SL) (Fig. 9.9(a)). Larger trout were more efficient manipulators of all stickleback.

Swallowing ability of a predatory fish is largely a function of its gape and the diameter of the prey (Werner 1974; Zaret 1980; Hoyle and Keast 1987). I measured the distance between the posterior edges of the upper jaw as an estimate of predator gape (MD) and the maximum diameter of stickleback with dorsal and pelvic spines erected (PD). Escape frequencies were re-examined in relation to PD/MD. These data (Fig. 9.9(b)) demonstrate that when PD/MD is less than 0.6, the prey very rarely escape. However, as PD/MD approaches unity, escape frequencies increase rapidly, reaching about 94 per cent when PD/MD = 1.4. A logistic curve describes 90 per cent of the variance. Therefore, large adult size of stickleback provides refuge during both pursuit and manipulation.

CONCLUSIONS

Is there any substantive ecological insight to be derived from this intensive study of a single population? The large number of predator species taking stickleback over a 10 yr period is much greater than that previously reported, and appears anomalously high in comparison with evolutionary studies of other taxa. However, within any short observation period such as a single season, the diversity and abundance of piscivores was usually low, and only with increased observational period did total diversity increase. Therefore, high diversity is the consequence of the extended time frame over which the observations were made, rather than a reflection of any elevated ecological complexity of the habitat. Because each of the predators has the potential to influence both demographic and selective processes in the life history, short-term investigations would have major conceptual limitations for evaluating life history evolution or functional morphology (see Greene 1986 for useful discussion).

What then are some evolutionary effects of this predator regime on variation in stickleback defences? A definitive assessment of this effect will not be available for a number of years because of the complexity and diversity of interactions in the Drizzle Lake population that have not yet been addressed. One interpretation of the available data is that the population is exposed to normalizing selection around a single optimal phenotype, a 'jack of all trades', representing an average response to salmonid, avian, and mammalian piscivores. Previous univariate assessment of variability in lateral plates (Moodie and Reimchen 1976*a*) indicates higher variance and higher asymmetry in adult stickleback from Drizzle Lake than in other populations where trout predators are abundant, a finding which is the inverse to that predicted on intense normalizing selection.

It seems more plausible that the population is subject to diversifying selection, as would be predicted in multiniche models (Van Valen 1965). Although direct evidence for this is rare in field studies (Schluter *et al.* 1985; Grant and Grant 1989*a*), there are many examples showing extensive variability of traits within populations associated with habitat complexity, for example, cladocerans (Kerfoot 1975), gastropods (Reimchen 1979), cichlids (Greenwood 1974), stickleback (McPhail page 422 this volume) and birds (Van Valen 1965), and the theory basic to the process of diversifying selection has been developed (see Endler 1986 and Wilson 1989 for reviews). In Drizzle Lake, there are numerous pursuit and manipulation methods among the predators, including plunging (kingfisher), wading (heron), diving (loon, grebe), surface feeding (merganser), laceration and puncture (salmonids), compression (birds), chewing (otter), and grappling (odonate naiads), as well as spatial differences in the distributions of the piscivores. Therefore, there may be multiple optima with different abundance of piscivores among habitats and among seasons, producing a shift in the phenotypic distributions proportional to the number of pursuit and manipulation failures for each predator. Although a causal relationship has not been demonstrated with predators, phenotype frequencies differ from one part of the lake to the next (Reimchen and Nelson 1987) and from season to season (Reimchen unpubl. data) consistent with spatial and temporal differences in fitness of phenotypes. Analyses of these issues are continuing.

A major attribute of stickleback for evolutionary investigation is that much of the observed population differentiation is recently derived, probably since the last ice advance (Moodie and Reimchen 1976*b*; Bell 1984*a*, 1988; Bell and Foster page 16; McPhail page 401 this volume). Consequently the accumulation of historical factors that may confound analyses of form will be less important. Ecological evaluation of this variation has the potential, therefore, of generating fundamental insight into the processes of evolutionary change. From the investigations on Queen Charlotte Islands stickleback populations, it is possible to formulate some minimum criteria which would be required to realistically partition the causes of intra- or interpopulation variation in prey defences. These are:

(1) causes and amount of age-specific mortality;

(2) temporal and spatial patterns of habitat use by predators;

(3) predator efficiencies during search, pursuit, and manipulation;

(4) phenotype spatial distributions;

(5) phenotype fitness against different predators.

Such criteria, which should in theory be basic to a study of prey defences, have not been determined for any population of any species, and consequently, their completeness cannot be assessed. This is not to imply that other selective forces, unrelated to predation, are not also operating on defence characters. Clearly, multiple factors such as drift, founder effect, linkage, ontogeny, and allometry can influence present population attributes. However, whether these are ever primary causes of variation in defensive characters can only be unambiguously determined if the role of predators can be empirically discounted.

These criteria offer considerable potential for addressing functional aspects of microevolutionary trends and may yield insight into larger-scale differences among related taxa. Gould (1984) has criticized evolutionary biologists for pursuing the causes of fine-scale differentiation among populations: 'It is time for the pendulum to swing back to a position at the pluralistic middle. I believe that both the great systematists of the 1930s and the great synthesists were correct—some geographic variation within a species is clearly adaptive, but much is a non-adaptive product of history.' (p. 236). I suspect that there is little relevant information in the primary literature to support or reject this belief. Continuing studies of Galapagos finches (Grant and Grant 1989*b*), which may be the most thorough assessment of variability in trophic traits in any species, indicate major temporal and spatial components to fitness among phenotypes. As yet, there exists no consensus to assess the relative proportion of adaptive and non-adaptive components to this variation. Studies on population differentiation in stickleback are probably more extensive than for most species, and we are not even close to a stage where the relative amount of adaptive variation in traits can be evaluated with any confidence. We may be greatly under-estimating the potential for endocyclic or diversifying selection within populations, and greatly under-estimating the selective differences among populations. Most studies, because of the constraints of time, have not been able to address the spatial and temporal complexities of life history interactions, and therefore full evaluation of variability remains impossible. Only when a rigorous assessment of fine-scale variability is obtained for different populations among representative taxa will it be possible to formulate significant generalizations on how much variation in prey defences in natural populations is a remnant of history, and how much is an adaptation to locality-specific selective pressures.

One of the major tenets of the last three decades of studies on stickleback is that predation intensity differs among populations and is associated with differences in morphological and behavioural traits. In virtually all populations, the occurrence of predatory fish is correlated with enhanced expression of spines and lateral plates, while the absence of predatory fish is associated with reduced expression of these traits. This relationship has been attributed to a relaxation of selection by predators (Hagen and Gilbertson 1972; Moodie and Reimchen 1976*a*; Gross 1978). From the investigations on the Queen Charlotte Islands populations, I suspect that categorizing localities according to the presence or absence of predatory fish or by compiling lists of potential predators, while providing useful insight into population differentiation, provides inaccurate information on relative predation and selection intensities. The reasons for this can be summarized as follows. (1) If multiple sources of prey are available, the presence of predatory fish in a locality does not in itself provide direct evidence that stickleback are subject to predation. It follows that locality differences in the abundance of predatory fish are not a meaningful description of relative predation intensity, since none or all may be consuming stickleback. (2) The absence of predatory fish does not indicate that predation levels on stickleback are low, because other piscivores, including avian and invertebrate taxa, which are present in most localities, may represent the major cause of mortality. (3) The presence of a single piscivore in a small stickleback population may produce the same predation level, relative to total population size, as that of multiple predators in a large population. An otter or a grebe need only briefly forage on a population, irregularly between years, to generate a substantial quantitative effect on that population.

(4) Extensive predation in a population does not provide evidence that selection is operating, because the strength of selection will be proportional, not to predation levels, but rather to the foraging inefficiency of the predators during search, pursuit, and manipulation (Vermeij 1982; Reimchen 1988). If all stickleback encountered by predators are consumed, there will be no selection on defensive traits, even if predation intensity is very high. Consequently, any evaluation of the importance of specific predators as selective agents requires a combination of data on foraging levels and failure rates during the life history of the prey. Estimates of the intensity of selection operating in the entire population become possible only if comparable data are collected for each piscivore in the locality.

In conclusion, there is mounting evidence that the morphology and behaviour of stickleback are associated with particular predator regimes. Lateral plate modes near seven are common among populations exposed to predatory fish and birds, stickleback with five lateral plates have higher survival than non-fives when individuals are exposed to garter snake predators in experimental tanks, modes of three or four occur where avian piscivores are prevalent, while lower modes, at least on the Queen Charlotte Islands, occur where macroinvertebrate piscivores are prevalent. I propose that it is the proportions of different predator species among habitats, rather than predation intensity, that constitute the driving selective force for population differentiation in lateral plates and spine morphology. Differences in search, pursuit, and manipulation efficiencies among the predators combined with amount of mortality will determine the selective pressures operating among populations.

ACKNOWLEDGEMENTS

I am particularly indebted to Sheila Douglas for comments on the manuscript and for extensive field assistance during the last 12 years of the study. I am also grateful to M. A. Bell, J. B. Foster, and J. S. Nelson for encouragement and discussion. These investigations were initially supported by grants from the Ecological Reserves Unit, Government of British Columbia, and subsequently by grants from the National Sciences and Engineering Research Council of Canada to J.S. Nelson (NSERC A-5457) and to the author (NSERC A-2354).

10

Adaptive variation in antipredator behaviour in threespine stickleback

F.A. Huntingford, P.J. Wright, and J.F. Tierney

A major aim in behavioural ecology is to understand how natural selection acts on behavioural variants. One way of unravelling the relationship between phenotype and selection is the comparative approach, in which behavioural variants are related to environmental features (Huntingford 1984; Krebs and Davies 1987). If a behavioural trait (for example, the removal of egg shells by parents after young gulls hatch) is regularly associated with a particular environmental factor (for example, risk of predation on chicks), it is likely that the former is an adaptation to the latter (Tinbergen *et al.* 1962). However, behavioural phenotypes are notoriously labile, so differences among populations may reflect flexible responses of the animals to environmental differences as well as evolved hereditary adaptations. Behavioural differences among populations can only be considered as the result of natural selection if a genetic basis for the behavioural differences can be demonstrated.

The comparative approach can be applied at the level of taxonomic units at the species level and above, or at the level of local populations within species. The latter is particularly valuable for testing adaptive hypotheses because the variance attributable to phylogenetic divergence (Harvey and Mace 1983; Ridley 1983; Lauder 1986) is limited. Regardless of whether interspecific or intraspecific comparisons are employed, it is important to show that the behavioural differences do not simply reflect common ancestry but are likely to be independent adaptations of the groups compared (Bell and Foster page 20 this volume). In this chapter we review the use of population comparisons to investigate the relationship between antipredator behaviour and local regimes of predation among populations of threespine stickleback, *Gasterosteus aculeatus*. The results suggest that different predation regimes can produce behavioural divergence. In some instances these may have a straightforward genetic basis, but in other cases their developmental origin is more complex, depending on interactions between innate behavioural predispositions and early social experience.

The chapter begins with a general overview of how threespine stickleback respond to predators, and then addresses the causes of variation in this context. In particular we evaluate the causes of the extensive, seemingly adaptive, differentiation of antipredator behaviour among populations. Finally we describe the early ontogeny of antipredator behaviour and the proximate factors that affect its expression in adults.

ANTIPREDATOR BEHAVIOUR OF THE THREESPINE STICKLEBACK

In spite of their protective armour, which includes the spines for which they are named, threespine stickleback fall prey to many kinds of piscivorous animals, among them mammals, birds, reptiles, fish, and insects (Reimchen Table 9.1 this volume). Although often effective (Hoogland *et al.* 1957), armour is a last line of defence, employed only after behavioural defences have failed.

Risk of predation can be reduced if potential prey either avoid encounters with predators or actively evade predators once encountered (the primary and secondary defences, respectively, of Edmunds 1974). Both kinds of defence are seen in stickleback. Fraser and Huntingford (1986) have shown that in the laboratory, stickleback (non-breeding adults) avoid feeding patches in which predatory fish (brown trout, *Salmo trutta*) are present. They also reduce the risk of detection by staying in or near aquatic vegetation or other forms of cover. For example, in the presence of a predator (brown trout), adult stickleback restrict their foraging to weed-beds (Ibrahim and Huntingford 1989*a*). In Crystal Lake, British Columbia, fry remain in or near vegetation until they reach 15 mm SL, a size at which they are no longer eaten by adult conspecifics (Foster *et al.* 1988). When predatory fish are rare or absent, stickleback avoid vegetation, and thus distance themselves from potential predation by dragonfly naiads that perch on vegetation (Foster *et al.* 1988; Fig. 10.1).

Once predators have been encountered, several evasive manoeuvres are possible. Threespine stickleback may form large schools (e.g. Wootton 1976; Bentzen and McPhail 1984; Foster *et al.* 1988), especially when feeding on plankton in open water. Small prey fish experience reduced individual risk of predation while schooling (e.g. Neill and Cullen 1974; Pitcher *et al.* 1983), as do individuals in smaller, less cohesive associations, such as those characteristic of stickleback on the breeding grounds (Neill and Cullen 1974; Pitcher *et al.* 1983). So stickleback may gain protection from group membership.

Effective evasive manoeuvres are also performed by solitary stickleback. On detecting a potential predator such as a pike, *Esox lucius*, the immediate response of a stickleback is to stop what it is doing, raise its spines, and fixate on the predator (Hoogland *et al.* 1957, Benzie 1965). Often, it will then move closer (Huntingford 1976*a*). This is predator inspection (a term coined by Pitcher *et al.* 1986 with reference to minnows, *Phoxinus*

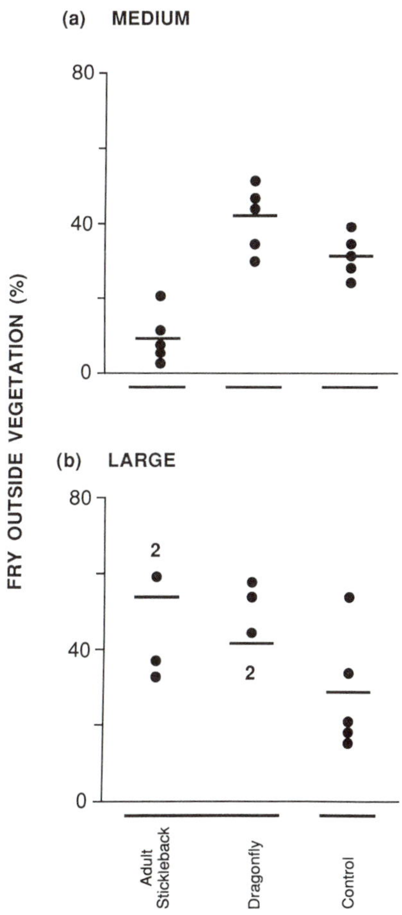

Fig. 10.1 Use of open water, as opposed to vegetation, by (a) medium-sized stickleback fry (11–15 mm SL) and (b) larger fry (21–25 mm) in pools with no predators (control), with potentially cannibalistic adult stickleback, and with dragonfly naiads. Dots represent means from replicates and lines represent group means. Lines below the horizontal axis connect treatments that do not differ significantly (after Foster *et al.* 1988).

phoxinus), during which the stickleback, seemingly acquires the information necessary for assessing predator identity and implicit level of risk, as they are known to do (Foster and Ploch 1990).

If the potential predator is deemed to pose a risk, the stickleback responds with one of a variety of evasive manoeuvres (slow retreat, unpredictable jumps, or rapid swimming to cover). Although there is no detailed information on this point, choice of response probably depends on some combinations of the species (Foster and Ploch 1990), hunting style (Hoogland *et al.* 1957), size (Foster and Ploch 1990) and proximity of the predator, its speed of attack, and the availability of cover (Hoogland *et al.* 1957; Benzie 1965). If cornered, the stickleback locks its spines in a raised position, and the manipulation required to swallow a stickleback that has its spines locked in this way affords it a chance to escape (Hoogland *et al.* 1957; Reimchen 1983, page 269 this volume). Stickleback that do escape at any point in the encounter may remain frozen or hidden for a period before resuming normal activities.

VARIATION IN ANTIPREDATOR BEHAVIOUR

Although all threespine stickleback have the capacity to perform these antipredator responses, the intensity and nature of the responses are extremely variable, even when the fish are exposed to the same kind of predator under standardized conditions (Fig. 10.2). Why do some individuals fail to show any response to a direct attack, others give a weak response followed by rapid resumption of normal behaviour, while still others respond vigorously and remain frozen in cover for many minutes? Some of the variability comes about because the same individual fish responds differently from one occasion to the next when presented with an identical predatory

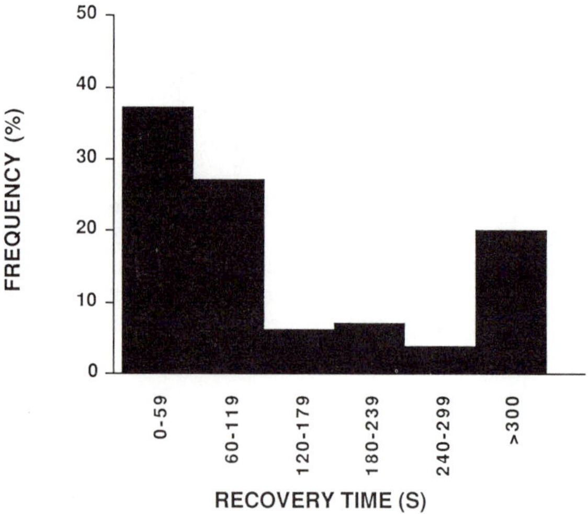

Fig. 10.2 Variability in antipredator responses of non-breeding, adult threespine stickleback. Recovery time is the interval between response to a simulated attack by an avian predator and recovery of normal activity in 5 min tests on 164 different non-breeding adult threespine stickleback (J.E. Tierney, F.A. Huntingford, and D.W.T. Crompton unpubl. data).

stimulus. In contrast to such within-individual effects, another source of variability is the existence of permanent between-individual differences in behavioural phenotype.

Within-individual variation

Effects of parasites

Some spectacular changes in the antipredator behaviour of individual stickleback are caused by parasitic infection. There is a large and expanding literature showing that the larval stages of many parasite species effect changes that render their intermediate hosts more vulnerable to predation by the definitive host; such changes potentially increase the parasites' chances of successfully completing their life cycle (see reviews by Milinski 1990; Moore and Gotelli 1990). Sometimes the changes are morphological, but often the parasite interferes with the antipredator responses of the intermediate host.

In threespine stickleback, changes of this kind occur as a result of

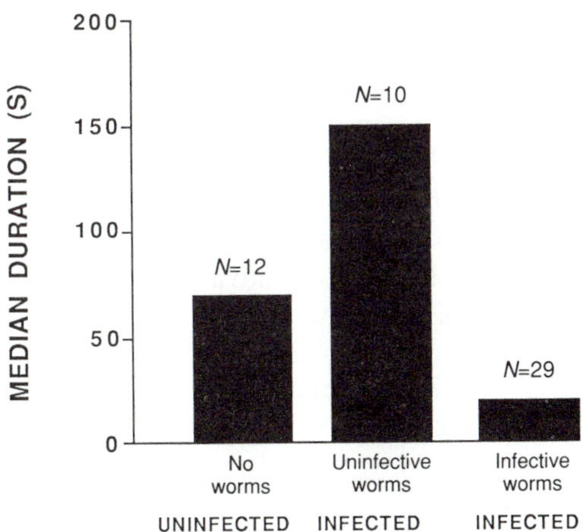

Fig. 10.3 Effects of parasites on one aspect of the antipredator responses of threespine stickleback. Median duration is the median interval for individuals between first response to a simulated attack by an avian predator and resumption of normal activity (recovery time) in non-breeding adult threespine stickleback. Recovery time is shown in relation to infection with the cestode *Schistocephalus solidus*. Uninfective worms are those weighing less than 50 mg; above this weight, the worms are fully capable of establishment in the definitive host, but below this weight establishment is effectively impossible. The three groups are statistically distinct at $P < 0.001$ ($H = 13.62$, $N = 51$) using a non-parametric ANOVA (Tierney *et al.* in press).

infection by the tapeworm (Cestoda) *Schistocephalus solidus* (Giles 1983*b*; see Milinski 1990 for review). During the early stages of infection, before the parasite is sufficiently mature to become established in its definitive host (a bird), various aspects of the responses of infected stickleback to simulated avian attack (including freezing and suppression of normal activity) are enhanced (Fig. 10.3; Tierney *et al.* in press). Although the following consequences remain to be documented, such a behavioural shift probably reduces the chances that the fish will be captured by a predator, and thus promotes the survival of the parasite.

Once the tapeworm is infective to the definitive host, the behaviour of the stickleback changes so that an attack elicits no more than a weak and short-lived response (Fig. 10.3; Tierney *et al.* in press). Once again the effect remains to be documented, but in nature this change in behaviour may have the effect of promoting transmission of the tapeworm from the intermediate host to the definitive host. These behavioural changes may be by-products of infection, perhaps caused by enhanced food and oxygen requirements in heavily infected fish (Giles 1987*a*,*b*), but may also be the result of a direct change in responsiveness to danger (Milinski 1990). So some of the variability in antipredator behaviour documented in Fig. 10.2 may depend on the acquisition of a parasitic infection and on the developmental stage of the parasite.

Reversible responses to altered environmental conditions

Other intra-individual differences are the result of flexible responses to changes in the balance of advantages and disadvantages of performing antipredator behaviour. Although the advantages of avoiding being eaten are self-evident, the benefits of showing a particular response may vary with circumstances. For example, adult sticklebacks in mixed shoals of *G. aculeatus and G. wheatlandi* gain protection from predation, so that individuals within the shoal are less vulnerable, especially if the group is large. FitzGerald and van Havre (1985) have shown that in such shoals the rate of recovery from disturbance is negatively correlated with group size, suggesting that the fish adjust their behaviour to the protective effect of group membership.

Antipredator responses can also have disadvantageous consequences that may reduce fitness, such as interference with other important activities like feeding or reproducing. For brevity, such disadvantageous consequences are referred to here as the costs of antipredator responses. These costs, and with them the risk that animals are willing to accept in order to perform other essential activities, can change as a result of alterations in internal state and external conditions (e.g. Carlisle 1982; Lima and Dill 1990).

For example, hungry threespine stickleback (non-breeding adults) are willing to accept greater risks of predation in order to feed than are recently fed individuals (Fraser and Huntingford 1986). The reproductive condition of males also influences their antipredator behaviour. In laboratory aquaria, non-territorial male sticklebacks flee more readily from a hunting pike (a predator at the site from which they were collected) than do those with territories but no young, which in turn flee more readily than those with young (Huntingford 1976*b*). Similarly, in the field, when confronted with a tethered trout, territorial males with young in their nests are slower to take flight, and hide for shorter periods, than do those without young (Kynard 1978*a*; see also FitzGerald and van Havre 1985). In some lakes in British Columbia, Canada, territorial males confronted with a large prickly sculpin, *Cottus asper*, hid for shorter periods and accepted greater risk in attacking the sculpin when they had young in the nest, especially if the young were numerous and/or well developed (i.e. of greater reproductive value; Pressley 1981; but see Foster and Ploch 1990).

Pike, trout, and sculpin are predators of adult stickleback, so that in this case, fleeing probably increases the chance of survival of the individual concerned. As sculpin also feed on young stickleback in the nest, one direct cost of flight by breeding males is potential loss of their brood (Pressley 1981; Foster and Ploch 1990). An additional, indirect cost to breeding males of responding to a predator is interrupted paternal care. When required to defend their territories against other males in the presence of a predator (brown trout), male stickleback with a brood of eggs maintain higher levels of attack towards the intruding conspecific than do those with an empty nest. The highest rates of attack were elicited when males were defending broods acquired early in the season. At the site used in this study, fish from late-hatched broods often fail to breed in the following summer, and few fish survive for more than 14 months; early-hatched broods are therefore more valuable than those acquired later. The parental males themselves die shortly after breeding, and thus the prediction is that they would invest heavily in later broods, as these represent their last opportunity for breeding. A low level of investment in late-hatched broods (which has also been reported both in the laboratory and in the field by Kynard 1978*b*) is therefore paradoxical, and may represent a relictual behaviour from the perennial condition (Ukegbu and Huntingford 1988). In any event, breeding male stickleback adapt their antipredator responses to the costs of keeping safe, adjusting accepted risk in relation to the presence and value of their brood. This adjustment presumably comes about because the mechanisms that govern the response to a predator are susceptible to internal cues (such as hormonal state) and external cues (such as the presence of young in the nest) that track the reproductive status of the father and the value of the brood.

Between-individual variation

Gender effects

In contrast to parasite-induced behavioural shifts and flexible responses to the changing costs of self-preservation, other differences in antipredator responses reflect permanent differences among individual stickleback. For example, especially during the breeding season, male stickleback show weaker responses than females, both during encounters with a pike and following a simulated attack by a model bird (Giles and Huntingford 1984). The reasons for these gender differences are obscure. In causal terms, they may be the result of shared physiological controls for aggression and predator avoidance in males (Bakker page 349 this volume). In functional terms, the value of ensuring protection to the brood may outweigh the cost of an enhanced predation risk.

Population effects

In a number of species, antipredator behaviour has been shown to covary with predation risk, for example in ground squirrels (Owings and Coss 1977; Towers and Coss 1990), in prairie dogs (Loughry 1988, 1989), in garter snakes (Arnold and Bennett 1984; Herzog and Schwartz 1990), in salamanders (Brodie *et al.* 1984; Dowdey and Brodie 1989), and in spiders (Riechert and Hedrick 1990). Such relationships have been particularly well studied in small freshwater fishes, which are vulnerable to a range of predators and which often exist in many localized populations exposed to varying predation regimes. For example, guppies, *Poecilia reticulata*, from sites where predatory fish are abundant show well-developed antipredator responses, including strong schooling (which provides protection against predatory fish), compared with guppies from sites at which predatory fish are rare or absent (Seghers 1974; Magurran and Seghers 1990; Reznick *et al.* 1990). Guppies that coexist with an invertebrate predator, the freshwater prawn, *Macrobrachium crenulatum*, are particulary responsive to this predator (Magurran and Seghers 1990). Similarly, various protective responses (including schooling and predator inspection) are well developed in minnows, *Phoxinus phoxinus*, from a site where pike are abundant compared with those from a site where pike are absent (Magurran 1986, Magurran and Pitcher 1987).

The predation regimes to which threespine stickleback populations are exposed also vary, and this appears to have influenced the evolution of a number of morphological traits (Reimchen Chapter 9 this volume). Is there any evidence that their antipredator behaviour has also diversified? Foster (1988, 1990) found that in populations of threespine stickleback in which cannibalism by groups is common, breeding males perform diversionary displays that distract the attention of potential cannibals from their nest; this behaviour is not shown by males from sites where such cannibalism does not occur.

The antipredator responses of sticklebacks do not always reflect extant predation regimes. Pressley (1981) demonstrated that male stickleback in Trout Lake, British Columbia, Canada, would inspect and attack a preserved prickly sculpin and adjusted their response to brood value, just as did stickleback sympatric with the sculpin, even though it was absent from Trout Lake. In contrast, in Crystal Lake, British Columbia, male stickleback approached a preserved sculpin directly and rapidly, without adjusting their response to the value of the brood. Absence of any adjustment of protective responses in relation to the status of a brood is probably a derived character, as prickly sculpin are native to shallow marine waters where the ancestral marine form is found (Foster and Ploch 1990). These examples suggest that in stickleback, as in guppies and minnows, population differences in antipredator behaviour may reflect local predation risk. This possibility has been studied most intensively in sticklebacks from a number of sites in the United Kingdom; and the rest of this chapter focuses on these studies.

VARIATION IN ANTIPREDATOR BEHAVIOUR AMONG UNITED KINGDOM POPULATIONS

Boldness and risk of predation

In an initial survey of 13 populations of freshwater stickleback (Huntingford 1982), sites were classified on the basis of gill netting and stomach content analysis of potential predators into sites with abundant piscivorous fishes

Table 10.1 The behaviour patterns with significant loadings on the 'boldness' factor identified by multivariate analysis in Huntingford (1982) and shown in Fig. 10.4. This comprised a principal components analysis followed by varimax rotation, using a total of 17 behavioural variables and 145 subjects. Initial analyses were run on the separate populations to confirm that the relationships between variables were consistent before combining data for a single analysis. The boldness factor accounted for 35 per cent of the total variance in the data.

Behaviour patterns with:	
Positive loadings	Negative loadings
Time pectoral sculling[a]	Time hiding in weed
Time feeding	Time motionless
Time facing/approaching[b]	Time continuous swimming[a]
	Time with spines raised
	Time to resume pectoral sculling

[a] Pectoral sculling is the mode of swimming typical of undisturbed threespine stickleback, whereas disturbed stickleback typically swim continuously.

[b] Predator inspection loads positively on the boldness factor because only the boldest fish showed any predator inspection towards the live, actively hunting pike.

of any species known to eat stickleback (high-risk sites) and those at which such fishes were rare or absent (low-risk sites). Non-breeding adult stickleback were collected from each site and maintained in the laboratory for at least one month, during which time they were fed the same rations. Their responses to a predatory fish were screened in a standard laboratory test, during which they were given 10 min exposure to a pike that was motivated to stalk but not to strike.

For the purpose of this initial survey, the behaviour of each fish was measured by a single compound score (derived from a principal components analysis) that summarized its 'boldness' during the encounter with the pike (Huntingford 1976*a*; Table 10.1). The results are unambiguous; wild-caught stickleback from high-risk sites were markedly less bold (i.e. they had better-developed antipredator responses) than those from low-risk sites (Fig. 10.4). Subsequent studies on other populations have confirmed this general conclusion and have also shown that stickleback respond differently to predatory birds in relation to local risk of avian predation (Giles and Huntingford 1984).

Behavioural differences between a high-risk and a low-risk site

To characterize more precisely the differences between threespine stickleback from high-risk and low-risk sites, more detailed behavioural studies

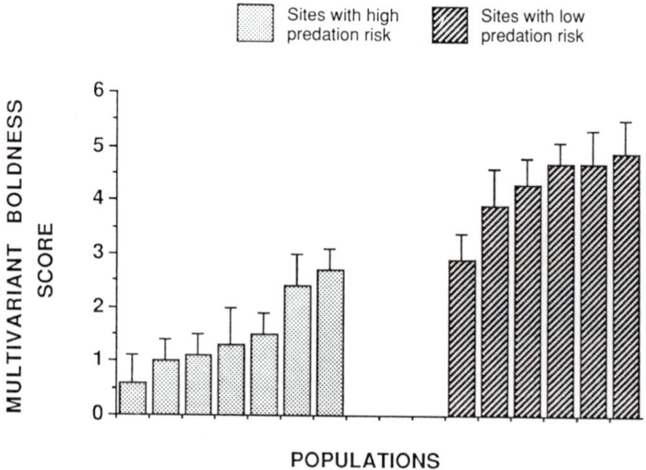

Fig. 10.4 Site-specific levels of antipredator responses in non-breeding adult stickleback. Mean values (± 1 SE) of a multivariate boldness score (see Table 10.1) are shown for stickleback from 13 sites in the UK ($N \geqslant 10$ fish per site) in which piscivorous fish are abundant (shaded bars) or rare hatched bars). The population means for low-risk sites fell above the overall mean significantly more often than did those for high-risk sites ($P < 0.001$, Fisher's exact test; after Huntingford 1982).

were conducted on two populations whose antipredator behaviour had been shown to differ markedly in the previous surveys. One of these, the River Endrick, is part of the Loch Lomond drainage system and contains a variety of predatory fish that are known to prey on stickleback at this site. In addition, piscivorous birds, especially heron, are abundant and are known to take stickleback from this river (Giles and Huntingford 1984). This site was therefore designated a high-risk site. The other site, Inverleith Pond, Edinburgh, is a small pond that has been in existence for several hundred years at least. The pond contains no fish other than stickleback. In the course of extensive sampling, no invertebrate predators of stickleback have been found, and piscivorous birds are rare (although black-headed gulls sometimes roost on the pond). This site was therefore designated a low-risk site.

Differences between wild-caught adults

Rather than using compound scores for comparison of these two populations, each aspect of the stickleback's reaction to predators was analysed separately. Wild-caught, non-breeding adult stickleback from the two sites were individually videotaped for an 8 min period during which they were exposed to a realistic fibreglass model of a brown trout (Tulley 1985). The model was moved according to a pre-arranged routine that included a

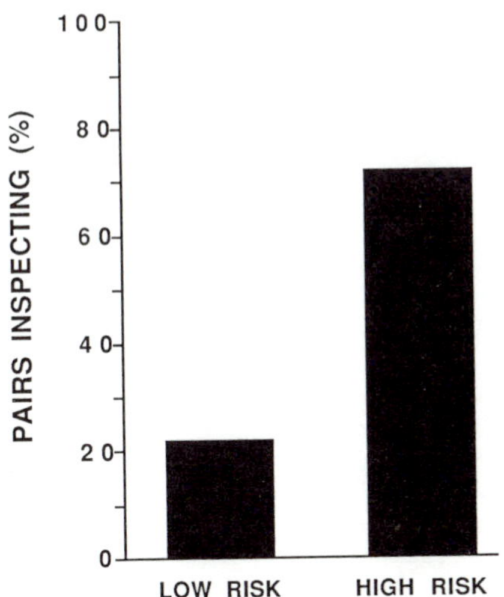

Fig. 10.5 The frequency of predator inspection by pairs of stickleback from a site without predatory fish (low risk) and from another site in which predatory fish are abundant (high risk) when confronted with a model trout ($N = 30$ pairs for the low-risk site and 50 for the high-risk site, $G = 19.40$, $P < 0.001$; Barrie and Huntingford, unpubl. data).

period when the trout was hidden in the weeds, a period when it moved out of the weeds and remained stationary in open water, and a period when it charged the stickleback at a fixed speed. In a separate study, inspection responses to the model by pairs of fish from the two study sites were recorded (Barrie and Huntingford unpubl. data).

Stickleback from the high-risk site inspected the predator far more often than did those from the low-risk site (Fig. 10.5), and responded by keeping still or freezing rather than by jumping (a less effective response), both to the first movement of the trout (Fig. 10.6(a)) and as an initial response to a direct attack (Fig. 10.6(b)). They subsequently made faster escape responses (median speed = 45 cm s^{-1} compared with 31 cm s^{-1}; Mann–Whitney $U = 7$, $N_1 = 7$, $N_2 = 8$, $P < 0.05$). In addition, when escaping, the stickleback from the high-risk site were significantly more likely to move at right angles to the line of attack than were those from the low-risk site (which tended to swim towards the approaching predator) and took longer to resume normal activity following an attack (Huntingford and Wright unpubl. data).

In brief, therefore, as compared with stickleback from a low-risk site, those that are naturally at risk of predation by piscivorous fish stand out as showing greater vigilance, more effective responses to direct attack,

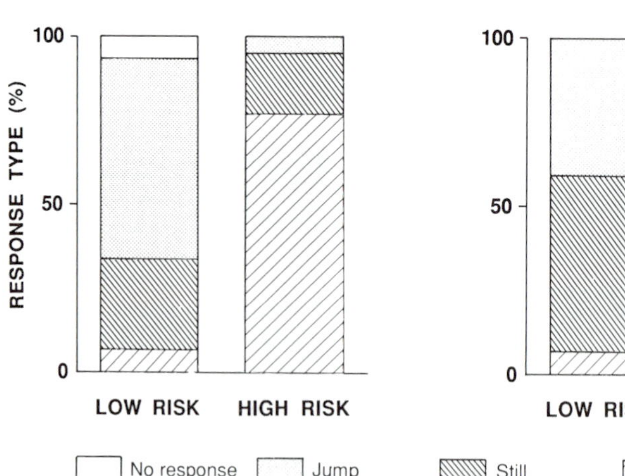

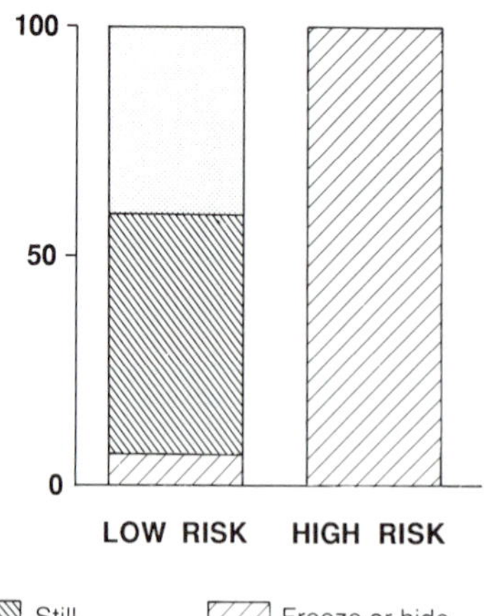

Fig. 10.6 The responses of stickleback from a site without predatory fish (low risk) and from another site in which predatory fish are abundant (high risk) (a) to the first movement of a model trout and (b) to a direct attack. Freezing is distinguished from still by complete cessation of respiratory movements. (a) $G = 10.83$, $N = 20$ for both sites, $P < 0.01$; (b) $G = 39.34$, $N = 20$ for both sites, $P < 0.001$ (Huntingford and Wright unpubl. data).

and a longer period of behavioural suppression following such an attack. Conversely, fish from low-risk sites give higher priority to other important requirements such as foraging. These behavioural traits are likely to promote survival in the habitat in which each population lives.

Ontogeny and plasticity of site-specific antipredator responses

Because stickleback can modify their behaviour after an encounter with a hunting pike, showing enhanced vigilance and stronger escape responses (Benzie 1965), the differences in antipredator responses between wild-caught fish from high-risk and low-risk sites could be the result of differential exposure to unsuccessful attacks prior to capture. On the other hand, they may represent inherited behavioural differences that develop regardless of experience. In guppies (Seghers 1974; Breden *et al.* 1987) and in minnows (Magurran 1990), the site-specific levels of response persist in fish raised in the laboratory without any encounters with predatory fish. Such results indicate that the behavioural differences are inherited traits.

To determine whether the differences in antipredator behaviour of threespine stickleback are also inherited, we compared the responses of laboratory-reared individuals from Inverleith Pond (low risk) and from a tributary of the River Endrick (high risk) to a simulated attack by a predatory bird. Little response to the model bird was elicited from small

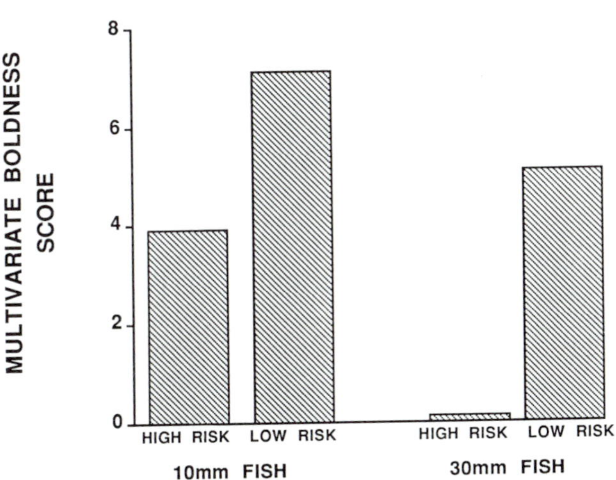

Fig. 10.7 Responses of predator-naïve stickleback of two sizes from a high-risk and a low-risk site to exposure to a model avian predator. Bars depict median value of a multivariate 'boldness' score (Table 10.1) in laboratory-reared stickleback from a high-risk and a low-risk site at standard lengths of about 10 mm and 30 mm. The 30 mm fish from the high-risk site (a tributary of the River Endrick, see text) have lower scores than do the other three groups (after Tulley and Huntingford 1987*a*).

individuals (about 10 mm in length) from either population (Fig. 10.7). However, strong and potentially effective predator avoidance developed in the offspring of fish from the high-risk site by the time they were 30 mm long (large enough to be eaten by a heron, *Ardea cinerea*, for example, Giles 1984*b*), even though they had never experienced an attack. In contrast, the offspring of stickleback from the low-risk site remained unresponsive as adults (Fig. 10.7; Tulley and Huntingford 1987*a*).

Similarly, when laboratory-bred stickleback were exposed to a model of a predatory fish, they showed the pattern of antipredator behaviour typical of their site of origin. Fish from the River Endrick (a high-risk site) that were reared by their fathers and exposed to a model piscivorous fish at an age of 8 w tended to be more vigilant, showing more pronounced responses to the first movement of a distant predator than did similarly reared fish from the low-risk, Inverleith Pond population (Fig. 10.8(a)). In addition, in response to a direct attack, high-risk fish were more likely

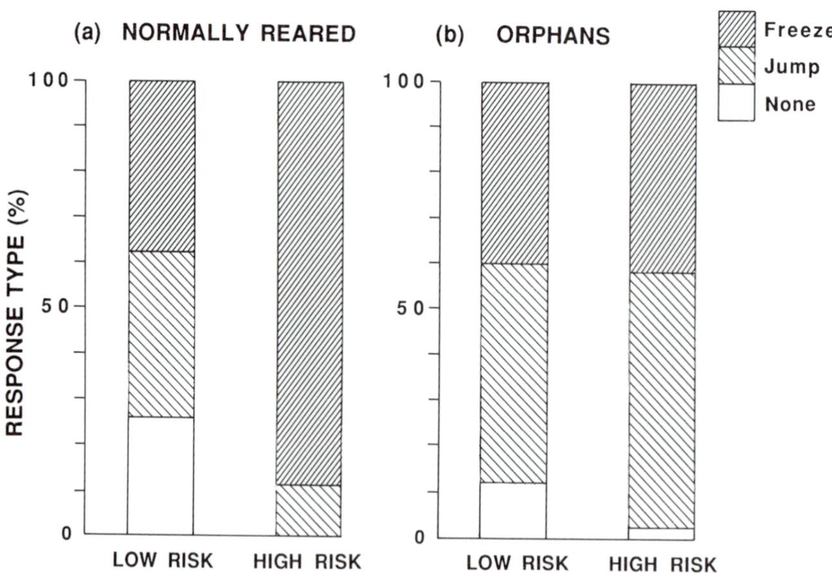

Fig. 10.8 The responses of laboratory-reared stickleback from the high-risk and low-risk populations elicited by the first movement of a model piscivorous fish: (a) normally reared, (b) orphans. Normally reared, high-risk fish are more likely to freeze ($\chi^2 = 7.10$, d.f. = 1, $P < 0.01$) than to jump or to do nothing (the last two being combined to avoid small expected values; $\chi^2 = 7.92$, d.f. = 1, $P < 0.01$) in response to the first movement of the model than are all other categories of fish. No other comparisons were significant. (Overall: $\chi^2 = 20.27$, d.f. = 3, $N = 110$, $P < 0.001$; Huntingford and Wright unpubl. data).

to escape at right angles to the line of attack than to move directly towards or away from the predator (Fig. 10.9(a)). Following an attack, the fish from the high-risk site underwent a longer period of behavioural suppression than did low-risk fish (median = 50 s compared with 8 s; Kruskal–Wallis ANOVA, $H = 11.54$, $N = 13, 31$, $P < 0.001$; Huntingford and Wright unpubl. data). These results suggest that the characteristic features of stickleback from high-risk and low-risk sites are inherited traits that develop in the absence of differential experience of attack by predators.

However, things are not so simple. In the same experiment, other broods of stickleback from these two populations were reared in the laboratory as orphans, without a period of paternal care, and at 8 wk were also subjected to a simulated encounter with a predatory fish. High-risk fish now showed poorly developed protective responses similar to those of their low-risk counterparts. The pattern of response to the first movement of the model predator is the same for high- and low-risk orphans (Fig. 10.8 (b)), as is the distribution of escape directions (Fig. 10.9(b)), and so is the

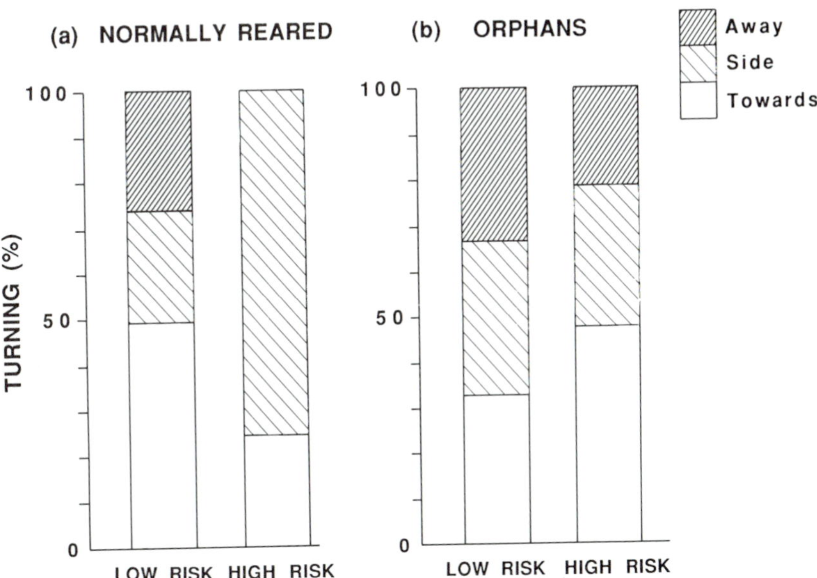

Fig. 10.9 The direction of escape responses of laboratory-reared stickleback from low-risk and high-risk populations when 'attacked' by a model piscivorous fish: (a) normally reared, (b) orphans. Normally reared, high-risk fish are more likely to escape to the side, as opposed to towards or directly away from the line of attack (the last two being combined to avoid small expected values; $\chi^2 = 3.46$, d.f. = 1, $0.1 > P > 0.05$), than are all other categories of fish. No other comparisons were significant. (Overall: $\chi^2 = 8.77$, d.f. = 3, $N = 54$, $P < 0.01$; Huntingford and Wright unpubl. data).

period of behavioural suppression following an attack (recovery times for high-risk and low-risk orphans are 10 s and 4.5 s respectively; Kruskal-Wallis ANOVA, $H = 1.52$, $N = 26$, 12, $P > 0.22$). Similar results were found for stickleback tested with a model piscivorous fish at an age of 4 months (Huntingford and Wright, unpubl. data; see also Tulley and Huntingford 1987*b*).

The explanation for this seemingly bizarre result lies in the nature of the interaction between stickleback fathers and their fry. When young stickleback have absorbed the yolk sac and start to disperse from the nest, their fathers swim after them, gather them up in their mouths, and return them to the nest (Rowland page 338 this volume). If the fry attempt to escape, as they often do, the fathers swim quickly after them and retrieve them. In these cases, young stickleback experience an encounter with a rapidly accelerating larger fish bent on snapping them up. Evidently, this sufficiently resembles an actual encounter with a predatory fish that the young learn to avoid attacks by larger fish more efficiently in consequence (a possibility originally suggested by Benzie 1965). Interactions between guppies and cannibalistic conspecifics enhance the predator avoidance abilities of experienced survivors (Goodey and Liley 1986), and there is little reason to expect that retrieval interactions between paternal stickleback and their young should not have the same effect.

Thus, interactions with their father promote the development of effective predator avoidance in stickleback from high-risk populations, but they do not have the same effect on low-risk fish. The duration of the paternal phase and the amount of time devoted to seeking straying fry was similar in the two populations, both in the laboratory and in the field (Huntingford and Wright unpubl. data). At a temperature of 15–16°C, the fry remained in or near the nest for 2–3 d after hatching, until absorption of the yolk sac was complete. After this time, they gradually dispersed from the nest area (on foraging trips), and started to experience retrieval attempts by the father. From 3 d to 6 d after the brood had hatched, in the laboratory, males of both populations spent approximately 40 per cent of their time searching for and retrieving fry; they then began to build a new nest and subsequently ignored their now-independent young.

Although there were no differences between the two sites in the duration of the father–fry relationship, there were differences in its nature. In particular, both retrieval by fathers and escape by fry were faster in fish from the high-risk population (Fig. 10.10). Retrieval and fry escape speeds were significantly correlated for both categories of fish, but fry from the high-risk site (unlike their low-risk counterparts) were potentially able to outswim their fathers. From hatching, high-risk fry reacted more strongly to their fathers' approaches than did low-risk fry, which consistently ignored their fathers when they swam near by (Fig. 10.11) and jumped up towards them as they approached. It is not yet clear whether this difference in

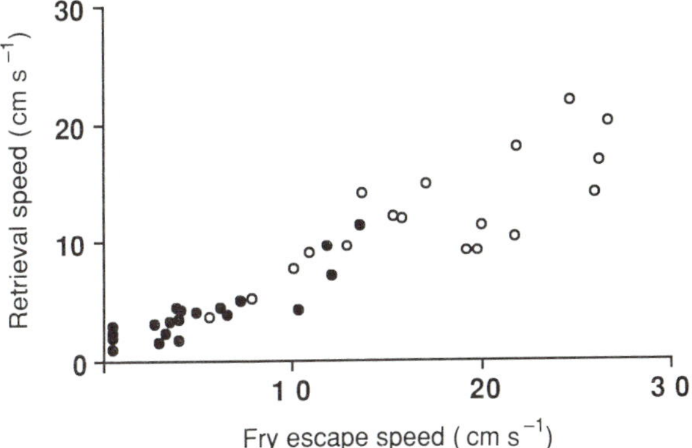

Fig. 10.10 Speed of retrieval attempts by paternal stickleback in relation to the escape speed of their fry in fish from a high-risk site (open circles) and a low-risk site (filled circles). Father and fry speeds are correlated in both populations (Spearman rank-order correlations, high-risk site $r_s = 0.73$, $P < 0.01$, $N = 18$; low-risk site $r_s = 0.86$, $P < 0.01$, $N = 20$). Speeds are significantly greater for high-risk fish than for low-risk fish, both for fathers (medians = 11.9 cm s^{-1} and 3.9 cm s^{-1} respectively; Mann–Whitney $U = 43$, $P < 0.001$) and for fry (medians = 17.7 cm s^{-1} and 3.6 cm s^{-1} respectively; Mann–Whitney $U = 9$, $P < 0.001$). High-risk fry have significantly higher speeds than do their fathers (Wilcoxon $T = 1.5$, $P < 0.001$), but this is not the case for low-risk fry (Wilcoxon $T = 65$, $P > 0.1$; Huntingford and Wright unpubl. data).

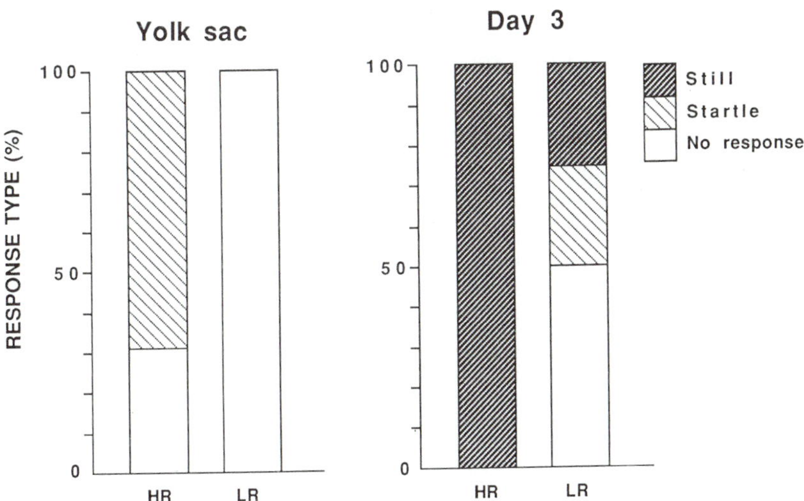

Fig. 10.11 Responses of stickleback fry from the high-risk (HR) and low-risk (LR) sites to close approaches by their father on hatching (yolk sac) and on the first day of the retrieval period (day 3). Comparison at the level of broods shows that high-risk fry are significantly more likely than low-risk fry to respond to their father (by either giving a startle response or remaining still) as opposed to showing no response (Mann–Whitney $U = 9$, $N_1 = 7$, $N_2 = 9$, $P < 0.05$; Huntingford and Wright unpubl. data).

responsiveness is a cause or an effect of the faster retrieval attempts.

Either way, in response to these faster chases, during the course of the paternal period the high-risk fry (but not their low-risk counterparts) became more vigilant and more effective at escaping from their fathers; their initial simple escape responses were amplified into a full spectrum of effective responses to a potentially dangerous larger fish. In nature, this early effect of paternal care may well make the difference between

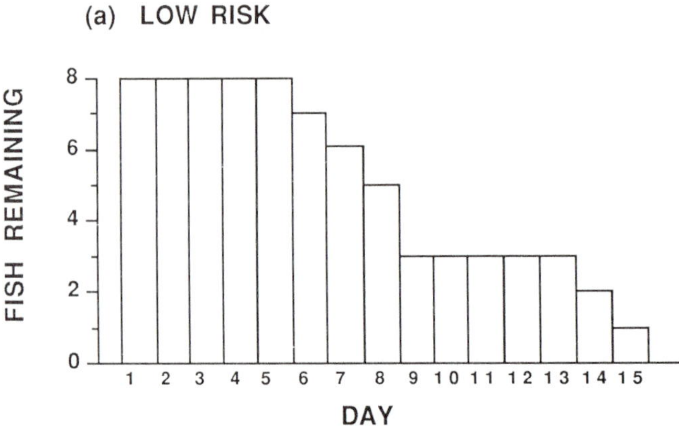

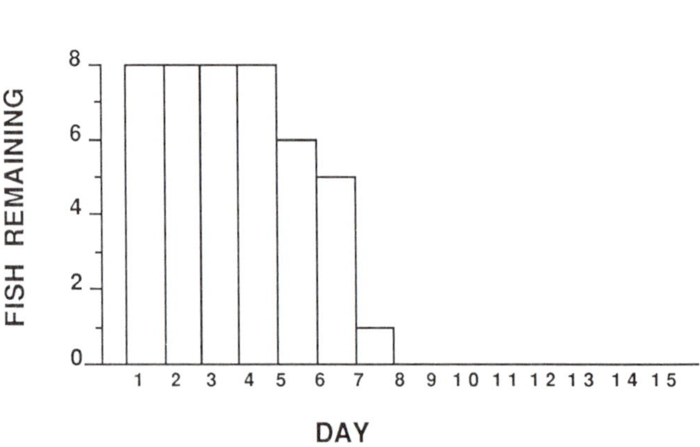

Fig. 10.12 The number of predator-naive stickleback from the two study sites that had failed to reach the criterion for learning to avoid a previously favoured but now-dangerous feeding patch on successive days of avoidance training. High-risk fish took significantly fewer days than low-risk fish to reach the criterion (median = 6 and 8 days respectively; Mann–Whitney $U = 11$, $N_1 = 8$, $N_2 = 8$, $P < 0.05$; after Huntingford and Wright, in press).

life and death when the young stickleback first encounter a real predator.

It is possible that the different intensity of paternal care experienced by young stickleback from the two study site interacts with a predisposition in fish from the high-risk site to modify their behaviour in the light of adverse experience. Laboratory-reared, predator-naive stickleback from the two study sites (aged 6–9 months and measuring 30–35 mm SL) were subjected to an avoidance conditioning regime. The fish were trained to feed in two equally profitable feeding compartments; they learned this quickly, but in every case developed a strong preference for one of the compartments. During the avoidance conditioning procedure, the fish were given one session per day in which each entry into this favoured compartment was followed by a single simulated attack from a model predator (a dark looming object approaching rapidly from above). When not participating in training sessions, the subjects were confined in a separate compartment that contained no food. The criterion for learning this task was failure to enter the previously favoured but now-dangerous compartment on three successive days (see Huntingford and Wright 1989 for details).

With two exceptions, all the fish reached this criterion within 15 d. However, stickleback from the high-risk site were markedly quicker at learning to avoid the dangerous patch (Fig. 10.12; Huntingford and Wright in press *b*). Laboratory-reared minnows from a high-risk site that had been given a single exposure to pike subsequently showed stronger responses than did naive fish from the same site; no such experience-induced adjustment was found in fish from a low-risk site, suggesting that these fish also have an inherited difference in readiness to adapt their behaviour in the light of adverse exerience (Magurran 1990).

CONCLUSIONS

The antipredator behaviour of threespine stickleback is very variable, and some of this variability can be ascribed to intra-individual changes in state. In particular, how a stickleback behaves in any given encounter with a predator is influenced by the likely adverse and beneficial consequences of the possible responses. Long-term, interindividual differences also contribute to the observed variability in antipredator behaviour in threespine stickleback, and some of these differences can be related to local environmental conditions, and in particular to the predation risk prevailing at different sites.

These site-specific levels of antipredator behaviour appear to have a rather complex developmental origin. Comparison of a pair of sites with markedly different predation regimes has shown that, as a result of a difference in the vigour with which paternal stickleback retrieve their fry, young from the high-risk site have more opportunity than do their low-risk counterparts to learn effective techniques for avoiding larger fish. They make good use of this opportunity, having a predisposition to modify their behaviour in the light of adverse experience. In consequence, a small initial divergence in behaviour (more vigorous response to the father's approach in young stickleback from the high-risk site) is amplified. Therefore when they subsequently encounter larger fish with predatory intentions, threespine stickleback from the site where predators are abundant have a full repertoire of effective antipredator responses that is effectively absent in those from the site where predators are uncommon.

ACKNOWLEDGEMENTS

We thank N.B. Metcalfe, D.H. Owings, and R.G. Coss for extremely helpful comments on earlier drafts of this manuscript.

11

Proximate determinants of stickleback behaviour: an evolutionary perspective

William J. Rowland

Many factors initiate and control behaviour in threespine stickleback. Photoperiod, temperature, and other environmental variables typically have pervasive effects on responsiveness and may be exerted indirectly over a long time scale. Their effects on processes such as migration, osmoregulation, and reproductive cycling are mediated largely through the endocrine system. This aspect of stickleback biology has provided fertile grounds for study by physiologists (Guderley Chapter 4 this volume). Here I concentrate on the more immediate external factors that influence reproductive behaviour of threespine stickleback on the breeding grounds. These proximate factors tend to operate on a shorter time scale and to affect neural control of behaviour.

I focus on the threespine stickleback, *Gasterosteus aculeatus*, although findings from other species are sometimes considered as they relate to stickleback evolution and behaviour. I have also included some of the very early references on a topic whenever appropriate. This work is often overlooked because it is not published in English, is otherwise inaccessible, or is written in a style that does not conform to modern standards and is therefore assumed to be unworthy of serious attention by biologists. Nevertheless, a fair number of these works include accurate and insightful observations, and the enthusiasm and clear writing style of their authors can serve as an inspiration to workers today.

I have organized this chapter chronologically, to reflect the order in which functionally related activities typically appear during the reproductive cycle of the stickleback and to emphasize how behaviour performed at one stage often brings the fish into a situation that elicits the next stage of its reproductive cycle.

Beginning with the fish's arrival at their breeding grounds, I first discuss the factors that influence aggression and the acquisition of a breeding territory by the male. I then discuss the factors involved in his selection of a nest site and in the construction of the nest. The completion of the nest leads to the courtship phase, in which mating occurs. I consider this from the perspective of both the male and the female stickleback, and conclude the chapter with a discussion of proximate factors involved in the male's care of his offspring to the point that they become independent.

Visual cues exert a dominant influence on behaviour of all gasterosteid fishes, and in this family vision has been the most studied sense. Although the discussion reflects this emphasis, I consider evidence for the involvement of other senses in mediating reproductive behaviour in stickleback.

This chapter is intended to serve several purposes: to describe the behaviour to which contributors refer in other chapters, to provide an up-to-date review of work on the stimuli that affect stickleback behaviour on the breeding grounds, and to present some preliminary or unpublished data that may suggest avenues for future research.

In keeping with the theme of this book, I try to maintain an evolutionary perspective in so far as this is practicable. Many of the experiments on stickleback behaviour that are discussed below, however, were conducted during a time when ethology was developing rapidly. It was therefore logical to interpret the proximate features that so dramatically influenced an animal's behaviour within the ethological context of sign stimuli and innate releasing mechanisms. These interpretations are not invalid, but another perspective, here an evolutionary one, can often provide insights that may otherwise be overlooked.

AGGRESSION AND TERRITORIALITY

The behaviour employed by male threespine stickleback to establish and maintain territories or to defend themselves and their spawn from predators has long interested biologists (e.g. Anonymous 1830; Warington 1855). The kind of intruder can profoundly influence the form of aggression. Territorial males typically direct ritualized head-down threats, paralleling, and overt attacks toward solitary individuals of the same (van Iersel 1953; Morris 1958; Li and Owings 1978*a*) or related (Rowland 1983*a*,*b*) species, but may respond very differently to groups of conspecifics or other predators (Huntingford *et al.* Chapter 10; Foster page 394 this volume). This variation suggests that aggression toward rivals and aggression toward predators may represent different phenomena, although at some level they may be causally linked (e.g. Huntingford 1976*a*).

In view of the central importance of rival-induced aggression for the establishment and maintenance of breeding territories, I now discuss the role of various proximate factors that are believed to play an important role in the initiation and control of conspecific aggression in threespine stickleback. These include male colour and size, the posture, movement, and location of the rival male, and previous experience with rivals.

Colour

Visual cues probably exert the strongest sensory influence on aggressive behaviour in sticklebacks. When males of closely related sympatric species were presented in watertight glass cylinders at various distances from a nest, each species evoked a different level of response from the nest-owner (Rowland 1983*a*,*b*; Fig. 11.1).

The conspicuous nuptial colour patterns that are unique to reproductive males of each species are presumed to serve as cues on which much of this species discrimination and conspecific-directed aggression is based. The most widely cited evidence of this is the experiment by Tinbergen and his co-workers (Tinbergen 1948, 1951). When a crude dummy with a red underside and a more detailed but non-red dummy were presented to territorial threespine stickleback males, the red dummy elicited the most attacks. Although the red dummy's effectiveness at eliciting attack was slightly enhanced by adding a light blue back and blue eyes, males neglected most other features (Tinbergen 1953). This led Tinbergen to conclude that it was the male colour pattern, particularly the red underside, that served as the primary feature or 'sign stimulus' eliciting territorial aggression in this species.

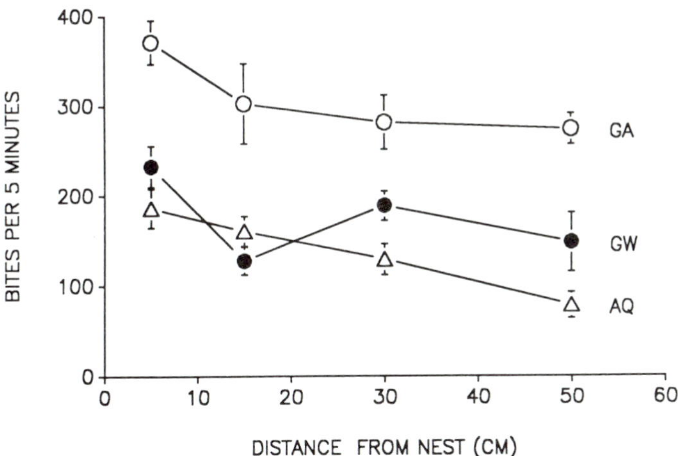

Fig. 11.1 Attack rate of five territorial male threespine stickleback to three species of intruder males presented at four distances from the territorial male's nest (mean ± SE). The intruder species were the threespine stickleback, *Gasterosteus aculeatus* (GA), the blackspotted stickleback, *G. wheatlandi* (GW), and the fourspine stickleback, *Apeltes quadracus* (AQ). There is a decrease in bite rate to each intruder species with increasing distance from the nest (Page's $L = 141$ for GA, $L = 138$ for GW, and $L = 147$ for AQ, where $k = 4$ and $N = 5$ for each intruder species; all $P < 0.05$; Page 1963).

Subsequent attempts to verify the effects of nuptial colour on stickleback aggression have yielded mixed results. When Muckensturm (1969) presented males from a European population of the threespine stickleback with dummies of different colours, the most effective colour for provoking attack varied among individuals. Males did not attack red more than other colours, and when males were tested under illumination from which red wavelengths were removed by a blue filter, their aggressive reactions did not decrease (Chauvin-Muckensturm 1979). Similarly, Peeke *et al.* (1969), studying a Californian (USA) population of threespine stickleback, and Wootton (1971*b*), studying a British Columbian (Canada) population, failed to detect differences in the mean rates of attack directed by territorial males at silver dummies versus dummies that were partially or entirely red.

In other studies, the rates at which males attacked non-red dummies versus dummies with a red underside differed, but in the opposite direction. My own research on stickleback from a Long Island (New York) marine population and a Dutch anadromous population provided results that contradicted those described by Tinbergen (1953). Overall, the non-red dummy received more attacks than did the dummy with a red underside, but a small minority of males showed the opposite response (Rowland 1982*a*; Rowland and Sevenster 1985). In contrast, Collias (1990) reanalysed results he had obtained from Tinbergen's laboratory in Leiden in 1947 and confirmed statistically that male threespine stickleback attacked dummies with a red underside more than those without red. But Collias' results also revealed that the attack-provoking effect of red was less robust than had often been implied in the literature: several of his males showed little tendency to attack the red dummy more than the non-red one, even though he presented the dummies close to (7–13 cm) the males' nests in an attempt to maximize aggression.

One plausible explanation for the different results from these studies is that red elicits fear as well as aggression in males, but that the relative increase in these two factors is context dependent (Rowland 1982*a*). Specifically, when the dummy is presented near the nest, as in Collias' study, it may elicit aggression more effectively than it elicits fear, and thus operate as a 'sign stimulus' for aggression (*sensu* Tinbergen 1951). But when the dummies were placed farther from the nest, as in the studies by Rowland (20 cm from the nest) and Rowland and Sevenster (30 cm from the nest), the consequent decrease in aggression (van Iersel 1958; Fig. 11.1 this chapter) might have led to relatively more fear of the red dummy, hence relatively more biting at the non-red one. Thus, a dual effect of red colour (i.e. provocation and intimidation) could explain the discrepancies among studies.

Discrepancies among studies of nuptial colour effects in threespine stickleback could also be accounted for by differences in illumination used in aquarium experiments (Reimchen 1989). Because differences in the relative amounts of horizontal, upwelling, and downwelling light affect hue or brightness contrast of objects viewed underwater (for reviews see Lythgoe 1979; Nicol 1989), the dimensions of aquaria and the source and placement of their illumination may affect the way nuptial colour is perceived by stickleback. Although this possibility cannot account for differences among individuals within a given study, it is an important consideration for future research design.

Another possibility is that differences in hue, saturation, or extent of red used to colour the dummies, in the procedure used to present the dummies, or in an individual's experience or genotype, affected the motivational balance between attack and avoidance behaviour in males. These differences could also have led to the conflicting results obtained in past studies. Subsequent findings that males more often abort attacks on dummies with a red underside, unusually large body size, or head-down threat posture, than on dummies lacking these features, are consistent with this hypothesis (Rowland 1983*a*; Rowland and Sevenster 1985).

Recent studies reveal wide interpopulational variation for many aspects of behaviour in threespine stickleback, including aggression (for reviews see chapters by Hart and Giles, Huntingford *et al.*, Bakker, Foster and McPhail this volume). This variation could also account for some of the inconsistencies that have arisen in studies of male response to nuptial colour.

In the threespine stickleback, there is a positive association between the male's colour score and tendency to attack intruders (Rowland 1984; McLennan and McPhail 1989*b*). This association could explain why the red underside should be intimidating to opponents and why brighter males tend to dominate duller ones (Bakker and Sevenster 1983). The observation that male threespine stickleback in some populations maintain intense nuptial colour, even when they are guarding offspring but no longer courting females (Moodie 1972*a*,*b*; Kynard 1978*a*), suggests that the intimidation effect may also play a role in the nest-owner's defence of nest and offspring.

Although some accounts probably overstate the effectiveness of red for provoking attack in *G. aculeatus* males, this does not necessarily invalidate a sign stimulus interpretation of nuptial colour, as some imply (Reiss 1984). On balance, the response of male stickleback to the sight of a red underside in an opponent is seemingly affected in a way more complicated than implied in the original reports and even in recent textbooks on animal behaviour (e.g. Grier 1984; McFarland 1985; Drickamer and Vessey 1992).

If the red underside is interpreted broadly as an index of male condition rather than as a releaser of aggression (Rowland 1982*a*, 1984; Milinski and Bakker 1990), then the inter- and intrapopulational variation reported for male response to this feature is not unexpected. Just as individuals differ in their tendency to attack or avoid dummies (Peeke *et al.* 1969; Wootton 1971*b*; Rowland 1982*a*; Bakker and Sevenster 1983; Rowland and Sevenster 1985) and live males (Sevenster 1961; Wootton 1971*b*;

Rowland 1984), so too will the provoking and intimidating effects of the red underside vary among them. If intrasexual selection has played a role in the evolution of male nuptial colour in stickleback, then it is probably the intimidating effect that maintains or elaborates red colour in *G. aculeatus* populations. Indeed, this effect may even lead brighter males to dominate duller ones in this species (Bakker and Sevenster 1983), although the possibility that brighter males dominate merely because they behave more aggressively or are physically superior needs to be tested.

Size

Initially, Tinbergen (1953) concluded that intruder size had little effect on aggression in stickleback, because territorial males reacted aggressively even to a red mail truck passing in the distance outside the laboratory window. However, he also discovered that a dummy three times the size of a stickleback would provoke an attack when presented at a distance, but not when presented in the aquarium. In consequence, he acknowledged that these abnormally large objects were probably effective in provoking attack because at a distance they subtended a small angle on the retina. Thus, what may be critical in this situation is the apparent size of the intruder, in much the same way that the apparent size of prey has been found to be important in stickleback foraging behaviour (Gibson 1980; but see Hart and Gill page 232 this volume).

Although in many fishes, larger body size may confer an advantage in aggressive interactions, it has been assumed that male stickleback are similar enough in size that this factor is unimportant (Wootton 1976). In fact, most studies that have looked for a size advantage in aggressive interactions between male *G. aculeatus* failed to find a statistically significant effect (van den Assem 1967; Sargent and Gebler 1980; Bakker 1986; FitzGerald and Kedney 1987). In some populations, however, males differ by 15 per cent or more in body weight, and in such cases heavier males tend to defeat their smaller opponents when the fish compete directly for the same territory (Rowland 1989*a*). Larson (1976), too, found that in a small British Columbia lake, a larger benthic form of *G. aculeatus* excluded a smaller limnetic form of the species from a vegetated area that both forms chose to occupy.

I examined the effect of body size on aggressive encounters more directly by presenting dummies of different sizes to territorial male threespine stickleback (Rowland 1983*a*). When the proportions of aborted attacks ('backoffs'; see Rowland and Sevenster 1985) to total attacks (backoffs + bites) were compared, it was found that males aborted a greater proportion of attacks to larger dummies than to smaller ones (Fig. 11.2). This result suggests that larger males are more intimidating than smaller males. If this is true, then size effects could confer behavioural as well as physical

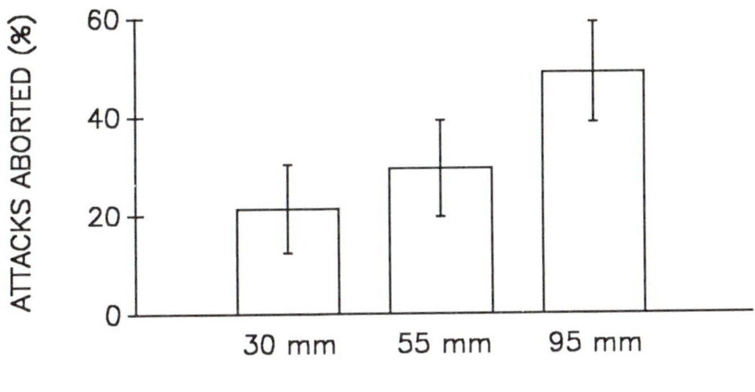

Fig. 11.2 Per cent of attacks (mean ± SE) by nine territorial male threespine stickleback that were aborted when males were presented with dummies of three different standard lengths. Each male was presented in random order with each dummy once at a constant distance from the nest. There is an increase in the proportion of attacks aborted with increasing dummy size ($P < 0.01$; Page's $L = 119$, $k = 3$, $N = 9$).

advantages to the larger opponent in aggressive encounters between stickleback, especially in interspecific interactions, where size differences are often considerable (Rowland 1983*a*,*b*; Gaudreault and FitzGerald 1985).

It is not known whether males fail to attack oversized dummies (Tinbergen 1953; Rowland 1983*a*) because they fail to recognize them as stickleback, possibly perceiving them as potential predators, or because they are merely intimidated by them. It is possible that when the dummy is only slightly larger than the male, intimidation is the primary factor, but that when the dummy is much larger, the male instead responds as it would to a potential predator. When the dummy is several times larger than the male, the male approaches the dummy in a cautious, exploratory manner (Huntingford *et al.* page 278 this volume), occasionally attacking it from behind on the tail or flank (pers. obs.). This behaviour is very similar to that observed when threespine stickleback attack large prickly sculpin, *Cottus asper* (Pressley 1981; Foster and Ploch 1990), suggesting that the males are responding to the dummy as if it posed a predation risk.

Posture

Posture is another feature that is commonly cited as a sign stimulus for aggression in stickleback (Tinbergen 1948, 1951). Ter Pelkwijk and Tinbergen (1937) reported that a dummy or a live male in a glass tube elicited more attacks from a male threespine stickleback in his territory when presented head-down than when presented horizontally. Because males threaten rivals by tilting their bodies head-down, this posture should be more intimidating than a horizontal one. If the above argument concerning the dual (i.e., provoking and intimidating) effects of nuptial colour is valid, then males should show a corresponding response to posture.

Baerends (1985), in reviewing the evidence for sign stimuli in stickleback, questioned why the responses of males to red nuptial colour were so much more variable than were their responses to dummies in the vertical posture. I suspect that the response to posture only appears to be less variable than the response to colour because posture effects have not been subjected to the scrutiny that colour effects have received. Indeed, a re-examination of posture effects on aggression in threespine stickleback revealed lower attack rates to a head-down dummy than to a horizontal one, presumably because the males perceived the head-down dummy as more threatening (Rowland and Sevenster 1985).

Males also attacked a dummy male less when it was presented head-up (Rowland and Sevenster 1985). Head-up postures, commonly assumed by defeated males and courting females, are associated with submissiveness and have been interpreted as appeasement displays serving to inhibit aggression (Morris 1958; Bastock 1967). This effect might therefore have been responsible for the reduced attack on the head-up dummy.

The original dummy experiments on stickleback behaviour may not be strictly comparable to recent studies, and the different experimental techniques used in them may have contributed to the seemingly contradictory results. For example, in a more recent series of experiments, we presented males with stationary dummies specifically to control for the effects of movement (Rowland and Sevenster 1985). The classic experiments, however, tested for posture effects in stickleback by comparing male response to dummies or dead fish presented in various postures and moved to imitate threatening or normally swimming males (ter Pelkwijk and Tinbergen 1937). Therefore, stronger aggressive responses toward the head-down dummy may better reflect a response to the combination of movement and posture than to posture alone. Indeed, ter Pelkwijk and Tinbergen cautioned that their experiments did not allow them to separate the relative importance of posture and movement for evoking aggression in males, but this point seems to have been overlooked in later interpretations of their results.

Movement

Movement is a powerful stimulus for aggression in stickleback. In the posture experiment discussed above, a dummy, or even a dead female, attached to a wire and made to imitate the threat display of a rival (or possibly the movements of a nest-raiding fish; Foster page 394 this volume) by thrusting it head-down into the bottom provoked intense attack from males. The same dummy or dead fish manipulated to imitate the normal swimming movements of a stickleback provoked much less aggression (ter Pelkwijk and Tinbergen 1937).

Fleeing stickleback also provoked chasing and biting in an opponent (van Iersel 1953). Sevenster (1961) used vertical movement and Peeke *et al.* (1969, pers. comm.) used horizontal movement to induce male stickleback to attack dummies. Even undirected movement increases aggression, as males intensified attacks to a stickleback held in a glass cylinder as soon as the latter moved (Rowland 1984). Thus, dummies were probably moved in the original stickleback experiments to increase the chance that males would respond to them (ter Pelkwijk and Tinbergen 1937; Baerends 1985).

Laboratory observations on freely interacting stickleback reveal that males monitor the behaviour of their opponents and respond accordingly; i.e. the response of one fish elicits a specific reaction in the other. This is especially evident for highly coordinated activities such as spine-fighting, where males carefully match their opponent's movements, rapidly chasing and circling each other head-to-tail, with a pelvic spine pointed toward their opponent (van Iersel 1953).

Location

Spatial context is a critical factor in animal aggression. Indeed, many of the aggressive activities of stickleback are related to territorial defence and are therefore strongly site specific. This dependence was convincingly demonstrated by an experiment in which two neighbouring males (A and B) were placed in separate glass cylinders and allowed to interact visually (Tinbergen 1953). When both cylinders were placed in A's territory, male A threatened and tried to attack male B, whereas B attempted to escape. The response of the males was reversed when the cylinders were moved into B's territory.

Van Iersel (1958) extended this experiment by presenting a stimulus male in a glass cylinder at various distances from the nest of a test male. He found that the closer the stimulus male was presented to the nest, the more intensely it was attacked by the test male. More recent work confirms this effect of distance, even on aggression provoked by intruders of another species (Huntingford 1976*a*, 1977; Rowland 1983*a*,*b*). This effect holds as well for other stickleback species (Fig. 11.1) and may serve to adjust male defence to the increasing danger of an intruder as it approaches the male's nest.

The differences in aggression that various species evoke from territorial males may be a question of species recognition, reflecting the degree of similarity that the territorial males share with intruders. Alternatively, the different responses to intruders of each species may reflect the kind and magnitude of threat posed to the male and his nest (Rowland 1983*a*; Foster and Ploch 1990).

Tactile cues

Threespine stickleback have receptor nerve fibres that end in the epidermis (Whitear 1971), and these provide the skin with tactile sensitivity (Wootton

1976). Thus, even though vision is the primary sense involved in elicitation and control of aggression in this species, tactile stimulation probably plays some role.

Tinbergen (1951) noted that touch receptors may have a specific releasing function in fighting in this species. He found that when a male strikes an opponent with his snout, this evokes a similar response in the opponent. Tinbergen, too, was able to evoke a retaliatory response from a male by imitating this tactile response with a glass rod. He therefore concluded that, although fighting as a whole is controlled by visual stimuli, the release of this specialized response in fighting stickleback is almost or entirely independent of vision.

Chemical cues

There is no evidence that external chemical cues directly elicit or control aggression in threespine stickleback. Indeed, olfaction appears to be poorly developed in *G. aculeatus* (Teichmann 1954; Wunder 1957), and the lack of response of this species even to food odours (Kleerekoper 1969) would suggest that chemical cues are unlikely to be involved in its aggressive behaviour, though they may function in group recognition (van Havre and FitzGerald 1988). Furthermore, the structure of the brain and olfactory receptors of threespine stickleback suggests that chemical stimuli play a minor role in this species' behaviour (for references see Wootton 1976). It is, nevertheless, inadvisable to rule out the possibility of direct or indirect (Segaar *et al.* 1983) involvement of chemical cues in the absence of relevant behavioural data.

Experience

The influence of territorial boundaries and nest site on the focus, form, and intensity of aggression in male stickleback is well established (see above). Experience will therefore influence the aggressive behaviour of males, because it is through experience that they come to recognize the spatial relationships of their territories and nest sites.

Social experience appears to exert considerable influence on aggressive behaviour and the outcome of fights in threespine stickleback. For example, males made to lose a fight by introducing them into the tank of another male lost subsequent contests when paired against a naive male in a neutral tank. Males made to win, however, by placing an intruding male into their territory tank, were victorious in subsequent contests with naive males (Bakker and Sevenster 1983). Bakker *et al.* (1989) also found that losing had a stronger and longer-lasting effect on subsequent fight outcomes than did winning, a finding that is consistent with studies on other fishes (e.g. McDonald *et al.* 1968; Frey and Miller 1972; Francis 1983). Bakker *et al.* (1989) suggested that this is so, either because winning and losing affect the physiological mechanisms that mediate agonistic behaviour differently, or because the loser receives immediate positive reinforcement (termination of the fight) as a consequence of its response (retreat). That is, it is the loser and not the winner whose response immediately terminates the fight.

Continuous or repeated presentation of a stimulus may reduce an animal's responsiveness to that stimulus through habituation. Thus, territorial male stickleback decreased their attacks to males viewed behind a glass partition (van den Assem and van der Molen 1969). They also habituated if they had only intermittent view of their neighbour or could freely interact with him. Peeke *et al.* (1969) also found that male stickleback habituated to repeated presentation of a rival male in a glass cylinder, but if the same rival was presented at a different location in the male's tank, or if a different rival was presented at the same location, there was a partial recovery of attack by the territory owner (Peeke and Veno 1973). Recovery of attack was greatest if a different rival was presented at a different location.

Agonistic interactions among stickleback often interfere with courtship (Wunder 1934; Li and Owings 1978*a*; Sargent and Gebler 1980; Borg 1985; Ward and FitzGerald 1987; Rowland 1988) or nesting activities (van den Assem 1967; Sargent 1985; Rowland 1988). The stimulus-specific nature of habituation does, however, provide a mechanism for reducing interaction among neighbours and the risk, time, and effort that such interaction incurs. Laboratory experiments reveal that reduction in aggression through habituation can increase the time that territorial males spend in courtship and nesting activities (Rowland 1988), but the extent to which habituation operates in nature is unknown.

Sevenster (1968, 1973) found that the opportunity to 'fight' another male separated by a glass partition served as a positive reinforcement for territorial male threespine stickleback. If this reinforcement was contingent on biting a rod suspended in their tank, males learned to perform this task. The response of males that had been conditioned to the rod suggests that some of the aggression-evoking effects of the rival become transferred to the rod during this process. Hollis (1990) has discussed how Pavlovian conditioning of aggression in male blue gouramies, *Trichogaster trichopterus*, could enhance their territorial defence by signalling impending aggressive encounters, and such effects may also occur in male threespine stickleback. Learning processes are therefore likely to play a role in modifying the proximate mechanisms that affect stickleback aggression in nature. A recent attempt to train threespine stickleback to decrease their use of head-down threat display, by punishing males with electric shock when they performed this response, proved unsuccessful (Losey and Sevenster 1991). In fact, seven of nine males so punished increased their use of head-down threats. This result emphasizes that, although learning provides a mechanism by which some stickleback activities can be adjusted to the environment, the biological constraints imposed on other activities will render them resistant to such modification.

NEST-BUILDING BEHAVIOUR

Once the male threespine stickleback has established a territory, he is ready to build a nest. Tinbergen (1951) stated that warm, shallow water and the presence of suitable vegetation play an important role in activating nest building. In the initial stages of this behaviour, the male picks up mouthfuls of sand from almost anywhere on the bottom and quickly ejects them. Scattered digging has been observed regularly in the field (Foster pers. comm.). In the laboratory, it results in the formation of a crater-like pattern over the bottom of the tank.

The male eventually focuses on a specific nest site and engages in more intense digging as he carries the sand farther and farther away before ejecting it. This results in the formation of a pit in which the nest will be built (Tinbergen 1951).

The choice of a nest site has been the subject of several studies. In aquaria, males often nest in the corner or near vegetation, rocks, or other objects (Hancock 1852; Warington 1852, 1855; van Iersel 1958; van den Assem 1967; Tschanz and Scharf 1971; Jenni 1972). In nature, *G. aculeatus* males may show similar attraction for nesting in cover (Black and Wootton 1970; Moodie 1972*a*; Kynard 1978*a*), but this tendency is not as marked as in some other sticklebacks (e.g. Morris 1958; Winn 1960; Courtenay and Keenleyside 1983; FitzGerald 1983). By nesting in cover, males may reduce nest predation and courtship interference by other males (reviewed by Whoriskey and FitzGerald page 192 this volume). The proximity of rocks, plants, and other objects may also serve as landmarks to facilitate the male's orientation to his nest (van Iersel 1958, Tschanz and Scharf 1971).

It is becoming clear that stickleback populations differ in the way that males respond to proximate factors affecting nest-site selection. For example, in one Japanese population some male threespine stickleback build upright nests on a vertical shore wall (Mori 1988), a habit not seen in other populations. Mori suggested that these unusual sites may be used when high population densities prevent males from obtaining territories on the bottom. Atypical nest-site selection has also been described in other sticklebacks (Reisman and Cade 1967; McKenzie and Keenleyside 1970; Griswold and Smith 1972; Rowland 1974*a*,*b*).

Male threespine stickleback often tend to favour certain nesting substrates over others, although there exists considerable variation among populations. For example, males from a stream-resident freshwater population in the Little Campbell River, British Columbia, Canada, typically nest on mud in still water, whereas those in a parapatric anadromous population nest predominantly on sand in mild currents. When offered a choice of mud or sand in aquaria, males from both populations built nests on the substrate typical of their population, suggesting that the interpopulation difference reflected active choice (Hagen 1967; see also McPhail page 408 this volume).

Threespine stickleback that nest in brackish marsh pools with mud bottoms on Long Island, New York, USA, will nest on sand in the laboratory. Given a choice between black and off-white sand of identical texture, these males always nested in the former (10:0, $P < 0.001$, binomial test). Offered a choice between fine and coarse black sand, the former was favoured (9:1, $P < 0.011$, binomial test), suggesting that the males sought a substrate of colour and texture similar to the mud of their natural breeding grounds. The extent to which this preference is influenced by genetic factors or prior experience with such substrates is an open question.

Even during the initial stages of nest building in aquaria, Long Island males spent little time digging in fine gravel but increased their digging in sand, suggesting that the preference for the finer substrate reflects the ease with which males manipulate the finer-textured sand. Moreover, if males and their nests are less conspicuous when viewed against a dark background, the preference that these males express for the darker substrates may render them less vulnerable to predators. In the absence of further information, the possible causes and functions of substrate preferences in threespine stickleback must remain speculative.

Blouw and Hagen (1990) described a population of 'white stickleback' that nest exclusively in plumes of filamentous algae in shallow marine and brackish waters of Nova Scotia, Canada. Even when eggs and sperm from these fish were combined in the laboratory and the resulting progeny were reared to maturity, the male offspring would nest only on algae, indicating that the preference for this unusual nesting site by the white stickleback is inherited.

Males start to search for nest material in the latter part of the sand-digging stage (van Iersel 1953). Males that were provided with a layer of sand in which they could dig, collected and incorporated other material into the nest sooner and more frequently than did males that could see sand but were denied access to it by a plate of glass (Schütz 1980). This finding suggests that the collection of nest material is brought about by digging as well as by the pit that this activity produces. Schütz also found that sand digging was reduced if males could bring nest material into the pit. Thus sand digging, pit formation, and the collection and placement of nest material all interact to synchronize nest building in stickleback.

In the appetitive stages of gathering nest material, the male spends much time seeking and testing material (van Iersel 1953). When the male finds a strand of algae, a loose stem, or other plant matter, he sucks it up and spits it out several times in rapid succession. If it is acceptable, he incorporates it into the nest. As nest building progresses, the searching and testing phase drops out (van Iersel 1953). The male now swims directly to a source of nest material, tears off a piece, and brings it quickly to the nest. Here it is stuffed into the developing nest structure and glued with a substance that is produced by the male's kidneys (Rinkel and Hirsch 1940).

The male applies this gluelike substance with a peculiar gliding movement as he presses his anal opening against the nest (Leiner 1929; Wunder 1930; van Iersel 1953). Each gluing bout lasts a few seconds and is performed repeatedly during the nest-building phase.

As the accumulating nest material becomes consolidated, it forms a matlike structure that is anchored to the substrate. The male continually shapes this structure by pushing into it, focusing on a point that becomes the entrance. The male bores into the nest with increasing intensity until he eventually tunnels through the entire structure. This initial act of 'creeping through' completes the nest and marks the transition to the courtship phase, when the male is now competent to court females (van Iersel 1953; Sevenster 1961). Here again, one can observe how internal and external factors interact to coordinate reproductive activities and produce a functional behaviour sequence in stickleback (Guiton 1960; Wootton 1976).

Visual cues

Visual stimuli play an important role in the initiation of nest-building behaviour in stickleback. For example, the sight of nest-building material under glass stimulated sand digging in males, whereas sandy ground under glass stimulated them to collect suitable nesting material (Schütz 1980). The selection of nest sites, too, was influenced by visual features. Tschanz and Scharf (1971) tested males in aquaria under various combinations of conditions and found that they chose subdued illumination, plants in the corner of the tank, and vertically striped background patterns.

These tests also revealed that males use objects as landmarks for more localized orientation close to the nest, but that they use differences in illumination levels to find their way back when they are farther from their nest site. When stones, plants, or other objects close to the nest are moved, males often search for their nest at a location corresponding to its position relative to the objects before they had been moved (pers. obs.).

That such males are fooled even though the nest is undisturbed and in full view emphasizes the role of visual cues for the recognition of nest site in threespine stickleback (see also Wootton 1976). Visual cues have also been implicated in selection of vegetation type in fourspine stickleback, *Apeltes quadracus* (Baker 1971), a species that in nature also selects certain plants on which to nest (Courtenay and Keenleyside 1983).

Males are selective in their choice of nest-building materials. For example, Wunder (1930) presented male threespine stickleback with nest-building material of various colours. He found that males tended to place the more brightly coloured fibres around the nest entrance, and suggested that this served to advertize the entrance. Leiner (1931) also found that a small number of males tended to place red and other brightly coloured material around the nest entrance, but was reluctant to conclude that this represented a marking mechanism.

Morris (1958) investigated this phenomenon more fully and concluded that the bright contrasting ring of colour with which males surrounded the nest entrance was a result of changes in colour selectivity during nest building. He documented interindividual differences in colour patterns, and showed that males tended to ignore red material early on but chose this colour more often as the nest neared completion and the entrance was being formed. This shift in selectivity may therefore have provided males with a mechanism to increase the conspicuousness of the nest. This could assist the male in orientating to the nest in an environment generally devoid of landmarks, and might even help induce females to enter the nest and spawn.

The limited colour range of nesting material normally available to males raises questions concerning the relevance of these preferences in nature (Wootton 1976). Material of contrasting shade, however, is more likely to be available in nature, and males might use this contrast to highlight the entrance to their nests, especially during the courtship phase. Of the ten males who built nests on black sand in the substrate colour tests discussed above, six had carried white sand from the opposite end of the aquarium and deposited it around the nest entrance, making the entrance conspicuous. This may, however, have been an inadvertent result of attempts by the males to hide the nest entrance; because only two dishes of sand were present during the substrate tests, males seeking sand from elsewhere in the tank to cover their nest could only select sand of contrasting shade (white).

If male threespine stickleback engage in nest marking, it is most likely that they would do so during the earlier stages of courtship, before they receive eggs. Field data reveal that the eggs and nest material of threespine stickleback are subject to attack from conspecifics (Whoriskey and FitzGerald page 202; Foster page 394 this volume), so it is not surprising that males in nature may try to conceal their nests by covering them with vegetation and debris (Wootton 1972*a*; Kynard 1978*a*). Indeed, nest marking has not been documented in nature, and in populations subjected to high levels of nest raiding by conspecifics, nests are cryptic even during the courtship phase (Foster 1988). Further experiments that address the question of male nest marking are clearly needed.

Males also accept or reject nest-building material on characters other than colour. Warington (1852) and van Iersel (1953) thought that males might evaluate the specific gravity of items by watching how fast they sink and rejecting those that were too heavy or too light. Perhaps by choosing material of a particular density, males are able to avoid incorporating into their nest decaying or otherwise unsuitable material. Schütz (1980) concluded, however, that the weight of material had little influence on its selection by males for nest building. Although Schütz did not investigate colour choice, he found that males did choose nesting material of certain lengths, diameters, and surface textures, and that males learned to distinguish these features visually during the course of the nest building period. Such features are evidently important for proper nest construction in stickleback.

Tactile cues

Little systematic study has been conducted on the role of tactile cues in nest-building behaviour, but the effect of sand digging in the early stages of this phase (Schütz 1980) and the preferences of Long Island males for nesting substrates of finer texture were probably mediated in part through such stimuli. Males also choose nest-building material of coarse surface texture and may use tactile cues to perceive such differences (Schütz 1980). Coarse texture may facilitate binding and adhesion when material is glued into the nest. Moreover, the male presses his anogenital opening against the surface of the nest when he glues. The tactile stimuli produced by this contact probably trigger and properly orientate glue secretion.

Another opportunity for tactile stimulation occurs when males creep through the nest, a behaviour that brings the male into intimate contact with the nest tunnel. Experimental manipulation of nest structure suggests that tactile feedback does provide information to the male. For example, when a male's nest was shortened by cutting off the hind part and removing it, the duration of creeping through decreased accordingly ('t Hart 1978). Furthermore, males whose nests were shortened began to rebuild quickly and usually restored their nests to the original length within 2 h. Although 't Hart was unable to identify the nature of the stimulation that enabled males to register the duration of creeping through, tactile and/or visual cues were probably involved in this process.

The damage sometimes sustained by nests during female entry (pers. obs.) and the importance of the nest in protecting eggs from predators (Potts 1984) emphasize the advantages of a properly constructed nest in this species. Tactile stimuli may provide essential information concerning the form, size, and strength of the nest, and can therefore affect the male's transition to the courtship phase.

Chemical cues

There is little evidence that chemical cues play a role in nest-building behaviour. As the broad range of items that males accept for nest building includes such varied substances as plant, animal, and synthetic material (Schütz 1980), it would appear unlikely that males rely on chemical cues for the selection process. Even the state of decomposition of plant material, which one expects might be distinguishable through chemical cues, had no effect on the acceptance or rejection of nest-building material by male threespine stickleback (Schütz 1980).

Chemical cues are likely to be involved in selection of breeding areas by anadromous stickleback populations, in so far as fish distinguish and choose particular salinities (reviewed in Guderley page 92 this volume). Furthermore, the annual return of breeding stickleback to specific marsh pools on Long Island (unpubl. data), and the avoidance of temporary pools by stickleback in Quebec, Canada (Whoriskey and FitzGerald 1989), raise the intriguing possibility that breeding individuals may return to their natal breeding grounds to spawn (Hagen 1967; Kynard 1978*a*). This phenomenon has been well documented in salmon and results from a process of chemical imprinting (e.g. Hasler *et al.* 1978). A similar process might occur in stickleback, but the technical difficulties involved with marking enough stickleback fry and recapturing them as adults have so far presented an insurmountable obstacle to biologists.

COURTSHIP AND MATING BEHAVIOUR

Courtship of the threespine stickleback is one of the best-studied examples of mating behaviour in the animal kingdom. Warington's (1852) accurate but incomplete description of the attempts of a male threespine stickleback to induce a female to spawn in its nest is among the earliest published accounts of courtship behaviour in this species. Since then, many detailed descriptions and analyses of courtship in the threespine stickleback have appeared in the literature (see Wootton 1976 for references).

Stickleback courtship comprises a sequence of modal action patterns (Barlow 1968) that alternate between the male and female. The performance of a given behaviour by one member of the courting pair elicits from its partner a behaviour constituting the next step of the courtship sequence. Signals from the male thus in large part depend on the female's behaviour, and vice versa. The response chain that results from this process insures that sperm and eggs are shed in the same place at the same time (Tinbergen 1951). This synchronization is essential in sticklebacks, because fertilization is external and gametes have limited viability outside the body.

Courtship is a dynamic process that depends on the internal state of the participants and thus varies among individuals from the same population. Courtship motivation also fluctuates within the same individual over time. This fluctuation can be easily observed in males responding either to conspecifics held in a clear flask or to stationary dummies (Rowland 1984). Certain populations of threespine stickleback differ consistently with respect to the details of male courtship behaviour (Wilz 1973; McPhail and Hay 1983; Ridgway and McPhail 1984; Foster page 391, McPhail Chapter 14 this volume). Nevertheless, the behavioural components that constitute courtship are qualitatively similar, and the basic courtship patterns are conserved enough among populations that homologous components are readily identifiable. The courtship of a marine population of threespine stickleback from Long Island is representative of the species and thus

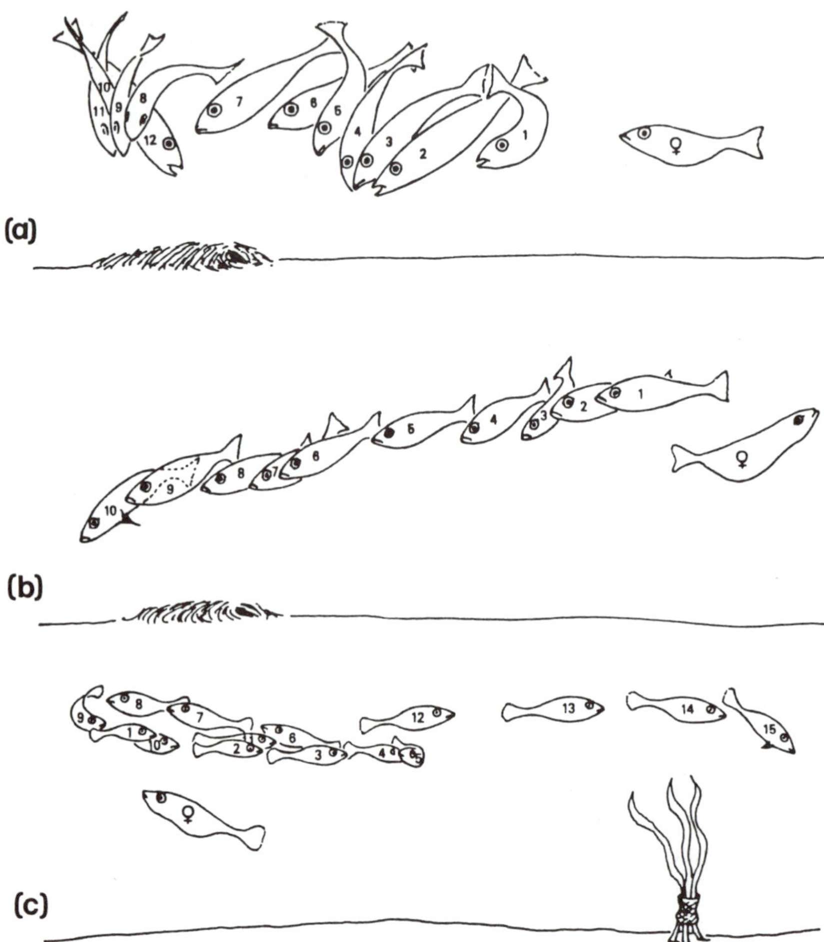

Fig. 11.3 Time–motion film analysis of male courtship dance patterns of (a) *Gasterosteus aculeatus*, (b) *G. wheatlandi*, and (c) *Apeltes quadracus*. Frames were taken at 1/24 s intervals, and are numbered consecutively.

serves as the basis for the description of courtship that follows.

When the male detects a female in his territory, he approaches her in a series of horizontal darting movements that alternate toward and away from the female (Fig. 11.3(a); see also Fig. 1.4). This motor pattern is called the 'zigzag dance' (Leiner 1929; Tinbergen 1951). Zigzagging may be interspersed with bites, butts, or other aggressive behaviour directed toward the female, and the frequency and intensity of these activities vary among males (Sevenster 1961; Wilz 1973) and within the same male over time (Sevenster 1961; Wilz 1975). If the female is receptive, the male swims to his nest, fans, glues, and performs other nest activities. The male eventually creeps through his nest, a behaviour associated with blanching of his body. This colour change makes his crimson throat and belly appear even brighter, as though the male '. . . were somewhat translucent and glowed with an internal incandescence . . .' (Warington 1852). The male returns to the female and courts her even more vigorously before returning to the nest (Sevenster 1961). Thus, the male oscillates between courting the female and performing nest activities. The process ceases when the pair spawns or the female departs.

When a sexually receptive female encounters a courting male she assumes a characteristic head-up posture. In the most intense form, the female lifts her tail and assumes a posture of lordosis, with her back forming a concave arch (see Fig. 3 of ter Pelkwijk and Tinbergen 1937). With rapid, nearly imperceptible fin movements, she turns to face the courting male as he circles and weaves around the tank (Sevenster 1973). Often, the female will follow the male before he is prepared to lead her to the nest. If this occurs, the male swims under the female, rolls his body to the side, and backs up against her so that his erect dorsal spines touch her abdomen (Leiner 1930). This 'dorsal pricking' (Fig. 1.5) causes receptive females to hold their position on the male's territory while the male returns to his nest and engages in nest activities (Wilz 1970*a*,*b*).

Ethologists have proposed several different functional explanations for the nesting activities that males perform during courtship. These explanations need not be mutually exclusive. Moreover, the wide behavioural and morphological variation so far documented among populations of threespine stickleback raises the possibility that these nest activities might serve different functions in different populations.

The nesting activities observed during courtship were originally interpreted as displacement behaviour (Tinbergen 1951; Sevenster 1961). This causal explanation maintained that aggressive and sexual tendencies in the male normally inhibit nesting and parental behaviour, and that when the two former tendencies are balanced during courtship, they cancel each other, thereby 'disinhibiting' nesting and parental behaviour (Sevenster 1961).

Wilz found, however, that territorial males that engaged in these nest activities underwent a motivational shift, from an aggressive to a more sexual state. Such males decreased their attacks on the female and increased their zigzagging behaviour. Wilz therefore concluded that dorsal pricking and its associated nest activities function to synchronize male and female courtship activities.

Several workers have proposed other functions for the nest activities performed during courtship in stickleback. For example, Sevenster-Bol (1962) and McFarland (1974) suggested that these activities maintain the nest in condition to receive receptive females. Rohwer (1978) predicted that female stickleback would spawn with males that already have eggs. He proposed that males could persuade females to spawn in their nests by behaving as though the nests contained eggs (i.e. by fanning the empty nest).

Data concerning the role of eggs in determining mate choice in threespine stickleback are equivocal (Ridley and Rechten 1981; Whoriskey and FitzGerald 1985*c*; Jamieson and Colgan 1989; Goldschmidt and Bakker 1990), although in other species females clearly choose to spawn in nests that contain eggs (e.g. Constantz 1985; Marconato and Bisazza 1986; Unger and Sargent 1988). The hypothesis that nest-directed behaviour is attractive to females remains to be tested.

When the male is ready to lead the female to his nest, he approaches her, turns abruptly, and rushes back to the nest. When the pair reach the nest, the male rolls his body to one side and with his dorsal surface facing the female, thrusts his snout into the entrance. In this way the male shows the nest entrance to the female (Leiner 1930; Wunder 1930; Tinbergen 1951). If the female enters the nest, she squirms into the tunnel until her head protrudes from the other end (but see 'Cues from the nest', page 336). When the female is in the nest, the male quivers his snout along her caudal peduncle several times, alternating from one side of her body to the other. Within a minute or so, the female bends her tail upwards, shivers, and deposits her eggs. After she swims out of the nest, the male squirms through, shivering momentarily over the eggs as he fertilizes them.

The act of fertilization and the presence of a fresh clutch of eggs in the nest causes a sudden decline in the male's willingness to court; this refractory period lasts for about an hour (Sevenster-Bol 1962). The refractory period probably provides an opportunity for the male to restore his nest, and perhaps permits sperm accumulation for subsequent spawnings. Fertilizations have an increasingly negative effect on the sexual tendency of males over the long term (van Iersel 1953). This decline may be associated with a gradual hormonal breakdown resulting from gonadal change or some other physiological mechanism (Sevenster-Bol 1962).

Following fertilization is a period of intense nest activity, when the male pushes the new clutch down into the nest, repairs any damage the nest may have suffered, and extends it forward to cover the newly spawned eggs. The male resumes his guard duties and drives off intruders, including females. The male's sexual tendency gradually recovers (but see below), and in about an hour he is again competent to court females and fertilize eggs (van Iersel 1953; Sevenster-Bol 1962).

The complex interactions that exist among the male, female, nest, and eggs provide numerous potential cues for eliciting and coordinating mating behaviour in stickleback. Because these cues and the mechanisms by which they are perceived play an important role in mate choice and sexual selection, their investigation is crucial to understanding the evolution of stickleback behaviour and morphology.

Colour

Preference for red

Conspicuous nuptial colour in threespine stickleback is restricted to males, and in most populations reaches maximum intensity during courtship. Nuptial colour may therefore serve an important signal function for courtship in this species. Leiner (1930) was one of the first to consider the role of intersexual selection in the evolution of nuptial colour in male threespine stickleback. By presenting receptive females to males in aquaria illuminated by monochromatic light, Leiner attempted to test the effects of male nuptial colour on females. Because males could induce females to spawn in their nests under green light (540 nm), Leiner mistakenly concluded that the red underside was unimportant to mating success and that only the zigzag dance played a significant role.

The ability of stickleback to mate in green light demonstrates only that the red nuptial colour of males is unnecessary, not that it is unimportant (ter Pelkwijk and Tinbergen 1937). Indeed, Wunder (1934) concluded earlier that female threespine stickleback selected males primarily on the basis of some aspect of nuptial colour, but this conclusion was confounded by the greater tendency for brighter males to court females and dominate duller rivals.

Experimental evidence for a direct effect of nuptial colour on female choice was first obtained by ter Pelkwijk and Tinbergen (1937), who found that receptive females followed a dummy to an artificial nest only if the dummy had a red underside. Taking advantage of a population that was polymorphic for male nuptial colour, Semler (1971) also documented an effect of red on male courtship success. He demonstrated that females choose males with red nuptial colour over a non-red (silver, mottled, or black underside) male, and that they choose non-red males with artificial red colour (lipstick or nail polish) over non-red males treated with clear lipstick or nail polish.

Although a significant proportion of females tested by Semler selected red males, 28 per cent selected non-red males. Because Semler only tested each female once, it is not known whether the difference in choice by females reflected female error or a behavioural polymorphism. The distinction is an important one for understanding the evolution of male nuptial colour.

Recent research on anadromous and resident freshwater threespine stickleback in several coastal rivers in Washington, USA, indicates that even when nuptial colour changes from red to black in a lineage, the females may still choose red over black males. In these drainages, resident freshwater stickleback in the upper reaches of the streams have diverged from the ancestral marine form (red colour) and evolved black nuptial colour, seemingly as a consequence of threat display convergence with the Olympic mudminnow, *Novumbra hubbsi* (McPhail 1969; Hagen and Moodie 1979;

Hagen *et al.* 1980). Female preference for red males has been partially lost, however, in the black species adjacent to a suspected zone of hybridization, suggesting that selection may favour prezygotic isolation in sympatry (McPhail 1969; Bell 1976*b*).

By functioning as a sign stimulus for female courtship, the red underside provides a mechanism by which female threespine stickleback recognize and choose prospective mates. In nature, however, virtually all males from which females are likely to choose (i.e. males with nests) will display some red colour. Thus, females should be able to distinguish differences in the relative expression of this trait.

A recent laboratory study of *G. aculeatus* from Europe revealed that females were capable of such discrimination and spent more time orientating to the brighter of two males presented simultaneously (Milinski and Bakker 1990). This preference was no longer statistically significant when males were presented under green light, suggesting that females chose males primarily by differences in colour rather than in behaviour.

McLennan and McPhail (1990) also found that females from an anadromous population in British Columbia orientated to and tracked the brighter of two reproductive males. When the females were released from the jars in which they were held during this assay, they typically mated with the brighter male, although duller males succeeded in mating in a minority of pairings. The duller males that mated did so by intruding into the territory of the brighter male and interfering with that male's courtship attempts, suggesting that male behaviour may play an indirect role in courtship success when males compete for females.

My own observations reveal that at least the early stages of courtship can be evoked in female threespine stickleback in the absence of nuptial colour. For example, when females from a Long Island population were held in large stock tanks without males, the gravid ones often assumed the head-up posture and concave back and followed female tankmates, as in heterosexual courtship. Thus, females that are sufficiently motivated may become less selective of mates.

Preliminary experiments using dummy presentations led to similar conclusions. When I presented moving dummies on a carousel apparatus to females, few followed the one with the red underside more. Contrary to expectations, most females divided their time between red and non-red dummies (Table 11.1). Although females willing to follow dummies might have been so receptive that they no longer discriminated among them, female response to nuptial colour may vary more than past studies have recognized.

Recent study reveals that there is considerable interpopulational variation in female preference for male colour in the guppy, *Poecilia reticulata* (Houde 1988; Houde and Endler 1990). It is possible that threespine stickleback populations vary similarly (Rowland 1982*a*). Differences in experimental methodology (e.g. Baerends 1985; Rowland and Sevenster

Table 11.1 Female courtship responses (measured as mean number of following bouts per 4 min trial) to moving all-silver and red dummy males presented on a carousel apparatus. The total number of trials in which females responded to either dummy is shown in the last column.

Female[a]	Mean follows per trial of		Total trials
	Silver belly silver irises	Red belly silver irises	
1a	2.8	5.8	4
1b	3.0	3.0	1
1d	1.5	0.3	4
2a	4.4	2.8	4
2b	1.2	2.6	5
2c	3.5	2.3	4
3a	0.5	1.0	2
4a	4.4	6.8	5
5e	10.6	4.0	7
6a	1.8	3.8	4
6b	0.0	2.0	1
6c	1.3	0.3	3
6d	17.7	15.2	9
Grand mean ± SE	4.05 ± 1.36	3.84 ± 1.08	

$P = 0.459$, for one-tailed Wilcoxon matched-pairs, signed-ranks test.

[a] Numbers refer to females, letters refer to experimental trial.

1985), experience or motivational state of the subjects, and other such factors may also give rise to differences in the response of female threespine stickleback to red colour, as discussed above for male response to this feature.

Significance of red colour

By choosing brighter red males over duller ones, females may be selecting males of high quality. Red nuptial colour in stickleback is based on carotenoids (Lonneberg 1938; Brush and Reisman 1965). Because carotenoids may be a limited dietary resource for many species, the extent or intensity of red colour may reflect an individual's ability to accumulate protein and other limited nutrients that are associated with them in nature (Rothschild 1975; Endler 1980). If bright males sire fitter offspring and provide superior parental care than do duller males, the reproductive success of females that mate with them will be greater than if they had chosen dull males.

Milinski and Bakker (1990) found that the brightness of the male's red nuptial colour and physical condition were positively correlated in a European population of threespine stickleback. The colour and condition of preferred males deteriorated after males were infected with parasites, and females no longer selected them (Milinski and Bakker 1990). These results support the hypothesis that animals select mates by traits that reliably indicate their bearer's condition (Andersson 1982, 1986; Hamilton and Zuk 1982; Kodric-Brown and Brown 1984). Milinski and Bakker (1990) did not state whether male courtship intensity was correlated with male condition. They noted, however, that males in poor condition might be able to muster enough energy to court when the need arises, but are unlikely to meet the long-term demands for maintaining bright colour.

Condition factors based on weight-to-length relationships (Bolger and Connolly 1989) and colour scores (Rowland 1984) of male threespine stickleback from Long Island were subjected to a regression analysis. In contrast to the freshwater population of threespine stickleback studied by Milinski and Bakker (1990), redness of Long Island males was not correlated with condition ($r^2 = -0.097$, $N = 32$, $P > 0.05$; unpubl. data). Perhaps the intensity of red in Long Island males is limited less by foraging ability, parasites, or general nutrition levels than by short-term (e.g. hormonal or motivational) differences among males. Under such circumstances, the conspicuousness or overall appearance of males, rather than their red colour alone, may lead females to choose them (see below).

An alternative interpretation for the association between male nuptial colour and female response is provided by the sensory exploitation hypothesis. This hypothesis maintains that the properties of signal receivers exert a major influence on the evolution of the signal (reviewed in Ryan 1990). Thus, selection is thought to act on the signaller to exploit any pre-existing bias in the perceptual system of the signal receiver. With respect to courtship in the threespine stickleback, red is regarded as having been selected from among a range of potential colours in the male because this colour is especially effective in stimulating the female's visual system. Over time, the courtship signal comes to match the perceptual properties of its intended receiver.

Investigators studying the psychophysics of vision in threespine stickleback implied that the visual system of the female was adapted to receive the signal emitted by the male (Cronly-Dillon and Sharma 1968). By projecting stripes of monochromatic light on the walls of a chamber and measuring the threshold for the optomotor response at various wavelengths, these investigators obtained photopic (light-adapted) spectral sensitivity curves for reproductive and non-reproductive stickleback. Fish exhibited peaks of sensitivity at 500–512 nm (blue) and at 594 nm (orange). Female (but not male) sensitivity to 594 nm increased with the onset of the reproductive season. This suggested that the eyes of female threespine stickleback may adapt seasonally to enhance detection of male nuptial colour.

A major problem here, however, is that the match between male colour and female sensitivity obtained for the threespine stickleback is only approximate and may therefore be spurious. If such associations could be extended to other species of sticklebacks, this would provide more convincing evidence for the coevolution of visual signal–receptor systems in this family of fishes.

Preliminary data suggest that light-adapted female blackspotted stickleback, *Gasterosteus wheatlandi*, and female *G. aculeatus* from Long Island, New York, USA, possess the same two spectral sensitivity peaks reported by Cronly-Dillon and Sharma (1968), but that the blackspotted stickleback is relatively more sensitive to blue-green wavelengths (Baube unpubl. data). Male blackspotted stickleback develop a yellow-green nuptial colour during the breeding season (Perlmutter 1963; Reisman 1968*a*; McInerney 1969; Rowland 1983*a*), which may be taken as further evidence of the coevolution of the receptor and signal systems in sticklebacks.

Because neither signal nor receptor exists without the other, it is reasonable to assume that selection acts on both components simultaneously. Visual receptors of reproductive females must do more than detect and assess potential mates. The properties of these receptors will be shaped by ecological factors that often require compromise solutions and constrain the structure and function of signalling systems. The extent to which ecological factors affect courtship and other forms of signalling in gasterosteid fishes is the subject of current research.

Role of other colours

The majority of studies of male nuptial colour in threespine stickleback have emphasized the importance of the red underside, but the effects of this feature are likely to be confounded by the blue iris and bluish white dorsal colour constituting the other components of male colour (Titschack 1922; Wunder 1930; Craig-Bennett 1931; Ikeda 1933; McLennan and McPhail 1989*a*). The expression of all three colour components varies with male behaviour, and both colour and behaviour reflect the male's reproductive state (Craig-Bennett 1931; Ikeda 1933; Reisman 1968*b*; Rowland 1984; McLennan and McPhail 1989*b*).

The blanched dorsal colour of courting male threespine stickleback provides a contrasting background that probably enhances the effectiveness of their red underside to attract females. By reversing the effects of countershading, a light dorsum alone will increase male conspicuousness and may in its own right serve as a sign stimulus for female courtship (van Iersel 1953).

This possibility finds support from a study of the 'white stickleback' that occurs sympatrically with the typical form of *G. aculeatus* in Nova Scotia, Canada (Blouw and Hagen 1990). Although the two forms mate assortatively and are reproductively isolated, females of both forms are initially attracted more to white males, even though the latter may exhibit little or no red on their ventral surface. Because breeding males develop a white dorsal and flank colour that renders them so conspicuous, the greater attractiveness of white males to females can probably be attributed to this trait.

Although the correspondence between the intensities of the red underside and blue eyes of threespine stickleback is only approximate (McLennan and McPhail 1989*a*), the widespread occurrence of the latter trait in reproductive males, even in populations where males develop little or no red (Semler 1971; Reimchen 1989), suggests that blue eyes may have an important signal function in courtship. Data from a Long Island population of *G. aculeatus* provide preliminary evidence of this. Gravid females presented with moving non-red dummies on a carousel apparatus followed the dummy with blue irises more than the one with silver irises (Table 11.2). Therefore, females may choose males with blue eyes, at least in the absence of a red belly.

McLennan and McPhail (1989*a*) found that blue eye colour varied less in intensity than red underside colour and suggested that red is potentially the more important signal by which females discriminate among males of their species. This is a reasonable assumption, but only choice tests can establish how much female choice depends on hue, intensity, etc. of the components of nuptial colour in male threespine stickleback. In view of the wide interpopulational variation in morphology (reviewed in Bell 1984*a*; Reimchen Chapter 9 this volume) and behaviour (reviewed in chapters by Hart and Gill, Huntingford *et al.*, Bakker, Foster, and McPhail this volume) of this species, one might expect to find corresponding variation in female response to different components of male nuptial colour. Such differences are likely to be related to local conditions (Reimchen 1989).

Table 11.2 Female courtship responses (measured as mean number of following bouts per 4 min trial) to moving all-silver and blue-eyed dummy male presented on a carousel apparatus. The total number of trials in which females responded to either dummy is shown in the last column.

Female[a]	Mean follows per 4 min trial of		Total trials
	Silver belly silver irises	Silver belly + blue irises	
1a	1.7	4.9	10
2a	2.8	6.3	6
2b	1.5	3.0	4
3a	1.0	2.9	9
4a	7.7	7.7	7
5a	1.4	1.6	5
6a	6.7	8.3	3
Grand mean ± SE	3.26 ± 1.04	4.96 ± 0.97	

$P = 0.016$, for one-tailed Wilcoxon matched-pairs, signed-ranks test.

[a] Numbers refer to females, letters refer to experimental trials.

Female colour

The pigmentation pattern of the breeding female stickleback, though less flamboyant than that of its male counterpart, has received some attention from biologists. Non-reproductive males and females usually appear olive tan to grey dorsally, with a silvery cast along their flanks and belly. In breeding females, especially those from marine populations, lateral colour often intensifies to a brassy silver. Van Iersel (1953; and reported in Sevenster 1961) concluded that the silver colour was a releasing stimulus for male courtship in *G. aculeatus* from the Netherlands. Results obtained from a Long Island, New York, population of *G. aculeatus* support this. When males were offered a choice between gravid dummy females painted silver or black, they courted the silver dummy more (Table 11.3). The possibility that males were reluctant to court the abnormally contrasting black dummy cannot, however, be excluded.

In the open water column, crypsis is best served by the uniform silvery flank colour (Edmunds 1974) characteristic of non-reproductive fish. When receptive females move to the shallow breeding areas to spawn, however, they become especially vulnerable to avian predators that frequent the marsh pools (Bull 1964; Whoriskey and FitzGerald 1985*b*; Williams and Delbeek 1989; pers. obs.).

Females in some populations develop a dark vertical barring on the upper half of their flanks when they become reproductively active (Wunder 1934; Williams and Delbeek 1989; Rowland *et al.* 1991). In the shallow breeding habitat, the barring pattern disrupts the fish's outline (Nikolsky 1963), so that against the dark, non-uniform background of the tidal pools the fish becomes less visible to predators. The barring pattern may therefore serve as protective colour for the females.

Wunder (1934) first suggested that this barring pattern could be used by males when selecting among gravid females. As Wunder had, we found that female threespine stickleback developed similar pigmentation that correlated with their sexual receptivity and intensified in the presence of

Table 11.3 Effects of female dummy colour on male courtship response, based on single 4 min pairwise presentations of dummies to 19 males.

Response		Mean ± SE per 4 min to	
		Silver dummy	Black dummy
Visits to dummy	**	5.53 ± 5.56	3.74 ± 4.09
Seconds at dummy	*	46.32 ± 40.31	23.63 ± 22.70
Zigzags to dummy	**	24.42 ± 33.09	5.42 ± 9.54

$*P < 0.02$; $**P < 0.01$, for two-tailed probabilities on Wilcoxon's rank-sum, matched-pairs test.

courting males. Based on my own experience that barring served as a reliable cue for recognizing receptive females in a stock tank, we tested whether males might attend to this cue when courting. Indeed, Long Island males presented with a barred and an unbarred dummy courted the barred dummy more (Rowland *et al.* 1991).

The barring pattern of the female could serve as a courtship signal that suppresses attack and arouses sexual tendencies in the male, thereby helping to synchronize mating activities of the courting pair (Morris 1956). If this is true of barring, male responsiveness to the pattern should increase and its conspicuousness should also increase until balanced by counter-selective pressure from predation. An alternative interpretation maintains that signals advertising female receptivity may facilitate female mate choice by inciting competition among potential mates (Cox and LeBoeuf 1977; Farr and Travis 1986). In so far as the second interpretation is likely to depend on the veracity of the first, the two possible interpretations should not be considered mutually exclusive.

In species where males compete, females may mate with superior males more often than expected by chance, simply by accepting those that dominate in intrasexual interactions. If brighter, more vigorous male stickleback dominate rivals (Bakker and Sevenster 1989), this dominance will increase access to females and further enhance their reproductive success. Cox and LeBoeuf (1977) suggested that female elephant seals may even incite competition among males, and by mating with the victor produce offspring of higher fitness. Female stickleback might similarly exercise such passive choice of mates (see below).

Behaviour

The species specificity of certain elements of courtship, particularly those that occur during the initial stages, suggests that movement is a critical component of stickleback courtship. The differences in courtship displays commonly observed among sympatric species may have evolved in large part through selection for species identification (Tinbergen 1952; Morris 1956). Figure 11.3 illustrates the courtship dances of three sticklebacks that occur sympatrically on the north-east coast of North America. These displays are performed by males soon after they encounter prospective mates, and in conjunction with the species-distinctive nuptial colours, probably help to maintain reproductive isolation among the species (Reisman 1968*b*; McInerney 1969; Rowland 1970).

Leiner (1930) recognized the importance of male courtship behaviour when he observed that *G. aculeatus* females spawned with males that were denied the advantage of nuptial colour through the use of monochromatic illumination. Manipulating dummies to imitate the courtship movements of the male, ter Pelkwijk and Tinbergen (1937) induced females to follow the dummies and even to enter an artificial nest. Similarly, females paid little attention to a stationary dummy male suspended in their tank, but often adopted a head-up posture and followed the dummy when it was moved through the tank in a manner that approximated male leading movements (Rowland 1989*c*). The effectiveness of movement is also evident from the sudden increase in attention females direct to males that start to zigzag, even if the pair is separated by a glass partition.

The zigzag dance of the threespine stickleback is often cited as a classical example of a fixed action pattern, even though it may vary in amplitude (Tinbergen 1951) and frequency (Sevenster 1961). If differences in courtship behaviour reflect differences in male condition, then they, like colour, might also serve as effective criteria for female mate choice. The total visual stimulation resulting from the interaction of the male's behaviour and colour may therefore provide the most important cue by which females detect and evaluate potential mates. Indeed, it is the combination of rapid movement and bright colour that provides the element of conspicuousness so characteristic of courtship display in animals (Bastock 1967).

It has been suggested that female threespine stickleback might choose to mate with males that perform the zigzag dance at relatively high frequency (Gross and Franck 1979; Ridley 1986). There exists, however, no empirical support for this hypothesis, and in studies on three populations no such association was found (Ward and FitzGerald 1987; Jamieson and Colgan 1989; Milinski and Bakker 1990).

If it exists, an effect of male courtship vigour on female choice of mates may be obscured by a positive association between courtship vigour and brightness of the male's red nuptial colour (Rowland 1984; Ward and FitzGerald 1987; McLennan and McPhail 1989*b*). Females may therefore choose mates on the basis of either attribute, although Milinski and Bakker (1990) have argued that courtship intensity is an unreliable indicator of male condition because even males in poor condition may be able to court vigorously. Perhaps when carotenoids differ little among males or in a way that does not reliably indicate differences in overall quality, females attend more to other cues.

In Milinski and Bakker's (1990) study, the zigzag dance may have been inhibited by the use of small tanks (Bakker and Sevenster 1989; see also below). Consequently, the role of male courtship behaviour in female choice under natural conditions may have been under-estimated. Moreover, McLennan and McPhail (1990) illustrate how behaviour can influence male mating success indirectly through male–male interaction (see above).

In view of the wide variation discovered among populations of stickleback in nature, it is not unlikely that the relative roles of nuptial colour, behaviour, and other traits affecting mate choice will also depend on water conditions (e.g. Reimchen 1989), presence of predators (e.g. Moodie 1972*b*; Foster pers. comm.; this chapter), and other ecological factors.

Elements of the aggression that male stickleback use to establish and defend territories from rivals also appear during courtship (Tinbergen 1951; van Iersel 1953; Morris 1956; Sevenster 1961). Moderate levels of aggression may have a stimulatory effect on receptive females (e.g. McPhail and Hay 1983; Ridgway and McPhail 1984), but excessive aggression may interfere with courtship (Wilz 1973; Ward and FitzGerald 1987; Rowland 1988). Moreover, the extent to which males express aggression during courtship may lead to consistent differences in courtship among populations (Wilz 1973), and these may play a role in mate selection by females.

For example, McPhail and Hay (1983; see also McPhail page 408 this volume) found assortative mating between populations of stream-resident freshwater and marine *G. aculeatus* from the Little Campbell River, British Columbia, Canada, and attributed it to quantitative differences in male courtship. Freshwater males more often zigzagged as a first response to a female, whereas marine males more often bit the female. Because this was true regardless of the type of female courted, the authors suggested that the two courtship responses reflected differences between males rather than between females from the two populations. Thus, females may assess males by comparing the ratio of zigzags to bites. McPhail and Hay also found that freshwater males zigzagged more to freshwater females than to marine females, but the mechanism by which males distinguished the two kinds of females was not ascertained.

Ridgway and McPhail (1984; see also McPhail page 421 this volume) investigated positive assortative mating between limnetic and benthic populations of threespine stickleback from Enos Lake, British Columbia, Canada. The two types of males differed in the way they approached and led females, with benthic males biting more and limnetic males zigzagging more during the initial phase of courtship. Benthic and limnetic females clearly distinguished between the two kinds of males, indicating that differences in the initial phase of courtship may play a role in the assortative mating in these two populations.

Ziuganov (1988) studied a completely plated and a low-plated species of threespine stickleback occurring sympatrically in the White Sea and Kamchatka River basins in Russia. He, too, found that males of the two species differed in the degree of aggression shown during courtship, and also in their initial approach to females, their leading behaviour, and the way they showed the nest entrance. These differences suggest that male courtship behaviour plays a role in maintaining their reproductive isolation.

Sexual selection theory predicts that in species such as the threespine stickleback, where both males and females make a substantial parental investment, both sexes will discriminate among potential mates when given a choice (e.g. Trivers 1972; Williams 1975). Tinbergen (1951) emphasized how successful reproduction in the threespine stickleback depended on sequential exchange of signals between the courting partners. The interplay of behaviour, colour, size, and other properties that occurs during the courtship sequence thus provides both male and female stickleback with ample opportunity to evaluate prospective mating partners and to exercise mate choice, either by terminating the courtship prematurely or by spawning (Hay and McPhail 1975; see also McPhail Chapter 14 this volume).

Posture

Courting males of other species of sticklebacks tend to adopt a head-down posture in courtship, particularly during the leading phase (Wootton 1976), but male threespine stickleback do not appear to do so. When the male leads the female to the nest, he typically retains a horizontal posture as he dashes back to his nest, only tilting downward as he swims toward the bottom to approach the nest entrance. At this point the male rolls his body to the side and briefly assumes the unique body posture (showing) that displays his red underside to the female and points out the nest entrance.

When zigzagging, males usually maintain a horizontal position also. A male may briefly adopt a head-down or head-up posture as well, but this seems to result primarily from the male attempting to orientate himself with respect to the female as he courts her.

For female threespine stickleback, posture appears to be an important sign stimulus for eliciting male courtship. Ter Pelkwijk and Tinbergen (1937) provided an accurate illustration of the head-up posture and concave back that females typically assume when courted. Tinbergen (1951) reported that a male stickleback even courted a dead tench, *Tinca tinca*, when it was presented in a head-up posture, but the tench was made to simulate female courtship movements when it was presented to the male (ter Pelkwijk and Tinbergen 1937). It is therefore difficult to determine from these observations the extent to which posture may contribute to the elicitation of courtship in male stickleback.

A re-examination of sign stimuli in male stickleback by using stationary dummies failed to confirm the effectiveness of head-up posture (Rowland and Sevenster 1985). When Long Island males were presented with stationary dummies of gravid females, they courted those in horizontal posture more than those presented head-up or head-down. Because biting, too, was less frequent to dummy males and dummy females presented head-up, we suggested that this posture may function in courtship more as an appeasement signal to suppress attack by males than as a sexual signal (Rowland and Sevenster 1985). A subsequent study by McLennan and McPhail (1990) offered a similar interpretation.

Body shape

Wunder (1934) reported that male threespine stickleback mated preferentially with females of greater girth, but this observation might have reflected the greater receptivity of such females rather than male preference for them.

By comparing a distended dummy with an undistended one, Tinbergen (1948) demonstrated that the swollen belly of a gravid female stickleback served as a sign stimulus for male courtship. However, males in nature may be courted by several gravid females simultaneously (Kynard 1978*a*). Because such females will vary in size or egg content, males that can distinguish among them and court accordingly could improve their reproductive success.

Indeed, threespine stickleback males presented with pairs of dummies of equal length but different degrees of abdominal distension chose the more distended one (Rowland 1982*b*), even when distension far exceeded normal limits (Rowland and Sevenster 1985). Although stationary dummies were used in these experiments to rule out any confounding effects that female behaviour might have on male choice, similar results were obtained with males presented with pairs of live females that differed in body weight (Sargent *et al.* 1986).

Interestingly, males did not court the more distended or larger dummies exclusively, but distributed courtship in proportion to the dummies' projection areas (Rowland 1989*b*). This indicates that males use the area of the image projected on the retina, or some closely correlated trait, as a proximate cue for comparing mates.

Even when males were presented with four dummies of different degrees

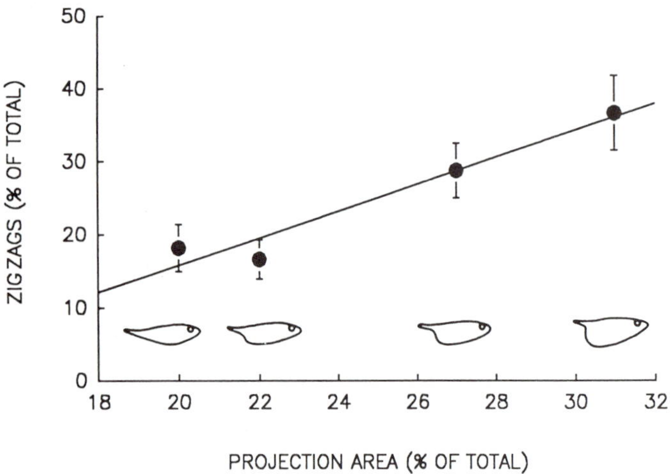

Fig. 11.4 Number of zigzags (mean ± SE) directed to each of four simultaneously presented dummies that differed in degree of abdominal distension. The size of each dummy is expressed as per cent of the total projection area of all four dummies. All dummies were presented to 27 males at an equal distance from their nest. There is an increase in the proportion of zigzags with increasing proportion of projection area of a dummy ($P < 0.001$; Page's $L = 722.5$, $k = 4$, $N = 27$).

of abdominal distension simultaneously, they divided courtship approximately in proportion to the projection area of each dummy (Fig. 11.4). Because abdominal distension in females is associated with sexual receptivity as well as fecundity (Rowland 1982*b*), a perceptual mechanism that causes males to select such females would confer a reproductive advantage to males.

Males evidently attend to the shape, and not just the increased lateral projection area that results when the distension of the female's abdomen is increased. When males were presented with a superdistended (supernormal) dummy (lateral projection area = 539 mm^2) and a larger but normally distended dummy (lateral projection = 661 mm^2) in a single trial, they courted the supernormally distended one 45 per cent more (23.00 ± 3.10 v. 15.82 ± 2.55 zigzags per 4 min trial; mean ± SE; $T^+ = 298$, $N = 27$, $P < 0.01$, Wilcoxon signed-ranks, matched-pairs test). This suggests that configurational cues (Tinbergen 1951; Ewert 1980) influence courtship in male threespine stickleback. Therefore it is probably not coincidental that courting females assume a posture that displays the distended abdomen to advantage.

The superdistended dummy also presents a greater cross-sectional area, which would be visible to males as they swim around the dummies. This could confound interpretation regarding the cues to which males attended in the above experiment. More convincing evidence for the importance of configurational cues in male courtship was obtained by disrupting the normal spatial relationships of the dummy. When two normally distended dummy females were presented horizontally to males, one in the normal belly-down position and the other in an abnormal belly-up position, males courted the belly-down dummy 70 per cent more (16.57 ± 2.87 v. 9.74 ± 2.50 zigzags per 4 min; $T^+ = 197$, $N = 21$, $P < 0.01$). Thus, despite the similarity in the orientation of the dummies' body axes, projection areas, etc., males distinguished the different spatial relationships of the dummies' body contours.

Female stickleback may also use body shape as a cue in courtship, but the meagre evidence that supports this is circumstantial. For example, Mori (1984) found a marked sexual dimorphism with respect to head size in resident freshwater *G. aculeatus* from Japan. The ratio of head length to body length was greater in males than in females, suggesting the possibility that this attribute might provide an additional cue for sexual identification in this population. McPhail (1984; see also McPhail page 418 this volume) found considerable differences in the body shape of males from two biological species of threespine stickleback from Enos Lake, British Columbia. Ridgway and McPhail (1984) felt it likely that females use shape as a cue for selecting mates.

Body size

The increase in fecundity with body size in teleosts (Williams 1966) is well documented in threespine stickleback (Baker page 164 this volume). Male stickleback should therefore choose larger females over smaller ones. Mate choice experiments revealed that Long Island males selectively court the larger of two dummy females when the dummies are presented simultaneously (Rowland 1989*b*). As in the shape experiment discussed above, courtship was distributed in proportion to the projection area of the two dummies, even if the dummies exceeded the size range of real females. The supernormality effect thus apparently overrode any reluctance males might have had to approach an object large enough to be a predator.

Courtship reflected the relative sizes of the two dummies, but males were more likely to perform ambivalent behaviour (e.g. backing off) when presented with dummies whose projection area was approximately ten times the projection area of a female stickleback (Rowland 1989*b*). This response could have occurred because the dummy was too large to be recognized as a potential mate, or because it was perceived as a potential predator. That very large dummies (several times larger than a stickleback) were perceived as potential predators was suggested by the cautious, exploratory approach (i.e. predator inspection; Pitcher 1986, Huntingford *et al.* page 278 this volume) of males to the dummies (pers. obs.). In the few instances in which the approach escalated to an attack, it was directed toward the dummy's tail or flank, initiated from behind the dummy, and followed by a quick retreat. This behaviour was very like that elicited by large prickly sculpin (Pressley 1981; Foster and Ploch 1990).

Although males courted dummies that were larger than any females in the population, the distance from which they courted increased with increasing dummy size (Fig. 11.5; Baube unpubl. data). Whether the increase in courtship distance resulted from increased fear of larger dummies, or whether males tend to match the image the female casts on the retina to some expected size when courting, is uncertain. In any case, we know that size and distance interact in stickleback, because the relative attractiveness of a dummy can be increased if it is presented closer to the male. Male stickleback may therefore estimate distance or size by the angle the female's image subtends on their retina.

Students of optimal foraging theory have already proposed this 'apparent-size hypothesis' to explain how prey selection is mediated in stickleback (Gibson 1980; Wetterer 1989; but see Hart and Gill page 232 this volume). But if males attend solely to the female's projection area, one would predict dummy attractiveness to be inversely related to the square of its distance. Further data are needed to test this prediction.

Females also court larger potential mates (dummies) more actively than they do smaller ones (Rowland 1989*c*). When normal- and supernormal-

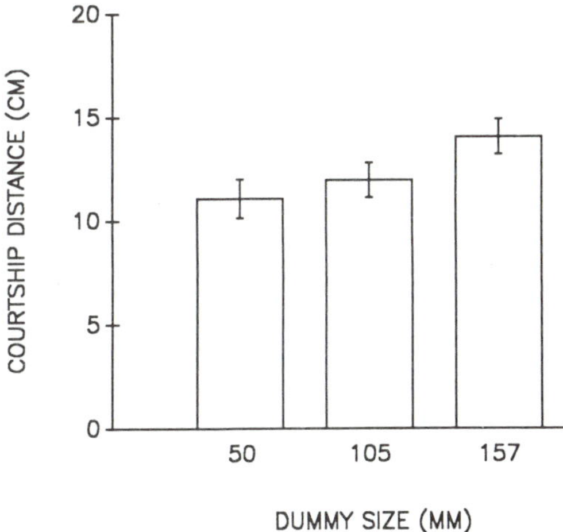

Fig. 11.5 Courtship distance (mean ± SE) of males presented with dummy females of identical shape but different body length, based on 19 males presented once with each dummy. The 50 mm dummy approximated the size of females from this population of stickleback. There is an increase in courtship distance with dummy size ($P < 0.01$; Page's $L = 244.0$, $k = 3$, $N = 19$).

sized dummy males were moved together through the tanks of receptive females to simulate courting males, the females spent more time following the larger one. Like those of males, the females' preferences were incomplete, and they divided their following time in proportion to the projection area of the two dummies. Similarly, females from a population of relatively small body size chose males from a population of unusually large threespine stickleback when offered a choice of males from the two populations (Moodie 1982). Moodie suggested accordingly that the large males served as a supernormal stimulus for females of smaller body size.

Distance

Courtship response in stickleback depends on how far from the nest the male encounters the female. For example, males zigzagged more to females presented at 100 cm than at 10 or 280 cm from the nest (Symons 1965). Symons concluded that the lower rate of zigzagging at 10 cm probably resulted from the correspondingly higher amount of showing that males performed when they encountered females so close to the nest. Unpublished studies by van Iersel and by Sierksma and Sevenster (cited in Sevenster-Bol 1962) also found a reduced rate of zigzagging to females close to the nest. They attributed this reduction to an increased inhibition of sexual tendency by the increase in aggression that results when males are close to the nest (Fig. 11.1).

An interesting implication of this finding is that the use of small aquaria in laboratory studies may lead to a somewhat distorted view of stickleback courtship in nature. Indeed, Bakker and Sevenster (1989) suggested that the small size of aquaria (34 × 17 × 20 cm) used in their behavioural genetic study of stickleback may have inhibited the full expression of courtship in males because females were forced to remain close to the nest. This possibility emphasizes the need for further study of stickleback in the field, where behaviour is less likely to be affected by such spatial constraints.

The limited field observations of courtship so far reported for *G. aculeatus* indicate that males often encounter several females at a time (e.g. Kynard 1978*a*; Borg 1985). Under such conditions, distance could become an important factor in choosing mates. If a female is less likely to follow the male to the nest the further away the nest is when he starts courting, or if being farther from the nest increases courtship costs or vulnerability to nest raiding for the male, then he should preferentially court females nearest to his nest.

To test this possibility, we presented dummy females at two distances from the male's nest. Preliminary trials revealed that males courted a dummy twice normal size (in all linear dimensions) more than a normal-size dummy when the dummies were presented equidistant from the males' nests (Table 11.4). Males presented with two normal-size dummy females, one at 22 cm and the other at 44 cm, still courted both dummies, but spent more time courting the closer one (Table 11.4).

We then tested whether an increase in mate size would compensate for increased mate distance. Indeed, males spent about the same amount of time courting the large dummy presented at 44 cm as they did the normal-size dummy presented at half that distance (Table 11.4). This result

Table 11.4 Effects of distance on female size preference of male stickleback, based on single 4 min pairwise presentations of dummies to 27 males.

Dummy pairing	Response (mean ± SE) to each dummy per trial		
	Visits	Seconds at dummy	Zigzags
Normal at 22 cm	*** 8.11 ± 0.78	*** 115.33 ± 12.40	** 35.41 ± 6.17
Normal at 44 cm	6.19 ± 0.66	55.56 ± 6.61	20.48 ± 3.51
Normal at 44 cm	7.74 ± 1.11	44.52 ± 5.93	12.11 ± 2.21
Large at 44 cm	***10.96 ± 1.46	*** 80.37 ± 8.35	** 24.00 ± 5.67
Normal at 22 cm	7.00 ± 0.93	67.41 ± 8.71	17.59 ± 3.97
Large at 44 cm	5.93 ± 0.82	53.33 ± 8.11	12.37 ± 2.90

**$P < 0.01$; ***$P < 0.001$ for two-tailed probabilities on Wilcoxon's rank-sum, matched-pairs test.

is consistent with the apparent-size hypothesis discussed above, because the size of the retinal image cast by the supernormal dummy would approximate that of the smaller, closer dummy.

This result also might imply that stickleback are unable to perceive absolute size differences between dummies presented at different distances. However, the binocular vision of fishes facilitates perception of absolute size differences between objects viewed at different distances (e.g. Herter 1930), and such size constancy is retained even in monocular fish (Douglas *et al.* 1988). The well-developed visual sense of stickleback (Wootton 1976) suggests that they, too, should have the capacity to perceive absolute size differences among objects irrespective of distance. Therefore, the equal allocation of courtship between the large dummy at 44 cm and the normal dummy at 22 cm may instead reflect a trade-off between the advantages of courting a larger mate and the disadvantages of having to go farther from the nest to do so.

The female's proximity to the nest may also reflect her readiness to spawn. If this is true, then the male should value the closer female more than a female farther from the nest, even if he recognizes that the closer one is of equal size or smaller. However, stickleback have not been studied fully enough to warrant conclusions concerning the limitations of size constancy of vision in this species.

The effects of distance also depend on the size, shape, and orientation of the territory, distance from neighbouring males, density of females, and other physical and biological factors. The relevance of such information can only be reliably established in conjunction with data collected in the field.

Experience

Sevenster (1968, 1973) demonstrated that male stickleback can be conditioned to perform a simple operant response (i.e. swimming through a ring) for the opportunity to court a female through a glass partition. There is, however, little information on how experience may influence stickleback courtship in more natural contexts. Males are often observed searching and zigzagging in the area of the tank where a live or dummy female was presented, long after the female has been removed.

Such observations suggest that stickleback learn where they are likely to encounter mates, much as they learn where they have encountered food (Thomas 1974; Hart and Gill page 220 this volume). Guiton (1960) reported observations on threespine stickleback suggesting that the nest site may itself come to acquire sexually stimulating qualities through conditioning. When a male was induced to move his nest to a new site by repeated destruction of his nest, he responded sexually (quivering) to a dummy presented at the old nest site, now devoid of nest or pit, but attacked the same dummy when it was presented at the site where the new nest was being constructed.

Males must associate social stimuli with their location in the field. Thus, by learning where females, rivals, and other intruders are likely to appear, males could adjust their behaviour, territories, and nest sites accordingly. Similarly, females could learn the location of territories and nest sites of courting males, and return to or avoid particular males or areas, depending on their experience there. If stickleback distinguish neighbours from strangers and relate this discrimination to specific locations within the territory (Peeke and Veno 1973), then perhaps they can extend this ability to the sexual context.

To what extent does learning influence courtship and mate choice in animals? Stickleback are often regarded as a species whose mating behaviour is largely innately determined (e.g. Cullen 1960), so it would be interesting to examine whether males can learn to modify their pattern of mate choice. The ability to vary mate choice might have a selective advantage. We are currently testing this hypothesis by presenting males with a pair of dummies that differ in abdominal distension (Rowland 1989*b*). We then attempt to shift the initial choice of males by immediately removing both dummies from the male's tank for 30 min (a presumably aversive stimulus) as soon as the male begins to court the dummy we have designated as 'negative'. This paradigm was chosen to simulate the case where a male, by focusing courtship on the fatter but non-receptive of two females (e.g. a parasite-distended female), may lose the opportunity to mate with either one because they leave the male's territory. Preliminary results so far indicate that male choice for the more distended dummy female is resistant to modification by experience.

Peeke and Peeke (1973) noted the lack of information regarding the role of habituation in sexual behaviour of fishes. Given the relatively short period during which male threespine stickleback actively court (i.e. the courtship phase), compared with the period when they actively attack, males are likely to maximize their chances of obtaining additional matings if they persist in courting prospective mates during the courtship phase. One might therefore expect courtship behaviour to be more resistant to habituation than is the case for aggressive behaviour in this species. Courtship behaviour of stickleback may rely more directly on hormonal changes and other physiological processes (reviewed in Guderley Chapter 4 this volume) than on experiential ones for its timing and control.

Tactile cues

Male threespine stickleback often initiate courtship by biting or butting the female. This behaviour quickly drives off unreceptive females, but receptive ones face and approach males that behave this way (e.g. Sevenster 1968). This activity may test the receptivity of prospective mates and sexually stimulate those that are ready to spawn. Violent contact behaviour at the beginning of courtship is typical of males from several populations of threespine stickleback (e.g. Wilz 1973; McPhail and Hay 1983;

Ridgway and McPhail 1984).

Male threespine stickleback also provide gentler forms of tactile stimulation to females during courtship, including dorsal pricking (discussed above) and the quivering that males direct to females that have entered the nest. If the male is removed just prior to quivering, the female fails to spawn. The female can be induced to spawn, however, by repeatedly tapping her caudal peduncle with a rod in a way that simulates the tactile stimulus normally provided by the male (Tinbergen 1951). This activity might have originated from the male's attempts to enter the nest and fertilize the eggs deposited by the female.

Tactile stimulation may also stimulate leading in threespine stickleback males, because receptive females sometimes contact the male with their snout during courtship. Indeed, contact by the female is characteristic of courtship in fourspine stickleback, *A. quadracus* (Hall 1956; Reisman 1963; Rowland 1974*a*), brook stickleback, *Culaea inconstans*, and blackspotted stickleback, *G. wheatlandi* (Reisman 1968*a*; McInerney 1969; Rowland 1970; Wootton 1976). In *G. wheatlandi*, the female repeatedly touches her snout between the male's erect pelvic fins as he vibrates his body and slowly leads her on a circuitous route to the nest. This vibration probably provides both visual stimulation to the female as the male's orange pelvic fins oscillate before her eyes, and tactile stimulation to the male and female when they contact. As dorsal pricking and the frequent nesting activities seen in *G. aculeatus* males when they court are absent in *G. wheatlandi*, the protracted leading behaviour of the latter species may represent an alternative mechanism for synchronizing male and female mating activities.

Reproductive male *G. wheatlandi* are yellow-green and inconspicuous in vegetation, so tactile signalling could provide an effective way to gain and hold the attention of a female in such habitats. Moreover, the female provides feedback to the male through such contact. If this contact is interrupted while the male is leading, he backs up and waits for her to re-establish contact or repeats the courtship sequence. Populations of *G. aculeatus* that differ in conspicuousness, or that inhabit waters that differ in visibility, should be compared to determine whether the relative contribution of tactile signalling varies accordingly.

Tactile cues are probably critical for the culmination of courtship: the fertilization of the eggs. Fertilization occurs when the male creeps through the nest just after the female has deposited her eggs there. The tactile stimuli that the male experiences during this final activity of the courtship sequence are probably necessary for the emission of sperm (Sevenster-Bol 1962).

Chemical cues

As noted above in the discussion of the role chemical cues in aggression, neither behavioural evidence nor the structure of the brain and olfactory receptors of threespine stickleback suggest that chemical stimuli play a major role in this species' behaviour. Nevertheless, the possible influence of chemical cues in stickleback courtship has not been ruled out. Leiner (1930) assumed that the glue which males secreted over the nest during courtship had a sexually stimulating effect on the male himself, rather than on the female. Leiner noted that males usually glued immediately before they led the female to the nest. Although the gluing act itself may effect a motivational shift in males that results in leading (Wilz 1973) and consequent nest entry by the female, the secretion may provide some chemical stimulation to the male or female.

In some fishes, chemical cues from receptive females may stimulate courtship in conspecific males (e.g. Crow and Liley 1979; Farr and Travis 1986). The discovery that mucus from freshly laid eggs can elicit fertilization in male threespine stickleback (van Iersel 1953) suggests that chemical stimuli could be involved in other aspects of courtship behaviour in this species. Perhaps leakage of mucus or other fluids from the cloaca of gravid females provides a chemical cue by which males recognize females whose spawning is imminent. This kind of cue could virtually ensure a successful mating by the male.

In a pilot study, I presented territorial male threespine stickleback with two visually identical dummies, each fitted with a fine plastic tube that permitted water to flow slowly from the ventral opening. When water from a jar containing a receptive, gravid female was allowed to flow from one dummy, males showed no obvious difference in response to this dummy compared with response to the dummy through which plain tank water flowed. Further study is needed, however, before any conclusions can be drawn concerning a possible role of chemical cues in courtship.

Cues from the nest

The vast majority of stickleback courtships do not result in spawning, and these are often terminated by the female before she reaches the nest (e.g. Tinbergen 1954; Borg 1985). These incomplete courtships could reflect nest site preferences of females. For example, females may prefer to spawn in concealed nests (Sargent and Gebler 1980; Sargent 1982).

It is also possible that females identify and select nests on the basis of specific visual cues such as long pieces of plant material at the entrance parallel to the nest tunnel. These 'leidstengels' (Dutch for 'lead stems') are a commonly observed feature in nests of stickleback from Europe and North America (Wunder 1930; pers. obs.), and may serve to mark the nest entrance. Bright or contrasting colours around the nest entrance could serve the same purpose.

On arriving at the nest, females in captivity often fail to spawn after examining and even after entering the nest (e.g. Warington 1852; Li and Owings 1978*a*; Ridley and Rechten 1981; Jamieson and Colgan 1989; pers. obs.). Some authors have interpreted this failure as an expression of selection by the female. Ridley and Rechten (1981) concluded that females preferred to spawn with males that already possessed eggs, and that females determined this by poking their snout into the nest. Others (e.g. Wootton 1976; Jamieson and Colgan 1989), however, invoked a motivational explanation for aborted nest entry and suggested that many of these incidents observed in the laboratory may have resulted from workers using females not yet ready to spawn. However, aborted nest entry has been observed in the field in stickleback from British Columbia, Canada (Foster pers. comm.), Long Island New York, USA (pers. obs.), and the Netherlands (Goldschmidt and Bakker 1990); it is not merely a laboratory artefact. Female stickleback may therefore place their snout into the nest entrance to obtain information about the condition of the nest to decide whether to abort or to complete nest entry.

PARENTAL BEHAVIOUR

Once the male has positioned the fertilized eggs in the nest and repaired whatever damage the nest may have sustained during spawning, he redirects his attention to courting. Beside courting females and guarding the nest, the male must now care for the spawn. Although Leiner (1960) described what may be rudimentary egg and nest tending by female fifteenspine stickleback, *Spinachia spinachia*, it is generally believed that it is exclusively the male that administers parental care in the Gasterosteidae. This phase of the reproductive cycle of *G. aculeatus* has been carefully analysed by van Iersel (1953).

In addition to guarding spawn, the primary paternal activity is fanning. The male hovers head-down above the nest with his head close to the entrance and oxygenates the eggs by fanning a current of water across the top of the nest with his pectoral fins (Hancock 1852; Tinbergen 1951; van Iersel 1953; Sevenster 1961; Wootton 1984*a*). This activity is repeated at regular intervals, depending on the number and developmental state of the eggs, the environmental conditions, and the internal state of the male (van Iersel 1953; Sevenster 1961; Wootton 1976; Whoriskey and FitzGerald page 197 this volume). The male also consumes dead or fungus-infected eggs, behaviour which may help to inhibit the spread of disease (van Iersel 1953).

Although the male's sexual tendency recovers about an hour after fertilization, recovery is less complete with each succeeding fertilization (van Iersel 1953). These internal changes, as well as those brought about by the increasing stimulation from the developing eggs, increase parental behaviour and decrease sexual behaviour and nuptial colour. This process continues until parental behaviour inhibits sexual behaviour and courtship is no longer expressed (van Iersel 1953). During the parental phase, males usually assume a duller, more cryptic colour (Craig-Bennett 1931), but in some populations red colour may be maintained or may reach a peak during this period (Moodie 1972*a*). This retention of red colour lends some support to the suggestion that nuptial colour may also serve as a warning signal to potential nest predators (Moodie 1972*b*).

As the eggs develop, their need for oxygen increases. To accommodate this change, the male increases both the frequency and duration of fanning bouts (van Iersel 1953; Sevenster 1961). A minimum amount of time is required to expel the water from a nest at the start of each fanning bout, and the increase in bout duration is well adapted to keeping the eggs supplied with oxygen throughout the parental phase (van Iersel 1953).

The demand for oxygen reaches a peak late in the parental phase. To facilitate oxygenation of the eggs, the male increases water exchange through the nest by boring holes in the top and rim of the nest (Wunder 1930; Leiner 1931). As the eggs near hatching, the male picks and tears at the nest and sucks out sand until the nest consists of little more than a heap of algae and debris with a depression in the centre (Wunder 1928; Leiner 1930; van Iersel 1953). Observations of stickleback in the wild also reveal that some males disassemble their nests and pile sticks and rushes around them just before the fry hatch (Foster 1988). This activity improves ventilation and produces a structure that may serve as a nursery in which the newly hatched young can seek refuge, as described for *G. wheatlandi* (McInerney 1969), *Pungitius pungitius* (Morris 1958), and the brook stickleback, *Culaea inconstans* (McKenzie 1974).

The parental behaviour of the white stickleback of Nova Scotia, Canada (Blouw and Hagen 1990) has become reduced to the dissemination of eggs. Shortly after the male has crept through the nest to fertilize the newly spawned eggs, he removes them from the nest and spits them into the filamentous algae that encompasses much of his territory (Jamieson *et al.* 1992*b*). The abundance of algae on the breeding grounds of these fish evidently provides an environment in which the separate eggs are cryptic and can obtain enough oxygen to develop without any further attention from the male. Whether emancipation from parental care in the white stickleback has increased the maximum reproductive output of males is unknown.

Reports of parental male stickleback retrieving young date back more than a century (e.g. Hancock 1852; Warington 1852, 1855). As young wander from the nest they are pursued, sucked up into the male's mouth, and spat back into the nest. This behaviour is performed intensely on the day of hatching and for some days thereafter (Feuth-de Bruijn and Sevenster 1983; see also Huntingford *et al.* page 292 this volume).

Leiner (1930) noted that parental male stickleback often retrieve and place another's fry into their own nest if the fry are placed nearby. Indeed, satiated parental male threespine stickleback may even retrieve fry of the ninespine stickleback, *P. pungitius* (Feuth-de Bruijn and Sevenster 1983). These observations suggest that males may have difficulty distinguishing their offspring from those of others. On the other hand, males may recognize alien fry as such, but adopt them as a way to reduce the probability that their own young will be taken if a predator attacks the nest (e.g. McKaye 1984).

The response of a parental male toward newly hatched fry is thought to depend on various factors, including his experience, point in the parental phase, hunger state, and the number of egg clutches he has acquired (van Iersel 1953; Feuth-de Bruijn and Sevenster 1983). The influences of the cues involved in retrieval of young and in other aspects of parental behaviour in *G. aculeatus* are therefore likely to vary accordingly.

Adult stickleback feed on a variety of prey, including conspecific eggs and fry (e.g. Hynes 1950; Semler 1971; Kynard 1978*a*; Worgan and FitzGerald 1981*a*; Hyatt and Ringler 1989*b*; Whoriskey and FitzGerald page 202, Hart and Gill page 232; Foster page 394 this volume). Leiner (1929) questioned why males do not devour their spawn and concluded that males must gradually develop an inhibition during the parental phase that prevents them from doing so. Leiner argued that the limited area and time available to parental males for foraging forces them to fast most of the time. But Feuth-de Bruijn and Sevenster (1983) found that parental males eagerly devoured prey (live *Artemia*), especially toward the latter part of the parental cycle, just when the eggs began to hatch. This indicated that the process that suppresses males from eating their own spawn was more specific than a general inhibition of feeding.

Feuth-de Bruijn and Sevenster (1983) also found that even though males started to retrieve young several days before the eggs hatched, hungry males would eat their own young if they had previous experience eating young. When satiated, such males switched from eating to retrieving young. Feuth-de Bruijn and Sevenster interpreted this behaviour as evidence for competition between eating and retrieving young as the parental phase progressed.

These findings are of particular relevance to the filial cannibalism hypothesis, which predicts that male stickleback may eat a small percentage of their spawn to maintain themselves in condition to care for the remaining spawn (Rohwer 1978 but see Sargent 1992; FitzGerald and Whoriskey 1992). Perhaps the interaction between hunger level and parental behaviour could limit filial cannibalism to periods of severe food deprivation, but field study is needed before any conclusions can be drawn.

The male is unable to keep up with the growing young as they stray farther and farther from the nest. Within a week after they hatch he ceases retrieving them, and instead chases and devours them as he would any other prey (van Iersel 1953). The breakdown of the nest, the fleeing response of the young to the approaching male, and other changes in external stimuli contribute to this change in behaviour, but internal factors also play a role (van Iersel 1953; Sevenster 1961; Wootton 1976).

Thus, the organization of parental behaviour in stickleback depends on an interplay between internal and external factors, and consequences of egg development may effect changes in the external stimuli that help coordinate fanning, nursery preparation, and other paternal activities. Although the experimental work on stickleback parental behaviour has dealt primarily with the role of chemical cues in controlling fanning and recognizing spawn, visual and tactile cues have also been implicated. I now consider external factors that are thought to play a role in the elicitation and control of parental behaviour in threespine stickleback.

Visual cues

The contribution of vision to other aspects of stickleback reproduction makes it likely that this sense is important for mediating parental behaviour, too. For example, fanning behaviour clearly depends on visual input. Tinbergen (1951) noted that fanning could be separated into a stereotyped motor component and a more variable orientation component. The latter component depended on the position and orientation of the male's nest; if it was tilted, the male tilted his body accordingly so that his body was orientated at the same angle relative to his nest when he fanned.

Vision also mediates egg retrieval. When males found a cluster of eggs outside their nest, they usually retrieved the eggs and replaced them in the nest (Leiner 1929; van Iersel 1953). Ter Pelkwijk and Tinbergen (1937) discovered, however, that the cluster had to comprise a minimum of five eggs to release retrieval. When one or two eggs were found lying outside the nest they were eaten immediately.

Clumps of gelatin placed outside the nests of males were also retrieved, provided that the clumps were at least as large as a cluster of five or six eggs (ter Pelkwijk and Tinbergen 1937). The effectiveness of these gelatin clumps in eliciting retrieval can probably be attributed partly to their resemblance to the transparent pale yellow of freshly laid eggs, because gelatin clumps coloured blue or red were not retrieved, but were carried far away from the nest. This experiment does not rule out a role for chemical cues, but if stickleback eggs produce some unique chemical stimuli, these appear to be unnecessary for egg retrieval in parental males.

Once the eggs are in the nest they are generally hidden from view, so it is unlikely that vision plays a major role in their care. The pursuit and retrieval of young by parental males do, however, involve visual cues. Males presented with day-old fry visually fixated on one of the fry as soon as it moved (Feuth-de Bruijn and Sevenster 1983). Visual cues may also help the male to distinguish young from prey or other moving objects when they are close by. In any case, the young must be taken into the mouth before they are retrieved or eaten, so the final decision regarding their fate may depend on chemical or tactile cues.

Chemical cues

Male stickleback accept and care for eggs that have been fertilized by other males (Evers 1878; Leiner 1930; van Iersel 1953). Thus male stickleback appear unable to distinguish their own eggs from those of others. Males in the laboratory sometimes steal eggs from the nests of rivals, carry the stolen eggs to their own nest, and care for them (van den Assem 1967; Wootton 1971*a*; Whoriskey and FitzGerald page 196; Foster page 390 this volume).

FitzGerald and van Havre (1987) directly tested for egg recognition ability in stickleback by presenting parental males with their own clutch and an alien clutch in separate mesh bags. Because males attacked their own eggs as readily as alien eggs, the authors concluded that male stickleback cannot recognize their own eggs. The possibility that the sack itself instigated attack on the eggs even though males had recognized them cannot be ruled out.

Females, however, did attempt to devour alien eggs more than eggs they had laid themselves, leading FitzGerald and van Havre (1987) to propose that egg recognition prevents females from preying on their own eggs when they participate in nest raiding. Smith and Whoriskey (1988) found subsequently that females readily attacked their own eggs if they were presented in the same bag with alien eggs. They therefore rejected FitzGerald and van Havre's hypothesis on the grounds that nests in nature usually contain clutches from several females. Whatever its function may be, the egg recognition shown by females is probably mediated by chemical cues that are easily masked or altered.

Chemical cues may also play a role in the recognition of young by parental male stickleback, but only to the species level. Male threespine stickleback readily retrieved unrelated conspecific young (Leiner 1929; Feuth-de Bruijn and Sevenster 1983) but treated young ninespine stickleback, *P. pungitius*, differently (Feuth-de Bruijn and Sevenster 1983). Ninespine young were quickly devoured if the males were hungry, but if the males were satiated they held the ninespine young in the mouth for some time and performed mumbling movements with their jaws. Several threespine males finally retrieved ninespine young, but others dropped them after testing them in this manner. This suggests that males may obtain tactile or chemical cues when they take young into their mouth, and that they accept or reject them depending on external stimuli and the males' internal state.

Perhaps the best-documented example of chemical cues in stickleback behaviour is in the control of parental fanning. When water from a bucket containing a large number of threespine stickleback was siphoned into a nest that contained eggs, the male increased the duration and frequency of fanning (van Iersel 1953). Because the rates of O_2 consumption and CO_2 production increase during development of embryos, van Iersel presumed that males gauged this when they poked their head into the nest entrance and then adjusted their fanning levels accordingly.

Sevenster (1961) extended van Iersel's 'bad water' experiments by charging tapwater with CO_2 and siphoning this into the nest of parental males. This, too, increased fanning, but Sevenster was unable to establish a clear correlation between the concentration of CO_2 and the extent to which fanning increased. He suggested that both increased CO_2 and reduced O_2 levels may elicit fanning and even show heterogeneous summation.

The use of direct chemical cues, rather than solely an internally generated cycle, to control fanning is advantageous because it allows for fine-tuning to local environmental conditions. Despite variations in substrate, temperature, water currents, and other environmental factors, males can maintain efficient fanning levels for a given set of conditions. Internal factors would lead the male to visit the nest periodically, increase responsiveness to eliciting stimuli, and help to suppress behaviour that might otherwise interfere with paternal care (van Iersel 1953; Sevenster 1961).

Finally, chemical factors may determine the time at which the nest is disassembled to provide increased ventilation for the developing embryos. Indeed, van Iersel (1953) found that males increased their boring and pulling at the nest more than fourfold during the 'bad water' experiments.

Tactile cues

Whenever the object of a response comes into close contact with the male's mouth, the possibility exists that tactile as well as chemical perception is involved in the control of that response. Therefore the prolonged holding in the mouth and eventual acceptance of ninespine young by some threespine males could be the result of similar tactile cues produced by young of the two species. The possible role of tactile cues in the male's perception of the eggs has already been noted.

Experience

The repeated performance of nesting activities and the general experience with the nest surroundings may enhance the efficiency with which the male performs his parental duties. If males complete multiple breeding cycles in a season, as laboratory studies show is possible (e.g. van Iersel 1953; Sargent 1985), improvement between cycles would be possible. The limited data indicate, however, that a large proportion of males in most (Whoriskey and FitzGerald page 193 this volume) but not all populations (Foster pers. comm.) fail to renest.

Particularly in populations in which males typically nest only once in the breeding season, the short duration of the breeding cycle and the rapidly changing factors associated with the developing progeny probably provide little opportunity for males to improve their parental behaviour through practice. Therefore, experience is less likely to be important for shaping parental behaviour than it is for shaping foraging and other activities that are repeated regularly throughout a fish's lifetime.

The experiments of Feuth-de Bruijn and Sevenster (1983) revealed, however, how experience could influence parental behaviour in stickleback indirectly by interfering with the parental male's response to fry. When non-parental males were kept hungry and then fed conspecific fry, they came to accept them as prey. When these males mated and entered the parental phase, they devoured rather than retrieved conspecific fry, including their own, unless they were kept satiated with *Artemia* or *Tubifex*.

In many populations, few if any fish that have spawned survive to the next breeding season (Wootton 1984*a*). Of those that survive, we do not know whether the males are able to breed more than one season. Moreover, although it might appear unlikely that if males were capable of breeding more than one season, they could modify their parental behaviour in accordance with experience from the previous season, this possibility cannot be ruled out. Such questions emphasize the need for further study of life history in stickleback.

CONCLUSIONS

The vast literature on threespine stickleback, dating back more than two centuries (e.g. Arderon 1746), bears testimony to the fascination the species has long held for ethologists. The wide availability, ease of maintenance, and well-developed social behaviour of this common little fish make it an ideal system for studying the causation, control, and evolution of behaviour. These qualities have led to a proliferation of studies on the species that continues to this day.

Many of the examples discussed in this review illustrate how careful laboratory observation and experimentation can reveal much about the factors that elicit and control reproductive behaviour in threespine stickleback. This research provides insight, not only into the behaviour of stickleback, but into behaviour in general. Indeed, stickleback research has had an enormous impact on the very foundation of ethology. Such fundamental concepts as sign stimulus, releasing mechanism, fixed action pattern, response chain, and displacement activity developed in large part from observations on this fish.

Despite our detailed knowledge of stickleback behaviour, many questions remain. For example, why do males fan the nest during courtship, even before they have obtained eggs? And why do males steal eggs from the nests of rivals, transfer them to their own nests, and care for them? Do these activities merely reflect a limitation or misfiring of the mechanism(s) controlling behaviour, or do they serve a purpose for the animal? The answer to such questions may be best answered in evolutionary—behavioural ecological terms, an approach that students of stickleback behaviour strongly advocate, and which has gained momentum in the past two decades.

An evolutionary approach is crucial to a complete understanding of stickleback behaviour and is bound to generate new and important questions.

Several authors in this volume presented evidence that the behaviour of threespine stickleback is at least as evolutionarily malleable as morphological traits. The wide interpopulation variation, even among stickleback populations in close proximity, is impressive and has forced us to alter our view of 'the' threespine stickleback and of speciation processes in general.

The relationship between behaviour and evolution is, however, bidirectional. Because selection acts on the whole organism, the constraints that behavioural mechanisms impose on the species play a critical role in its evolution. It is therefore imperative that we continue to investigate the proximate determinants of stickleback behaviour at all levels of organization, from neurophysiology to ethology and animal learning. Only by adopting this broad approach can we gain a complete understanding of the stickleback's evolution.

ACKNOWLEDGEMENTS

I wish to acknowledge the help of the many students who assisted me in the research reported here and shared in my enthusiasm for studying sticklebacks. I especially thank C. Baube and D. Kostka for assisting me for several seasons in the laboratory and field. S. Beeching, T. Horan, V. Savage, S. Schwen, and D. Winslow provided additional assistance.

I thank G. Barlow for his many thoughtful comments and helpful suggestions on this chapter. T. Bakker, C. Baube, S. Beeching, T. Horan, and M. Hosking provided additional helpful comments.

I also thank my wife, Ineke, and my daughters, Marijke and Sylvia, for their patience and understanding when I spent several spring vacations in the laboratory or salt-marsh.

This chapter is dedicated to the memory of my brother, R. George Rowland, Jun., who tended the other end of the seine when I started studying sticklebacks.

12

Evolution of aggressive behaviour in the threespine stickleback

Theo C.M. Bakker

Threespine stickleback, *Gasterosteus aculeatus* L., are highly polygynous; a male stickleback may collect as many as 20 clutches of eggs from different females in a single breeding cycle (e.g. Kynard 1978*a*). In addition, he may complete several breeding cycles during a single season (Wootton 1976, 1984*a*). We can therefore expect (Trivers 1972) strong competition to occur among male stickleback for access to females and females to be selective in their choice of mates.

Indeed, reproductive male stickleback compete directly for females, often interfering with one another's courtship (e.g. Ridgway and McPhail 1987; Goldschmidt and Bakker 1990). Competition for territories may also represent indirect competition for females, in that territory sites may be differently attractive to females (e.g. Kynard 1978*a*). In both instances, competition takes the form of intense aggressive interactions (Rowland Chapter 11 this volume).

In contrast, competition among females for mates is less often observed, presumably because throughout most of the reproductive season there is an ample supply of potential mates (e.g. Kynard 1978*a*). Late in the breeding season, however, simultaneous solicitation of males by multiple females is more common (Kynard 1978*a*; Borg 1985; Foster, pers. comm.), and aggressive interactions between courting females can be observed (Foster, pers. comm.).

These sexual selection processes, combined with the possibility that juvenile stickleback and adults outside the breeding season may maintain territories for feeding and/or shelter from predators (MacLean 1980; Bakker and Feuth-de Bruijn 1988), result in the occurrence of stickleback aggression in nearly every social situation. Aggression in the stickleback and the selective forces that act on it are thus diverse. This diversity can also be expected to be reflected in the underlying causal mechanisms and genetic bases of aggression. For a complete understanding of the evolution of aggression this complexity must be taken into account.

This chapter concentrates on intraspecific aggression. Antipredator tactics, including aggression directed at predators, are discussed by

Huntingford *et al.* (Chapter 10 this volume). The chapter is organized in three main parts. In the first, the forms of aggression exhibited by threespine stickleback (aggression of juveniles, aggression of adult females, territorial aggression of males, dominance of reproductive males) are described, and methods of measuring aggression are discussed. The central, second part reviews evidence for the existence of genetic variation in aggressiveness, both within and among stickleback populations. Attention is paid to genetic correlations between different forms of aggressiveness. Additionally, endocrine influences of the pituitary–gonadal axis in controlling aggressiveness in different phases in the life cycle of stickleback are reviewed, and the endocrine bases of genetic correlations between different forms of aggressiveness are discussed. The final section evaluates the evolution of stickleback aggression. Natural and sexual selection processes that act on a complex character set of different forms of aggressiveness, sexual behaviour, red breeding coloration, and several life history characters, are analysed. Topics that are discussed in the final part include the costs of aggression, aggression and reproductive success, and aggression and life history mode.

STICKLEBACK AGGRESSION: A MULTIFARIOUS PHENOMENON

The aggressive behaviour of threespine stickleback is strongly influenced by immediate context and by past experience (Rowland Chapter 11 this volume). Nevertheless, there are systematic differences in aggressive behaviour between life history stages and sexes, and between comparable life history stages of males and females across individuals and populations of threespine stickleback. The aggressive behaviour of reproductive (territorial) males is better studied than that of non-reproductive males, females, and juveniles. Territorial males are usually far more aggressive than are individuals in the other three groups (e.g. Wootton 1984; Bakker 1986).

Aggression of juveniles

Juvenile threespine stickleback typically, but not always, form large schools (Foster *et al.* 1988). Aggression among schooling juveniles has not been investigated in the field but has often been observed in the laboratory (Leiner 1929; Muckensturm 1969; Sevenster and Goyens 1975; Goyens and Sevenster 1976; Bakker 1985, 1986; Bakker and Feuth-de Bruijn 1988). Aggressive interactions first appear about 4 wk after hatching (Bakker 1986). During the first weeks after the onset of aggression, juveniles that are attacked typically flee without reciprocating. However, about 3 wk after the onset of aggression, attacked juveniles begin to counter-attack, a response that can lead to roundabout fights (rapid circling with spines erect). At approximately the same age, territorial behaviour first appears

(Bakker 1986; Bakker and Feuth-de Bruijn 1988). In the laboratory, aggression changes during the juvenile stage in a characteristic pattern that is not fully understood (Bakker 1985, 1986). Both sexes display similar levels of aggressive and territorial behaviour during the juvenile stage (Sevenster and Goyens 1975; Bakker 1985, 1986; Bakker and Feuth-de Bruijn 1988).

Aggression of adult females

Adult females typically forage in groups composed of adult females and non-reproductive males, leaving only to spawn (Keenleyside 1955; Black and Wootton 1970; Symons 1971; Kynard 1978*a*; FitzGerald 1983; Bentzen and McPhail 1984; pers. obs.). The aggressive behaviour of adult females in the laboratory resembles that of juveniles, typically involving direct attacks on one another (Leiner 1929, 1931; Wunder 1934; Sevenster and Goyens 1975; Li and Owings 1978*a*; Bakker 1985, 1986; Wootton 1985*b*). Females tend to be less aggressive when gravid, and as they age, aggressive interactions decline in frequency (Bakker 1986).

Although much less pronounced than in males, dominance relationships may also develop among females in groups (Leiner 1931; Wunder 1934; Wootton 1976; Li and Owings 1978*a*; Bakker 1986). For example, adult females from a Californian (USA) freshwater population established dominance relationships in the laboratory in which status was positively correlated with home range size (Li and Owings 1978*a*). Females of higher status also had greater access to males and were more effective than subordinates at disrupting courtship by other females. Although neither phenomenon has been studied in detail in the field, foraging territories are sometimes defended by females during the breeding season (MacLean 1980; Foster pers. comm.), and aggressive interactions frequently occur between females that are courting the same male in Crystal Lake, British Columbia, Canada (Foster pers. comm.). Defence of foraging territories has also been observed outside the breeding season in one Canadian lake (MacLean 1980).

Aggression of adult males

Pre-breeding aggression

During and outside the reproductive season, non-reproductive adult males participate in sexually mixed schools (e.g. Kynard 1978*a*; MacLean 1980). In the laboratory, their aggression resembles that of juveniles and females (e.g. Sevenster 1961; Wai and Hoar 1963; Baggerman 1966, 1968; Huntingford 1979). At the beginning of the breeding season, males leave the schools and establish a nesting territory. Males often must compete with other males for territory sites (e.g. Mori 1990*b*), and aggressiveness increases markedly at this time (as established in laboratory studies, e.g. Sevenster 1961; Baggerman 1966, 1968; Bakker 1986). In consequence, aggressive interactions, including biting, bumping, threatening, and roundabout fighting, are often observed. Males that are attacked by another male may counter-attack, flee, or assume threatening postures. The aggression shown by males in the period preceding nest building is often called 'pre-breeding aggression'. The ability of a male to obtain a territory in competition with other males is termed 'dominance ability'.

Territorial aggression

Once a territory is established, the male defends it vigorously against intruders ('territorial aggression'). Usually, the territory owner attacks rival males with a direct charge that ends in a bite or a bump, and roundabout fighting is rare. Similar attacks are directed at large juveniles, females, and often, other fish species. Once the nest is completed, the male typically courts females rather than attacking them, although aggressive behaviour may still break through at various points in the courtship sequence ('courtship aggression').

Breeding male threespine stickleback are well known for their high levels of aggression against territorial intruders, particularly against rival males. Among reproductive males of the Gasterosteidae, *Gasterosteus aculeatus* males achieve the highest levels of aggression (Morris 1958; Wilz 1971; Huntingford 1977; Rowland 1983*a*,*b*; FitzGerald 1983; Gaudreault and FitzGerald 1985).

Male aggressiveness changes during the reproductive cycle in a pattern thought to reflect changes in the value of the resource (territory and nest) being defended. Initially, the presence of a nest appears to increase the value of the resource. This increase is indicated by defence of a larger territory when a nest is present (Stanley and Wootton 1986), by a drop in aggression after removal of the nest (Symons 1965), and by a decrease in aggression with increasing distance from the nest (van Iersel 1953; Symons 1965; Black 1971; Huntingford 1976*a*,*c*, 1977; Rowland 1983 *a*,*b*, page 299 this volume). Acquisition of eggs also appears to enhance the value of the resource, and the level of aggression directed toward conspecific rivals continues to increase with the acquisition of additional clutches (Sargent and Gross 1986).

Territorial aggression during the parental phase

There has been considerable disagreement in the literature as to whether aggressive defence of the nest subsequently increases (Wunder 1928; Huntingford 1976*a*,*b*,*c*, 1977; Kynard 1978*a*) or decreases (van Iersel 1953; Segaar 1961; Sevenster 1961; Wootton 1971*b*) during the parental phase. This inconsistency seems to be due to differences in the distance from the nest at which aggression was measured in different studies and to changing spatial patterns of aggression by males as the embryos mature.

After egg acquisition, the size of the territory defended by the male decreases over the course of the parental phase (Black 1971; Kynard 1978*a*), leading to a decrease in levels of aggression at distances greater than about 30 cm from the nest. During this period, however, levels of aggression remain high closer to the nest (Symons 1965; Black 1971). During the parental phase a distinction must therefore be made between the value of the nest and that of territory size. It is clear that the value of the nest increases once it contains eggs, but the advantage of maintaining a large territory that seems to be associated with intersexual selection (van den Assem 1967; Black 1971; Li and Owings 1978*a*; Goldschmidt and Bakker 1990) disappears once the male enters the parental phase. The decrease in territory size during the parental phase may reflect a trade-off between the increasing need to care for the embryos as they mature (Rowland page 338 this volume) and the need to defend the nest from egg raiders (Whoriskey and FitzGerald page 202; Foster page 394 this volume). Aggressiveness increases sharply after the eggs hatch (Sevenster 1961; Black 1971; Wootton 1971*b*; Huntingford 1976*a,b,c*, 1977; Kynard 1978*a*). This makes sense functionally, both because of the increased reproductive value of offspring, and because of the increased vulnerability of newly hatched young to predation by conspecifics (but see Worgan and FitzGerald 1981*b*).

The relationship between territorial aggressiveness and resource value is even more evident in the antipredator behaviour or boldness that males show towards predators of both adults and offspring (Huntingford *et al.* page 283 this volume). In the threespine stickleback, boldness towards a hunting pike and aggressiveness towards conspecifics covaried over the course of the breeding season, suggesting that these traits share internal causal factors (Huntingford 1976*a,b,c*, 1982; Giles and Huntingford 1984; Tulley and Huntingford 1988). Further evidence of this relationship was provided by Kynard (1978*a*), who demonstrated that the boldness of males confronted with a rival male paralleled that of males confronted with a trout (predator of the male) over the course of a brood cycle, increasing from the empty-nest stage to that in which males were defending fry.

In some studies such as Kynard's (1978*a*), the reproductive value of the guarded progeny has been shown to affect male risk-taking. Males with embryos were bolder when confronted with a preserved sculpin (Pressley 1981) or overhead stimulus (FitzGerald and van Havre 1985) than were those without embryos, and boldness toward a preserved sculpin was also correlated with embryo age. In contrast, Foster and Ploch (1990) detected no effect of embryos on male risk-taking, and FitzGerald and van Havre (1985) found no difference between male responses to an overhead stimulus when the males had embryos versus fry in their nests.

THE MEASUREMENT OF AGGRESSIVENESS: METHODS

Individual levels of aggression can be quantified in a number of ways, but the test developed by van Iersel (1958) is most widely used to measure aggressive motivation of territorial male stickleback. This test employs a 'rival' male enclosed in a glass tube. He is introduced into the territory of an isolated experimental male at a specific distance from his nest for 5 min or so, and the numbers of bites and bumps directed at the intruder are counted. If appropriate opponents are used, this method can also be used to measure the aggressive motivation of territorial males during courtship (e.g. Sevenster 1961; Sevenster-Bol 1962; Bakker 1986), or of juveniles or adult females (e.g. Sevenster and Goyens 1975; Bakker 1986).

Variously simplified models, or 'dummy' males, have also been used to measure territorial aggressiveness. This method is valuable in that it permits alteration of specific features (for instance, nuptial coloration) of the stimulus male to ascertain the effect of the character on aggression elicited from the experimental male (Rowland page 298 this volume). The method can prove problematic, however, as there appears to be considerable inter-individual and interpopulation variation in responsiveness to dummies (e.g. Wootton 1971*b*; Baerends 1985; Rowland and Sevenster 1985).

A final method involves scoring of aggressive interactions between two territorial males separated by a glass partition (e.g. Baggerman 1966, 1968; Wootton 1971*b*). Although this method is relatively simple to carry out, interpretation of results can be complicated by habituation (Peeke 1969, 1983; Peeke *et al.* 1969, 1979; Peeke and Veno 1973, 1976; van den Assem and van der Molen 1969; Rowland 1988).

Each of these methods can produce different absolute levels of territorial aggression (Wootton 1971*b*, 1972*b*). Consequently, it is impossible to compare levels of aggression between studies that employ different methods of measuring territorial aggressiveness. Even when the same general method is used, minor methodological differences can have profound effects (but see Giles and Huntingford 1985). For example, the distance from the nest at which a male or a dummy is presented can significantly affect the level of aggression that is measured (see above). These problems make it hard to distinguish between interpopulation differences in aggressiveness and methodological effects, except when identical methods are applied to multiple populations (e.g. Huntingford 1982, but see Giles and Huntingford 1985). All of these methods can, however, be used to test for comparable changes in aggressiveness over the breeding cycle (Wootton 1971*b*, 1972*b*).

Establishing the frequency of aggressive interactions among individuals in standardized groups is an alternative method of quantifying interpopulation differences in aggressiveness when interindividual variation is not a primary concern. The method has been applied to juvenile stickleback (Goyens and Sevenster 1976; Bakker 1986; Bakker and Feuth-de Bruijn 1988) and was as successful as the standardized tube test in discriminating differences in the levels of juvenile aggression between genetically differentiated lines (Bakker 1986). It is probably most useful for measuring aggressiveness in juveniles, females, and non-reproductive males.

Unfortunately, controlled methods of measuring territorial aggressiveness have rarely been applied in the field or in naturalistic laboratory settings (exceptions: Black 1971; Kynard 1978*a*; Gaudreault and FitzGerald 1985). Instead, individual levels of aggression are usually deduced from observations of natural encounters among focal males and other individuals. These observations are relatively uncontrolled, and do not measure a standardized index of the male's aggressive motivation because the aggressive state of a male may be influenced by factors such as the number and quality of near neighbours or distance from the nest. Although it may be possible to examine associations between aggressiveness and reproductive success through field studies of this type, one must be cautious in drawing conclusions about the evolution of aggressiveness from such research. These concerns have been taken into account in the ensuing discussion. There obviously is a need for a general, standardized index of aggressive motivation in field studies. Other methodological concerns will be discussed as appropriate throughout.

CAUSES OF VARIATION IN AGGRESSIVENESS

As is probably the case for all behavioural traits, aggressiveness of threespine stickleback in all life history stages is influenced by immediate environmental and social conditions. In particular, the effects of the rival male's phenotype (e.g. nuptial coloration, body size, behaviour) on the aggressive behaviour of reproductively active males have been extensively investigated over the last 40 yr (reviewed by Rowland page 299 this volume). In contrast, study of the genetic bases of variation in aggressiveness has only recently been initiated. Because a knowledge of the extent of genetic variation underlying variation in aggressiveness is a prerequisite for evolutionary interpretation, I will begin by discussing the genetics of aggressive behaviour in threespine stickleback.

Genetic influences on aggressive behaviour

The presence of genetic variation within populations can be inferred by breeding and rearing individuals in the laboratory under identical conditions. The individuals can be scored for aggressive phenotypes at different stages in their development and, if appropriate experimental designs are used (e.g. Falconer 1981), heritabilities and genetic correlations can be calculated. A similar design can be used to examine differences among populations in the genetic bases of behaviour. The threespine stickleback is an excellent subject for research in behavioural genetics because many environmental sources of variation can be controlled in a laboratory setting (Bakker 1986). For example, fertilized eggs can be incubated in the laboratory to avoid the effects of differences in the quality of paternal care. Recent research has established that there is a genetic basis for variation in aggression within and among populations of threespine stickleback.

Genetic differences among populations

Intriguing interpopulation differences in levels of aggression have been reported for juvenile and adult threespine stickleback when aggression has been assayed using standardized methods across all populations. As in most other aspects of phenotype, threespine stickleback populations appear to differ substantially in aggressive characteristics.

Territorial aggressiveness of reproductive males has been shown to vary among freshwater populations in Scotland (Huntingford 1982; Giles and Huntingford 1985; Tulley and Huntingford 1988). Additionally, aggressiveness during standardized courtship experiments differed between a marine population from Rhode Island, USA, and a freshwater population from Oxford, UK (Wilz 1973), between the sympatric Enos Lake species pair (Ridgway and McPhail 1984, 1987; McPhail page 421 this volume), and between parapatric anadromous–freshwater species pairs in British Columbia, Canada (McPhail and Hay 1983; McPhail page 408 this volume), and Washington, USA (McPhail 1969). Finally, juveniles, adult females, and males from Scottish freshwater populations differed in their boldness towards live and dummy predators in standardized laboratory tests (Huntingford 1982; Giles 1984*a*,*b*; Giles and Huntingford 1984; Huntingford and Giles 1987; Tulley and Huntingford 1987*a*, 1988; Huntingford *et al.* page 285 this volume), and this character is correlated with aggressiveness. Huntingford (1976*a*,*b*,*c*) postulated shared internal causal factors as an explanation for the correlation of these two aspects of behaviour, although this idea has yet to be tested directly. There is a consensus that predation has played a major role in moulding the behaviour and morphology of threespine stickleback (Reimchen Chapter 9; Huntingford *et al.* Chapter 10 this volume). To the extent that there exist correlations between aggressiveness and behavioural and morphological defences against predators, predation may have acted indirectly to mould the levels of aggressiveness displayed by threespine stickleback.

Suggestions concerning genetic involvement, and thus adaptive variation, can be deduced from the above-mentioned interpopulation differences. Strictly speaking, this is only justified if the populations concerned have been bred and raised under identical conditions to eliminate potential influences of environmental variation on variation in aggressiveness. Almost all the above-mentioned studies do not come up to these strict requirements and are therefore inconclusive with respect to genetic variation, although they may suggest adaptive variation in stickleback aggressiveness.

The best-documented interpopulation variation involves genetic differences in juvenile aggression. Bakker and Feuth-de Bruijn compared juvenile aggression and territoriality in laboratory-bred offspring from wild-caught parents originating from an anadromous population (Den Helder, Netherlands) and a freshwater population (Emst, Netherlands) (Bakker *et al.* 1988; Bakker and Feuth-de Bruijn 1988, unpubl. data). Aggressive interactions among groups of five juveniles (mixed sexes) were measured during weekly 5 min observation periods until the first individual in the group reached reproductive maturity. During much of the juvenile stage, juveniles from the anadromous population were less aggressive than were those from the freshwater population (each week from 10 wk after fertilization onwards: Mann–Whitney U test, one tailed, $U \leq 104$, $Z \geq 2.898$, $P < 0.002$) (Fig. 12.1). Juveniles from the freshwater population were also more likely to establish territories than were juveniles from the anadromous population.

The progeny of both reciprocal crosses exhibited aggression levels similar to those of the freshwater population (Bakker and Feuth-de Bruijn, unpubl. data), suggesting dominance of alleles for higher aggression levels. Sustained directional selection is expected to lead to dominance of genes controlling the expression of the trait in the direction selected for (e.g. Broadhurst and Jinks 1974). Assuming that freshwater stickleback populations have been derived from anadromous or marine ancestors (Bell 1984*a*, 1988; Bell and Foster page 14 this volume), the genetic dominance of higher juvenile aggression levels is in the direction expected if selection has favoured greater aggressiveness in freshwater populations, as suggested by these studies.

Honma and Tamura (1984) qualitatively described parallel differences in aggressiveness between laboratory-bred juveniles of an anadromous and

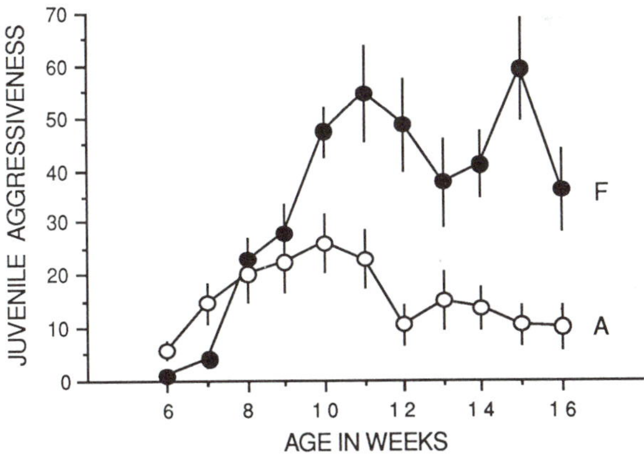

Fig. 12.1 Mean levels of juvenile aggression as a function of age (weeks after fertilization) in groups of five juveniles belonging to a Dutch freshwater (F) population (19 groups) and a Dutch anadromous (A) population (23 groups). The groups were made 3 wk after fertilization, and aggressiveness was scored as the total number of bites and bumps in the group during weekly 5 min observation periods. Error bars represent one standard error of the mean.

a freshwater population in Japan. One of the reciprocal crosses showed hybrid sterility (Honma and Tamura 1984; Honma *et al.* 1986).

Bakker and Feuth-de Bruijn also documented small, but significant differences in the levels of territorial aggression exhibited by laboratory-bred, isolated males from the Dutch anadromous and freshwater populations they studied (Bakker et al. 1988; Bakker and Feuth-de Bruijn, unpubl. data). Males from the anadromous population spent less time in aggressive behaviour toward an enclosed rival ($\bar{X} =$ 26.6 per cent biting-bumping time during 5 min tests; $N = 44$ males) than did males from the freshwater population ($\bar{X} =$ 34.5 per cent; $N = 44$ males; Mann–Whitney U test, two tailed, $U = 721.5$, $Z = 2.057$, $P < 0.04$).

Additionally, the males from these two populations differed in levels of sexual activity during tests of territorial aggression, but the relationship was the inverse of that for territorial aggression. During these tests, the number of zigzags performed by each male was recorded. The zigzag dance is a prominent component of male courtship (Rowland page 314 this volume) and is considered a reliable measure of male courtship readiness (e.g. van Iersel 1953; Sevenster 1961). Although this behaviour is displayed most often to ripe females, it is also elicited, although at a lower frequency, by male intruders. In both situations, there exists a mutually inhibitory relationship between the aggressive and sexual tendencies (Sevenster 1961; Symons 1965; Rowland 1984). Males from the anadromous population performed zigzags more often ($\bar{X} = 16.6$ per 5 min; $N = 44$) than did those from the freshwater population ($\bar{X} = 2.4$ per 5 min; $N = 44$; Mann–Whitney U test, two tailed, $U = 256.5$, $Z = 5.943$, $P < 0.001$).

Finally, the differences in territorial aggression between these populations were paralleled by apparent differences in dominance ability between the males (Table 12.1). If two reproductive, isolated males are simultaneously introduced in a tank unfamiliar to both and just large enough for the settlement of one territory, then one of the males usually dominates the

Table 12.1 Outcomes of dyadic combats between inexperienced laboratory-bred, individually isolated reproductive males from a Dutch freshwater (F) and a Dutch anadromous (A) population.

Dominance test characteristics	F dominant	A dominant	Undecided	P[a]
Round-robin; small tanks[b]	71	27	2	<0.001
First tests[c]; small tanks	10	2	1	<0.05
First tests; larger tanks	9	1	0	<0.02

[a] χ^2 test, two tailed.
[b] Tank sizes: small, 34 × 17 cm; larger, 69 × 40 cm.
[c] First tests: results are for the first tests to which each male was exposed, avoiding effects of experience (see text). Round robin: results include the outcomes of all possible pairwise combinations of males (see text).

other after a short and intense fight (Bakker and Sevenster 1983). The dominant male begins nest building, while the inferior male remains quiet at the water surface or hidden between plants, where he is attacked by the dominant male upon movement.

When all pairwise comparisons of relative dominance are made among a group of individually isolated males, the males can be arranged in a linear order of dominance based on the probability of winning the dominance contests (Bakker and Sevenster 1983; Bakker 1985, 1986). Males from the freshwater population displayed greater dominance ability, regardless of whether mean dominance ability was deduced from the outcomes of all possible pairwise combinations of males or only on the basis of the first test to which each male was exposed (Table 12.1). The latter assessment avoids possible confounding effects of experience (Bakker and Sevenster 1983; Bakker *et al.* 1989). Population differences in territory size (e.g. Giles and Huntingford 1985) might also affect dominance ability in small (34 × 17 cm) tanks. This possibility was ruled out by assessing dominance in larger (69 × 40 cm) tanks (Table 12.1).

Genetic variation within populations

In some freshwater stickleback populations, there exists an association between morphology (lateral plate number) and aggressiveness (Huntingford 1981) or competitive abilities (Moodie 1972*b*; Kynard 1979*a*). Because lateral plate number is partly under genetic control (e.g. Hagen 1973), this association indicates, albeit indirectly, genotypic variation of aggressiveness within stickleback populations.

The first evidence of genetically based differences in aggression within threespine stickleback populations was provided by Goyens and Sevenster (1976), who demonstrated differences in juvenile aggressiveness between laboratory-bred progeny of different parents from a freshwater population in the Netherlands. This research stimulated more detailed and extensive studies designed to evaluate both intra- and interpopulation variation in aggression (Bakker 1985, 1986; Bakker and Feuth-de Bruijn 1988; Bakker *et al.* 1988; Bakker and Sevenster 1989; Bakker and Feuth-de Bruijn, unpubl. data). These studies provide the conclusive evidence for genetic variation in aggressiveness of threespine stickleback.

To assess the heritability (in the narrow sense) of aggressiveness, Bakker (1985, 1986) conducted directional selection experiments in which separate lines were artificially selected for high and low values of several forms of aggression. The base population was derived from laboratory-bred progeny of 25 males and 25 females collected in freshwater streams flowing into the Apeldoorns kanaal in the Netherlands. Independent selection lines, one each for enhanced and reduced levels of aggression, were established for each of three forms of aggression. Aggressive phenotypes of juvenile males and females were scored, and those with the most extreme phenotypes were used to establish lines with high (JH) and low (JL) levels of juvenile aggression. Similarly, adult males were scored for territorial aggressiveness, and adult females for female aggressiveness, in establishing the high (TH) and low (TL) territorial aggression lines. In establishing the high (DH) and low (DL) dominance lines, males were scored for dominance ability and females were selected at random. In addition to these six selection lines, an unselected control (C) line was maintained by breeding randomly selected adults. The selection lines were maintained for three generations. The control line was tested against the selected lines after two generations of random mating. Juveniles used to establish each generation were isolated

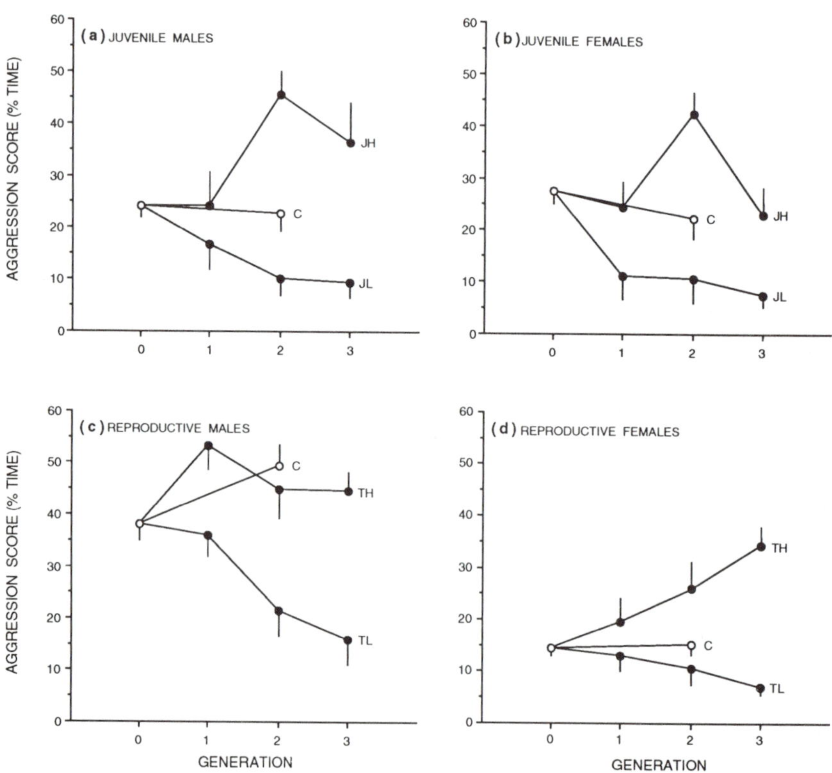

Fig. 12.2 Responses to selection for high and low levels of juvenile (JH and JL lines) and territorial (TH and TL lines) aggression. Aggressiveness in an unselected control line (C) is also presented. Juvenile aggression was measured in (a) juvenile males and (b) juvenile females. Territorial aggression was measured in (c) reproductive males and (d) reproductive females. Aggressiveness is expressed as the mean per cent biting-bumping time during weekly 5 min aggression tests. Error bars represent one standard error of the mean.

from their sibs 21 d after fertilization, well before the onset of juvenile aggression (Fig. 12.1).

Selection for reduced juvenile aggressiveness produced significant divergence from the control line in both sexes after one generation of selection. The differences persisted for the ensuing two generations (Figs. 12.2(a),(b); Bakker 1986). Selection for enhanced aggressiveness was less successful, producing significant divergence from the control line only in the second generation. In the third, levels of aggression were similar to those in the control line in the second generation. Similarly, selection for reduced territorial aggression produced significant divergence from the control line in reproductive males, but selection for enhanced aggression did not (Fig. 12.2(c); Bakker 1986). In the females, however, selection in both low and high territorial aggression lines produced significant differences from the control line by the third generation of selection (Fig. 12.2(d); Bakker 1986). In the six lines in which realized heritabilities (h^2) could be calculated for each sex (JL, TH, TL), five were significant and ranged from 0.29 ± 0.04 ($h^2 \pm$ SD) to 0.64 ± 0.07 (Bakker 1986).

Analysis of the dominance data was more complex because dominance had to be measured in contests between two individuals. The outcome of any contest therefore depends on the phenotypes of both. Within each generation of the DH and DL lines, all possible pairwise combinations of males were tested. The criterion of selection was based on the number of tests that each male had won. In each generation, the joint response to two-way selection for dominance was determined from interline dominance tests with males randomly chosen from both lines. Selection for low and high dominance ability produced significant divergence between the two lines by the third generation (Fig. 12.3), at which time males from the DH line won 19 of 24 dominance contests against males from the DL line ($\chi^2 = 8.17$, d.f. $=1$, $P < 0.01$; Bakker 1986). In the second generation, DH males won 5 of 10 contests against control males, while the DL males won only 3 of 10 contests. This, and the results of dominance tests between both dominance lines and the other selection lines in the third generation (see below), suggested that the divergence in the DH and DL lines was due to a decrease in the dominance ability of DL males rather than an increase in that of the DH males. The estimated realized h^2 for dominance ability (combined two-way response) was 0.34 (Bakker 1986).

These results demonstrate that there is substantial heritable variation for aggressiveness in at least one population of threespine stickleback. The seeming lack of response of males to selection for enhanced dominance ability (Fig. 12.3) and territorial aggressiveness (Fig. 12.2(c)) can probably best be explained as a consequence of long-term selection for high levels of territorial aggression and dominance ability in the natural population (Bakker 1986). In contrast, the seeming lack of response to selection for increased levels of juvenile aggression in both males and females

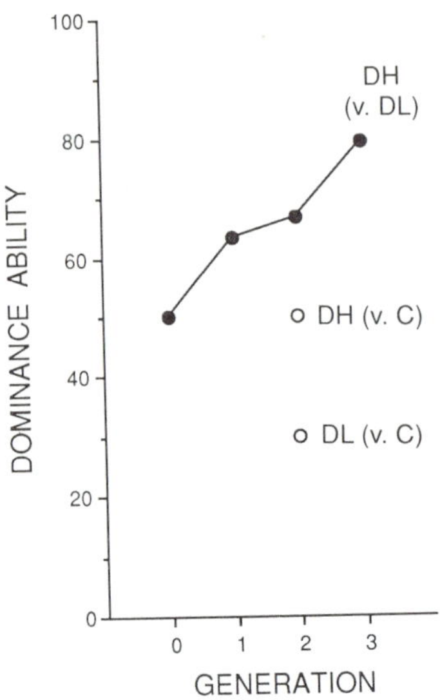

Fig. 12.3 The response to selection for high (DH line) and low (DL line) dominance ability, measured as the proportion of cases in which DH males won dominance contests against DL males (including the first unselected generation). The dominance ability of DH and DL males was also scored in tests against the control line (C) in the second selected generation.

(Fig. 12.2(a),(b)) may result from reduced embryonic viability and correlated increases in female aggressiveness during courtship that reduce the probability of successful spawning. Both effects were observed in the JH line (Bakker 1986).

Genetic correlations between forms of aggression

Character correlations can profoundly affect evolutionary responses to selection, because selection on a particular trait will produce changes in correlated characters (e.g. Falconer 1981; Lande and Arnold 1983; Endler 1986). The evolutionary response to directional selection will be constrained if it produces disadvantageous changes in correlated characters because the net fitness differential may be negative. Although we certainly do not have a complete understanding of how or why different forms of aggression are correlated with one another and with other aspects of phenotype, recent research has provided some valuable insights.

Imposition of artificial selection for several forms of aggression (Bakker 1985, 1986; see above) provided evidence of genetic correlations between some but not all forms of aggression. The study consisted of a series of double selection experiments (that is, line X was selected for trait x and screened for trait y, while line Y was selected for trait y and screened for trait x, etc.), because the fish in each line were screened for all the investigated forms of aggression. This design permitted estimation of genetic correlations (Falconer 1981), which express the extent to which the variation of two characters is influenced by common genes. One has to realize, however, that estimates of genetic correlations are subject to large sampling errors (e.g. Falconer 1981).

Each female in each line was assayed for both juvenile and female aggressiveness. Males were assayed for juvenile aggressiveness, territorial aggressiveness, dominance ability, and aggressiveness during courtship. From these data, the mean score for each form of aggression was calculated for each generation in each line. Thus, for example, as a direct response to selection, mean female aggression was determined for each of three generations of the TH line. Three mean scores were also calculated for the correlated response of juvenile aggressiveness in the same line, so that the correlation between the two forms of aggression could be examined. This was repeated in each line for all appropriate forms of aggression. A single mean for each form of aggression was calculated for each sex in the control line, using the combined scores of all individuals in generations 0 and 2 (this gave the most reliable values of control line fish in calculating genetic correlations; Bakker 1986).

The genetic relationship between juvenile aggressiveness in females and adult female aggressiveness was assessed using the control, JH, JL, TH, and TL line means (Fig. 12.4(a)). There was a strong correlation among generation means between the two forms of aggression, suggesting that the same loci affect both (genetic correlation: 0.98). In contrast, juvenile and territorial aggression in males were not strongly correlated among generation means of the JH, JL, TH, and TL lines (Fig. 12.4(b)), suggesting that in males, juvenile aggressiveness is only partly governed by the same genetic factors as is territorial aggression (genetic correlation: 0.50). The correlation between territorial and courtship aggression among generation means of the TH and TL lines also points to common genetic influences between these two forms of aggression (Fig. 12.4 (c); Sevenster 1961).

Selection for dominance ability produced little correlated change in other forms of aggressiveness; DH and DL line fish did not differ significantly in any of the other forms of aggressiveness (Bakker 1986). This result suggests that dominance ability is affected by other genetic factors than are juvenile, territorial, or courtship aggressiveness. The zero genetic correlation between juvenile aggressiveness and dominance ability was further substantiated by the outcomes of interline dominance tests in the third generation of selection. JH and JL males had similar dominance abilities to those of DH males (y-axes of Fig. 12.5(a),(b); Bakker 1986). Bidirectional selection for territorial aggressiveness resulted, however, in parallel changes in dominance ability; TH males had dominance abilities that were similar to (or even higher than) the dominance abilities of DH

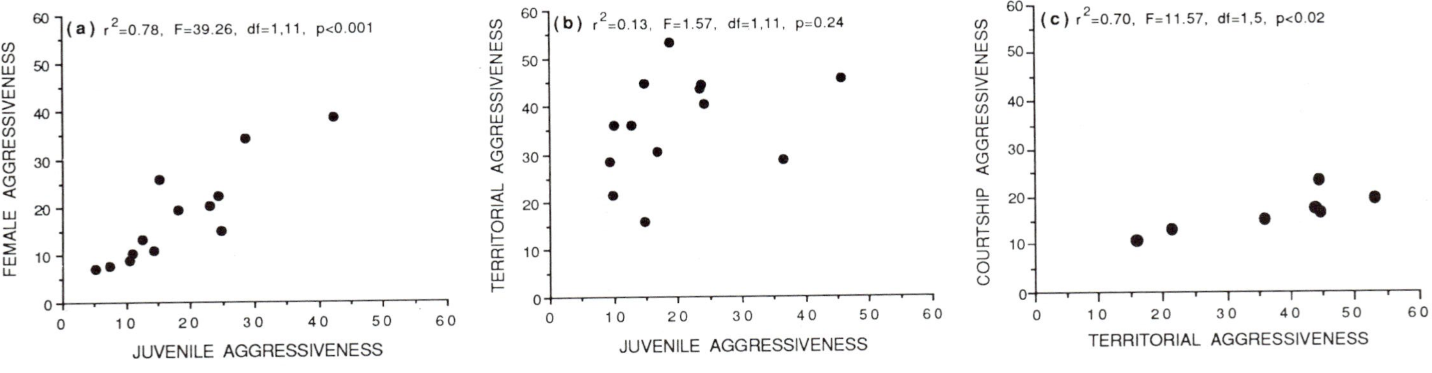

Fig. 12.4 Correlations between (a) juvenile and female aggressiveness, (b) juvenile and territorial aggressiveness, and (c) territorial and courtship aggressiveness. The data points on each graph represent the means of the aggression scores of all individuals in a single generation in a selected line. Three generations are represented for each of four selected lines: those selected respectively for high and low juvenile aggression (graphs (a) and (b)) and those selected respectively for high and low territorial aggression (graphs (a), (b), and (c)). A single mean was calculated for the control line using scores of individuals in the base population and in the second generation of the control line. See the text for additional details.

males, while the dominance abilities of TL males were similar to those of DL males (*y*-axes of Fig. 12.5(a),(b); Bakker 1986). This outcome seems to contradict a zero genetic correlation between territorial aggressiveness and dominance ability (see above). In this population, dominance ability correlates positively with the degree of red breeding coloration (Bakker and Sevenster 1983; Bakker 1986; see below). Also in the interline comparisons of dominance ability, differences between lines in the degree of red coloration explain the greater part of variation in dominance ability (Fig. 12.5(a),(b)). These results suggest a different genetic causation of colour changes in the territorial aggression lines and the dominance lines, leading to a zero genetic correlation between territorial aggressiveness and dominance ability in the dominance lines, but a positive one in the territorial aggression lines.

In combination, these studies demonstrate the existence of substantial genetic variation in aggressiveness of threespine stickleback. Natural and sexual selection acting on this variation has the potential to change the levels of aggression within and between populations. Indeed, parallel differences in the aggressiveness of juveniles from two anadromous and two freshwater populations suggest that natural selection may have done so. The genetic correlations among different forms of aggression are comparable in sign and magnitude with the corresponding phenotypic correlations (Bakker 1985, 1986), as usually is the case (e.g. Falconer 1981; Cheverud 1988). Stickleback aggression is thus characterized by a complex genetic correlation structure among different forms of aggression which may constrain the evolutionary response of specific forms of aggression. Evolution of stickleback aggression can best be understood by taking this complexity into account. Analyses of hormonal influences on stickleback aggression (see below) make clear that the genetic correlation structure of aggression is part of a larger complex involving reproductive behaviours and several life history characters.

Hormonal influences on aggression

The hormonal influences of the pituitary–gonadal axis on aggression by male threespine stickleback have been studied extensively. Gonadotropins are pituitary hormones, whose secretion is triggered by light (e.g. Slijkhuis 1978; Borg *et al.* 1987*c*). Secretion of gonadotropins in turn stimulates the production of gonadal hormones (androgens). Under winter conditions (short photoperiod, low temperature), androgens have a positive feedback effect on gonadotropin synthesis (Borg *et al.* 1986). This also holds when male stickleback that have been maintained under short photoperiods are stimulated into breeding condition by long photoperiods. It is only during the breeding season (long photoperiod, relatively high temperatures) that androgens inhibit gonadotropic cells (Borg *et al.* 1985; Guderley page 105 this volume).

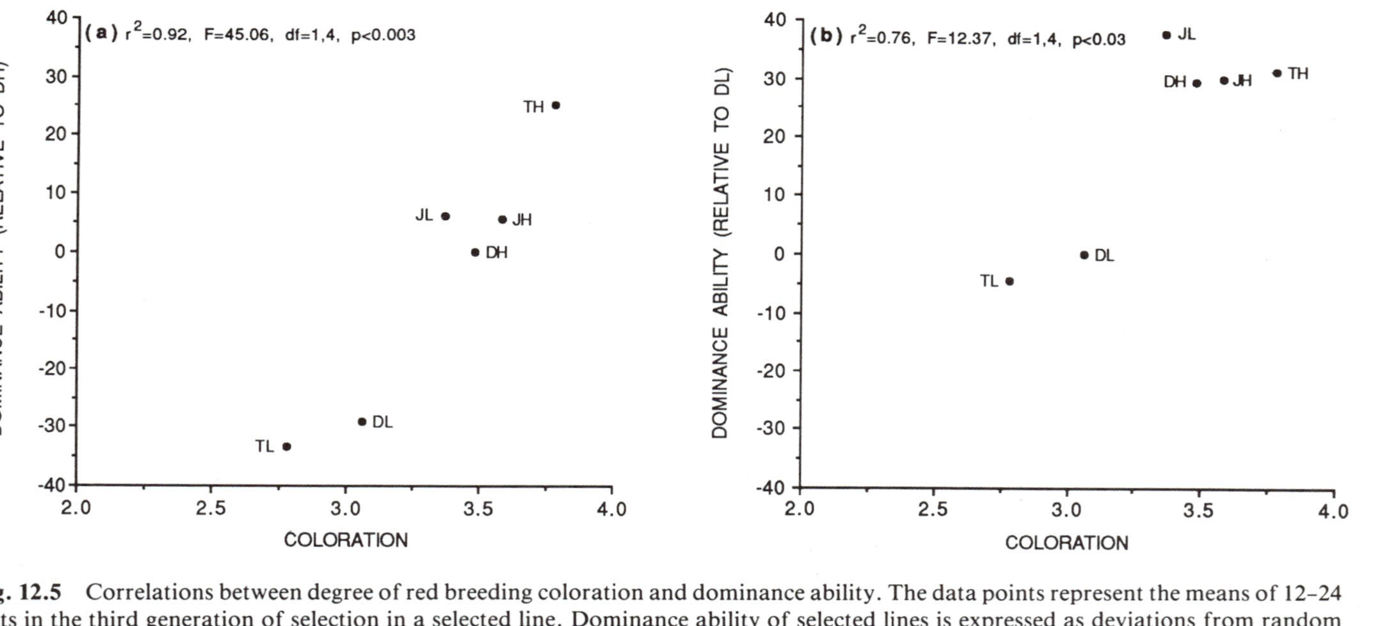

Fig. 12.5 Correlations between degree of red breeding coloration and dominance ability. The data points represent the means of 12–24 tests in the third generation of selection in a selected line. Dominance ability of selected lines is expressed as deviations from random (50 per cent of tests won by males of a selected line) in dominance tests against males of (a) the high dominance line (DH) and (b) the low dominance line (DL). The degree of red coloration of both contestants was scored on a four-point scale after the test. Scoring was conducted in the males' home tanks after presentation of a ripe female in a tube. JH, JL: high and low juvenile aggression lines, respectively; TH, TL: high and low territorial aggression lines, respectively.

In pre-breeding males, the level of aggression is related to the level of gonadotropins rather than to that of androgens (Hoar 1962*a*,*b*; Wai and Hoar 1963; Baggerman 1966, 1968; Wootton 1970). A combination of several methods (manipulation of photoperiod, castration of males, administration of testosterone or gonadotropins) was used to show this. For example, castrated, pre-breeding males (long photoperiod) are as aggressive as their intact counterparts, indicating that androgens are not important determinants of male aggressiveness at this stage.

Hormonal control of territorial aggression: controversial data?

The hormonal control of territorial aggression in the sexual phase (after nest building but before the start of the parental cycle) seems less clear. Males that are castrated or treated with anti-androgens in this phase show a drop in aggressiveness (Fig. 12.6; Baggerman 1966, 1968; Wootton 1970; Rouse *et al.* 1977), suggesting that androgens influence territorial aggressiveness. The variances within the sham-operated and castrated groups were high in Wootton's experiments, however, and the difference was not significant (Wootton 1970). In additional castration experiments, Wootton (1970) did document a decline in bites directed at stimulus males by empty-nest males measured at weekly intervals following gonadectomy. Unfortunately, no statistics are provided, but the error bars (Fig. 3 in Wootton 1970) suggest

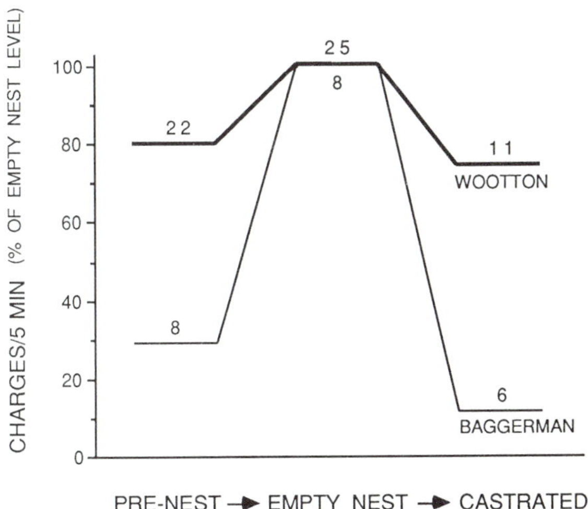

Fig. 12.6 Mean levels of aggression (charges per 5 min) of pre-nest, empty-nest, and castrated (empty-nest) males relative to those of empty-nest males in studies by Baggerman (1966) and Wootton (1970). The numbers of males tested are indicated in the figure. Data were estimated from Wootton, Table I and Fig. 2, and from Baggerman, Fig. 2, 5, and 7 (pre-nest: average of four tests during the 12 d period before nest building; empty nest: average of seven tests in the first 21 d after nest building; castrated: average of seven tests during the 21 d period after castration).

a significant effect of castration on aggressiveness. Baggerman (1966, 1968) did not test the significance of the decline in her experiments, but as her castrations had a greater effect than did those in Wootton's study (Fig. 12.6), the decline is probably significant in spite of the small sample size. Thus, the castration experiments suggest that androgens play a role in regulating aggressiveness at the empty nest stage. It is clear that additional experiments are needed to further test this hypothesis. Gonadectomy can be accomplished with minimal injury because the stickleback male's testes are more accessible than in most other fish species; this makes the threespine stickleback ideal for behaviour–endocrinological research (Borg pers. comm.).

An alternative interpretation of the decrease in aggressiveness following castration is that it could have been caused by loss of the nest rather than changes in androgen levels (Wootton 1970). Removal of a male's nest in the empty nest stage can cause aggression to decline (Symons 1965; Stanley and Wootton 1986), and castrated males do not maintain their nests. Although the nest may be a causal factor in determining the levels of aggression in nesting males, other findings counter the interpretation that the nest is the only causal factor involved. The establishment of dominance relationships between reproductive males in tanks unfamiliar to the contestants can be accompanied by intense fights (Bakker and Sevenster 1983; Bakker 1986). Additionally, parallel changes in kidney size in threespine stickleback males selected for high or low levels of territorial aggression suggest that androgens play a role as well (Bakker 1985, 1986; see below).

The effect of castration was much more pronounced in Baggerman's (1966, 1968) than in Wootton's (1970) study, leading to different interpretations as to the role of gonadotropins in controlling aggressiveness of empty-nest males. In Baggerman's study, the level of aggression of empty-nest males following castration was reduced to very low levels (Fig. 12.6). Her interpretation of this was that there had been a switch in the hormonal control of aggressiveness, such that androgens took over the role of gonadotropins and directly affected aggression during the sexual phase. Wootton (1970) pointed out a problem with this interpretation. He noted that the level of aggression of the castrated males did not drop below that of pre-nesting males (Fig. 12.6), suggesting that the same causal factors that determined pre-breeding aggressiveness were still in effect. Baggerman's data also fit in with this interpretation, because it is unlikely that in her study the levels of pre-breeding aggression and aggression after castration differed significantly.

The disagreement as to whether gonadotropins influence aggressiveness during the sexual phase arose because levels of pre-breeding aggression differed between studies. Apart from some other methodological differences, Baggerman also used fish from a Dutch anadromous population whereas Wootton's fish came from a British freshwater population. It has been suggested several times that population differences caused the discrepancy between the studies (Liley 1969; Munro and Pitcher 1983; Villars 1983). Indeed, parallel differences in the levels of aggression before breeding have been reported for anadromous and freshwater populations from the Netherlands and Japan (see above). It will be argued in the next section that in the case of the Dutch populations, the differences are due to different levels of gonadotropins.

The most plausible overall picture that emerges is that a stickleback male's aggression is controlled by gonadotropins outside the reproductive period, and by both gonadotropins and androgens during the reproductive period. For a proper evaluation of the evolution of aggressiveness in threespine stickleback, it is crucial that we understand the causal mechanisms leading to variation in aggressiveness. The pituitary–gonadal axis plays a key role in this, not only by controlling aggressiveness in different phases in the life cycle, but also by controlling sexual behaviour, the physiological capability for reproduction, and the male's secondary sexual characters (Guderley page 108 this volume). The evolution of aggressiveness may therefore be influenced indirectly through the pituitary–gonadal axis as a consequence of selection on different aspects of phenotype.

The endocrine basis of genetic correlations: constraints on the evolution of aggression?

Differences in life history and morphological traits among the lines selected for different forms of aggression provided evidence of the involvement of two classes of hormones in control of aggression (Bakker 1985, 1986). Juvenile aggressiveness was negatively correlated with the age at sexual maturity among generation means of the low and high juvenile aggression lines (Fig. 12.7; Bakker 1986). Thus, selection for juvenile aggressiveness was accompanied by a change in the age at sexual maturity, such that JH fish matured earlier than JL fish (age at sexual maturity in the third selected generation in days after fertilization: JH males, $\bar{X} \pm \mathrm{SD} = 117.8 \pm 13.8$, $N = 15$; JL males, 130.7 ± 17.3, $N = 11$; JH females, 121.9 ± 15.8, $N = 17$; JL females, 134.6 ± 19.3, $N = 12$; Mann–Whitney U test, one tailed, $U = 46$ and 61, respectively, $P < 0.04$ for both sexes) (Bakker 1986). Additionally, the onset of juvenile aggression (in d after fertilization) was later in JL fish, as assessed in the third selected generation in standardized groups of juveniles (JH, $\bar{X} \pm \mathrm{SD} = 34.7 \pm 1.9$, $N = 6$; JL, 41.8 ± 3.4, $N = 9$; Mann–Whitney U test, two tailed, $U = 0$, $P < 0.002$) (Bakker 1986). Finally, the incidence of female ripeness was lower in JL fish than it was in JH fish (Fig. 12.8). These results suggest that selection for juvenile aggressiveness has acted on the (effective) level of gonadotropic hormones because teleost gonadotropins induce spermatogenesis, spermiation, and testicular steroidogenesis in males, and vitello-

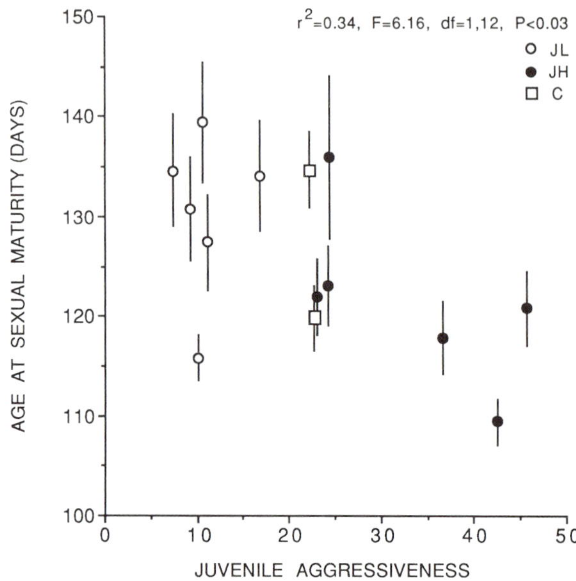

Fig. 12.7 Correlation between juvenile aggressiveness and age at sexual maturity (in days after fertilization). The data points represent the means of all males or females in successive generations of the juvenile high (JH) and low (JL) aggression lines and the second generation of the control (C) line. Sexual maturity was defined for males as completion of the first nest and for females as spawning readiness for the first time. Error bars represent two standard errors of the mean.

genesis, ovarian oestrogen secretion, oocyte maturation, and ovulation in females (reviewed in e.g. Idler and Ng 1983; Ng and Idler 1983). Some of these functions of gonadotropins are impressively demonstrated in the platyfish, *Xiphophorus maculatus*, in which several alleles at a single locus affect the development of the gonadotropic cells in the pituitary gland (e.g. Schreibman and Margolis-Nunno 1987; Schreibman *et al.* 1989). Various combinations of alleles at this locus determine, for instance, the age at which sexual maturity occurs (between 8 and 104 wk).

Attainment of sexual maturity was delayed in the JL line, suggesting that selection on this aspect of life history could produce a correlated response in juvenile aggressiveness (and correlated forms of aggression). Similarly, selection on body size could indirectly influence juvenile aggressiveness because growth slows or stops after attainment of sexual maturity (e.g. Wootton 1976, 1984*a*, Crivelli and Britton 1987; Mori and Nagoshi 1987); later-maturing individuals tend to be larger.

Finally, females in the JH line appeared to mature clutches more rapidly than did females in the JL line. This could indicate a genetic correlation

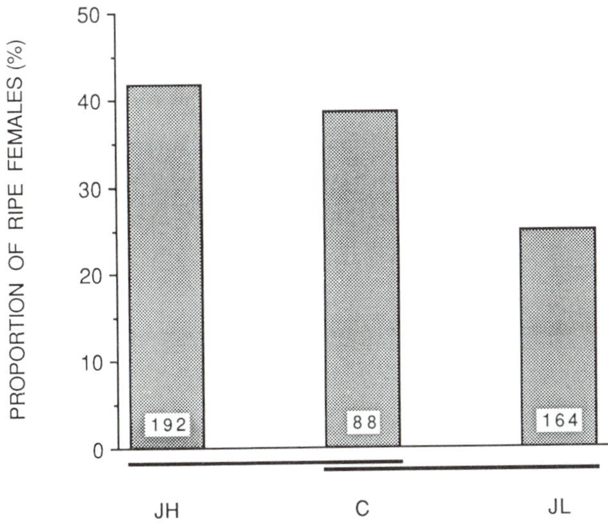

Fig. 12.8 The proportions of ripe females in the high (JH) and low (JL) juvenile aggression lines and in the control (C) line. The proportions were based on the number of ripe females in four female aggression tests (one per week) after the females became ripe for the first time. Data from females in all three generations were pooled within each selection line. The numbers of aggression tests are indicated at the bases of the histogram bars. The overall data set was heterogeneous ($G = 11.7712$, d.f. = 2, $P < 0.005$). Lines below the *x*-axis connect treatments that did not differ significantly at the 0.05 level (STP procedure of Sokal and Rohlf 1981; $G_H < 5.701$, d.f. = 1, $k = 3$).

between aggression and clutch maturation rate, or could simply result from smaller females producing smaller clutches at a higher frequency than larger females (but see Wootton 1973*b*). Data from the JH, JL, and C lines in the second selected generation (Bakker, unpubl. data), where differences among lines in the level of juvenile aggression were greatest (Fig. 12.2), give some support to the former explanation. Females selected to mother the next generation matured later and tended to be larger in the JL line than in the JH line (age at sexual maturity ± SD: JL, 157.0 ± 19.1 d after fertilization, $N = 3$; JH, 108.3 ± 9.7, $N = 3$; Mann–Whitney *U* test, one tailed, $U = 0$, $P = 0.05$; length ± SD: JL, 6.1 ± 0.3 cm; JH, 5.6 ± 0.4; Mann–Whitney *U* test, one tailed, $U = 1.5$, $P = 0.15$) or C line (age at sexual maturity 131.8 ± 17.4 d, $N = 9$, Mann–Whitney *U* test, one tailed, $U = 3.5$, $P < 0.05$; length 5.8 ± 0.2 cm, Mann–Whitney *U* test, one tailed $U = 5$, $P = 0.097$). Although there was some variation in age at which lengths were assessed (JL, 288.7 ± 4.6 d after fertilization; JH, 271.3 ± 18.5; C, 307.9 ± 12.7), this does not account for the observed differences. One to two months before the length assessments (age: JL, 241.0 ± 10.1 d after fertilization; JH, 217.8 ± 28.1; C, 246.7 ± 21.1), JL females also produced larger clutches than did JH females (JL, $\bar{X} \pm$ SD = 180.0 ± 26.6 eggs; JH, 143.5 ± 5.1; Mann–Whitney U test, one tailed, $U = 0$, $P = 0.05$) or C females (132.4 ± 27.1 eggs, Mann–Whitney U test, one tailed, $U = 2$, $P = 0.025$). The smaller clutches produced by JH females seemed to be more than compensated for by a higher clutch maturation rate (Fig. 12.8). In the second selected generation, the difference between per cent ripe JH and JL females during the tests was even greater (Bakker 1986). Thus JL females enjoyed a greater clutch size, but lifetime fecundity was probably greater in JH females.

Selection on territorial aggressiveness seems to have affected the level of androgen production rather than the level of gonadotropin production. TH line males have enlarged kidneys relative to control line males, a condition indicative of elevated levels of androgen production (Wai and Hoar 1963; Moufier 1972; de Ruiter and Mein 1982), whereas TL males have smaller kidneys than those of the controls (Fig. 12.9). Parallel changes in the degree of red breeding coloration in TH and TL males (Fig. 12.5) further support this hypothesis.

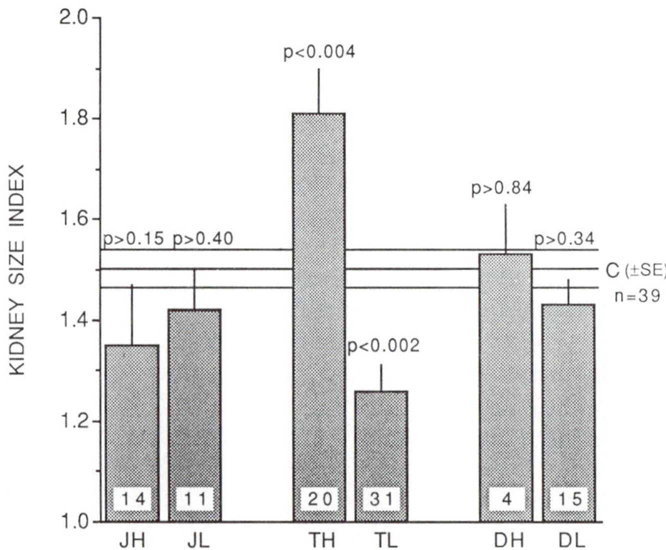

Fig. 12.9 Mean kidney size index of males in the fourth selected generation of the juvenile high (JH) and low (JL) aggression lines, territorial high (TH) and low (TL) aggression lines, and in the high (DH) and low (DL) dominance lines. The numbers of males examined are indicated at the bases of the histogram bars. Error bars indicate one standard error of the mean. Relative kidney sizes in the selected lines were compared with that of the control (C) line by the Mann–Whitney U test, two tailed.

Although genetic correlations between aggressiveness and most aspects of reproductive behaviour have yet to be investigated, their common hormonal control and signalling system (red nuptial coloration of the male) make it likely that some aspects of aggressive and reproductive behaviour will prove to have evolved in concert. The probable hormonal causes of such character correlations are the androgens, which have been implicated as determinants of aggressiveness, nest building, nest-directed activities, courtship behaviour, and the secondary sexual characteristics of the male (Craig-Bennett 1931; Ikeda 1933; Hoar 1962*a*,*b*; Wai and Hoar 1963; Baggerman 1966, 1968; Wootton 1970; Rouse *et al.* 1977; Borg 1981, 1982*a*, 1987; Borg *et al.* 1987*c*, 1989*c*; Andersson *et al.* 1988; Mayer *et al.* 1990*a*; Guderley page 105 this volume). Thus, androgen levels, levels of some forms of aggression, nuptial coloration, and reproductive behaviour tend to covary over the life cycle of the male. It also follows that differences in androgen levels among individuals or populations can also lead to positive correlations between these characters (Rowland 1984; Giles and Huntingford 1985; McLennan and McPhail 1989*b*).

Although some forms of aggression are affected by androgen levels, gonadotropins may affect levels of most forms of aggression and exert a primary effect on some (see above). For this reason, not all forms of aggression are positively correlated with reproductive behaviour. For example, in lines selected for low levels of juvenile aggression (JL) and low levels of territorial aggression (TL), males of the third and fourth selected generations displayed similarly low levels of aggressive activity during courtship. However, JL and TL males differed significantly in direct (courtship intensity expressed as the number of zigzags) and indirect (particular nest-directed activities) measures of sexual tendency; compared with males of the control line or of the corresponding high line, JL males tended to display an enhanced sexual activity, whereas the sexual activity of TL males tended to be reduced (Bakker and Sevenster 1989).

These findings support the interpretation that levels of juvenile aggression are affected primarily by gonadotropin levels, whereas territorial aggression is also affected by androgen levels. Only when territorial aggression was selected against, was there a parallel correlated response in sexual activity, a character known to be affected by androgen levels (see above). Because there exists a mutually inhibitory relationship between the tendency to behave aggressively towards a stimulus and the tendency to behave sexually over a short time period and within an individual male stickleback (Sevenster 1968, 1973; Peeke 1969; van den Assem and van der Molen 1969; Wilz 1972; Rowland 1988; Bakker and Sevenster 1989), a one-sided reduction of the aggression level as in JL males causes an increase in sexual activity. Thus, although males from both selection lines were equally aggressive during courtship, the different selection regimes had opposite effects on sexual activity.

A comparison of the Dutch freshwater and anadromous populations (see above) gives some insight into the evolutionary implications of the different involvement of gonadotropins and androgens in the aggressive and sexual behavioural systems. The laboratory-bred freshwater fish were more aggressive and territorial during the juvenile stage (Fig. 12.1), and freshwater males had higher levels of territorial aggression but lower levels of courtship activity (see above). The differences in levels of territorial aggression did not appear to be attributable to differences in androgen levels, as mean kidney size in the anadromous population (kidney size index ± SD = 2.06 ± 0.32; $N = 25$) did not differ significantly from that in the freshwater population ($\bar{X} = 2.07 \pm 0.22$; $N = 37$; Mann–Whitney U test, one tailed, $U = 455$, $Z = 0.108$, $P > 0.45$). Because laboratory-reared individuals from the two populations differed markedly in their levels of aggression during the juvenile stage (Fig. 12.1), it is likely that differences in the level of (or sensitivity to) gonadotropins were responsible for the difference in territorial aggressiveness between the populations. The higher sexual activity of the anadromous males agrees with this interpretation. These results suggest that selection for reduced or enhanced aggression in juvenile stickleback has implications for aggression and sexual activity of mature fish.

Aggressive and reproductive behaviour also appear to be linked by the dual signalling role of the male's red breeding coloration in at least some populations. Early research in which dummies were presented inside the territories of males demonstrated an aggression-releasing effect of the red undersides of male threespine stickleback (ter Pelkwijk and Tinbergen 1937; Tinbergen 1948, 1951; see also Baerends 1985; Collias 1990), causing Tinbergen to conclude that this coloration was the primary sign stimulus for territorial aggression in this species. This interpretation has been questioned, however, because several recent studies failed to detect an aggression-releasing effect of the red belly (Rowland page 300 this volume).

One explanation for the differences among studies may lie in the intimidating effect of the male's red coloration. Tinbergen (1948) found that a dummy with a red belly was more effective in inhibiting rival males from entering the territory than was a non-red dummy. More recent research has suggested that dominance ability of males may also be correlated with the development of the red coloration (Bakker and Sevenster 1983; Bakker 1985, 1986), but again, not all studies have documented such a relationship (FitzGerald and Kedney 1987; Rowland 1989*a*).

Intriguingly, the successful selection for low dominance ability in male stickleback (Fig. 12.3) was paralleled by a significant reduction in the degree of red breeding coloration (Fig. 12.10(a),(b); Bakker 1986) (high versus low in generation 3, Mann–Whitney U test, one tailed, $U = 43.5$, $N_1 = 14$, $N_2 = 13$, $P < 0.01$ after aggression; $U = 56$, $N_1 = 13$, $N_2 = 15$, $P < 0.03$ after courtship). The DH line showed no increase in dominance

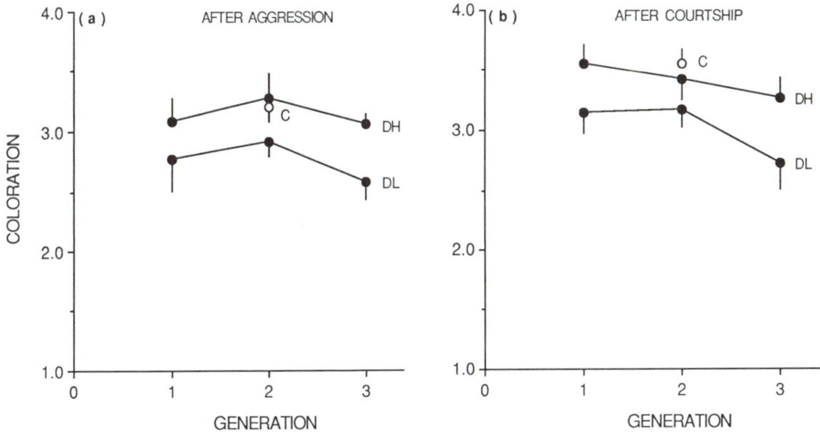

Fig. 12.10 The mean degree of red breeding coloration of males in three successive generations selected for low (DL line) and high (DH line) levels of dominance ability, and in the second generation of the control (C) line. Degree of red coloration was scored (a) after presentation of a male in a tube and (b) after presentation of a ripe female in a tube. Error bars indicate one standard error of the mean.

ability (Figs. 12.3, 12.5(a),(b)), or in brightness (high versus control in generation 2, Mann–Whitney U test, two tailed, $U = 78.5$, $N_1 = 11$, $N_2 = 16$, $P > 0.63$ after aggression; $U = 73$, $N_1 = 11$, $N_2 = 16$, $P > 0.44$ after courtship). Notice also the difference in the degree of red coloration of all males, irrespective of their origin, after presentation of a ripe female (Fig. 12.10(b)) and of a rival male (Fig. 12.10(a)); females make males flush more strongly (Wilcoxon matched-pairs signed-ranks test involving generations of the DH, DL, and C lines, $T = 629$, $N = 83$, $Z = 5.070$, $P < 0.001$). This difference points to the role of red coloration in intersexual selection (see below).

Bidirectional selection on territorial aggression also produced parallel, bidirectional differences in degree of red coloration (Fig. 12.5(a),(b)), suggesting that this character can be modified through two genetically distinct physiological pathways (Bakker 1986). The colour change in the low dominance line is likely due to androgen-independent factors (Fig. 12.9) that, for example, influence the concentration and/or distribution of red and other pigments, rather than to the probably androgen-dependent factors that influence coloration in the territorial aggression lines.

Some of the discrepancies among studies on the aggression-releasing effect of red coloration are probably due to methodological differences, because the red coloration on a male or on a dummy can either elicit aggression or inhibit attack. Differences in the quality of the red colour or in presentation could alter the response of territorial males (Rowland page 300 this volume). Alternatively, the discrepancies could reflect population differences. The only study that compared the response of males from different populations towards dummies with and without a red belly did find, however, a similar intimidating effect of a red belly in males from different populations, i.e. in males from a marine (brackish) population sampled on Long Island, New York, USA, and in those from a mixture of Dutch freshwater and anadromous males (Rowland and Sevenster 1985).

In contrast, there is evidence of a difference in the intimidating effect of red coloration between a Dutch freshwater and a Dutch anadromous population. Laboratory-reared anadromous males tended to be slightly, but significantly brighter than laboratory-reared freshwater males, suggesting a genetic basis to the difference. Their dominance ability was, however, less than that of their freshwater counterparts (Table 12.1). When linear orders of dominance were constructed for the males in each population, there existed a significant positive correlation between the degree of red coloration and dominance ability in the freshwater population (Fig. 12.11(a)) but not in the anadromous population (Fig. 12.11(b)). The former correlation was not attributable to particular males (jackknifing correlation coefficients ranged from 0.61, $N = 9$, $P < 0.05$, to 0.85, $N = 9$, $P < 0.01$). Body size and aggressiveness were both uncorrelated with dominance ability in these populations (Bakker and Feuth-de Bruijn unpubl. data).

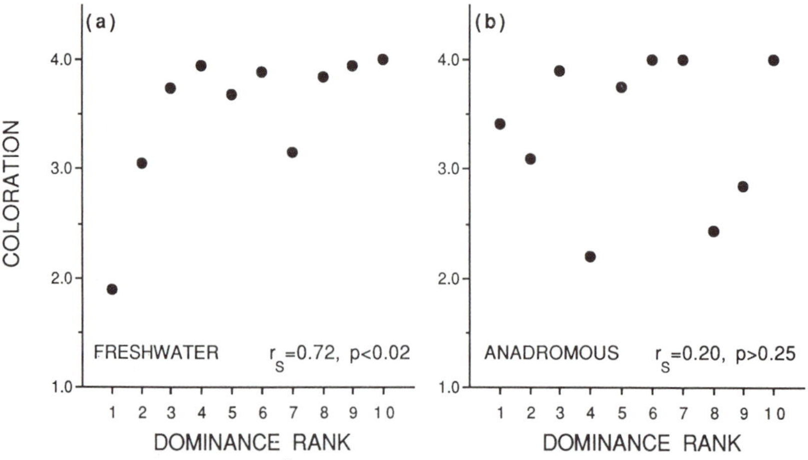

Fig. 12.11 The relationship between dominance rank and degree of red breeding coloration among ten males from (a) a Dutch freshwater population and (b) a Dutch anadromous population. Dominance rank was assessed in dyadic combats between isolated males, and mean degree of red coloration was quantified on a four-point scale after each dominance test upon stimulation with an enclosed ripe female. r_s is the Spearman rank correlation coefficient.

The male's red coloration has also been shown to function in intersexual selection in threespine stickleback (see also Rowland page 317 this volume). Ter Pelkwijk and Tinbergen (1937) noticed that there was little response of ripe females to male dummies that lacked a red belly. Subsequently, Semler (1971) demonstrated that females from Wapato Lake, Washington, USA, a population in which males were polymorphic for the development of red nuptial coloration, preferred red over non-red males. It was only recently shown that female stickleback prefer the more intensely red-coloured male in simultaneous (McLennan and McPhail 1990; Milinski and Bakker 1990) and sequential choice situations (Bakker and Milinski 1991). Using fish from a Swiss freshwater population (near Roche/ Montreux; probably anadromous fish introduced at the end of the 19th century, Bakker unpubl. data) in choice tests under two light conditions, Milinski and Bakker (1990) showed that female mating preference is based almost exclusively on the intensity of the male's coloration, which reveals his physical condition.

Clearly then, red male nuptial coloration functions as a signal both in aggressive interactions with conspecifics and in mate choice. Although it has been suggested that the evolution of colour patterns in the Gasterosteidae is more strongly correlated with intersexual selection than with intrasexual selection (McLennan *et al.* 1988), the apparent signal function of the red nuptial coloration in aggressive interactions among males suggests that in threespine stickleback, intrasexual selection could also have affected the evolution of this trait. This conclusion obviously applies only to the intimidating effect of red, because it is very unlikely that its aggression-releasing effect has played a role in the evolution of red breeding coloration in threespine stickleback; it may be rather disadvantageous to its bearer to provoke aggression of rivals. Equally clearly, because red nuptial coloration serves a signal function in both intrasexual aggression and courtship, selection on either could produce a correlated change in the other. As a result, it may prove difficult to discriminate the relative importance of selection on the two signal functions of this character.

NATURAL SELECTION, SEXUAL SELECTION, AND THE EVOLUTION OF AGGRESSIVENESS

Among the sticklebacks, *G. aculeatus* has the best developed morphological defence mechanism against vertebrate predators (e.g. Hoogland *et al.* 1957; Nelson 1971; Bowne Chapter 2 this volume) and at the same time the highest levels of aggression and the most pronounced breeding coloration (e.g. McLennan *et al.* 1988). There exists a consensus that predation is a major selective force in both morphological and behavioural evolution of the threespine stickleback (Reimchen Chapter 9; Huntingford *et al*, Chapter 10 this volume). Because of the well-developed morphological and behavioural defences observed in many populations, it has been argued that the intensity of predation may be lower in this species than in other species of sticklebacks (but see Reimchen page 254 this volume, for evidence of heavy predation loads in some populations). On this basis it has been suggested that the relative freedom from predation might have facilitated the emancipation of reproductive males from nesting in areas of dense vegetation to the open (Morris 1958; Wilz 1971; Wootton 1976, 1984*a*) where competition for females would be more intense.

If correct, this scenario suggests that relative freedom from predators, in some populations at least, could have permitted the evolution of male traits that enhance their competitive abilities (e.g. high aggression levels) and attractiveness towards females (e.g. red breeding coloration). As these characteristics impose temporal and energetic costs and increased predation risk, there must be a net advantage to nesting in breeding aggregations in open habitats. Under laboratory conditions, males with near neighbours develop more intense red breeding coloration (Reisman 1968*b*, Sargent 1985), and are more motivated to court females late in the brood cycle (Sargent 1985) than are solitary males. Furthermore, males in nesting aggregations in nature may be able to attract more females (Kynard 1978*a*; Mori 1990*b*) and to detect sneakers at a greater distance (Mori 1990*b*) than can solitary males. These factors could result in higher lifetime reproductive success for gregarious males (Sargent 1985; Sargent and Gross 1986).

The costs of aggression

There can be little question that aggressiveness should be subject to natural selection, at least in part because aggression is energetically costly. Chellappa and Huntingford (1989) demonstrated that participation in a fight of more than a few seconds' duration left both males with depleted liver glycogen. In short fights, the level of liver glycogen was lowest in the defeated male. As the fights increased in length, the liver glycogen levels decreased, as did the difference in levels between the males. Similarly, under laboratory conditions, males that held territories in isolation lost weight less rapidly during the brood cycle than did males with neighbours (males were separated by a transparant partition), suggesting a cost of territorial aggression (Sargent 1985; Sargent and Gross 1986). Males that were fed more expanded the territories they defended (Stanley and Wootton 1986).

So far as involvement in aggressive interactions affects physical condition, it may affect the intensity of breeding coloration (Milinski and Bakker 1990) and hence courtship success (see above). The fitness of reproductive males should be enhanced by minimizing the energetic costs they incur during aggressive interactions. The frequency and intensity of aggressive interactions could be reduced through habituation of aggression against familiar rivals (neighbours), allowing males to maintain intense red breeding coloration and high levels of sexual activity (see below; Rowland 1988).

Habituation is probably only possible after neighbouring males have assessed the fighting abilities of one another when these abilities are signalled reliably (e.g. through the intensity of breeding coloration, which is positively correlated with physical condition; Milinski and Bakker 1990) and when aggressive intentions are signalled honestly (Losey and Sevenster 1991).

For other reasons as well, high levels of aggression are not always advantageous. Under laboratory conditions, when territories are being established, highly aggressive males are not more likely to obtain territories than are less aggressive males (Muckensturm 1969; Bakker and Sevenster 1983; Bakker 1986; FitzGerald and Kedney 1987; Rowland 1989*a*). Red breeding coloration (Bakker and Sevenster 1983; Bakker 1986) or large body size (Rowland 1989*a*) may be more important determinants of the outcome of competition for territories.

Furthermore, costs of aggression probably include increased risk of predation owing to reduced vigilance and increased conspicuousness, and interference with courtship activity (see below).

Territorial aggression and determinants of reproductive success

Territorial aggression and territory size

High aggression levels at relatively great distances from the nest do appear to be necessary for maintenance of a large territory under laboratory conditions (van den Assem 1967; Black 1971; Stanley and Wootton 1986; Ward and FitzGerald 1988), although evidence for a positive association between territory size and aggressiveness is mixed. In laboratory studies, the relationship was positive (van den Assem 1967; Black 1971). In the only field study of this relationship, the frequency of aggressive interactions between rival males was inversely related to territory size (Goldschmidt and Bakker 1990).

The differences in the relationship between aggressiveness and territory size that have been documented in different studies could be due to differences among the study populations. This is impossible to determine, however, because of the variation in research methods. Differences in methods of measuring aggression can substantially alter the results. For example, Black (1971) found that aggressiveness was positively associated with territory size, and that changes in aggressiveness paralleled changes in territory size during the parental phase, when aggressiveness was scored as the number of aggressive interactions between males sharing an aquarium. In contrast, tests in which a rival male was presented in a tube 20 cm from the nest and aggression was measured as the number of bites and bumps per 5 min revealed no correlation between territory size and aggressiveness (Black 1971).

Additionally, the measure of aggressiveness differed among studies, as did the accuracy with which these different measures were likely to reflect male aggressiveness. For example, counts of the number of aggressive interactions between males and freely moving conspecifics are likely to be affected by immediate environmental conditions, such as the amount of cover near a nest or the number of foraging individuals moving through the area at the moment. A final problem with comparison of these studies is that the relationships between aggression and territory size were scored at different stages in the reproductive cycle. A further complication is that a large body size and intense red breeding coloration probably contribute to the maintenance of a large territory by signalling superior fighting ability, in the case of coloration through its positive correlation with physical condition (Milinski and Bakker 1990; also implicitly suggested by studies of Rowland 1984; McLennan and McPhail 1989*b*). The relationship between male aggressiveness and territory size under natural conditions could best be resolved with controlled field experiments.

Reproductive tactics

In homogeneous habitats, superior reproductive success is associated with large territories, both in the laboratory (van den Assem 1967; Black 1971; Li and Owings 1978*a*) and in the field (Goldschmidt and Bakker 1990). Males with large territories experienced greater mating success (van den Assem 1967; Li and Owings 1978*a*) and reared more progeny (van den Assem 1967; Black 1971) in the laboratory than did males with smaller territories. Similarly, males with large territories in a trout pond in the Netherlands experienced greater mating success and suffered less egg raiding than did those with smaller territories (Goldschmidt and Bakker 1990).

The advantages of large territories include lower rates of aggressive interactions initiated by neighbours and lower rates of sneaking by rival males (van den Assem 1967; Goldschmidt and Bakker 1990). These interactions decrease reproductive success directly by interrupting courtship and through loss of fertilizations to sneaking males that enter the nest and release sperm over the eggs. In addition, these interactions stimulate aggressive behaviour in the owner of the territory, thereby slowing courtship (van den Assem 1967; Black 1971; Li and Owings 1978*a*,*b*; Ridgway and McPhail 1987). As there exists a mutually inhibitory relationship between the tendencies to show aggressive and sexual behaviour (see above), a high aggression level decreases a male's willingness to lead a ripe female to his nest. Recent aggressive interaction causes a prolongation of courtship and an increased risk that the female will depart before spawning (e.g. van den Assem 1967; Wilz 1970*a*,*b*, 1972; Borg 1985; Ward and FitzGerald 1987).

This negative relationship between aggressive and reproductive behaviour was also evident in tests of female mate choice when a small territory size was imposed by the design of the experiment. Aggressive males experienced lower spawning success than did those that were less aggressive, because females departed before spawning (Ward and FitzGerald 1987). It is clear not only that aggression is energetically costly, but that during the courtship phase of the reproductive cycle, high levels of aggressiveness are also costly in terms of lost mating opportunities. As a result, selection against high levels of courtship aggression could produce decreases in other correlated aspects of aggression, holding them at lower levels than would otherwise be favoured by selection.

In homogeneous habitats, selection thus favours large territory size because it assures relatively undisturbed courtship. In contrast, FitzGerald (1983) and FitzGerald and Whoriskey (1985) documented an inverse relationship between internest distance (an indirect measure of territory size) in the field and the number of eggs in nests in a salt-marsh population in Quebec, Canada, which they interpreted as evidence of an inverse relationship between territory size and reproductive success. The negative correlation was, however, established in a heterogeneous habitat where the nests of males with small territories were protected by the banks of the tidal pools. Several studies have shown that concealment of nests increases reproductive success (Moodie 1972*b*; Kynard 1978*a*; Sargent and Gebler 1980; Sargent 1982; Mori 1990*b*). Nest concealment can seemingly also reduce the frequency of aggressive interactions during courtship and paternal care (Sargent and Gebler 1980). Nest concealment also decreases the frequency of courtship interference and the probability of stolen fertilizations and egg raids by sneakers (Sargent and Gebler 1980). Under such circumstances, selection for high levels of territorial aggression may be reduced.

The above review of the relationship between territorial aggressiveness and reproductive success opens the possibility of the existence of two reproductive tactics: a 'nest concealment' tactic and a 'large territory size' tactic, both leading to superior reproductive success by way of reduced interference from rivals. The nest concealment tactic would require neither such great dominance abilities (large body size and/or intense red breeding coloration) nor such high aggression levels as the large territory size tactic would. The nest concealment tactic would likely have a relative advantage in populations that are subject to a high vertebrate predation pressure (Huntingford *et al.* Chapter 10 this volume), whereas insect predators would likely reverse this advantage (Foster *et al.* 1988). Additionally, in populations where parasites exert a major influence, the large territory size tactic would possibly be favoured through intensified intersexual selection for bright red breeding coloration (Hamilton and Zuk 1982; Milinski and Bakker 1990). The presence of both strategies within stickleback populations could be one of the causes for a sustained genetic variation of aggressiveness.

Some support for the possible existence of two reproductive tactics comes from observations on a benthic–limnetic species pair of threespine sticklebacks in Enos Lake, British Columbia, Canada (McPhail page 421 this volume). In this lake, territorial males of the limnetic species, which nests in open habitats, are frequently involved in aggressive interactions with conspecific neighbours. Such interactions are rare between benthic males that nest in vegetation. Consequently, courtship in the limnetic species is protracted relative to that in the benthic species because of a high frequency of interruptions by rival males (Ridgway and McPhail 1987).

In order to judge the impacts of different selection regimes on aggression, field studies that directly compare the fitness of different aggressive phenotypes in different contexts or in different populations are essential. No such studies have been conducted yet, although the threespine stickleback should be an excellent subject for such research. This small fish can readily be observed *in situ* in many aquatic habitats (e.g. Kynard 1978*a*; FitzGerald 1983; Borg 1985; Foster 1990), and different populations are exposed to differing selection regimes known to have produced divergence in other behavioural phenotypes (e.g. Huntingford *et al.* Chapter 10; Foster Chapter 13 this volume). The results discussed above on interpopulation differences in aggression suggest that at least some aspects of aggressive behaviour have also diverged among populations, but more research, involving field and laboratory observations on more populations, is needed.

Life history mode and the evolution of aggression

Differences in aggressiveness of individuals from two pairs of anadromous and freshwater populations suggest that there are differences in aggressiveness among populations that reflect the selection regimes to which they are exposed (Honma and Tamura 1984; Bakker and Feuth-de Bruijn 1988). The lower levels of juvenile aggression measured in both anadromous populations relative to the freshwater populations could be favoured because juveniles in anadromous populations migrate to the sea in large shoals (e.g. Daniel 1985). High levels of intraspecific aggression might be disadvantageous in shoals, and this disadvantage could account for the seemingly lower aggression of anadromous stickleback (Bakker and Feuth-de Bruijn 1988), although freshwater juveniles are also known to participate in large feeding schools (e.g. Foster *et al.* 1988). Additional field and laboratory studies are necessary to determine the robustness of this seeming dichotomy between freshwater and anadromous threespine stickleback and to ascertain the causes.

In the Dutch comparison (Bakker and Feuth-de Bruijn 1988), juveniles in the freshwater population were also more likely to establish territories in the laboratory. Although territoriality of juveniles has not been documented in the field, MacLean (1980) has observed defence of feeding territories in a Canadian lake outside the breeding season. Certainly, juvenile territoriality is more likely to be favoured in non-migratory than in migratory populations. Parallel differences in aggressiveness and territoriality have been observed in juvenile salmonids that migrate to sea immediately after hatching and those that delay migration (Keenleyside 1979).

Anadromous and freshwater populations should exhibit differences in life history and behavioural characters, owing to the higher costs of a migratory lifestyle (e.g. Stearns 1976; Roff 1988). There is evidence for the existence of a so-called migratory life history syndrome, in which the relevant life history characters show positive genetic correlations (e.g. Dingle 1988). Several life history characters have been compared between anadromous and freshwater stickleback populations and they support the predictions (wild-caught fish: e.g. Hagen 1967; laboratory-bred fish: Snyder and Dingle 1989, 1990; Snyder 1990; Bakker and Feuth-de Bruijn unpubl. data). Lines selected for enhanced and reduced levels of juvenile aggression suggest that aggressiveness is genetically correlated with this migratory life history syndrome through gonadotropins (see above). The evolution of juvenile aggressiveness (and correlated forms of aggressiveness) may therefore be influenced indirectly through selection on life history traits, such as body size and age at reproduction.

Comparisons of marine (or anadromous) and freshwater populations have the potential to provide insight into the directions of evolutionary change, because freshwater populations are thought to have been derived from the marine form (e.g. McPhail and Lindsey 1970; Bell 1976*a*; Bell and Foster page 14; McPhail page 401 this volume). For example, all of the populations in which the intensity of red nuptial coloration proved not to be an important determinant of dominance ability among males were anadromous populations (FitzGerald and Kedney 1987; Rowland 1989*a*; Bakker and Feuth-de Bruijn unpubl. data). The only population that has been studied in which coloration played a prominent role in determining dominance relationships was a freshwater population (Bakker and Sevenster 1983; Bakker 1985, 1986; Bakker and Feuth-de Bruijn unpubl. data). It is therefore possible that in marine populations, nuptial coloration functions primarily to determine patterns of mate choice (intersexual selection), and that the signal function of this coloration in dominance contests is a derived condition. This possibility could be tested in a systematic study of the signal function of the male's coloration across a large number of marine and freshwater populations.

CONCLUSIONS

Although a limited number of studies have investigated genetic influences on variation in aggressiveness, there is clear evidence for genetically based variation in aggressiveness within and among threespine stickleback populations. Aggressiveness is part of a complex character suite in which different forms of aggressiveness, sexual behaviour, the intensity of red breeding coloration, and several life history characters are genetically correlated to varying degrees. The genetic correlations in this complex are partly based on the multiple influences of hormones of the pituitary–gonadal axis. The evolution of aggressive behaviour can only be understood by taking this complex into account.

Different determinants of variation in reproductive success have been established, both in laboratory and in field studies. They suggest the possible existence of two reproductive tactics: a 'nest concealment' tactic and a 'large territory size' tactic, both leading to superior reproductive success by reducing interference from rivals. The tactics probably require different dominance abilities and territorial aggression levels, and their relative advantages are likely to depend on the presence and abundance of vertebrate and insect predators.

In addition, selection processes acting on aggressiveness in the juvenile stage and of adult females also have some impact on the evolution of aggressiveness in stickleback. The importance of aggression in juveniles and females leads to the notion that different life history modes, which are accompanied by differences in life history and behavioural characters (e.g. juvenile aggressiveness), may have hitherto unrecognized evolutionary consequences for stickleback aggression.

The proposed influences of different reproductive tactics and life history modes on the evolution of stickleback aggression generate testable hypotheses for future research.

ACKNOWLEDGEMENTS

I thank Gerry FitzGerald, Bill Rowland, and two anonymous reviewers for their comments. Enja Feuth-de Bruijn kindly permitted me to include our unpublished data. Manuscript preparation was supported by the Swiss National Science Foundation.

13

Evolution of the reproductive behaviour of threespine stickleback

Susan A. Foster

Ethologists have been interested in understanding evolutionary processes that structure behavioural patterns since the turn of the century (e.g. Whitman 1899; Heinroth 1911), or even earlier if Charles Darwin may be included among the ranks of ethologists (Darwin 1872). Typically, comparisons have been made among closely related species in an effort to examine historical sequences in phenotypic evolution and to determine the extent to which behavioural phenotypes have been influenced by phylogenetic history. These efforts have demonstrated clearly that at least some behavioural phenotypes are strongly conserved over long time periods in phylogeny and can therefore reflect the patterns of phylogenetic diversification within a clade (reviewed in Lauder 1986; McLennan *et al.* 1988; Burghardt and Gittleman 1990; Brooks and McLennan 1991).

It is only recently that comparisons of behavioural phenotypes have been extended to include comparisons among allopatric populations of a single species. Because geographic variation often reflects present adaptation to local conditions (e.g. Turesson 1922; Endler 1977, 1986; Riechert 1986) these comparisons have most often been used to infer the adaptive values of particular behavioural patterns (e.g. Arnold 1977; Riechert 1986; Hedrick and Riechert 1990; Huntingford *et al.* Chapter 10 this volume, and references therein). Additionally, they have been used to test evolutionary models (Riechert 1986) and to gain insight into the causes of character displacement (e.g. Bell 1976*b*; Littlejohn and Watson 1985; Schluter and McPhail 1992) and speciation (Verrell and Arnold 1989; Foster 1990, unpubl. data; Houde and Endler 1990).

Although most interpopulation comparisons have assumed that the populations were adapted to local conditions, in several instances in which the historical relationships among the populations could be inferred (using geological evidence, evidence from faunal distributions, or molecular data) it has been possible to reconstruct the evolutionary history of specific behavioural patterns (e.g. Cruz and Wiley 1989; Goldthwaite *et al.* 1990; Thompson 1990; Towers and Coss 1990; Foster unpubl. data). Such comparisons are valuable not only because they strengthen adaptive inference, but also because, ultimately, they should allow us better to understand how complex behavioural phenotypes evolve and how they dissipate under conditions of disuse.

The threespine stickleback, *Gasterosteus aculeatus*, and the sticklebacks (Gasterosteidae) as a group are superb subjects for research on the evolution of reproductive behaviour. Although the threespine stickleback is the best known, each of the sticklebacks has been the subject of at least one detailed study of reproductive behaviour (reviewed in Wootton 1976, 1984*a*; Rowland Chapter 11 this volume). Because the phylogenetic relationships among the gasterosteids have been examined using osteological, biochemical, and karyotypic data (Bowne Chapter 2 this volume), it has been possible to examine the evolution of reproductive behaviour in this group in a phylogenetic context (McLennan *et al.* 1988).

Additionally, there is substantial variation in behaviour among threespine stickleback populations (see below; Chapters by Hart and Gill, Huntingford *et al.* and Bakker), and in some cases, speciation has occurred (see below; McPhail Chapter 14 this volume). In several regions where the morphological and behavioural diversity of freshwater populations is particularly great, interpopulation differences are thought to have evolved postglacially (Bell and Foster page 16; McPhail page 401 this volume). As freshwater populations must have been derived from the relatively uniform marine or anadromous populations, the character states in the latter can be inferred to be the primitive (plesiomorphic) states for characters that are divergent in freshwater populations. It should thus be possible to interpret the directions and patterns of behavioural evolution in the freshwater radiation (see Bell and Foster page 20 this volume).

In this chapter, I review the evolution of reproductive behaviour in the Gasterosteidae, with emphasis on the threespine stickleback. I have not attempted to review every published speculation on the evolution of threespine stickleback reproductive behaviour. Many suggestions have been derived exclusively from laboratory studies of behaviour in small aquaria. Particularly in discussing adaptive value, I have included only those hypotheses for which there is empirical support from field studies as well. In consequence, this review is somewhat selective, and some interesting and plausible hypotheses have not been considered.

THE REPRODUCTIVE BEHAVIOUR OF THE THREESPINE STICKLEBACK

Here I present a brief overview of the reproductive behaviour of the threespine stickleback that should enable the reader to follow the ensuing discussion. Additional details and references can be found in Rowland (Chapter 11 this volume), and Bakker (Chapter 12 this volume).

Breeding of the threespine stickleback typically is initiated at age 1 or 2 yr and occurs during a few spring and summer months (Baker page 155 this volume) when males move into shallow water from aggregations feeding in deeper water. At the onset of the breeding season, males in most populations develop red throats and fore-bellies, and blue colour in the eyes and on the dorsolateral surfaces. However, expression of this nuptial coloration varies among populations of *G. aculeatus* (Reimchen 1989), and radical departures from this pattern occur (e.g. McPhail 1969; Moodie 1972*a*; Blouw and Hagen 1990). Males establish territories in which they build tubular nests comprising vegetation glued with a secretion from the kidneys. Establishment of territories typically entails chasing and biting as well as frequent ritualized interactions, including a head-down threat, in which the male turns his body broadside to his opponent with spines erect, and circular fighting, in which the males circle one another rapidly.

Courtship is a complex ritual in which behavioural cues are exchanged between the male and the female. It is often initiated when the male swims toward the female with a rapid series of side-to-side jumps known as the zigzag dance (Fig. 1.4). If the female is receptive, he leads her to his nest, inserts his snout in the nest, and rolls on his side (showing behaviour). The female enters the nest beneath him and releases all her eggs before emerging visibly thinner. Finally, the male passes through the nest, fertilizing the eggs, and then chases the female from his territory.

This sequence can be interrupted or replaced by dorsal pricking, in which the male is positioned below the female as both move forward (Fig. 1.5). The male periodically backs slightly and pricks the female's belly with his erect dorsal spines. This typically deters her from approaching the nest. Instead, she waits in a head-up posture until the male leads her to the nest or she leaves without spawning. Occasionally, spawning occurs without the zigzag dance or dorsal pricking. In these instances, the male swims straight to the female and bites her or hits her on the side before leading her to the nest.

Males court and spawn with multiple females over a period of 1–4 d, and then shift to the parental phase, in which they provide all care for the young. During this time they fan with their pectoral fins, providing oxygenated water to the developing embryos, and defend them from predators. They also modify the nest as the embryos mature, finally producing a nest pit, or tangled mass of vegetation in which the newly hatched fry rest. They may continue to defend the fry up to 2 wk after hatching.

INTERGENERIC COMPARISONS AND PHYLOGENETIC ANALYSIS

The infrequent use of behavioural phenotypes for addressing evolutionary questions may stem from the widespread perception of behaviour as ephemeral and highly sensitive to environmental variation. Although concerns of excessive behavioural lability are retained by some investigators (e.g. Parsons 1972; McClearn and DeFries 1973), others have argued that if behavioural traits are appropriately selected and analysed they can even be used as valuable taxonomic and systematic characters (e.g. Whitman 1899; Heinroth 1911; Lorenz 1950; Eberhard 1982; Dobson 1985; McLennan *et al.* 1988).

Because behavioural characters had fallen into disfavour as sources of information for systematic analysis by the middle of this century, it was not until recently that a cladistic analysis of the Gasterosteidae was conducted using behavioural characters alone (McLennan *et al.* 1988). The analysis was based primarily on reproductive and aggressive behaviour patterns that are expressed on the breeding grounds, although nest structure (an expression of nest-building behaviour) and male nuptial coloration (a more morphological character) were also included. The behavioural characters used were those expressed during male–male agonistic encounters, nest building, courtship, spawning, and parental care. All the genera of sticklebacks were included in the analysis, as were both long-established species of *Gasterosteus* (Fig. 2.13). *Aulorhynchus flavidus* represented the evolutionary outgroup, the Aulorhynchidae (tube-snouts), because it was the only species for which behavioural data were available.

The behavioural analysis yielded a cladogram that was congruent with the best-supported cladistic analysis of the Gasterosteidae based on osteological, karyotypic, and allozyme data (McLennan *et al.* 1988; Fig. 2.13 of Bowne page 45 this volume). McLennan *et al.* argued not only that this congruency demonstrated the value of behavioural characters for systematic analysis; but also that the behavioural data provided a more complete picture of the phylogenetic relationships among the gasterosteids than did other kinds of data because they resolved a trichotomy involving *Pungitius*, *Culaea*, and *Gasterosteus*. Bowne (page 50 this volume) has, however, argued that it is prudent to regard this group as unresolved, because the behavioural analysis conflicts with karyotypic data (Chen and Reisman 1970) and *Culaea* was not included in the only allozyme study conducted to date (Hudon and Guderley 1984).

Although behavioural data can certainly provide information useful to systematists, as argued by McLennan *et al.* (1988), it would not be surprising if the behavioural data failed to provide an accurate rendition of the phylogeny of the Gasterosteidae. A fundamental assumption of cladistic analysis is that most of the characters used to construct a cladogram are free to evolve independently of one another (e.g. Felsenstein 1985; Mayr and Ashlock 1991). This assumption is exceedingly unlikely to be met when all the characters are functionally related reproductive characters, and indeed, McLennan *et al.* (1988) argue that a number of the characters used in their analysis have evolved in concert.

An additional problem is that reproductive behaviour has been characterized in very few populations in the majority of stickleback species (references in Wootton 1976, 1984*a*; McLennan *et al.* 1988). If the other species are as variable in reproductive behaviour as the threespine stickleback is (see below), it is possible that sampling error has led to an incorrect characterization of the primitive reproductive behaviour of some species. *Pungitius*, at least, varies considerably among populations in several reproductive characters used by McLennan *et al.*, including nuptial coloration, nest construction, and courtship behaviour (Morris 1958; McKenzie and Keenleyside 1970; Griswold and Smith 1972, 1973).

Once the phylogenetic relationships among the gasterosteids are more highly corroborated and behavioural data have been collected in the field for a number of populations of each species, it will be possible to obtain a clear picture of the patterns of evolutionary change in behavioural phenotypes within this clade. Even without this resolution, comparative analyses of many aspects of the reproductive behaviour of the gasterosteids have provided insight into the evolution of specific behavioural characters (Hall 1956; Wootton 1976; McLennan *et al.* 1988; McLennan 1991). For example, all gasterosteids have polygynous mating systems in which the male is territorial, builds a nest, and provides all parental care. In contrast, nuptial coloration and some aspects of nest structure and of courtship and parental behaviour vary considerably among *G. aculeatus* populations (see below).

The gasterosteids should be particularly interesting in this context because so many aspects of their behaviour are highly ritualized. Some ritualized behaviours, such as the head-down threat, are found in all the genera except *Spinachia*, the primitive sister group to the other gasterosteids (Wootton 1976; McLennan *et al.* 1988). This suggests that, once evolved, this behaviour has been retained relatively unchanged in all the derived species. In contrast, the zigzag dance, observed in all but *Spinachia* and *Culaea*, has been modified structurally in each of the species (McLennan *et al.* 1988). It is only through such comparative analyses that we will be able to ascertain how highly structured behavioural patterns change over evolutionary time. It is also through analyses such as these that we will be able to develop an understanding of the origins and histories of the assemblage of primitive and derived behavioural character states that constitute the behavioural repertoire of the threespine stickleback.

INTERPOPULATION COMPARISONS

Despite the extensive early research on the threespine stickleback by ethologists (Rowland Chapter 11 this volume), the remarkable interpopulation diversity of some aspects of reproductive behaviour has been recognized only recently. Perhaps it was not noted earlier because, by chance, the populations that were studied were similar behaviourally. Alternatively, the search for species-typical behavioural patterns, combined with the masking effect of a common laboratory environment, may have caused researchers to overlook subtle differences among populations in the structure of particular behavioural patterns and in the frequency with which they were expressed. Nevertheless, recent field and laboratory work has clearly documented extensive variation in many aspects of the reproductive behaviour of threespine stickleback.

Inference of adaptive value

When interpopulation comparisons are used to infer the adaptive value of a particular trait, the inference is strongest when (1) a large number of localities is sampled, (2) there is a very close correlation between the character states and a complex spatial pattern of environmental variation, and (3) the seeming adaptation can be inferred to have evolved independently in multiple localities (Endler 1986; Bell 1988). These conditions can all be met readily in interpopulation comparisons of threespine stickleback (Bell and Foster page 20; McPhail page 401 this volume). Although associations between environment and phenotype are not adequate to demonstrate the action of natural selection, they can provide unique insights into possible causes of adaptive diversification and can lead to the development of experimental designs that can test for an effect of natural selection in moulding the phenotypes (Endler 1986).

An assumption implicit in such interpopulation comparisons is that the behavioural differences among populations reflect underlying differences in genotype (Endler 1986). Although interpopulation differences in other aspects of behaviour of the threespine stickleback have been shown to have a genetic basis (Huntingford *et al.* page 289 this volume; Bakker page 351 this volume), the appropriate rearing studies have yet to be performed in investigations of reproductive behaviour. Clearly, such studies must be included in future research efforts.

Male courtship behaviour

Wilz (1973) provided the first evidence of interpopulation differences in the courtship behaviour of male stickleback. In a comparison of a British freshwater population and a brackish water population from the eastern USA, he found that the males from the North American population incorporated more nest-directed activities and dorsal pricking in courtship than did the British males. Consequently, courtship by North American males was more protracted than that by British males.

Because he conducted his research exclusively in the laboratory, Wilz (1973) did not suggest a cause for the differences between the populations. My comparative field studies of six lacustrine populations of threespine stickleback in British Columbia, Canada, revealed parallel differences in male courtship behaviour among populations and suggested an adaptive explanation for this variation (Foster 1990, unpubl. data).

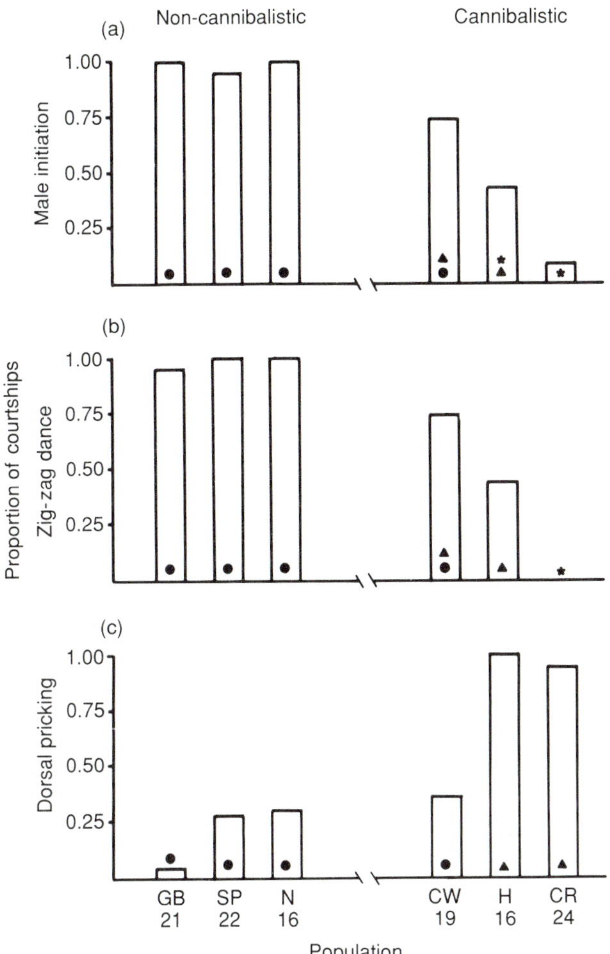

Fig. 13.1 Proportions of courtship interactions in six populations of threespine stickleback in British Columbia, Canada, in which (a) the male initiated the interaction, (b) the male performed a zigzag dance and (c) dorsal pricking occurred. Within rows, populations that share the same symbol at the base of the histogram bar form subsets that do not differ significantly (G statistic < 8.572, d.f. $= 1$, $P > 0.05$). Populations: CR, Crystal Lake; CW, Cowichan Lake; GB, Garden Bay Lake; H, Hotel Lake; N, North Lake; SP, Sproat Lake. Numbers along x-axis denote sample sizes.

In three of the six populations (i.e. in Garden Bay, North, and Sproat lakes), males typically initiated courtship with a zagzag dance and rarely performed dorsal pricking (Fig. 13.1). Courtship was extremely conspicuous, with the zigzag dance often extending several metres toward a female high in the water column. Following the dance, the male immediately attempted to lead the female to the nest.

In contrast, in two other populations (i.e. in Hotel and Crystal lakes), the female typically initiated courtship by swimming rapidly to the male, positioning herself above him, and initiating dorsal pricking. The zigzag dance was never observed in Crystal Lake, but occurred in 44 per cent of all courtships in Hotel Lake. When displayed by Hotel Lake males, the zigzag dance was slow, extended 1 m or less, and was performed near the substratum. Courtship in these populations was much less conspicuous than in the others and lasted a longer time. The structure of male courtship was intermediate in the sixth population in Cowichan Lake (Fig. 13.1; Foster unpubl. data).

These differences in courtship behaviour were associated with differences in cannibalistic behaviour among the populations. In Hotel and Crystal lakes, where courtship was relatively inconspicuous, groups of up to 300 females and immature males attacked nests guarded by males and cannibalized the defended young. During 700 h of observation in one breeding season in Crystal Lake, 68 per cent of the 38 focal nests known to contain embryos were destroyed by cannibalistic groups (Foster 1988). Cannibalism was never observed in any of the three lakes in which courtship was most conspicuous. In Garden Bay Lake, where the most extensive observations were made, laying occurred at 39 nests during 400 h of observation and none was observed to suffer cannibalism (Foster 1990, unpubl. data). In Cowichan Lake, where courtship characteristics were intermediate, small groups of up to ten adults attacked nests occasionally, but the risk to the young was evidently lower than in Hotel and Crystal lakes.

The advantage of inconspicuous courtship in cannibalistic populations is probably that foraging stickleback visually detect the nests they attack, using nest-directed behaviours of males and courting females as cues to nest location (Foster 1990 unpubl. data). Where cannibalistic foraging groups are common, inconspicuous courtship reduces risk to the nest and the offspring it contains. Dorsal pricking may also be advantageous in this context because it enables males to survey their surroundings before leading a female to the nest (Sargent 1982).

The interpopulation differences in courtship behaviour described here appear to represent a complex case of ecotypic differentiation. The threespine stickleback in the three non-cannibalistic populations phenotypically and ecologically resembled McPhail's limnetic form (McPhail 1984, page 418 this volume). As is typical of the form, they are slender-bodied, small fish that feed primarily on plankton in open water. The stickleback in Crystal and

Hotel lakes are larger, deeper bodied, and feed primarily on benthic material (benthic form, *sensu* McPhail page 418 this volume). They often fed in large groups in the territories of males, cannibalizing embryos when discovered. The Cowichan fish are intermediate in morphology and feed primarily on plankton. Occasional large individuals fed on benthic material among the territories held by reproductive males. They were attracted to disturbances on the breeding grounds and would consume guarded young when they discovered them (Foster 1990, unpubl. data).

Lavin and McPhail (1985) have demonstrated that the limnetic form of the threespine stickleback has typically evolved in large, deep lakes with extensive pelagic zones, whereas the benthic form is typical of small, shallow lakes with extensive littoral zones. It seems that, because of differences in foraging habitat, adults of the limnetic form typically do not cannibalize young defended by males (they do not encounter them), whereas adults of the benthic form encounter them regularly and are cannibals. Perhaps as a consequence, the courtship behaviour of limnetic stickleback tends to be more conspicuous than that of benthic stickleback, suggesting that ultimately, the evolution of courtship behaviour in freshwater stickleback has been influenced by the trophic habit of the stickleback (Fig. 1.6).

These observations on the courtship behaviour of allopatric benthic and limnetic stickleback may provide insight into the evolution of reproductive isolation between sympatric benthic and limnetic species of threespine stickleback found in six lakes in British Columbia (McPhail page 418 this volume). In one of these lakes, Enos Lake, Ridgway and McPhail (1984) described the reproductive behaviour of each species and documented complete assortative mating between the benthic and limnetic species. The differences between the species mirror those between the allopatric benthic and limnetic forms. If the differences in courtship behaviour between benthic and limnetic species pairs are in the same direction in all lakes (a possibility that has not been examined), a plausible hypothesis is that selection imposed by cannibalism has caused divergence in courtship behaviour that has, in turn, contributed to reproductive isolation between the benthic and limnetic forms (Foster unpubl. data).

Although the species pairs could have been derived from secondary contact between freshwater limnetic and benthic populations, geological evidence suggests instead that each lake was colonized by the marine form twice (McPhail page 424 this volume). According to this hypothesis, the first invasion gave rise to the benthic form because the lakes are small and shallow, and the second to the limnetic form, possibly as a consequence of competition for food with the benthic form (Schluter and McPhail 1992; McPhail page 425 this volume). If the courtship behaviour of marine stickleback resembles that of freshwater limnetics, a second invasion could have had much the same effect as secondary contact between freshwater benthic and limnetic forms.

One difficulty with attributing courtship divergence and reproductive isolation to cannibalism is that cannibalism by the benthic form should favour inconspicuous courtship in both forms in sympatry. If, however, hybrids suffer reduced fitness in areas of secondary contact, selection against hybrids could maintain differences in reproductive behaviour that promote assortative mating even at the cost of increased risk of cannibalism of young of the limnetic species.

The hypothesis that cannibalism has caused divergence in courtship behaviour that has contributed to reproductive isolation between the benthic–limnetic species pairs is intriguing, but remains to be tested. If this interpretation is correct, speciation in the benthic–limnetic species pairs must be seen as an incidental consequence of adaptive divergence driven by factors other than the efficacy of intersexual communication. This would provide one of the first clear counterexamples to Paterson's (1981, 1985) argument that isolating mechanisms are primarily a consequence of adaptations that facilitate intersexual communication under prevailing ecological conditions.

Male sneaking behaviour

Particularly in polygynous species, males can exhibit two or more kinds of mating behaviour that result in fertilization of eggs (e.g. Hamilton 1979; Thornhill and Alcock 1983; Dominey 1984*a*). In some instances the alternative behavioural phenotypes appear to represent evolutionarily stable strategies (e.g. Maynard Smith 1982; Austad 1984; Gross 1984), whereas in other instances they reflect facultative, opportunistic variation in which males that adopt secondary (conditional) tactics achieve lower reproductive success than do those adopting primary tactics (e.g. Dawkins 1980; Dunbar 1982; Kodric-Brown 1986). In either case, environmental and social contexts can affect the relative success of alternative strategies. Thus, the predictions of specific models can often be effectively tested by comparing populations that differ in exposure to the relevant environmental variables (e.g. Gross and Charnov 1980; Riechert 1986; Austad 1989).

Sneaking behaviour of male stickleback appears to be an example of a conditional, alternative mating tactic. Disruption of normal courtship behaviour by sneaking males has been reported in a number of laboratory studies of threespine stickleback (van den Assem 1967; Li and Owings 1978*a*; Sargent and Gebler 1980; Sargent 1982; Jamieson and Colgan 1989). Sneaking males freeze, assume drab coloration, sink to the bottom, and swim slowly to the entrance of the nest of a courting male. If undetected, the sneaker may enter the nest as the female leaves. Sneaking males release sperm over the eggs and then typically steal eggs which they carry to their own nests.

Recent field studies suggest that the incidence of sneaking behaviour varies dramatically among populations. In two Dutch freshwater populations, sneaking occurred during approximately 10 per cent of all courtship interactions (Goldschmidt *et al.* 1992). In both populations, sneaking was a typical early stage in the ontogeny of courtship by males that subsequently courted females with the full display repertoire. In contrast, sneaking was observed in only one of the six lacustrine populations I studied in British Columbia. During observations of > 200 courtships in which the female reached the nest entrance in Sproat Lake, only three males were ever observed to sneak toward the nests of courting males, and none reached the nest entrance. Each of these males had recently lost a nest, and they all ceased sneaking once they had built a new nest (Goldschmidt *et al.* 1992).

At present we do not know why sneaking is a prominent mating behaviour in some populations but not in others, although the distance between courting males appears to affect the probability of sneaking within populations, whereas sex ratio does not (Goldschmidt and Bakker 1990; Goldschmidt *et al.* 1992). Through interpopulation comparison, it should be possible to gain insight into the causes of variation in male mating behaviour among populations, thereby contributing to our understanding of the conditions under which alternative mating tactics are likely to evolve.

Female courtship behaviour

When female threespine stickleback are ready to spawn, they cease feeding and begin to court males. Their role in courtship differs among populations; in some cases the female usually initiates courtship and in others the male does.

Receptive females typically swim slowly over the breeding grounds, courting a number of males in sequence. In populations in which males typically initiate courtship, the behaviour of the female is usually passive at first (pers. obs.). At the male's approach she assumes a head-up posture and waits for the male to lead to the nest. If he does lead, she may quickly follow his lead, or instead, may swim away without following him to the nest. The basis of this decision is not known, although both body size and nuptial coloration have been implicated in mate choice (Rowland page 313 this volume; McPhail Chapter 14 this volume).

In other populations, the female plays a more active role. She swims from male to male, aggressively initiating dorsal pricking with each. The females in such populations (e.g. those in Crystal and Hotel lakes) appear to solicit males relatively indiscriminately, as they often initiate dorsal pricking with males that have incomplete nests or are already in the parental phase of the reproductive cycle. Between bouts of dorsal pricking, these females wait in the head-up posture while the male engages in nest-directed activities. If the male does not lead to his nest, the female may try to force entry while he is at the nest, or may depart to solicit other males (Foster unpubl. data).

In both kinds of population, the female often aborts entry to a nest once the male has led her to it and shown her the entrance. This behaviour has been reported in all populations that have been studied, whether the research was conducted in the laboratory (e.g. Leiner 1930; van den Assem 1967; Wootton 1974*b*; Ridley and Rechten 1981; Jamieson and Colgan 1989) or in the field (Wootton 1972*a*; Goldschmidt and Bakker 1990).

Wootton (1974*b*) suggested that the aborted nest entries could reflect intermediate stages in a developmental sequence leading to spawning readiness in females. If this hypothesis is correct, the consistency of this behaviour across populations may simply reflect an ontogenetic constraint on behavioural variation. Alternatively, the aborted nest entries may represent the rejection of particular nests by females. Ridley and Rechten (1981) originally suggested the latter hypothesis, arguing that the behaviour reflected preference for nests that already contained eggs. Support for the egg-preference hypothesis is equivocal (Jamieson and Colgan 1989), and it certainly cannot explain all instances of rejection, as the females often aborted entry into nests that contained eggs (Ridley and Rechten 1981; Goldschmidt and Bakker 1990). Nevertheless, this behaviour could be a mechanism of female choice, and the criteria upon which the females may base their decisions are potentially quite diverse (Rowland page 313 this volume).

In some populations, a high frequency of aborted nest entries is part of a cannibalistic foraging tactic of females. Females that are ripe court males for a protracted period before spawning. When shown the entrance to a nest that already contains eggs, the females grasp a mouthful of the eggs and either twist them loose from the others in the nest (stickleback eggs cohere) or pull the entire clump from the nest (Foster and Baker unpubl. data). This cannibalism has a devastating effect on the reproductive success of the male and presumably provides the female with a head start in provisioning her next clutch.

Deceptive courtship as a foraging mode is not found in all populations. Manzer (1976) found no embryos in the stomachs of any threespine stickleback he collected in Great Central Lake, British Columbia, during the breeding season, and there was no evidence of this form of cannibalism in Garden Bay, North, or Crystal lakes (Foster and Baker unpubl. data), or in either of two Dutch freshwater populations studied by Goldschmidt (pers. comm.). It thus appears that, although females may abort entry to nests in all or most populations of threespine stickleback, the behaviour may vary in function. Through a combination of experimentation and comparative observation, it should be possible to discriminate the functions of this behaviour in different populations and to gain insight into the causes of interpopulation variation in the expression of a remarkable form of intraspecific deception.

Inferring the direction of evolution

When primitive (plesiomorphic) character states can be recognized, interpopulation comparisons can provide unique insights not only into the way(s) in which historical events have affected behavioural phenotypes, but also into the directions and minimum rates of change of particular phenotypes (e.g. Cruz and Wiley 1989; Goldthwaite *et al.* 1990; Thompson 1990; Foster unpubl. data). Knowledge of character polarity can, in addition, help researchers avoid erroneous adaptive interpretations of patterns of geographic variation. Here I present two examples in which knowledge of character polarity has made it possible to evaluate the validity of specific hypotheses and to interpret comparative data correctly.

Female mate choice

Kaneshiro (1976) first argued that asymmetric mating preferences should reflect the direction of evolution. He suggested (1976, 1980) that during founder events, elements of ancestral courtship displays would be lost as a consequence of genetic drift and the associated genetic revolution. He predicted that as a result, females from the ancestral species would fail to respond to males of the derived species because those males would not display the full behavioural repertoire of the ancestral species. In contrast, females from the derived species would still recognize males from the ancestral population as appropriate mates because they would display the courtship elements retained in the derived species.

One of the examples that Kaneshiro (1980) used to support his hypothesis was McPhail's (1969) comparison of the mating preferences of a species pair of threespine stickleback from the Chehalis River, Washington, USA. The ancestral form possessed typical red nuptial coloration whereas the derived stream-resident form was jet black during breeding, seemingly as a consequence of convergence with the threat display of the sympatric mudminnow, *Novumbra hubbsi* (Hagen *et al.* 1980). Females of both species were shown to prefer males of the ancestral species (McPhail 1969), as predicted by Kaneshiro.

In a critique of Kaneshiro's hypothesis, Moodie (1982) pointed out that although the asymmetric preference was in the predicted direction, it was unlikely that the mechanism that generated the asymmetric mating preference was that outlined by Kaneshiro. He argued instead that selection for convergence with the threat display of the mudminnow had favoured the evolution of black nuptial coloration, but that selection had yet to alter the ancestral female preference for red nuptial coloration. In support of his argument, Moodie noted that McPhail (1969) had observed that derived females adjacent to the zone of overlap of the two species had lost the preference for red nuptial coloration (McPhail 1969), suggesting initiation of the evolution of prezygotic isolation.

Moodie (1982) cited a second example, in which he tested for asymmetry in mating preferences between females from a morphologically distinct, large-bodied population of threespine stickleback from Mayer Lake, British Columbia, and a parapatric population of stream-dwelling stickleback which were phenotypically similar to most freshwater threespine stickleback in the region (see also McPhail page 411 this volume). Although in nature these populations are reproductively isolated, he demonstrated that females from the stream-dwelling population responded more often to the lacustrine males than they did to males from their own population (Moodie 1972*b*). This was seemingly due to a preference for large males on the part of females in both populations. In this instance, if the stream-resident population is ancestral to the derived lacustrine species, the direction of asymmetry is the reverse of that predicted by Kaneshiro.

Although there is some uncertainty concerning the phylogenetic relationship between the populations studied by Moodie, there can be little doubt that Atlantic marine threespine stickleback are ancestral to the sympatric 'white stickleback', an undescribed species recently discovered along the coast of Nova Scotia, Canada (Blouw and Hagen 1990). Although Buth and Haglund (page 76 this volume; see also Haglund *et al.* 1990) have questioned the specific status of the white stickleback, I believe that Blouw and Hagen (1990) provided compelling evidence that the white stickleback is a biological species separate from *G. aculeatus*. The white stickleback is characterized by shimmering white male nuptial coloration. Blouw and Hagen (1990) demonstrated that there is complete assortative mating between *G. aculeatus* and the white stickleback in the laboratory. They have also shown that *G. aculeatus* females are strongly responsive to white males, and that white females are much less responsive to *G. aculeatus* males than to white males. In this case, females of both species prefer males of the derived species, yielding a mating asymmetry opposite in direction to that predicted by Kaneshiro (1976).

The data from studies of threespine stickleback suggest that, at least in this taxon, it would be unwise to attempt to discern the polarity of species derivation from present-day asymmetries in the mating preferences of females. These results do not invalidate the mechanism by which Kaneshiro (1976, 1980) suggested that mating asymmetries could evolve, but they do call into question the advisability of inferring the direction of evolution from mating asymmetry alone, a conclusion also reached by Wasserman and Koepfer (1980). Given the large number of species that must have evolved in boreal habitats of the Northern Hemisphere since deglaciation (Bell and Foster page 16; McPhail page 401 this volume), the threespine stickleback appears to be one of the most promising sources of information on the evolution of mating preferences and mechanisms of reproductive isolation available.

Diversionary display behaviour

A primary role of the male threespine stickleback during the parental phase is to defend his offspring from a diversity of predators (Whoriskey and FitzGerald page 202; Huntingford *et al.* page 292 this volume). In some lakes, the greatest risk to defended offspring is posed by conspecific foraging groups. This threat can be so great that Hyatt and Ringler (1989*a*) suggested that such cannibalism may play a primary role in population regulation in lacustrine populations. Although males can defend their nests against solitary conspecifics by attacking them directly, this tactic is ineffective against large groups of conspecifics (Foster 1988).

In habitats where cannibalism by groups occurs, males defend their nests by deception, using displays that divert conspecifics from approaching their nests. Such 'diversionary displays' have only been observed in cannibalistic populations, and only in this context (Whoriskey and FitzGerald 1985*c*; Foster 1988, 1990, unpubl. data; Ridgway and McPhail 1988; Hyatt and Ringler 1989*b*). In my observations on six lacustrine populations in British Columbia, Canada, males in the three cannibalistic populations (i.e. Crystal, Hotel, and Cowichan lakes) were the only ones that performed diversionary displays at the approach of groups of conspecifics. In the other three populations, males courted females in the groups or ignored them, depending on the male's reproductive state (Foster unpubl. data).

In the simplest form of the diversionary display (rooting), males swim rapidly out of their territories and dig vigorously in the substratum, seemingly mimicking a feeding individual (Fig. 13.2). If the display is effective, the group joins the male, apparently attempting to feed. The group subsequently departs without entering the territory. If the display fails to divert the group, the group may enter and feed in the territory, increasing risk to the brood. Rooting is the primary form of diversionary display exhibited by Cowichan Lake males. In Crystal Lake, most of the displays are similar, although they often include an erratic swim that causes the silvery sides of the fish to flash. In Hotel Lake, the displays are even more elaborate, with the male turning on his side and swimming erratically and

Fig. 13.2 Rooting diversionary display behaviour of male threespine stickleback.

conspicuously away from his nest. In some cases he descends and taps his snout along the substratum as if mimicking courtship showing (snout tapping), and in other instances, he remains well above the substratum, 'shimmering' through the water column conspicuously.

Diversionary displays have been described in three other lacustrine populations in British Columbia (Ridgway and McPhail 1988; Hyatt and Ringler 1989*b*), indicating that these behavioural patterns are widespread among lacustrine populations in this region. These lakes are each in separate river drainages from one another and from those I studied, suggesting independent origins from the marine ancestor. Diversionary displays could have evolved independently in each lake, and snout tapping and shimmering could represent progressive elaboration of the display by incorporation of conspicuous elements of courtship behaviour, as has been suggested in avian diversionary displays (e.g. Armstrong 1949*a*,*b*; Skutch 1955).

This explanation is unlikely, however, as all three forms of the diversionary display have been observed in an Atlantic salt-marsh population in Quebec, Canada (Whoriskey 1991, pers. comm.). Because Atlantic marine stickleback can be considered an outgroup for analysis of the evolution of character states in the freshwater populations of north-western North America (Buth and Haglund page 78 this volume), the limited displays of males in Enos, Paxton (Ridgway and McPhail 1988), Crystal, and Cowichan lakes (Foster 1988, unpubl. data) must be considered derived, phenotypes in which evolutionary change has occurred through loss of parts of the ancestral repertoire. Therefore, the more complex displays of males in Kennedy (Ringler pers. comm.) and Hotel lakes probably represent the ancestral character states. Recent observation of the full range of diversionary displays in a marine population breeding in a tidal lagoon on the Sechelt Peninsula, British Columbia, lends credence to this hypothesis (Foster unpubl. data).

CONCLUSIONS

As a subject for research on the evolution of behavioural phenotypes, the threespine stickleback is without parallel among the vertebrates. Not only can it be compared with its congeners, but postglacial endemic radiation makes possible the inference of directions and causes of recent evolutionary change through interpopulation comparisons. Because the threespine stickleback can be observed easily in both the laboratory and the field, functional interpretations of particular behavioural patterns are often possible and can be tested subsequently through experimental manipulation. Finally, the genetic and environmental bases of behavioural phenotypes can be ascertained because this small fish can easily be reared in the laboratory.

The potential value of behavioural differences between the genera of threespine stickleback as a source of phylogenetic information was recognized early (Reisman and Cade 1967; Rowland 1974*a*; Wootton 1976), although it was not until 1988 (McLennan *et al.*) that an effort was made to reconstruct the phylogenetic relationships among the gasterosteid genera on this basis. Unfortunately, the majority of behavioural phenotypes that have been carefully documented in all the species are associated with reproduction and aggression (McLennan *et al.* 1988), and therefore cannot be assumed to have evolved independently of one another. Although lack of independence among characters weakens confidence in the phylogenetic reconstruction (e.g. Felsenstein 1985; Mayr and Ashlock 1991), the degree of congruence with the best-supported cladogram for this group based on other phenotypic characters certainly suggests that elements of reproductive behaviour can be conserved during cladogenesis.

In species like the sticklebacks, which display substantial population differentiation, identification of the character states appropriate for phylogenetic analysis of the genera (i.e. primitive character states for each genus) can be exceedingly difficult. In the threespine stickleback, the phenotypically stable marine form is more likely to possess primitive character states than are freshwater populations (Bell and Foster page 14 this volume). Certainly, symplesiomorphies of Pacific and Atlantic marine populations, which probably were separated before postglacial freshwater populations were isolated (Buth and Haglund Chapter 3 this volume), are likely to represent the ancestral states for postglacial populations. The same inference can possibly be made for *Pungitius*, which has a similar distribution. However, phylogenetic relations within the genus *Pungitius* are more complex (Haglund *et al.* 1992*b*). The problem of inferring primitive behavioural phenotypes is likely to be more difficult in the other genera, and will probably require intergeneric outgroup analysis.

Certainly, there can be little doubt that comparison between character states in the derived freshwater forms and those in marine populations can provide evidence of character state polarity (e.g. Bell and Foster page 20 this volume). Knowledge of character state polarity not only permits testing of specific hypotheses, such as Kaneshiro's (1976) concerning the polarity of species derivation, but also facilitates inference of the causes of character state transitions.

Recognition of the value of population differentiation in *Gasterosteus* as a source of information on behavioural evolution has developed relatively recently (e.g. Huntingford 1982; Giles and Huntingford 1984; Ridgway and McPhail 1984). The value of judicious comparison among populations is evident from the examples in this chapter and those in Huntingford *et al.* (Chapter 10 this volume). Given that so few populations have been examined to date, it seems likely that we have uncovered only a small fraction of the behavioural diversity that occurs among threespine stickleback populations. Exploration of the value of this radiation as a source of information on the evolution of behavioural phenotypes has only begun.

ACKNOWLEDGEMENTS

I thank J.A. Baker, W.J. Etges, A.E. Houde, and W.J. Rowland for their very helpful comments on an earlier draft of this chapter. My very special thanks go to M.A. Bell for support throughout my rather prolonged postdoctoral career in his laboratory. His insights and discussions were critical to the development of many of the ideas presented here. My research and the writing of this chapter have been supported variously by USA National Institute of Mental Health National Research Service Award # F32 MH09244 (to S.A.F.) and by USA National Science Foundation Grants BSR91-08132 (to S.A.F.) and BSR89-05758 (to M.A. Bell).

14

Speciation and the evolution of reproductive isolation in the sticklebacks (*Gasterosteus*) of south-western British Columbia

J.D. McPhail

Over a century after publication of the *Origin of Species* (Darwin 1859), what species are, and how they are formed, remain contentious issues in evolutionary biology. This uncertainty poses a problem for anyone attempting a short review of the process of species formation in a specific group of animals: how to discuss speciation without becoming mired in the sometimes subtle philosophical and biological arguments that surround the species concept. The course I have chosen is to ignore issues that do not bear directly on speciation in stickleback. Consequently, this review deals only with the division of single species into two or more species (cladogenesis) and the processes involved in the acquisition of the main biological characteristic of species (reproductive isolation).

Three major models have been proposed to explain how single species divide into two or more new species. These are the geographic, parapatric, and sympatric models (for reviews see Mayr 1942, 1963, 1982; Bush 1975; Endler 1977). These models are not equally probable for all organisms, but in sexually reproducing animals the weight of evidence strongly favours the geographic model (Futuyma and Mayer 1980). Although the three models differ in important ways, they all invoke genetic divergence as a primary step in the process of speciation. Indeed, in recent years the genetic aspects of divergence have become a focus in speciation research (Carson 1975, 1987; Templeton 1981, 1987; Barton and Charlesworth 1984; Carson and Templeton 1984).

Research into the ecological, genetic, and behavioural changes that occur during speciation is inherently difficult, simply because the process usually is identifiable only after the fact. Thus most speciation studies compare closely related species (sister species) and, from these, draw inferences about the process of species formation. A danger of this approach is that it is more likely to provide information about the characteristics of species than about the processes that give rise to species. Often, however, no other option is available, and inferences derived from comparing closely related species have provided most of the conceptual foundation of our present view of speciation. Presumably, though, speciation is not a process that occurred only in the past. New species must still be in the process of formation, and if we can recognize these incipient species they should provide the best possible material for the study of the genetic, ecological, and behavioural changes that occur during species formation. If the current geographic view of speciation is correct, the majority of such incipient species must involve isolated populations. Unfortunately, there are no objective criteria for determining when divergence between allopatric populations has reached, or almost reached, the species level. In sympatric populations, however, the ability to coexist without interbreeding is usually accepted as the criterion for biological species (Dobzhansky 1940; Mayr 1942). In theory, this criterion is unambiguous; in fact, it is often difficult to apply. For example, in temperate freshwater fishes, sympatric species often hybridize (Hubbs 1955), and in such cases the capacity to maintain separate gene pools is the appropriate criterion for biological species. Therefore if divergent populations can coexist and, in spite of persistent hybridization, still maintain separate independent gene pools, they are biological species. Clearly, this criterion can be applied only in sympatry and requires detailed genetic and ecological information. This restriction is inconvenient for systematists, but it in no way lessens the importance of the biological species concept to research into the process of species formation. As Carson (1987) stressed, the goal of speciation studies is to understand how coexisting populations come into being, and it is unimportant whether or not systematists consider such divergent populations as species. This review deals exclusively with genetically divergent populations of the threespine stickleback, *Gasterosteus aculeatus*, that are in contact during their breeding season. The degree of contact varies from parapatric populations that abut in only a small area, through to fully sympatric populations. In all cases, however, some breeding individuals in these different populations come into contact, and thus the populations have the opportunity to exchange genes.

Geographically, this review covers only a small portion of the extensive Holarctic range of *G. aculeatus*: the Strait of Georgia and Vancouver Island on the south-west coast of British Columbia, Canada (Fig. 14.1). For convenience, this area is referred to simply as the Strait of Georgia region, but in the context of this review the region includes all of Vancouver Island. There are three reasons for restricting discussion to this geographic area. One is that the postglacial history of the region is reasonably well known (Mathews *et al.* 1970; Armstrong 1981; Clague 1981; Howes 1983), and consequently, divergent populations can be discussed within a reliable historical context. Second, in this area, some genetic and ecological information is available on a variety of divergent populations. And third, one important set of divergent populations is restricted to this area.

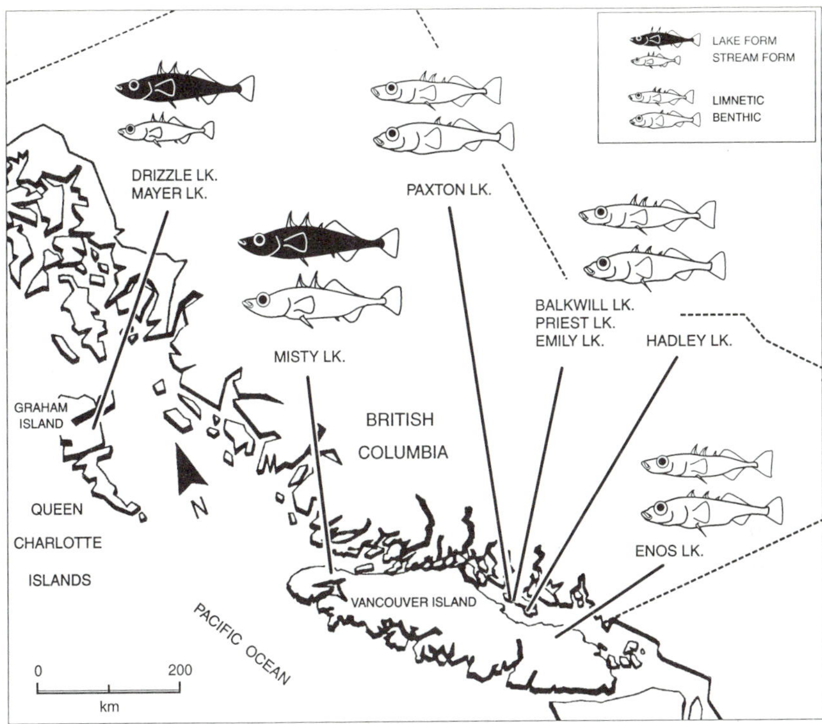

Fig. 14.1 The coast of British Columbia, western Canada: for this review the Strait of Georgia region includes Vancouver Island, the adjacent mainland, and the intervening islands. The stickleback cartoons indicate the distribution of the lake–stream and benthic–limnetic pairs.

Because of the glacial history of the Strait of Georgia region, stickleback divergence in this area is inextricably entangled with the problem of the postglacial colonization of fresh water. The entire region was covered in ice during the last (Fraser) glaciation. As the ice retreated about 13 000 yr ago, the sea swept in and flooded lowland areas. This submergence was followed by a period of rapid isostatic rebound, and by about 11 000 yr ago the land attained nearly its present conformation (Clague 1981). Most freshwater populations of *Gasterosteus* in the region are now found in areas that were postglacially submerged by the sea. This distribution, and the presence of freshwater populations on islands and isolated headlands, argues that postglacial colonization was through the sea; however, colonization probably was a complex process involving multiple invasions and more than one life history form of *Gasterosteus* (see also Bell and Foster page 16 this volume).

Extant freshwater populations probably were founded in two ways: (1) directly, either by isostatic rebound trapping marine populations in coastal lagoons and fiords that then became lakes, or by the active colonization of newly formed freshwater habitats by marine or anadromous stickleback, and (2) indirectly, by invasion first of one freshwater habitat and then secondary dispersal into another, different, freshwater habitat. The essential distinction between these colonization scenarios is that direct invasion involves only a single episode of selection in a new environment, whereas indirect invasion involves at least two different episodes of selection. For example, most rivers, streams, and lowland lakes that contain stickleback were accessible historically from the sea, and thus open to direct invasion. In contrast, many inland lakes were never directly accessible from the sea, and probably were colonized by stickleback that first invaded, and adapted to, streams, and then secondarily dispersed into lakes. Such historical differences in the mode of colonization may have contributed to the stickleback diversity now found in the fresh waters of the region.

Traditionally, freshwater populations are assumed to be derived from anadromous stickleback (Heuts 1947*a*; Münzing 1963; Hagen 1967; McPhail and Lindsey 1970; Bell 1976*a*), and anadromous stickleback are still abundant throughout the Strait of Georgia. If stickleback similar to extant anadromous fish were present at, or immediately after, the postglacial marine submergence, they could have founded freshwater populations. It is unlikely, however, that anadromous fish were the sole colonists of fresh water. Non-anadromous marine populations also occur throughout the Strait of Georgia, and many rivers that do not have anadromous runs have year-round estuarine populations of stickleback. The fully marine and estuarine life history forms of *Gasterosteus* probably were present in the early postglacial Strait of Georgia and also may have colonized fresh water. The genetic relationships of the fully marine, estuarine, and anadromous populations are unknown. They may represent separate gene pools or, alternatively, a single gene pool containing a complex life history polymorphism. Regardless of their genetic relationships, however, the three forms consistently differ in their life histories and in the frequency of the expression of morphometric and meristic traits (McPhail unpubl. data). Consequently, differences in the original source of colonists also may have contributed to the diversity of stickleback now found in the fresh waters of this region.

Certainly, the array of forms that now exist in the fresh waters of the Strait of Georgia region is remarkable, and the morphological diversity is comparable to that on the Queen Charlotte Islands, British Columbia, Canada (Moodie and Reimchen 1976*a*; Reimchen *et al.* 1985; Reimchen page 248 this volume). Among lakes, stickleback differ strikingly in body size, body shape, body colour, vertebral number, spine lengths, spine numbers, lateral plate expression, pelvic girdle expression, mouth size, eye position, gill raker number, gill raker length, and male nuptial colour. For most of these traits, this interpopulation variation is not random; instead, it is strongly associated with physical and biotic factors (Hagen and Gilbertson 1972; Moodie and Reimchen 1976*a*; Lavin and McPhail 1986; Reimchen and Nelson 1987; Reimchen 1989, Chapter 9 this volume). In addition, genetic studies (Hagen 1967, 1973; Hagen and Gilbertson 1973*a*; McPhail 1969, 1977, 1984, 1993) have established that many of these traits are inherited, and for a few traits there is experimental evidence that differences in expression influence the probability of survival or reproduction (Bentzen and McPhail 1984; Lavin and McPhail 1987). This evidence all argues that, although initially historic factors may have contributed to the observed variability, most of the interpopulation variation in freshwater stickleback in this region is a response to local selective regimes. Although interpopulation divergence is common in the region, divergent sympatric or parapatric populations are relatively rare. These latter instances, however, are of special interest to evolutionary biologists. The remainder of this review deals with the origin, and biological consequences, of genetic divergence in such stickleback populations.

Three sets of genetically divergent stickleback populations are known from the Strait of Georgia region. The best-studied set are the anadromous and stream-resident stickleback (Hagen 1967). During the breeding season, anadromous and stream-resident stickleback are in contact in many low-gradient streams throughout the area. This situation is not restricted to the Strait of Georgia region: the same, or similar, forms come into contact along much of the Pacific coast of North America, as well as along the coasts of Europe and Asia.

The second set of divergent populations involves lake- and stream-dwelling stickleback. Morphologically differentiated lake and stream forms of *Gasterosteus* are known from northern Vancouver Island and from Graham Island in the Queen Charlotte Islands (Moodie 1972*b*; Reimchen *et al.* 1985; Lavin and McPhail 1992). On these islands, divergent lake and stream populations are parapatric where streams enter, or exit, certain lakes. At present, the divergence between contiguous lake and stream forms appears to be confined to these two islands.

The third set of divergent populations involves sympatric plankton-feeding (limnetic) and littoral-foraging (benthic) stickleback (Larson 1976; McPhail 1984, 1992, 1993). Of the three divergences, this set has the most restricted geographic distribution. Sympatric limnetic–benthic pairs are confined to five small lakes on two islands in the Strait of Georgia and one small lake on Vancouver Island (McPhail 1992).

I discuss each of these sets of divergent populations below. First, I describe their morphology and life histories, and the nature of their contact. Then follows an examination of the evidence for separate gene pools, and for either hybridization or gene flow between populations. Next, I examine the nature and extent of reproductive isolation between the members of each set, and lastly, I discuss hypotheses concerning the origins of the different divergences. I then attempt to synthesize these data into a general discussion of speciation in *Gasterosteus*, with particular emphasis on the origins of reproductive isolation in these divergent parapatric and sympatric populations.

ANADROMOUS AND STREAM-RESIDENT STICKLEBACKS

Description of the two forms

Over much of the extensive Holarctic range of *G. aculeatus*, anadromous and stream-resident stickleback come into contact during the breeding season. In the Strait of Georgia region, the only published description of this contact is that of Hagen (1967) for the Little Campbell River. The following account is extracted from Hagen (1967) and supplemented with original observations of other nearby populations.

Morphologically, anadromous and stream-resident stickleback are strikingly different. The anadromous form is large, with a slim, silvery body, a complete set of lateral plates, numerous, long, slim gill rakers, and many dorsal and anal rays. In contrast, the stream-resident form is smaller, with a stocky, olive-drab body, a few anterior lateral plates (i.e. low morph), a low number of short, stubby gill rakers, and a small number of dorsal and anal rays (Table 14.1). With minor exceptions, these morphological differences between anadromous and stream-resident stickleback hold not only for populations throughout the Strait of Georgia region, but also for anadromous and stream-resident populations from over most of the geographic range of *G. aculeatus* (Heuts 1947*a*; Okada 1960; Hagen and Moodie 1982; Ziuganov 1983; Ziuganov *et al.* 1987).

During June and July, anadromous stickleback are abundant in the Little Campbell River from the estuary upstream to approximately the limits of tidal influence (a distance of about 2.5 km). Above this point, the nature of the stream gradually shifts from a slow-moving, soft-bottomed slough into a swift, gravel-bottomed trout stream. In this transition zone, anadromous stickleback are rare, but both stream-resident and hybrid stickleback are abundant (Hagen 1967). Above the transition zone, a high-gradient section extends upstream for several kilometres, and here only small numbers of resident stickleback are found in a few of the larger pools. Above this high-gradient section the stream changes again into a slow-moving, heavily vegetated slough containing dense populations of resident stickleback.

This pattern of anadromous fish in the lower reaches of streams and resident fish in the upper reaches is repeated throughout the Strait of Georgia region. In many streams, however, the distributions of anadromous and resident stickleback are not as neatly separated as in the Little Campbell River. In these streams, the distributions of anadromous and resident stickleback often overlap for several kilometres. Consequently, during the breeding season most of the anadromous population, and much of the

Table 14.1 Morphology[a] of a laboratory-reared anadromous–stream-resident pair (and their hybrids) from the Salmon River, near Fort Langley, British Columbia.

Form	Character								N
	Left plates	Dorsal rays	Anal rays	Gill raker number	Body depth (mm)	Jaw width (mm)	Spine length (mm)	Gill raker length (mm)	
Anadromous	33.02 (0.87)	12.26 (0.64)	9.23 (0.60)	22.10 (1.17)	10.77 (0.54)	2.81 (0.24)	8.64 (0.56)	1.71 (0.14)	60
Hybrids (F_1)	18.18 (3.11)	10.90 (0.54)	7.81 (0.56)	20.13 (1.06)	11.89 (0.42)	3.38 (0.24)	6.69 (0.44)	1.13 (0.10)	30
Hybrids (F_2)	19.33 (8.02)	10.91 (0.70)	7.80 (0.60)	19.80 (1.14)	11.99 (0.57)	3.32 (0.91)	6.49 (0.48)	1.17 (0.17)	30
Stream residents	4.14 (1.05)	10.04 (0.53)	7.51 (0.47)	16.11 (1.01)	15.54 (0.42)	3.53 (0.20)	4.42 (0.40)	0.82 (0.07)	60

[a] Counts and measurements (after McPhail 1984) are means with measurments standardized to a 50 mm standard length; numbers in parentheses are standard deviations.

stream-resident population, are sympatric rather than parapatric as in the Little Campbell River.

Evidence for separate genomes

Anadromous and stream-resident stickleback, and their hybrids, from the Little Campbell River and several adjacent streams (Salmon River, Kanaka Creek, and Nathan Creek) were reared in the laboratory. Both forms breed true: crosses made among anadromous fish produce only offspring with the anadromous phenotype, and crosses made among stream-resident fish produce only fish with the stream phenotype. Laboratory-reared F_1 hybrids are intermediate in their morphology (Table 14.1). These results imply that the major differences between the two forms (body shape, adult size, body colour, lateral plate number, and gill raker number) are inherited. In addition, there are allele frequency differences between the two forms at three electrophoretically detectable loci (Withler and McPhail 1985). These morphological and allozymic differences appear to be stable over time, and thus indicate that the anadromous and stream-resident forms of *G. aculeatus* represent separate gene pools. Similar results from genetic studies in Europe (Münzing 1959; Ziuganov 1983) and Japan (Ikeda 1934) suggest that this division of *G. aculeatus* into anadromous and resident gene pools is geographically widespread.

Evidence for hybridization and gene flow

In the Little Campbell River the transition zone is about 2.0 km long, and Hagen (1967) estimated that approximately 21 per cent of the adult stickleback in this area are hybrids (F_1, F_2, or backcrosses). In many streams in the Strait of Georgia region, however, there is no clear hybrid zone. This does not imply that there are no hybrids, but only that in most streams there is no sharply defined area where hybrids are abundant. Indeed, in most streams, hybrids appear to be less common than they are in the hybrid zone in the Little Campbell River; however, attempts to estimate the abundance of wild hybrids are often unreliable. This is because the identification of wild hybrids commonly rests on a single character, lateral plate number. Unfortunately, the presence of individuals with intermediate numbers of plates does not necessarily constitute evidence of hybridization (Hagen and Moodie 1982). Many stream-resident populations in the Strait of Georgia region, that in other aspects of their phenotype are unequivocally stream fish, are polymorphic for lateral plates. Without detailed morphological and biochemical data such populations are easily mistaken for hybrids; however, genuine hybrids (at least F_1 hybrids) are unambiguously intermediate in most morphological and biochemical traits. In the Strait of Georgia region, some hybrid individuals occur wherever anadromous and stream-dwelling stickleback come into contact. This suggests that there is persistent hybridization between the two forms, but except in ecologically intermediate areas like the Little Campbell hybrid zone, hybrids are relatively rare.

The restriction of hybrids to an ecologically intermediate area, and the temporal stability of morphological and biochemical differences above and below the hybrid zone, suggests that little or no gene flow occurs between the anadromous and stream-resident sticklebacks in the Little Campbell River. Although not analysed in the same detail, other streams in the Strait of Georgia region give a similar picture. Thus, in this area, the anadromous and stream forms of *Gasterosteus* appear to represent separate and independent gene pools.

Nature of reproductive isolation

There are no developmental or sterility barriers between the two forms in the Little Campbell River or in other streams in this area (Hagen 1967; McPhail pers. obs.). Crosses between the two forms, and their hybrids (F_1, F_2, and backcrosses), are interfertile and viable. Thus hybrids have the potential to act as a bridge between the two gene pools, and if hybridization and backcrossing were extensive, the gene pools ultimately should fuse. Since, in the Strait of Georgia region, there is no evidence of introgression in streams where the two forms coexist, there must be mechanisms that prevent, or greatly restrict, gene flow. The main factors restricting gene flow appear to be differences in the time of reproduction, differences in nest location, strong positive assortative mating, and perhaps hybrid inferiority. None of these mechanisms is capable in itself of totally preventing gene flow, but taken together, they appear to effectively isolate the two forms (Hagen 1967; Hay and McPhail 1975; McPhail and Hay 1983).

In the Little Campbell River, the stream-resident population starts breeding in April and reaches a peak in May. Breeding slows during June and ceases by mid-July. In contrast, the anadromous form begins in late May or early June and reaches a breeding peak in July. Thus, in the Little Campbell River the breeding seasons of the two forms only partially overlap. The breeding season outlined for the resident form in the Little Campbell River is typical of stream-dwelling sticklebacks in this area, but the timing of the anadromous run is unusual. At this latitude (49°–50° N), anadromous populations begin entering fresh water in April and the runs usually peak in May (see also Baker page 156 this volume). Each stream, however, has a characteristic run time that remains remarkably consistent from year to year. Thus, although temporal differences in breeding season may contribute to reproductive isolation in some anadromous and stream-resident populations, the differences in breeding times in the Little Campbell River are greater than usual in this region.

The within-stream distribution of anadromous stickleback is similar in all streams in the area. This form penetrates upstream either to the approximate limits of tidal influence or until a velocity barrier is reached. Above this point, if there are low-gradient areas, only stream-resident stickleback occur. Below the point of maximum upstream penetration of anadromous stickleback, both forms usually occur, but occasionally (as in the Little Campbell River) the stream-dwelling form is rare in the area of tidal influence. The restriction of the anadromous form to the lower reaches of streams is a consistent pattern, even when there is no physical barrier to upstream movement. This distribution implies that this form has different ecological requirements from those of stream-resident stickleback. Hagen (1967) showed that the two forms differed in nesting habitat. Given a choice, the anadromous form prefers sand substrate, a slight current, and relatively open nesting sites. In contrast, stream-resident fish prefer a mud substrate, no current, and heavily vegetated nesting sites. In the Little Campbell River, the distribution of the two forms mirrors the within-stream distribution of the two kinds of nesting site. Thus, differences in habitat preferences probably contribute directly to reproductive isolation. In the lower reaches of many streams in the Strait of Georgia region, however, the breeding areas for anadromous and stream-resident stickleback overlap for several kilometres. Nothing is known about nest sites in these streams, but if the two types of nest site found in the Little Campbell River also exist in these streams, they are not spatially separated as they are in the Little Campbell.

Hagen (1967) tested for positive assortative mating in the two forms and found no evidence of homogamy. Unfortunately, he used an inappropriate criterion for female choice (the 'head-up' display; see Rowland page 327 this volume). Later, Hay and McPhail (1975) used female nest entry as the criterion for choice and demonstrated that males and females of both forms show a strong preference for their own kind. They also demonstrated that although courtship in anadromous and stream-resident stickleback is essentially the same, the two forms perform the same behaviours at different frequencies and in different sequences. In particular, males behave differently during their first approach to females. Typically, anadromous males bite first, whereas stream-resident males usually zigzag. McPhail and Hay (1983) suggested that this difference in the initial stages of courtship may function as an interpopulation recognition signal. They also demonstrated changes in courtship behaviour in 'no choice' situations (tests involving males of one form and females of the other form). These behavioural changes imply that the forms perceive differences between fish from their own gene pool and fish from the other gene pool. This recognition may be primarily behavioural, but at least one aspect of morphology (body size) is also implicated. Borland (1986) investigated mate preferences in stickleback in an adjacent, but separate, stream (the Salmon River) and again found that differences in body size contributed to positive assortative mating between anadromous and stream-resident *G. aculeatus*.

Anadromous stickleback migrate to and from the sea, whereas stream-resident stickleback remain year-round in fresh water and are relatively sedentary, even within their stream. These different ways of life are reflected in differences in body shape and in sustained and burst swimming performances (Taylor and McPhail 1986). They also differ in trophic adaptations: the anadromous form has long, numerous gill rakers and a narrow mouth, whereas the stream-dwelling form has less numerous, short gill rakers and a relatively wide mouth. In *Gasterosteus* these differences are associated, respectively, with plankton feeding and foraging on benthic invertebrates (Hagen 1967; Bentzen and McPhail 1984; Lavin and McPhail 1986). These differences in trophic morphology and body form suggest that anadromous and stream-resident sticklebacks are adapted to different ways of life. Because hybrids between the forms are intermediate in most morphological traits, it is argued (Heuts 1947*a*; Hagen 1967) that they are competitively inferior to the appropriate parental form in either habitat. The assumption that intermediate morphology in hybrids translates into ecological inferiority has not been tested in anadromous and stream-dwelling sticklebacks.

In summary, the absence of developmental and sterility barriers in hybrids, combined with sympatric or parapatric contact during the breeding season, suggest that the two forms should introgress wherever large numbers of hybrids are produced. In most streams, however, hybrids are either rare or confined to areas of intermediate habitat. Circumstantial evidence suggests that outside of such intermediate areas, hybrids are competitively inferior to the parental forms. In the Little Campbell River, differences in breeding season may contribute to reproductive isolation, but in most places the breeding seasons of the two forms overlap extensively. Premating isolation appears to be mostly a product of positive assortative mating, and this is achieved partly through differences in nest location but also by differences in morphology and behaviour.

Origin of the divergence

Any evolutionary model attempting to explain the origin of anadromous–stream-resident pairs must address two observations. First, it must explain the extensive geographic distribution of the divergence: such pairs occur, in contact, in lowland streams along the temperate coasts of three continents. Second, it must explain the presence of the pairs in recently glaciated areas, particularly in streams on islands and other areas accessible only through the sea. Two models attempt to explain these observations. Münzing (1959, 1963) suggests that in Europe, anadromous and freshwater stickleback diverged before the last glaciation and survived the ice in separate refugia. They then came into postglacial contact and now hybridize along a broad secondary contact zone. In Europe, the geographic distributions of allopatric populations of both forms, and the location of the presumptive contact zone, are consistent with this hypothesis (Münzing 1963). However, to explain the presence of both forms around the northern periphery of

Europe (e.g. northern Norway, the Kola Peninsula, and Iceland), Münzing (1963) found it necessary to hypothesize a series of recent (postglacial), and independent, evolutions of freshwater stickleback from marine populations.

In western North America, the geographic distribution of anadromous and freshwater stickleback does not fit easily into Münzing's hypothesis of survival in separate refugia (McPhail and Lindsey 1970). As an alternative hypothesis, Bell (1976*a*) suggested that marine stickleback have colonized fresh water many times. He argued that, although the divergence between marine and freshwater stickleback is ancient (Bell 1977), the presence of freshwater populations in recently glaciated areas all along the Pacific coast of North America requires multiple, independent, postglacial origins of the freshwater form.

Thus, both Münzing (1963) and Bell (1976*a*) concluded that in recently glaciated areas, anadromous stickleback have given rise to freshwater populations on several independent occasions. In such areas, two models might explain these recent, multiple origins of stream-resident stickleback: an allopatric model postulating geographic barriers that isolate the two forms and allow divergence, or a parapatric model that postulates the evolution *in situ* of stream-resident populations from the anadromous form. The essential difference between the models is that the parapatric model does not require complete geographic isolation. At present there are not sufficient data to test between these models, and therefore both must be viewed as highly speculative.

One obvious way that anadromous stickleback might found a stream-dwelling population is by becoming land-locked through the action of some extrinsic barrier (e.g. a waterfall or a velocity barrier associated with isostatic rebound). Such accidental land-locking is common in diadromous fish (McDowall 1988), and once fish are isolated from the anadromous population, selection should favour individuals adapted to stream life. Consequently, a stream-resident form will be selected above the barrier and eventually, when the barrier breaks down, the original anadromous population and the new stream-dwelling population will come into secondary contact. The hybrid zone in the Little Campbell River (Hagen 1967) could be such a secondary contact zone. A weakness of the allopatric hypothesis, however, is the extensive geographic distribution of parapatric anadromous and stream-resident sticklebacks. They occur in low-gradient rivers and streams along the coasts of Europe, Asia, and North America. Thus, the allopatric hypothesis requires not only the accidental land-locking of the anadromous form in a large number of rivers and streams throughout the Northern Hemisphere, but also the subsequent replacement of the barriers with secondary contact zones, and all of this in the relatively short time since the last glaciation. Admittedly, such a scenario is possible, but it does not seem likely.

The other way in which stream-resident stickleback might have diverged from the anadromous form is through different selection pressures operating on opposite sides of an ecotone. This model requires some of the offspring of anadromous parents to remain voluntarily in fresh water, and that there be major differences in the selection regimes encountered by migratory and non-migratory individuals. Under the parapatric model, the hybrid zone in the Little Campbell River would be interpreted as a primary contact zone. Unfortunately, it is not possible at present to distinguish between primary and secondary contact zones (Endler 1977, 1982; Barton and Hewitt 1985). A strength of the parapatric model, however, is that it avoids the necessity of postulating a large number of accidental land-lockings. Instead, it requires the widespread existence of abrupt ecotones in low-gradient, coastal rivers and streams. This is a more realistic scenario than accidental land-locking, since ecotones at the freshwater–marine boundary are a universal feature of estuaries. Indeed, if multiple evolutions of stream-resident stickleback have occurred, then parapatric divergence across just such a physical boundary may be the simplest explanation for their presence along the glaciated coasts of three continents.

LAKE–STREAM DIVERGENCE

Description of the two forms

Morphologically, anadromous stickleback are remarkably uniform throughout their extensive geographic range; in contrast, even in small geographic areas, freshwater populations (especially those in lakes) are remarkably variable (Hubbs 1929; Heuts 1947*a*; Berg 1949; Okada 1960; Münzing 1963; Penczak 1965; Hagen 1967; Paepke 1968, 1970, 1971; Miller and Hubbs 1969; Igarishi 1970*b*; McPhail and Lindsey 1970; Aneer 1973; Bell 1976*a*; Ziuganov 1983; Francis *et al.* 1985; Mori 1987*a*; Ziuganov *et al.* 1987; Reimchen Chapter 9 this volume). Variability in freshwater populations is not random; there are often strong associations between specific environmental factors and morphology (Heuts 1947*a*; Hagen and Gilbertson 1972; Moodie and Reimchen 1976*a*; Gross 1977, 1978; Baumgartner and Bell 1984; Lavin and McPhail 1985; Reimchen *et al.* 1985; Reimchen 1989). One particularly striking association between habitat and morphology is found when lake- and stream-dwelling stickleback are compared. Both in Europe and in western North America, lake-dwelling stickleback have slimmer bodies, smaller mouths, and more, and longer, gill rakers than stream-dwelling stickleback (Hagen and Gilbertson 1972; Gross and Anderson 1984). The existence of similar patterns of variation in several inherited traits on two continents is compelling evidence that the association between morphology and habitat is a result of natural selection.

Morphologically divergent lake and stream stickleback normally are allopatric, but Moodie (1972*a*) briefly described a parapatric pair of lake–stream stickleback in the Mayer Lake system on Graham Island in the Queen

Charlotte Islands. More recently, Stinson (1983) and Reimchen *et al.* (1985) investigated a similar situation in the Drizzle Lake drainage, also on Graham Island, and found substantial morphological differences between stream fish and lake fish. In both the Mayer and Drizzle lake systems, the stream form is smaller than the lake form and has a deep body with a mottled colour pattern on the back and sides, and relatively few short gill rakers. In contrast, the lake form is larger, and has a slim body with uniformly dark back and sides, and numerous, long gill rakers (Table 14.2). These divergent lake and stream forms come into contact where streams enter and leave lakes.

Recently, Lavin and McPhail (1992) described a similar lake–stream pair in the Misty Lake system on northern Vancouver Island (Table 14.2). Again, the lake form is larger than the stream form and is slim bodied with uniformly dark back and sides and numerous, long gill rakers. The stream form is smaller and deep bodied, with a mottled colour pattern on the back and sides, and with relatively few short gill rakers. In the Misty Lake system, a discriminant function was used to classify fish as lake or stream type (Lavin and McPhail 1992). Not surprisingly, lake fish were found mainly in the lake, but small numbers of lake fish were collected in the stream at distances up to 300 m above the lake. In contrast, no stream fish were collected in the lake, but both forms were found in a swamp at the mouth of the inlet stream. In this transition area, stream fish were numerically dominant. Thus, although the two forms predominate in different habitats, there is a transition zone where they overlap (in breeding condition) and so they are in a position to exchange genes.

Table 14.2 compares the morphology of the Misty Lake pair with the two sets of parapatric lake and stream stickleback known from the Queen Charlotte islands. The morphological parallels are striking, particularly when the variability of stickleback from the same, and adjacent, river systems are considered (Moodie and Reimchen 1976*a*; Reimchen *et al.* 1985; McPhail and Haas unpubl. data). At the three sites, the members of the lake–stream pairs differ from each other in the same suite of traits, and the differences are in the same direction and of similar magnitude. It is unlikely that such parallels arose by chance.

Evidence for separate genomes

The Misty lake–stream pair, and their hybrids, were reared together in the laboratory. The major differences between the two forms (body shape, adult size, adult colour, and gill raker number) are retained in laboratory-reared fish (Lavin and McPhail 1992). In addition, F_1 hybrids reared under the same conditions are intermediate; this suggests that the major differences between the two forms are inherited. Also, the two forms breed true, and this result argues that the Misty lake–stream pair represent two distinct gene pools rather than some complex polymorphism.

In contrast to the morphological and breeding data, the available bio-

Table 14.2 Morphology[a] of a wild-caught lake–stream pairs from the three known sites: Misty Lake (Vancouver Island), Mayer and Drizzle Lakes (Graham Island), Canada.

Site	Form	Character				
		Spine length (mm)	Body depth (mm)	Gill raker number	Left plates	Colour of back and sides
Misty	Lake (N = 100)	5.9 (0.49)	4.2 (0.20)	19.8 (1.18)	6.1 (0.78)	Uniform black
	Stream (N = 100)	6.0 (0.46)	4.1 (0.23)	16.9 (1.23)	6.7 (0.45)	Mottled brown
Mayer	Lake	5.3	4.6	21.2	6.8	Uniform black
	Stream	6.4	4.4	16.6	4.7	Mottled brown
Drizzle	Lake	6.2	5.2	21.3	4.9	Uniform black
	Stream	6.7	4.6	17.4	3.6	Mottled brown

[a] Mayer data from Moodie (1972*a*); Drizzle data from Stinson (1983) and Reimchen *et al*. (1985). Spine length and body depth are ratios obtained by dividing the measurement by standard length. The values given are means with standard deviations (where available) in parentheses.

chemical data provide no evidence for differentiation between the forms. Sixteen enzymes representing 25 presumptive loci were examined in the two forms (Lavin and McPhail 1992). In the Misty populations, only two of these loci were polymorphic, and their allele frequencies were the same in both forms. This absence of allozymic differences between the forms does not necessarily imply a single gene pool; however, it does suggest the possibility of gene flow between the two forms.

Evidence for hybridization and gene flow

Barton and Hewitt (1985) point out that gene flow in areas of contact between divergent populations often involves only loci that are not subject to selection for alternative alleles in the two populations. Thus, in the Misty Lake system, if the allozymes assayed are effectively neutral in the two habitats, the absence of allele frequency differences between the two forms might indicate gene flow. In contrast, the morphological differences between the forms are clearly adaptive. The low gill raker number in the stream form is typical of stream-resident fish and is associated with a high frequency of macroinvertebrates in the diet (Hagen 1967; Gross and Anderson 1984; Hart and Gill this volume, Table 8.2, 8.3). A similar association exists for jaw length, and Lavin and McPhail (1986) demonstrated experimentally that large jaws and low gill raker number contribute to foraging success in stickleback feeding on benthic prey. In contrast, Bentzen and McPhail (1984) demonstrated that narrow jaws and increased gill raker number contribute to the foraging success of stickleback feeding on plankton. Similar cases can be made for the functional significance of other morphological traits that distinguish the two forms (e.g. body shape, Taylor and McPhail 1986; pelvic spine length, Gross 1978; and male colour, Reimchen 1989). Thus, it is possible that the morphological differences between the two forms are maintained by strong selection on opposite sides of the lake–stream ecotone, while gene flow occurs at other loci that are essentially neutral in the two habitats.

The seeming rarity of hybrids, however, suggests that gene flow between the two forms, if it occurs at all, is low. Two hundred and seventy-four fish were examined along a transect from the lake, through the inlet swamp and into the inlet stream. With one exception, the discriminant function unambiguously classified individuals as either lake or stream fish (Lavin and McPhail 1992). One individual from the lake had an intermediate score and was assigned with approximately equal probability to both groups. The morphology of this fish was similar to that of laboratory-reared F_1 hybrids, and it probably represents a hybrid between the two forms. This analysis suggests that occasional hybrids are produced. They are rare, however, indicating that either the incidence of hybridization is low or the survival of hybrids is poor. In either case, the net effect is that the two forms remain as separate ecological and genetic entities.

Nature of reproductive isolation

Little is known about reproductive isolation in these parapatric lake–stream pairs. In the Misty Lake system, there is no indication of major differences in breeding season between the forms. Gravid females of both kinds are abundant in the appropriate habitats during May and June. Breeding in both forms starts in April and is completed sometime in July. Most of the breeding appears to occur either in the lake or in the stream habitat, but some gravid females of both forms were collected in the transition area (the inlet swamp). Deeply stained water in both habitats makes observation difficult, and the only nests observed were a few located along the lake margin. No nests were observed in the stream, so it is not known whether the two forms nest in different microhabitats.

There are no obvious developmental or sterility barriers between the two forms. Both forms, and their F_1 hybrids, are interfertile, and there is no evidence of hybrid breakdown in F_2 zygotes (Lavin and McPhail 1992). Although this notion is untested, it is possible that adult F_2 hybrids are sterile or show reduced fertility. This is unlikely, however, since F_2 sterility does not occur in crosses between other, equally divergent, stickleback in the Strait of Georgia region (McPhail 1992).

The only attempt to examine premating isolation in a parapatric lake–stream pair is that of Stinson (1983). She investigated premating isolation in the Drizzle Lake lake–stream pair. As indicated earlier, there are remarkable parallels in morphology between the Drizzle lake and stream forms and those found on northern Vancouver Island. This similarity suggests that Stinson's findings may also be applicable to the Misty populations. Stinson (1983) used two field techniques to examine male mate preference: she simultaneously presented nesting males of the lake form either with two live gravid females (one of each form) in separate jars or with two models of gravid females suspended on threads. The number of approaches made by males to each jar was counted and found to differ. Males of the lake form showed a slight, but significant ($P < 0.05$), bias towards females of the lake form over females of the stream form. Hay (1969) used a similar jar technique with anadromous and stream-resident stickleback and obtained similar results (i.e. males showed a slight but significant bias towards females of their own kind). He followed his jar experiments with a series of more sophisticated male choice experiments in which there was direct contact between a male and two females, and in which the entry of a female into the nest was the criterion for choice. The results of these experiments confirmed the earlier jar tests. This suggests that field indications of positive assortative mating, even when obtained by relatively simple procedures, probably are valid.

Stinson's (1983) model experiments were an attempt to determine what aspects of the female phenotype elicited the male preference. Her experiments implicate female colour as a major factor and female size as a lesser, but still contributing, factor. Because females of the two forms of stickleback in the Misty system also differ in colour and body size, these same traits may contribute to positive assortative mating in the Misty lake–stream pair.

In summary, the absence of developmental and fertility barriers in laboratory-reared hybrids suggests that in nature, hybrids could act as a conduit for gene flow between Misty Lake stream and lake stickleback. The rarity of adult hybrids, however, argues that little, or no, gene flow occurs between the two forms. Since breeding adults of both forms are in contact in the transition zone between lake and stream, this scarcity of hybrids also suggests that some form of positive assortative mating probably occurs within the system.

Origin of the divergence

Any evolutionary model proposed to explain the origins of these lake–stream pairs must reconcile three observations: first, the restricted geographic distribution of the pairs (evidently they are confined to the central coast of British Columbia), second, the disjunct distribution of the pairs (they occur on two widely separated islands; Fig. 14.1), and third, the remarkable parallels among sites in the morphology of the divergent forms.

Two models might explain these observations. One is a historical model that postulates a single origin for the lake- and stream-dwelling forms. An alternative model postulates at least two parallel origins of the lake–stream pairs. The historical model envisages lake and stream forms evolving separately in the two habitats and then, after a period of divergence, coming into secondary contact. Some support for this hypothesis comes from the observation that, on average, allopatric populations of stream-dwelling and lacustrine stickleback differ morphologically. These differences involve the same traits, and are in the same directions, as those that distinguish the lake–stream pairs (Hagen and Gilbertson 1972; Moodie and Reimchen 1976*a*; Gross and Anderson 1984; Reimchen *et al.* 1985; Lavin and McPhail 1986). Consequently, the morphological differences observed between the members of lake–stream pairs are exactly the morphological differences expected from an allopatric origin in the two habitats. All that is needed to produce a parapatric pair is a change in drainage that brings previously isolated lake and stream populations into secondary contact. The parallels among the three sites are then explained by dispersal from this centre of origin.

A major problem for this model is the sea barrier between Vancouver Island and the Queen Charlotte Islands. Under this model, the lake–stream pairs had to diverge before the last (Fraser) glaciation and then either survive on both islands or postglacially disperse between islands. Both northern Vancouver Island and Graham Island contained small ice-free refugia during the last glaciation (Warner *et al.* 1982; Howes 1983), therefore stickleback could have survived on either island. However, inferential evidence argues against survival on Vancouver Island. Misty Lake is well outside the unglaciated region on northern Vancouver Island (the Brooks Peninsula) and is separated from the peninsula by a mountain range. Also, the terrain on the Brooks Peninsula is precipitous and the small drainages on the peninsula seem to lack freshwater stickleback (Peden pers. comm.). Consequently, if the resemblances between the lake–stream pairs on the two islands reflect common ancestry, the pairs probably survived glaciation on the Queen Charlotte Islands (or some central coast refugium) and dispersed postglacially to northern Vancouver Island. Postglacial dispersal of freshwater sticklebacks from the Queen Charlotte Islands to northern Vancouver Island is not impossible. Indeed, the endemic plant taxa shared between the islands (Ogilvie 1989) provide compelling evidence for just such a biogeographic link. Thus, the historic model provides plausible explanations for both the origin of the lake–stream pairs and the striking parallels found among the three pairs.

The parallel model envisages the evolution *in situ* of the lake–stream pairs. Here, a single population is envisaged as colonizing a lake–stream system, and then diverging into two forms in response to different selection pressures on either side of the ecotone that occurs wherever streams enter or leave lakes. Such parapatric divergence is theoretically possible (Endler 1973, 1977; Lande 1982), and was suggested as a potential explanation for the parapatric lake–stream pairs on Graham Island (Reimchen *et al.* 1985). The previously described pattern of morphological divergence typical of allopatric lake and stream stickleback is equally compatible with this model. The existence of such differences clearly establishes that both the selection regimes, and the evolutionary responses of stickleback, differ in the two habitats. Thus, the parallel model postulates separate origins for at least the Vancouver Island and Graham Island lake–stream pairs, and attributes the morphological similarities shared by the pairs to similar selection regimes on the two islands.

The parallel evolution model, however, fails to explain the restricted geographic distribution of the lake–stream pairs. *Gasterosteus aculeatus* occurs in temperate and subarctic coastal lowlands throughout the Northern Hemisphere. Over much of this vast area, contiguous stream and lake habitats must be common. If differential selection across the lake–stream boundary is a sufficient explanation for the origin of the lake–stream pairs, they should occur throughout the species' range. Seemingly, they do not. So far, they are restricted to Graham and northern Vancouver islands; however, they may turn up elsewhere on the central coast, especially in areas that may have been associated with a coastal refugium (McPhail and Lindsey 1986; Ogilvie 1989). Whether they occur outside the central coast area is unknown. On southern Vancouver Island and the adjacent mainland, however, over 50 lake–stream ecotones have been examined, and none contains clear lake–stream pairs. Consequently, it is unlikely that the lake–stream pairs were overlooked on southern Vancouver Island. If the restriction of the lake–stream pairs to the central coast is real, it implies that there is some factor unique to this area that is involved in their evolution. This factor could be either common ancestory or a unique, geographically restricted ecological situation. To distinguish between these alternatives requires differentiating between resemblances caused by similar selection regimes and resemblances caused by common ancestory. As Endler (1982) stressed, for relatively recent divergences, this can be difficult and often is impossible. Nonetheless, the historical (common ancestory) model requires an older (pre-Fraser glaciation) divergence than the parallel evolution model. Thus, molecular comparisons such as those reported by Gach and Reimchen (1989) might solve this dilemma. For now, however, the origin of the lake–stream pairs remains an open question.

INTRALACUSTRINE DIVERGENCE–THE BENTHIC-LIMNETIC PAIRS

Description of the two forms

As noted earlier, lacustrine stickleback usually have slimmer bodies, narrower mouths and more, and longer, gill rakers than stream-dwelling stickleback. Amongst lacustrine populations, however, there is considerable variability. This variability is not random, and among allopatric lacustrine populations there is an association between lake size and morphology (Lavin and McPhail 1985). In particular, gill raker number (a trophic trait) tends to increase with lake size. Usually, stickleback living in small shallow lakes have relatively few, short gill rakers, whereas stickleback living in large, deep lakes have relatively large numbers of long gill rakers. As gill raker number is heritable in *Gasterosteus* (Hagen 1973), and high numbers of gill rakers are associated with planktivory (Bentzen and McPhail 1984), this relationship probably reflects an increase in the relative importance of plankton in the diets of stickleback inhabiting large lakes.

There are, however, six small lakes (on three islands in the Strait of Georgia region) that provide a striking exception to this generalization about lake size and trophic morphology. Each of these lakes contains two trophic forms of *Gasterosteus*: a littoral benthic-foraging form, and a limnetic plankton-feeding form. For convenience, these trophic forms are referred to as benthics and limnetics (Larson 1976; McPhail 1984, 1992). The limnetic stickleback are slim bodied, with numerous, long gill rakers, narrow mouths, and long slim snouts. In contrast, the benthic stickleback are deep bodied, with a few short gill rakers, wide mouths, and short broad snouts. There also are differences between benthic and limnetic stickleback in other meristic and morphometric traits; however, these differences are not as consistent from lake to lake as the differences in trophic traits and body shape (Table 14.3). Thus, although benthic–limnetic pairs occur in six lakes, and the trophic roles of the members of each pair are similar in the six lakes, these benthic–limnetic pairs are not all identical. This variation suggests that either the local selection regimes differ among the lakes, or the founding populations were different in each of the lakes.

Evidence for separate genomes

Two of the lakes containing benthic–limnetic pairs have been studied intensively: Paxton Lake on Texada Island, and Enos Lake on Vancouver Island. Benthic and limnetic stickleback from these lakes, and their hybrids, have been reared in the laboratory. For both lakes, the major morphological differences between the forms are retained in laboratory-reared fish. In addition, F_1 hybrids reared under the same conditions are intermediate in most morphological traits (McPhail 1984, 1992). Also, the two forms breed true in both lakes: limnetic crosses produce only limnetics, and benthic crosses produce only benthics. This implies not only that the morphological differences between the benthic–limnetic pairs in each lake are inherited, but also that the two forms represent separate gene pools rather than some complex trophic polymorphism. Further evidence for separate gene pools is provided by allozyme data. Sixteen enzymes representing 25 presumptive loci were examined in the two forms. Five of the loci are polymorphic and there are significant, and temporally stable, allele frequency differences between the sympatric forms (McPhail 1984, 1992). These morphological and biochemical differences argue strongly that in both Paxton and Enos lakes the benthic–limnetic pairs represent distinct gene pools. Fewer data are available for the pairs in the four other lakes, but they appear to be similar in most details to the pairs in Paxton and Enos lakes.

Evidence for hybridization and gene flow

In Paxton and Enos lakes, the morphological evidence suggests some hybridization between the forms. Discriminant function analysis indicates that about 1 per cent of the adults in both lakes probably are hybrids (McPhail 1984, 1992). In Paxton Lake, the incidence of adult hybrids has remained remarkably stable over 20 yr (McPhail 1992). This temporal stability implies that although the forms hybridize, there is no ecologically significant gene flow and the integrity of the two gene pools remains intact. The biochemical evidence supports this contention. Where there are data for several years, the allele frequencies within each form remain in Hardy–Weinberg equilibrium and the frequency differences between the forms remain stable over time (McPhail 1992). In addition, in Enos Lake, one locus (s*MDH-1**) is fixed at the common allele in benthics, but limnetics show an almost 18 per cent frequency of a variant allele (McPhail 1984). Once again, this result implies that there is no significant gene flow between these sympatric benthic and limnetic stickleback. Indeed, all the available

Table 14.3 Morphology[a] of benthic–limnetic pairs from six lakes.

Island[b]	Lake	Form	Meristic character					Morphometric character			*N*
			Left plates	Dorsal rays	Anal rays	Dorsal spines	Gill raker number	Body depth (mm)	Jaw length (mm)	Gill raker length (mm)	
Vancouver I.	Enos	Limnetic	4.69	11.27	8.60	0.64	24.25	8.38	2.54	1.78	50
		Benthic	5.02	10.45	7.23	0.95	18.48	10.64	3.14	0.93	50
Texada I.	Emily	Limnetic	6.27	11.33	8.90	1.0	23.93	8.53	2.62	1.52	50
		Benthic	3.63	10.54	7.29	0.90	19.09	10.86	3.69	0.62	50
	Priest	Limnetic	6.70	11.80	9.48	0.98	23.65	8.69	2.93	1.54	50
		Benthic	3.98	10.54	7.43	0.93	18.71	10.92	3.62	0.62	50
	Balkwill	Limnetic	6.47	11.87	9.53	0.67	23.23	8.58	2.80	1.57	50
		Benthic	4.10	10.90	7.63	0.90	18.30	10.63	3.63	0.67	50
	Paxton	Limnetic	5.55	11.88	9.12	0.98	23.72	8.76	2.55	1.52	50
		Benthic	0.63	10.78	7.23	0.56	18.64	10.61	3.77	0.77	50
Lasqueti I.	Hadley	Limnetic	6.13	11.07	8.33	0.97	24.03	8.09	2.27	1.27	39
		Benthic	4.83	10.10	6.87	0.77	19.10	11.04	3.46	0.92	50

[a] Counts and measurements (after McPhail 1984) are means, except for dorsal spines which is the proportion of three-spined individuals in each population; measurements are standardized to a 50 mm standard length.
[b] See Fig. 14.2.

evidence suggests that the members of these benthic–limnetic pairs represent separate gene pools.

Nature of reproductive isolation

There are no developmental or sterility barriers between the benthics and limnetics in either Paxton or Enos lakes (McPhail 1984, 1992). Both forms, and their F_1 and F_2 hybrids and backcrosses, are interfertile and produce fully viable offspring. Since there is no evidence of gene flow between the forms, the absence of postmating isolating mechanisms implies the presence of premating mechanisms that restrict gene flow. One possible mechanism is nest location. The two forms breed at the same time (April through June) and often occupy adjacent territories; however, males differ in nesting microhabitat. Benthic males nest in, or close to, dense cover, whereas limnetic males tend to nest in open areas (Table 14.4, and Ridgway and McPhail 1987). Thus, nest microhabitat may provide females with clues about male genotype.

Although field observations indicate a high degree of positive assortative mating in Paxton Lake, it is only in Enos Lake that courtship in a benthic–limnetic pair has been thoroughly investigated. In a series of mate choice experiments, Ridgway and McPhail (1984) established that the members of the Enos Lake benthic–limnetic pair recognize individuals of their own form and, given a choice, prefer to mate with their own kind. Evidently both morphological and behavioural traits facilitate this recognition. For example, when courting, Enos Lake benthic males are entirely black. In contrast, courting limnetic males have blue backs and sides, and bright red throats. These differences in male nuptial colour provide a strong visual indication of genotype. Interestingly, equivalent differences in male nuptial colour are absent in the other benthic–limnetic pairs. In addition to male colour, there are behaviours that also may signal genotype. Courtship behaviour in both forms is similar, but the sequence, frequency, and vigour of some behaviours differ between the two forms. In particular,

Table 14.4 Association between nest cover and trophic form in the Paxton Lake benthic–limnetic pair.

Form	Cover (ranked)[a]				N[b]
	1	2	3	4	
Benthic	2	12	16	20	50
Limnetic	62	36	28	5	131

[a] Cover at nest is ranked: 1, nest completely exposed; 2, sparse cover at nest; 3, moderate cover at nest; 4, nest completely hidden.

[b] The difference in sample size reflects the difficulty of locating benthic nests and does not necessarily imply a difference in the density of the two kinds of males.

Enos Lake benthic and limnetic males differ in their first approach to females, and females differ in their response to the male's first approach (Ridgway and McPhail 1984). Together, these morphological and behavioural differences produce a high level of positive assortative mating. Presumably, similar levels of positive assortative mating occur in the other lakes containing benthic–limnetic pairs.

Origin of the divergence

Any evolutionary model attempting to explain the origin of these benthic–limnetic pairs must address three observations. First is their restricted geographic distribution: sympatric benthic–limnetic pairs are found only in the Strait of Georgia region (Fig. 14.1); nothing similar has been reported from elsewhere in the extensive range of *Gasterosteus* (McPhail 1993). Second is their presence on three islands (Vancouver, Lasqueti, and Texada; Fig. 14.2), and third is the morphological differences within the set for each trophic form among lakes.

Two speciation models might account for these observations: the allopatric model (Mayr 1942), and the sympatric model (Bush 1975). Although the conditions necessary for sympatric divergence are restrictive (Futuyma and Mayer 1980), many small lakes fulfil one of the major requirements of sympatric speciation models: they provide two discrete habitats over which disruptive selection could operate. Typically, there are major differences in the size distribution and types of prey available in the littoral and limnetic zones of small lakes (Mittelbach 1981*b*). Thus, these different habitats provide adjacent, but strikingly different, foraging opportunities, and fish clearly use them differently. For example, when Werner *et al.* (1981) introduced bluegills, *Lepomis macrochirus*, into a pond containing primarily littoral and limnetic prey, they found that individual fish rapidly became trophic specialists. Some individuals foraged mainly in the limnetic zone, whereas others foraged predominantly in littoral areas. What is important, however, is that trophic specialists did better (in terms of food intake) than individuals that foraged in both habitats. This suggests that at high population densities, and consequently intense intraspecific competition, individuals that specialized on either littoral or limnetic prey would be at an advantage relative to individuals that used both habitats. If the tendency to specialize is partially inherited, the frequency of specialists in the population would rise. Thus, even among trophic specialists, the most efficient phenotypes would be favoured and disruptive selection based on competitive interactions between genotypes could lead to divergence (Wilson 1989) and, perhaps, to sympatric speciation (Rosenzweig 1978; Rice 1987).

With disruptive selection, however, positive assortative mating does not necessarily accompany morphological divergence (Felsenstein 1981). Most models of sympatric speciation require either a genetic association between mate preference and the traits under disruptive selection, or mating in

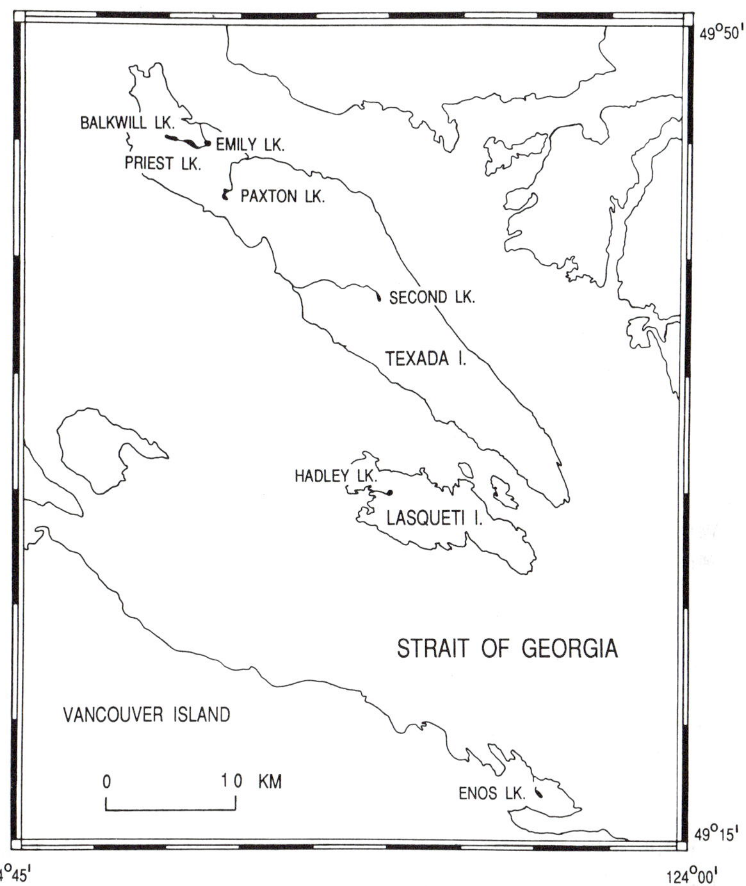

Fig. 14.2 The central Strait of Georgia region, showing lakes that contain benthic-limnetic pairs (note that they occur on three islands). The one lake without a pair is Second Lake; it contains an introduced population of hybrid origin.

separate habitats (Maynard Smith 1966; Pimm 1978; Halliburton and Gall 1981; Rice 1987). Although sympatric benthic and limnetic stickleback are adapted to exploit alternative foraging habitats (Larson 1976; Bentzen and McPhail 1984; Bentzen *et al.* 1984), there is no strong evidence that selection for foraging efficiency in these different habitats influences mate choice. In addition, even though benthic and limnetic males nest in different microhabitats, the nests of both forms occur only in the benthic foraging habitat. Thus, the benthic–limnetic pairs fail to meet a major requirement of sympatric speciation models: reproduction in different habitats.

The strongest argument against a sympatric origin for the benthic-limnetic pairs is their restricted geographic distribution. Over the vast geographic range of *G. aculeatus*, there must be many small, productive lakes that contain discrete benthic and limnetic foraging habitats. If disruptive selection operating across these habitats were sufficient to produce divergence, benthic–limnetic pairs should occur elsewhere. They do not. Indeed, there are lakes on Texada, Lasqueti, and Vancouver islands that contain only a single trophic form of stickleback but otherwise appear to be ecologically similar to adjacent lakes that contain benthic–limnetic pairs (McPhail 1993). This disparity argues that some factor, besides the availability of alternative habitats, is involved in the origin of these pairs.

All of the lakes containing benthic–limnetic pairs were ice-covered during the last (Fraser) glaciation, and then postglacially submerged beneath the sea. The landforms in the central Strait of Georgia only reached their present conformation about 11 000 yr ago. Curiously, pairs of sibling fish species are especially common in areas with histories of recent glaciation (Svärdson 1961; Lindsey 1963; McCart 1970; Henricson and Nyman 1976; Copeman 1977; Bodaly 1979). The usual explanations for such species pairs are multiple invasions. For example, Svärdson (1961) used two invasions from the sea to explain sibling species pairs in the Baltic region. He postulated an initial colonization followed by a period of isolation and divergence, and then a second invasion by the same species. A similar double-invasion hypothesis provides a plausible explanation for the origin of the sympatric pairs of benthic and limnetic stickleback. Like most lakes in the Strait of Georgia region, the lakes containing benthic–limnetic pairs probably were colonized from the sea during, or immediately after, the period of postglacial submergence. Later, isostatic rebound isolated these lakes and initiated a period of divergence in fresh water. Populations of *G. aculeatus* isolated in small lakes usually diverge from marine populations in trophic morphology and body shape (Hagen and Gilbertson 1972; Reimchen *et al.* 1985; Lavin and McPhail 1986), and this divergence typically is in the direction of benthic sticklebacks (i.e. a reduction in gill raker length and number, an increase in mouth width, an increase in body depth, and a reduction in median fin ray numbers). If, after many generations isolated in fresh water, these lakes were reopened to colonization from the sea, the original colonists may have diverged sufficiently that they would no longer mate randomly with the second set of colonists. Since the original colonists already had diverged from the plankton-feeding marine form towards littoral foraging, this option no longer would be available to the second set of colonists. If, however, there was sufficient limnetic production in the lake, the second colonists might persist as plankton feeders. Thus, double invasion could give rise to two trophic forms in lakes with adequate littoral and limnetic production.

In the central Strait of Georgia, there are several reasons why double invasion is an attractive explanation for the origin of the benthic–limnetic pairs. First, the hypothesis may be testable. It predicts a specific sequence of colonization and divergence: that the benthics are derived from the first colonists, and that the second set of colonists gave rise to the limnetics. Consequently, the benthic members of each pair have had longer to diverge from the ancestral marine form than the limnetics, and this difference in divergence time might be detectable at the molecular level. Second, although situations analogous to a double invasion are rare (i.e. where marine or anadromous stickleback gain entry to a lake that already contains freshwater stickleback), two cases are known (Narver 1969; Ziuganov *et al.* 1987). In both cases, the different forms appear to maintain themselves as separate reproductive units. Thus, a primary assumption of the double-invasion hypothesis—positive assortative mating between successive sets of colonists—appears to be confirmed. Third, and most important, there is geological evidence for an event in the Strait of Georgia that may have provided the opportunity for a second invasion from the sea. Mathews *et al.* (1970) postulate a second, smaller marine submergence in Georgia Strait about 2000 yr after the first submergence. It seems that this second submergence was short lived, and restricted to this region. It involved a change in sea level of about 50 m. Such a change would flood the barriers separating most of the lakes containing benthic–limnetic pairs from the sea and reopen them to colonization; however, it would not submerge these lakes. This local event provides a plausible explanation for both the origin of the benthic–limnetic pairs and their restricted geographic distribution. It also provides opportunities to corroborate the hypothesis. If a second marine submergence is necessary to explain the origin of the benthic–limnetic pairs, the pairs should be confined to a relatively narrow altitudinal range. Lakes below 50 m would have been resubmerged, and lakes above 100 m probably would have remained isolated from the sea. On Texada Island, the lakes with and without benthic–limnetic pairs roughly fit this pattern (McPhail 1993).

Two other important assumptions are implicit in the double-invasion hypothesis. The first is parallel evolution. The hypothesis assumes that the benthic–limnetic pairs evolved separately, but contemporaneously, on three islands. The morphological differences among the pairs on different islands (Table 14.3) are consistent with this assumption but do not constitute compelling evidence for parallel evolution. Again, however, this assumption may be testable at the molecular level. The second assumption is that there were competitive interactions between the two sets of colonists. McPhail (1993) discussed the evidence for trophic character displacement in the benthic–limnetic pairs. He argued that competitive interactions with the original colonists forced the second set of colonists to specialize in a trophic niche (plankton feeding) that normally is marginal for stickleback in small, shallow lakes (see also Schluter and McPhail 1992).

DISCUSSION

The genetics of speciation

The division of a single species into two or more species involves two interrelated processes: genetic divergence, and the evolution of reproductive isolation. Under the geographic speciation model, genetic divergence is a consequence of the splitting of a species' range into two or more isolated fragments, which then diverge genetically in response to different local selection regimes and eventually evolve into distinct species. Traditionally, geographic speciation is viewed as a gradual process, but this is not necessarily true. In particular, Mayr (1954, 1963, 1982) has addressed the problem of rapid geographic speciation. He argued that divergence can occur swiftly in populations isolated on the periphery of the range of widespread species. Such peripheral isolates often are founded by a small number of individuals representing only a fraction of the genotypes present in the main body of the species. Small numbers, complete isolation, and new selection pressures combine to initiate a 'genetic revolution' in some of these peripheral populations, and this, in turn, allows the fundamental genetic changes that facilitate rapid speciation. Mayr (1982) suggested that this process is sufficiently common that it deserves recognition as a special form of geographic speciation—peripatric speciation.

Carson (1975, 1982, 1987) used a similar model to explain the remarkable speciation found in Hawaiian *Drosophila*. The crux of Carson's model is that a species' genome consists of 'open' and 'closed' genetic systems. The 'open' system generates the normal genetic variability open to selection; the 'closed' system is strongly coadapted and contains polymorphisms that limit free recombination. Thus, the 'closed' system maintains the integrity of the species and is resistant to change. Carson (1987) argued that for rapid speciation, this 'closed' system must be destabilized, and that destabilization is most likely to occur during colonization. For Hawaiian *Drosophila*, Carson envisaged colonization by a small number of individuals (often a single inseminated female), followed by a 'population flush' during which selection is relaxed. Under these conditions novel recombinant phenotypes survive, and some may even flourish, in the new environment. Eventually, these novel forms may give rise to a new species.

Templeton (1981, 1987) also discussed the importance of founder effects in speciation, and proposed a genetic mechanism different from those proposed by either Mayr or Carson. The genetic details of these theories are beyond the scope of this review (see Barton and Charlesworth 1984 and Carson and Templeton 1984 for discussions); however, as explanations for rapid speciation, they all invoke colonization of a new area by a small number of founders, followed by some fundamental change in the genetic architecture of the new population.

The tempo of speciation of sticklebacks in the Strait of Georgia region

Apparently, the threespine sticklebacks, *G. aculeatus*, found in lowland lakes along the coast of British Columbia have gone through a process similar to that envisaged in models of rapid speciation. These lakes contain a remarkable array of morphologically and genetically distinctive stickleback populations (Hagen and Gilbertson 1972; Moodie 1972*a*; Larson 1976; Moodie and Reimchen 1976*a*; McPhail 1977, 1984, 1992, 1993; Reimchen *et al.* 1985; Reimchen Chapter 9 this volume). Since the entire Strait of Georgia region was glaciated, and the lowland areas were later flooded by the sea (Clague 1981), all the freshwater populations in this area are thought to be of postglacial origin. Therefore the maximum age of these divergent populations is about 11 000 yr and some could be younger. Certainly, the level of allozyme divergence among lacustrine populations, and between lacustrine populations and their putative marine ancestors, is consistent with a hypothesis of recent origin. In the Strait of Georgia region, the average genetic distance (Nei 1978) among populations in lakes that contain a single form of *Gasterosteus* is 0.024 ± 0.014 (30 lakes, 25 loci; see McPhail 1992 for the proteins and electrophoretic conditions). In freshwater fish, this level of genetic divergence is typical of relatively recent, geographically separate populations of the same species (Avise and Smith 1977; Kornfield *et al.* 1981).

Evolution of isolating mechanisms

Most divergent stickleback populations are allopatric and occur in isolated lakes. Consequently, it is difficult to determine whether barriers to gene flow have accompanied their morphological divergence. In the three combinations reviewed earlier, however, pairs of genetically divergent stickleback are in contact and so can be tested for reproductive isolation. In one set (the benthic–limnetic pairs), the two forms are not only reproductively isolated but also show well-developed resource partitioning. Because they are completely sympatric but maintain distinct and separate gene pools, they are 'good' biological species. This conclusion applies in spite of the fact that their morphologies are entirely encompassed within the range of variation typical of allopatric populations of *G. aculeatus*. The situation in the other two sets of pairs is not as clear. In both, contact between members of the pairs is parapatric, and a consequence of parapatry is that only a small proportion of the individuals in each population ever encounter individuals of the other population. Thus, unless the young remain near their birthplace, or return to the area as breeding adults, dispersal and recombination will continually erode any selection for positive assortative mating that might occur at the point of contact (Felsenstein 1981; Barton and Hewitt 1985). As a consequence, although parapatric contact zones can persist for long periods of time, complete reproductive isolation is unlikely to evolve (Sanderson 1989). In theory, then, neither the lake–stream pairs nor the anadromous–stream-resident pairs are likely to achieve complete reproductive isolation. Nevertheless, in both sets of pairs, although there is the opportunity for gene exchange, there is no evidence of introgression. Instead, these parapatric populations appear to maintain themselves as separate, independent gene pools, and this implies the existence of barriers to gene flow.

How barriers to gene flow evolve is a contentious issue in evolutionary biology. Typically, such barriers are divided into two types of isolating mechanisms: premating, and postmating (Dobzhansky 1937, 1940; Mayr 1963). Although this classification is oversimplified (Littlejohn 1981), the distinction is useful. There is general agreement that postmating barriers to gene flow (e.g. developmental incompatibility, sterility, and hybrid inferiority) arise as fortuitous by-products of genetic divergence. There is less agreement, however, regarding the evolution of premating barriers to gene flow. The argument centres on whether or not premating barriers to gene flow between divergent forms evolve only in isolation, or whether such barriers can be strengthened in areas of contact (Paterson 1978, 1981, 1982; Felsenstein 1981; Butlin 1989; Coyne and Orr 1989; Sanderson 1989).

The phenotypic and ecological basis of reproductive isolation

A striking feature of reproductive isolation in all the divergent pairs of stickleback is the absence of developmental or sterility barriers. Laboratory crosses, including F_2 and backcrosses, between members of divergent pairs typically produce viable, fertile offspring. The absence of such barriers to gene flow indicates that the overall genetic divergence between these pairs is so slight that it has not affected the compatibility of their ontogenetic and reproductive processes. Again, this implies a relatively recent origin for these divergent forms. Although there is no evidence of developmental or sterility barriers to gene flow in stickleback in the Strait of Georgia region, it is clear that such barriers between divergent populations of *Gasterosteus* can exist. In Japan, Honma and Tamura (1984) reported that male and, to a lesser extent, female hybrids between allopatric populations of anadromous and freshwater (but completely plated) stickleback are sterile. The existence of hybrid sterility in these interpopulation crosses demonstrates that postmating barriers to gene flow can arise as incidental by-products of genetic divergence in isolated stickleback populations. Presumably, however, these Japanese populations have been separated for a longer time than equivalent populations in the Strait of Georgia region. The relatively great genetic distances reported by Haglund *et al.* (1992*a*; Buth and Haglund page 78 this volume, Fig. 3.4) for some pairs of Japanese populations are consistent with a long period of isolation.

One postmating barrier that has both ecological and behavioural components, and appears to restrict gene flow between divergent populations in the Strait of Georgia region, is hybrid inferiority. In this area, some hybrids occur wherever there is contact between divergent pairs of stickleback. The presence of hybrids clearly indicates that premating isolation between members of divergent pairs is not complete. Although hybrids exist, however, they do not provide a bridge for significant gene flow between pairs. There may be gene flow between members of pairs at loci that are not under strong selection (Barton and Hewitt 1985; Harrison *et al.* 1987; Gach and Reimchen 1989; Rand and Harrison 1989), but if such gene flow occurs, it has not altered the basic morphological and ecological integrity of the divergent pairs. Thus, even in the face of persistent hybridization, divergent pairs of stickleback maintain their identity and there is no strong evidence of introgression. This implies that their gene pools are separate and effectively independent.

Typically, divergent pairs occupy sharply different habitats, and most of the morphological differences between members of pairs appear to fit them for life in different habitats (Bentzen and McPhail 1984; Taylor and McPhail 1986; Lavin and McPhail 1992). Because hybrids between divergent forms usually have intermediate phenotypes, hybrids are inferred to be competitively inferior to the parental forms, at least in the parental habitats (Heuts 1947*a*; Hagen 1967; Bell 1976*a*). This inference has been tested only in the benthic–limnetic pair. In this pair, Bentzen and McPhail (1984) experimentally demonstrated that the intermediate trophic morphology of hybrids translates into an intermediate foraging performance. That is, hybrids are not as efficient as benthics when foraging on bottom organisms or limnetics when foraging on plankton.

Aside from foraging behaviour, little is known about the behaviour of hybrid stickleback. However, the territorial, nest-building, courtship, and parental behaviours of *Gasterosteus* are complex, and these reproductive behaviours are an obvious area where intermediate, or aberrant, hybrid behaviour might be a disadvantage. McPhail (1969) observed that male hybrids between two phenotypically divergent stream-resident species exhibited all the normal *Gasterosteus* paternal behaviour, but their fanning was greatly reduced, and egg mortality was sharply increased compared with non-hybrid controls. In Japan, Honma and Tamura (1984) also suggested that fanning efficiency is reduced in hybrids between allopatric stickleback populations, and this might result in zygote mortality that is high relative to the parental forms. Also, courtship behaviour has been studied in two of the sets of divergent pairs (the anadromous–stream-resident pairs, and the benthic–limnetic pairs), and the members of these pairs differ only slightly in their courtships (Hay and McPhail 1975; McPhail and Hay 1983; Ridgway and McPhail 1984). In most stickleback populations, there is competition for mates and not all reproductive individuals breed (Kynard 1978*a*; Li and Owings 1978*a*; Borland 1986; McLennan 1989). Consequently hybrids, with their intermediate appearance and perhaps slightly aberrant behaviours, may be at a courtship disadvantage relative to the 'pure' parental forms.

This notion emphasizes an important aspect of hybrid inferiority in sticklebacks: it is relative. On their own, hybrids are capable of sequestering sufficient resources from the environment to survive, grow, and reproduce. They are competitively inferior only relative to the 'pure' parental forms. For example, first-generation hybrids between Paxton Lake benthic and limnetic stickleback were introduced into a small, stickleback-free lake on Texada Island (McPhail 1993). These hybrids founded a population that has survived for nine generations; clearly, they are capable of surviving and reproducing in the wild. This observation implies that hybrid inferiority in stickleback is a function of competitive interactions with the parental forms and, presumably, the degree of hybrid disadvantage varies with the density of the parental forms. If one parental form or both, declines in numbers, the survivorship of hybrids probably would increase and this could lead to introgression (Hubbs 1955). Thus, a major barrier (hybrid inferiority) to gene flow between divergent stickleback pairs is density dependent and, in *Gasterosteus*, population density is sensitive to changes in both the physical and the biotic environment (Eggers *et al.* 1978; Wootton 1984*a*). Consequently, hybrid inferiority, in itself, is too fragile to be an effective long-term barrier to gene flow. Nevertheless, hybrid inferiority may be important in the evolution of premating isolating mechanisms in some divergent pairs of stickleback. For example, Sanderson (1989) suggests that reinforcing selection may be possible in cases of complete sympatry. Thus, in the benthic–limnetic pairs, hybrid inferiority could provide the penalty for mismating that is necessary to reinforce selection for homogamy.

Except for the anadromous–stream-resident pair in the Little Campbell River, temporal isolation appears to be unimportant as a premating isolating mechanism in the Strait of Georgia region. In this area, premating isolation in divergent pairs of stickleback results primarily from ecological and behavioural factors. For example, in all of the divergent pairs there are microhabitat differences between forms in the location of nests (Hagen 1967; Stinson 1983; Ridgway and McPhail 1987). Although little is known about factors that influence mate choice in sticklebacks, differences in nest sites could contribute to reproductive isolation. In the wild, particularly in clear lakes, females probably make visual contact with potential mates at a distance. If females assess the nest microhabitat, and there is experimental evidence that they do (Sargent 1982), this may be the first in a series of choices that eventually leads to selecting a mate from the right gene pool.

Courtship is potentially an important premating isolating mechanism in sticklebacks. In *Gasterosteus* both sexes are involved in mate choice (Hay and McPhail 1975; McPhail and Hay 1983; Rowland page 313 this volume; Borland and McPhail unpubl. data), and courtship provides an opportunity not only for the choice of a mate from the right gene pool but also for the choice of the 'best' available mate from within the appropriate gene pool. For reproductive isolation, only the choice of a mate from the right gene pool is relevant; however, for the evolution of reproductive isolation, factors that influence mate choice within a population are important. This is because the components of the phenotype (behavioural or morphological) that influence mate choice within a population probably also affect mate choice between populations. Both morphology (Bell 1984*a*, page 470 this volume) and behaviour (Foster Chapter 13 this volume) in *Gasterosteus* can diverge rapidly in isolation. Thus, if mate choice within two allopatric populations is based on different morphological traits (or on traits selected in different directions), the populations probably would not mate randomly if they came into contact. This conclusion suggests that the traits involved in homogamy in stickleback are initially subject to selection (natural or sexual) for their effects on mate choice in allopatry, although their efficiency as isolating mechanisms may later be reinforced in areas of contact.

In divergent pairs, both behaviour and morphology appear to contribute to positive assortative mating. For two of the pairs there are data on courtship behaviour, and in both cases the differences in courtship between the forms are remarkably similar. There are no unique behaviours that might act as a genomic recognition signal; however, the sequence, frequency, and vigour of behaviours often differ between members of a pair. In the Little Campbell anadromous–stream-resident pair, and in the Enos Lake benthic–limnetic pair, the behaviour of males on their first approach to gravid females entering their territories differs between forms (McPhail and Hay 1983; Ridgway and McPhail 1984). The response of females to this first approach also differs, and forced courtships (where neither sex is given a choice) between members of a pair often break off at this point. Clearly, if a courtship behaviour is to serve as a species recognition signal, the signal would be most efficient at the beginning of courtship (Liley 1966). Still, in the laboratory, forced courtships between members of different gene pools eventually succeed. This suggests that although behavioural differences may contribute to positive assortative mating between divergent forms, they are not sufficient in themselves to completely separate the gene pools.

Even when both sexes are members of the same gene pool, a surprising number of courtships, both in the laboratory and in the field, are broken off before egg deposition. In the laboratory, this cessation is partly due to errors in estimating the female's physiological condition (Borland 1986), but in the field, where entry into courtship is voluntary, the high proportion (up to 70 per cent, McPhail pers. obs.; S.A. Foster pers. comm.) of aborted courtships suggests that both sexes 'shop around' for appropriate mates. Courtship brings males and females into close proximity, and thus provides the opportunity for a scrutiny of the morphology of potential mates. How morphology influences mate choice in stickleback has received little attention; however, there is experimental evidence that at least two morphological traits are involved in mate choice. These are male nuptial colour (McLennan and McPhail 1989*a*) and body size (Moodie 1982; Ridgway and McPhail 1984; Rowland page 330 this volume; Borland and McPhail unpubl. data).

If reinforcing selection increases the efficiency of reproductive isolation in areas of contact between divergent stickleback, it might be detectable in these morphological characters. This possibility was examined in the Salmon River, near Fort Langley, British Columbia. Here, Borland (1986) investigated the influence of female body size on male mate choice in stream-resident stickleback. In the Salmon River, stream-resident stickleback occur throughout the stream, whereas anadromous stickleback occupy only the lower 5 km of the stream, and then only during the reproductive season. Above the point of maximum inland penetration of anadromous stickleback there is an abrupt change in the size of resident females. In the area of sympatry, resident females average about 45 mm SL, while in the area of allopatry resident females average about 50 mm SL. In contrast, stream-resident males are the same size throughout the stream. Borland (1986) presented stream-resident males with a choice between two gravid females: one larger, and one smaller, than the male. The males came from two areas: sites above, and sites within, the area of sympatry with the anadromous form. Males from upper sites showed a strong preference for females larger than themselves, while males from the area of sympatry showed a strong preference for females smaller than themselves (Table 14.5). Since fecundity in *Gasterosteus* increases with female size (Baker page 164 this volume), the preference of lower river males for small females is contrary to expectation (Bell 1984*b*; Rowland 1989*b*, page 330 this volume). In the lower river, however, female size is a reliable indicator of genome (anadromous females range from 58 to 64 mm SL). This shift in male

Table 14.5 Male preference[a], relative to female size, in stream-resident stickleback from within (lower river) and above (upper river) the region of overlap with anadromous stickleback in the Salmon River, Fort Langley, British Columbia (after Borland 1986).

Male provenance	Female size		*P*
	Female > male	Female > male	
Lower river	2	11	<0.05
Upper river	11	0	<0.05

[a] The criterion for male choice was a successful courtship (i.e. female entry into the nest); probability values are from the binomial test (one tail).

preference within the stream-resident population favours positive assortative mating in the area of sympatry and thus suggests reinforcing selection.

In summary, gene flow between divergent populations of *G. aculeatus* is not restricted by any single factor. Instead, there is a constellation of factors, each of which contributes to restricting gene flow between the sympatric or parapatric forms but does not, in itself, constitute a total barrier to gene exchange. Some of the components of reproductive isolation (e.g. habitat preferences, spawning times, and the behavioural and morphological traits that result in hybrid inferiority) clearly are adaptive. Differences in these traits probably first evolve, in response to local selection regimes, either in geographic isolation or on opposite sides of ecotones dividing parapatric populations. Some sexually dimorphic traits, such as male nuptial colour, may be the product of sexual selection, although in *Gasterosteus* it is not easy to separate sexual and natural selection (McLennan and McPhail 1989*b*; Reimchen 1989). What is clear, however, is that there is active mate choice in stickleback and this is likely to result in sexual selection on some traits. If either the intensity of sexual selection, or the traits under sexual selection, differ in geographically separate populations or on opposite sides of an ecotone, this difference will result in some positive assortative mating at contact. The 'specific mate recognition systems' postulated by Paterson (1978, 1981, 1982) may arise in this way. Although most isolating mechanisms in divergent pairs of stickleback probably start to evolve in isolation, there is evidence that selection against hybrids in areas of contact enhances the effectiveness of barriers to gene flow (Borland and McPhail unpubl. data).

Imprinting and rapid speciation in threespine stickleback

Regardless of their mode of origin, effective barriers to gene flow have evolved remarkably rapidly in these divergent pairs of stickleback and, in the Paxton and Enos benthic–limnetic pairs, divergence seems to have reached the level of biological species. For fish, the evolution of species in less than 12 000 yr is unusually fast but not unique. There is evidence (Greenwood 1965) for the evolution of five species of *Haplochromis* (Cichlidae) in Lake Nabugabo (an isolated lagoon off Lake Victoria, Uganda) in 4000 yr. These cichlids, however, are still geographically isolated from their putative parents. Consequently, their status as species is based solely on their level of morphological divergence. Although morphological divergence also is a striking attribute of isolated stickleback populations, it is the rapid evolution of reproductive isolation in *Gasterosteus* that is remarkable and requires explanation.

A mechanism that may account for the rapid evolution of reproductive isolation in stickleback is imprinting. Immelmann (1975) notes that imprinting is widespread amongst vertebrates, and he drew attention to the potential evolutionary importance of the phenomenon. Typically, imprinting occurs at a restricted time of life (usually when individuals are young); it is persistent, and often involves species-specific characters (Immelmann 1975). The reproductive biology of stickleback appear to predispose them to imprinting. Young stickleback remain in the male's territory for up to 3 wk (pers. obs.), although in most populations the residence time is closer to 1 wk. If imprinting occurs at this early age, it could profoundly affect future mate choice. During their time in the male's territory, the young have an opportunity to imprint on the nest microhabitat. In addition, young females could imprint on their father's phenotype (e.g. body shape, colour). In a colonizing species like *G. aculeatus*, such imprinting could greatly accelerate the development of a new mating system. Certainly, the invasion of fresh water by stickleback usually must be followed by an episode of strong directional selection in this new environment. If individuals showing a greater than average development (or reduction) in some inherited trait that effects their external phenotype are at a selective advantage in the new environment, they will increase in frequency. If, in addition, female offspring imprint on their father's external phenotype, the proportion of females preferring that phenotype will also rise in the population. Consequently, with a combination of strong directional selection and imprinting, it might be possible to establish a new population-specific mating system without a genetic correlation between male phenotype and female preference. Also, when two previously isolated, and morphologically slightly differentiated, populations come into contact, imprinting alone should assure a significant level of positive assortative mating, even at first contact. If, in addition, there is a hybrid disadvantage, reproductive isolation should rapidly evolve.

In the Strait of Georgia region, *Gasterosteus* appears to be basically a marine animal with a large population reservoir in the sea. From this sanctuary the species has repeatedly colonized what can be viewed as a peripheral environment, fresh water. Given the large populations of stickleback in the sea and in estuaries, as well as the dense shoals of anadromous stickleback that migrate into fresh water, it is unlikely that freshwater populations founded directly from the sea were colonized by small numbers of individuals. Indeed, there is no evidence of either founder effects or 'genetic revolutions' in most freshwater populations. The same electrophoretically detectable loci are polymorphic in marine and freshwater populations, and all common alleles (present at a frequency of at least 1 per cent) are found in both environments (Withler and McPhail 1985). The same is true of morphology. Although some isolated freshwater populations show extreme expression, or more commonly extreme reduction, of morphological traits, there are no fundamental differences in morphology between marine and freshwater populations. The morphological differences that exist are in degree rather than in kind, and all populations clearly are recognizable as threespine stickleback.

The adaptive nature of speciation in threespine stickleback

If rapid speciation in stickleback is not based on founder effects and major genetic changes, what drives speciation in *Gasterosteus*? The answer appears to be natural selection. On recently deglaciated, geologically active coasts, freshwater environments are ephemeral: they appear and disappear with land movements and changes in sea level. In such areas, marine stickleback quickly colonize new freshwater environments. Many of these newly founded freshwater populations rapidly diverge from the marine populations. Typically, the freshwater fish faunas of such areas are depauperate and contain only a few euryhaline species. Therefore, predatory and competitive interactions in recently deglaciated areas probably are reduced relative to both the marine environment and freshwater areas with more diverse faunas. Perhaps this reduction allows *Gasterosteus* to exploit ecological opportunities that are not available in other areas. Certainly, in fresh water, most of the morphological divergence from the marine phenotype involves traits associated with ecological factors (e.g. foraging: gill raker number and length, mouth size, and body shape; predation: spine number and length, lateral plate number, and pelvic girdle structure; reproduction: fecundity, size at first maturity, and male nuptial colour). All these traits are inherited and therefore can respond to local selection. The repeated appearance, under similar ecological circumstances, of similar phenotypes in widely separate geographic areas (e.g. Hagen and Gilbertson 1972; Hagen and Moodie 1982; Reimchen 1983, 1989, page 248 this volume; Baumgartner and Bell 1984; Bell 1987) argues that divergence in *Gasterosteus* is adaptive. In all three sets of divergent pairs of stickleback there is some evidence of parallel evolution, and for the anadromous–stream-resident pairs, parallel evolution is the only compelling explanation for the circumboreal distribution of both forms.

In summary, rapid divergence in *Gasterosteus* in the Strait of Georgia region is associated with the colonization of depauperate coastal freshwater environments. There is no evidence that either founder effects or major genetic reorganizations are involved in this divergence, although some isolated populations may have gone through population bottlenecks (Withler and McPhail 1985). Instead, divergence in *G. aculeatus* appears to be driven by natural selection, and the available evidence indicates that this divergence is unrelentingly adaptive (Heuts 1947*a*, Hagen 1967; McPhail 1969, 1984, 1992, 1993; Hagen and Gilbertson 1972; Moodie 1972*a*,*b*; Moodie and Reimchen 1976*a*, Bell 1976*a*, 1984*a*; Gross 1977, 1978; Gross and Anderson 1984; Reimchen 1983, 1989). Most of the morphological traits involved are polygenic (Hagen and Gilbertson 1973*a*; McPhail 1977, 1984, 1992; Lavin and McPhail 1986); however, some traits that commonly diverge in freshwater populations (e.g. spine number, pelvic girdle structure, and lateral plate expression) appear to have thresholds (McPhail unpubl. data). In these cases, the genetic basis of the trait is polygenic but the phenotypic distribution is discontinuous. With such traits, small shifts in allele frequencies can produce major changes in phenotype and, thus, rapid morphological change. The expected pattern of divergence with threshold traits (striking morphological shifts but only minor genetic change) is typical of divergence in *Gasterosteus*. Interestingly, threshold traits also have been implicated in the rapid divergence of cave fish from their epigean ancestors (Wilkens and Hüppop 1986).

The paradox of rapid speciation and infrequency of old species of *Gasterosteus*

Most of the divergent populations of *Gasterosteus* in the Strait of Georgia region have evolved in geographic isolation; however, for two sets of divergent pairs a parapatric origin is possible, and for the anadromous–stream-resident pairs, the repeated *in situ* evolution of the stream form is the simplest explanation for the wide geographic distribution of both forms. There is no compelling evidence for sympatric speciation in stickleback from the Strait of Georgia region; however, speciation in *Gasterosteus* is something of a paradox. The evidence from the Strait of Georgia indicates that *Gasterosteus* is prone to isolation and rapid adaptive divergence. In addition, some divergent populations come into contact in this area, and where this happens, barriers to gene flow have evolved in a remarkably short time: in the benthic–limnetic pairs, well-defined biological species have evolved in less than 11 000 yr. Given this propensity to diverge, and the fact that the lineage has been present on the Pacific coast of North America since at least the Miocene (Bell 1977, page 441 this volume), one might expect *Gasterosteus* to be a speciose genus. It is not. There are only two well-marked (old?) species in the genus: the widely distributed *G. aculeatus* and a species confined to a narrow area on the Atlantic coast of North America, *G. wheatlandi*. From the evidence presented in this review it is clear that in the fresh waters of western North America, *G. aculeatus* is a complex of species; however, it is a complex of recently evolved species! Presumably, *Gasterosteus* has been invading and diverging in fresh water for millions of years (Bell page 446 this volume). Why, then, are there not more species, especially in areas that were not covered by the last glaciation? The answer may lie in the suggestion that *G. aculeatus* is basically a marine species. For a marine species, even one that regularly invades fresh water, lakes and streams are marginal environments. In such situations, *Gasterosteus* may resemble a weed: an excellent colonizer but a poor competitor. Thus, evolutionarily significant divergence may be possible only in areas that lack a primary freshwater fish fauna and are completely isolated from the sea. Certainly, most known divergent populations occur in ephemeral, isolated habitats (e.g. in lakes, especially lakes on islands, along recently glaciated coasts). Populations in such localities are particularly vulnerable, either to extinction or to periodic recolonization from the sea. Consequently, most divergent populations and biological species that evolve under these conditions are doomed either to genetic swamping or to extinction. They flourish briefly and then disappear without appreciable impact on the evolutionary trajectory of the main body of the species.

Although this scenario provides a plausible explanation for the concentration of divergent populations and biological species along the glaciated west coast of North America, as always with *Gasterosteus*, there are exceptions. In this case, a puzzling exception is the Atlantic coast of North America. Here, divergence in fresh water is minor relative to that found on the Pacific coast; instead, there is evidence (Blouw and Hagen 1990) of divergence, up to the level of biological species, in the sea!

ACKNOWLEDGEMENTS

So many people helped along the way that it is not possible to list them all; however, to all those field assistants who endured the wet and muck, and the laboratory assistants who fed, recorded, and coddled my sticklebacks, thanks for your patience and help. Without you this study would never have reached even its present state. My thanks also to the many graduate students whose original research projects form the foundation of much of this work; it was a pleasure and a privilege to work alongside them. Bob Carveth did the figures and helped in many other ways. Special thanks are due to a set of old friends and colleagues: Cas Lindsey for getting me started and keeping me going, Don Hagen for showing me how it was done, Ric Moodie for his ideas and quiet enthusiasm, and Tom Reimchen for his penetrating insights and regular reminders of the pitfalls of 'urban biology'. Over the years I discussed sticklebacks and evolution with these colleagues to the point that I no longer remember which ideas came from what researcher; however, any good ones probably are theirs and the errors certainly are mine. From its inception this research has been supported by the Natural Sciences and Engineering Research Council of Canada (operating grant A3451).

15

Palaeobiology and evolution of threespine stickleback

Michael A. Bell

The fossil record provides a unique chronological perspective on evolutionary processes. Although it is fragmentary and interpretation of palaeobiological evidence is difficult (Levinton 1988; Hoffman 1989), the fossil record is the only source for most information on rates and patterns of evolutionary change, timing of origin and extinction of taxa, past existence of extinct forms, and previous geographical and ecological distributions. Information from the fossil record has an important role to play in our understanding of evolutionary processes.

The *Gasterosteus aculeatus* species complex offers an excellent opportunity to apply palaeontological information to investigation of evolutionary processes. The wealth of information on the biology of extant threespine stickleback creates the potential for realistic inference of processes responsible for variation of fossil sticklebacks over geological time. Unlike most phenotypic variation studied in population biology (reviewed in Endler 1986), many well-studied, variable traits of *G. aculeatus* are osteological (reviewed by Bell 1984*a*; Wootton 1984*a*; Reimchen Chapter 9 this volume) and easily fossilized. Variation of lateral plate number in low morphs and of lateral plate morphs is ubiquitous and has been studied extensively in extant threespine stickleback (e.g. reviewed in Bell 1984*a*; Reimchen Chapter 9 this volume). Variation of dorsal spine number (Gross 1978) and pelvic girdle structure (Bell 1987) has also been studied in extant populations. Information on these fossilizable traits in extant populations provides the basis for interpretation of variation in fossils.

Other osteological traits that have received less attention in extant *Gasterosteus* have a good potential for analysis in fossils. Numerous landmarks for morphometric analysis are bony (Reimchen *et al.* 1985; Baumgartner *et al.* 1988; Baumgartner 1992) and readily observable in articulated fossils. New methods for acquisition and analysis of morphometric data (Rohlf and Bookstein 1990) present exciting opportunities for study of fossil sticklebacks. Dorsal and pelvic spine serration in extant threespine stickleback has received little attention, but greater serration appears to be associated with robust armour (Gross 1978). Thus, interesting

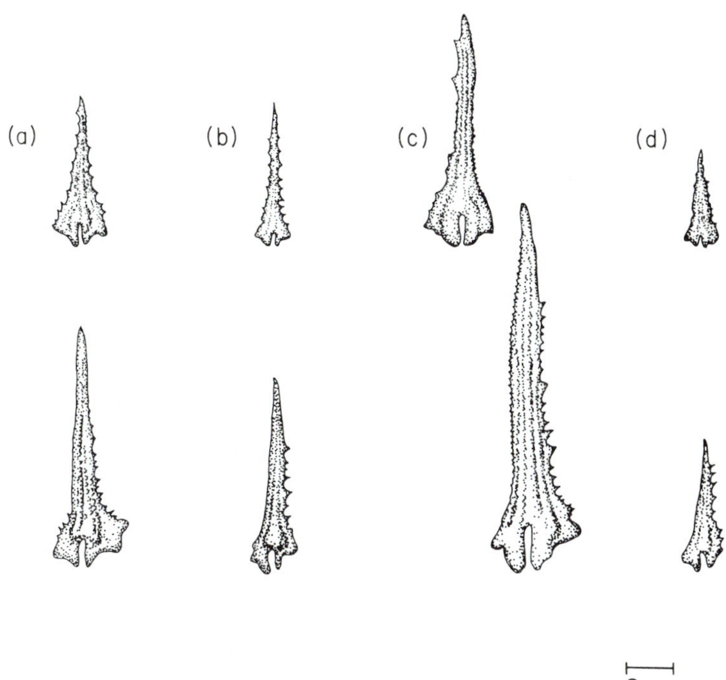

Fig. 15.1 Dorsal and right pelvic spines (*upper and lower rows, respectively*) from modern *Gasterosteus aculeatus* showing their diagnostic structure, including the broad base, proximal condyles, and variable marginal serrations. Sources of specimens are (a) Birch Cove, Cook Inlet, Alaska (71.3 mm standard length, anadromous), (b) Big Lake, Cook Inlet, Alaska (55.3 mm SL), (c) Bakewell Lake, Portland Canal, Alaska (73.3 mm SL), and (d) Santa Clara River, California (43.8 mm SL, *G. a. williamsoni*).

inferences could be made from isolated spines, which are distinctive and easily recognized (Fig. 15.1).

Wake and Larson (1987) proposed use of 'multidimensional analyses' that combine ecological, functional, developmental, genetic, and phylogenetic perspectives to study evolution of one phenotypic trait. The availability of phenotypes that are observable in the fossil record and can be studied from these diverse perspectives in extant threespine stickleback allows multidimensional analysis of stickleback evolution, with the added temporal dimension of the fossil record (Bell 1987, 1988). *Gasterosteus aculeatus* is one of the few taxa that can be used to bridge the gap between population biology and palaeobiology (Bell 1987, 1988).

OVERVIEW OF THE FOSSIL RECORD OF THE GASTEROSTEIDAE

Most fossil sticklebacks belong to the *G. aculeatus* species complex (Table 15.1). *Pungitius* has been reported from three deposits and *Culaea* from only one Holocene site. The other stickleback genera, *Spinachia* and *Alpeltes*, have no fossil record. In this section I summarize occurrences of fossil sticklebacks and consider their morphological variation in relation to palaeozoogeography and palaeoecology.

Taxonomy of fossil *Gasterosteus*

Most nominal species of fossil *Gasterosteus* do not warrant separation from the *G. aculeatus* species complex. This species complex includes numerous extant phenotypically diverse populations (Bell 1984*a*). Most threespine stickleback populations are allopatric, and therefore their specific status is uncertain. However, some populations within the complex have evolved isolating mechanisms and are biological species (McPhail Chapter 14 this volume). Although these biological species sometimes represent phenetic extremes for the complex (Schluter and McPhail 1992; McPhail Chapter 14 this volume), others exhibit more subtle distinctions (McPhail 1969; Blouw and Hagen 1990). The inconsistent relationship between morphological divergence and speciation in *Gasterosteus* provides a poor standard for inference of the specific status of fossil stickleback.

Gasterosteus apodus (Mural 1973) clearly is part of a continuum for armour reduction within *G. doryssus* (Bell *et al.* 1985*b*; see below), with which it was synonymized (Bell 1974). However, *G. doryssus* itself represents an extreme in armour reduction that is rarely equalled by extant populations of *G. aculeatus* (e.g. Bell 1974, 1987; Moodie and Reimchen 1976*a*; Reimchen 1980; Campbell 1984; Francis *et al.* 1985), and therefore I recognize *G. doryssus* in this review. Detailed comparisons with similar extant populations of *G. aculeatus* are needed for a more well-informed decision on the taxonomic status of *G. doryssus*.

Pungitius haynesi (David 1945) was synonymized with *G. aculeatus* (Bell 1973*a*). It may parallel *G. doryssus* in armour reduction, but its reference to *G. haynesi* (M.L. Smith 1981; Sychevskaya and Grechina 1981; Welton and Link 1982) seems premature because only two specimens are known. It is impossible to determine whether this stickleback's morphology exceeds the range of variation in extant *G. aculeatus* populations (pers. obs.).

Sychevskaya and Grechina (1981) described two nominal species of *Gasterosteus* from eastern Siberia. The characters they used to diagnose *G. abnormis* are highly variable among extant threespine stickleback, and its validity is suspect. Similarly, presence of four dorsal spines and three pelvic fin rays appears to place *G. orientalis* outside the range of variation for *G. aculeatus* (Nelson 1971; Bell 1984*a*), but its diagnosis is based on only one complete specimen and fragments of four others. Thus, the validity of this nominal species is also suspect.

Dawson (1871, 1872, 1894) reported a 'two-spined stickleback', but subsequent references to stickleback from Dawson's Green Creek site and adjacent peri-Champlain Sea deposits refer them to *G. aculeatus* (e.g. Gardiner 1966; McAllister *et al.* 1989). However, McAllister *et al.* (1989) noted that it is possible that the Green Creek stickleback represents the extant 'white stickleback' recently reported from Nova Scotia (Blouw and Hagen 1990).

Unfortunately, extensive morphological variation within and among extant *G. aculeatus* populations, uncertainty over the specific status of its divergent allopatric populations, plus small sample size and poor preservation of some nominal fossil species of *Gasterosteus* obscure their status. In this chapter, *G. doryssus* is the only extinct nominal taxon of *Gasterosteus* that will be recognized, though formal synonymy of other fossil *Gasterosteus* species must await careful reexamination of type specimens. Even *G. doryssus* clearly is a member of the *G. aculeatus* complex, sharing extreme armour reduction with many extant populations and differing from them only in the magnitude of such reduction. Other fossil *Gasterosteus* will be referred to *G. aculeatus* in the absence of compelling evidence that they are separate species. Reasonably large samples of well-preserved specimens with character states that exceed the broad range of variation in extant *G. aculeatus*, or at least novel combinations of unusual character states, are needed to justify description of new fossil *Gasterosteus* species.

Taxonomy and distribution of other fossil sticklebacks

It is important to clarify the taxonomic status of other fossil sticklebacks for zoogeographic considerations. *Gasterosteops hexicanthus* (Schtyl'ko 1934), which Berg (1949) listed without comment in synonymy with *Pungitius*, is from an inland Arctic drainage in western Siberia. Two other *Pungitius* records are from coastal freshwater deposits in the Pacific basin (Liu and Wang 1974; Rawlinson and Bell 1982). Cvancara *et al.* (1971) reported *Culaea inconstans* from a Holocene deposit in North Dakota, USA, well within its current range (Wootton 1976). Locations of these records are consistent with the modern geographical distributions of these genera.

Chronology of *Gasterosteus* fossils

Reynolds (1991) reported 'Gasterosteidae' from a late middle Miocene deposit in southern California, USA, based on a *Gasterosteus* pelvic spine (Reynolds pers. comm.). This deposit is approximately 16 Ma (1 Ma = one million years) old (Woodburne 1991; Reynolds pers. comm.), making this specimen the earliest precisely dated record of *Gasterosteus*. *Gasterosteus aculeatus* from deposits on the Kamchatka Peninsula and Sakhalin Island,

Table 15.1 Summary of the fossll record of sticklebacks.

Nominal species	Location[a]	Age[b]	Depositional environment	Material[c]	Reference[d]
A. Nominal species of Gasterosteus					
Gasterosteidae	S California, USA	16 Ma	Fluv–lac	Dis	1
G. abnormis[e]	Sakhalin I., Russia	M. Mio	Marine	Art	2
G. orientalis[e]	Kamchatka, Russia	M. Mio	Fluv or lac	Art	2
G. aculeatus	WC Nevada, USA	11.2 Ma	Lacustrine	Dis	3
G. aculeatus	WC Nevada, USA	LM Mio	Fluvial	Dis	4
Meriamella doryssa[f]	WC Nevada, USA	10 Ma	Lacustrine	Art	4–16
G. apodus[g]	WC Nevada, USA	10 Ma	Lacustrine	Art	8
G. aculeatus	C California, USA	L Mio	Marine	Art	17
Pungitius haynesi[h]	S California, USA	L Mio	Lacustrine	Art	18–20
G. aculeatus	S California, USA	2 Ma	Lacustrine	Art	21
G. aculeatus	C California, USA	Plio–Pleist	Fluvial	Art	22
G. aculeatus	Netherlands	Plio–Pleist	Fluvial	Dis	23, 24
G. aculeatus	NW Germany	E Pleist	Fluvial	Dis	25
G. aculeatus	C California, USA	M Pleist	?	Dis?	26
G. aculeatus	S California, USA	19–500 ka	Fluv–lac	Dis	27
G. aculeatus	C England	42 ka	Fluv–lac	Dis	28
G. aculeatus	S California, USA	25–33 ka	Fluvial	Dis	29
G. aculeatus	S California, USA	12 ka	Fluvial?	Dis	30
Gasterosteus sp.	S California, USA	L Pleist	Dune	Dis?	31
G. aculeatus	SE Ontario, Canada	Pleist–Holo	Lac or marine	Art	32–37
G. aculeatus	NE Greenland	Holo	Peat	Dis	38
G. aculeatus	W Greenland	9 ka	Lacustrine	Dis	39
G. aculeatus	NW Greenland	6.7 ka	Peat	Dis	40

B. Nominal species of Pungitius					
P. nihowanensis	Hopei, China	L Plio	Fluv–lac	Art	41
Gasterosteops hexicanthus[i]	W Siberia, Russia	Neogene	Fluvial?	Art	42
Pungitius sp.	S Alaska, USA	7 Ma	Fluvial	Art	43
C. Nominal species of Culaea					
Eucalia inconstans[j]	N Dakota, USA	9.5 ka	Lacustrine	Art	44

[a] N, north; S, south; E, east; W, west; C, central.
[b] Age is given in years, when available, or as Cenozoic epochs, which are sometimes abbreviated. L and M refer to late and middle; ka is thousand years and Ma is million years.
[c] Articulated (Art) skeletons or isolated disarticulated (Dis) elements.
[d] References: 1, Reynolds (1991); 2, Sychevskaya and Grechina (1981); 3, Bell unpublished data; 4, Bell (1974); 5, Jordan (1907); 6, Hay (1907); 7, Jordan (1908); 8, Mural (1973); 9, Bell (1984*a*); 10, Bell (1987); 11, Bell (1988); 12, Bell and Haglund (1982); 13, Bell *et al.* (1985*b*); 14, Bell and Legendre (1987); 15, Bell *et al.* (1987); 16, Bell *et al.* (1989); 17, Bell (1977); 18, David (1945); 19, Bell (1973*a*); 20, Welton and Link (1982); 21, Bell (1973*b*); 22, Fierstein *et al.* unpubl. data; 23, Gaudant (1979); 24, Gaudant (1981); 25, Obrhelova (1977); 26, Casteel and Hutchison (1973); 27, Jefferson (1991); 28, Coope *et al.* (1961); 29, Swift (1989); 30, Roeder (1985); 31, Reynolds and Reynolds (1991); 32, Dawson (1871); 33, Dawson (1872); 34, Dawson (1894); 35, Gardiner (1966); 36, McAllister *et al.* (1981); 37, McAllister *et al.* (1989); 38, Fredskild and Roen (1982); 39, Fredskild (1983); 40, Fredskild (1985); 41, Liu and Wang (1974); 42, Schtyl'ko (1934); 43, Rawlinson and Bell (1982); 44, Cvancara *et al.* (1971).
[e] Probably junior synonym of *G. aculeatus* because diagnostic characters given are highly variable in *G. aculeatus*.
[f] Junior synonym of *G. doryssus* (Jordan 1908).
[g] Junior synonym of *G. doryssus* (Bell 1974).
[h] Junior synonym of *G. aculeatus* (Bell 1973*a*).
[i] Referred to *Pungitius* by Berg (1949).
[j] Synonym of *Culaea*.

Russia, were also reported to be middle Miocene (Sychevskaya and Grechina 1981), but it is uncertain whether they predate the 16 Ma record. The next precisely dated occurrence is based on a single pelvic girdle with articulated spines and an associated dorsal spine from a lacustrine diatomite at Clark Siding, Nevada, USA (unpubl. obs., University of Michigan Museum of Paleontology number UMMP74785). Krebs and Bradbury (1984) reported a K/Ar date of 11.2 ± 0.3 Ma for a volcanic ash located stratigraphically above this specimen in the same quarry. Isolated *Gasterosteus* spines from fluvial deposits of the lower member of the Truckee Formation, Nevada, USA (Bell 1974) may be about the same age. Specimens of *G. doryssus* from the middle member of the Truckee Formation occur a few metres stratigraphically below a volcanic ash with a K/Ar date of 9.79 ± 0.12 Ma (Brown pers. comm.). A marine *G. aculeatus* from California, USA (Bell 1977) is approximately the same age (Brown pers. comm.). Numerous fossil *Gasterosteus* occur in much later deposits, mostly Pleistocene. Many of these later records come from California, where threespine stickleback must have become widely distributed during the Miocene.

Conflicting evidence for location of the generic radiation of the Gasterosteidae

Distribution of extant and fossil Gasterosteidae

Gasterosteus appears in five Tertiary Pacific basin formations and dates back at least 11 Ma or possibly even 16 Ma. Similarly, *Pungitius* from a 7 Ma-old deposit (Rawlinson and Bell 1982) is one of two Tertiary records from the Pacific basin. In contrast, *Gasterosteus* does not seem to have been reported from Atlantic basin deposits older than about 1.9 Ma (Gaudant 1979, 1981), and *Pungitius* has no fossil record whatsoever there. Although the time of appearance of taxa in the fossil record must be interpreted with extreme caution (Marshall 1990), the large difference between first occurrences of *Gasterosteus* in the Atlantic and Pacific basins is probably significant. The palaeogeographical distribution of *G. aculeatus* is consistent with Buth and Haglund's (page 82 this volume; Haglund *et al.* 1992*a*) inference from allozyme data that Atlantic basin *G. aculeatus* has been derived recently from Pacific populations. (Unfortunately, allozyme data for *Pungitius* (Haglund *et al.* 1992*b*) are inadequate to make similar inferences for this genus.) The late arrival of *G. aculeatus* in the Atlantic basin also conforms to a general bias in the direction of exchange of invertebrate species between the North Pacific and Atlantic during the past 3.5 Ma (Vermeij 1991). Buth and Haglund (page 82 this volume) postulate that *Gasterosteus wheatlandi* has been in the Atlantic much longer than *G. aculeatus*. Failure of *G. wheatlandi* to produce a fossil record may reflect its restriction to north-eastern North America, where Neogene and Holocene deposits are scarce (Savage and Russell 1983).

Today all five gasterosteid genera occur in the Atlantic basin or its tributary fresh waters, and three of them, *Spinachia*, *Apeltes*, and *Culaea*, as well as *G. wheatlandi*, are endemic to the Atlantic (Wootton 1976, 1984*a*; Lee *et al.* 1980). The distribution of extant stickleback genera indicates that the Gasterosteidae radiated in the Atlantic. Thus, restriction of early records of *Gasterosteus* and *Pungitius* to the Pacific basin appears to conflict with the zoogeographic evidence for generic radiation of the family in the Atlantic.

A palaeontological test for the geographic origin of *Gasterosteus aculeatus*

Absence of sticklebacks before the Pliocene in Atlantic basin deposits may represent absence of appropriate deposits, failure to sample them, or failure to identify fossil stickleback bones in existing collections. Nearly half of the fossil *Gasterosteus* records in Table 15.1 are from California, USA, but Long (pers. comm.) has found ten additional occurrences in Miocene to Pleistocene marine and freshwater deposits from California. Long's new freshwater records were found by re-examining assemblages that included the western pond turtle, *Clemmys mormorata*, which is commonly sympatric with extant threespine stickleback and is large, abundant, and distinctive enough to be well represented and easily recognized in samples from appropriate palaeoenvironments. He concluded that the distinctive dorsal and pelvic spines of *Gasterosteus* (Fig. 15.1) are commonly overlooked in fluvial deposits because they are too small to be collected without wet screening, and unlike turtle bones, they are likely to be unidentified by palaeomammalogists, who most often study such fluvial deposits.

Unfortunately, Long's approach cannot be applied to eastern North American deposits. Savage and Russell (1983) indicated that coastal north-eastern North America is virtually devoid of fossiliferous Miocene (0.23 per cent of localities) and Pliocene (0.0 per cent) terrestrial deposits, and glaciation severely limited preservation of Pleistocene fossils there. However, about a quarter of the Miocene deposits and numerous Pliocene and Pleistocene deposits that Savage and Russell (1983) listed occur in coastal western Europe (Germany, Belgium, France, Spain, Portugal) within the present range of *G. aculeatus* (Fig. 1.1).

Long's method could be applied to Neogene European deposits, but absence or infrequency of warm-water piscivorous fishes should be an additional criterion for selection of assemblages to analyse. Threespine stickleback do not appear to coexist with warm-water piscivorous fishes. The freshwater distribution of extant *G. aculeatus* in north-eastern North America is complementary to that of many piscivorous warm-water fishes (e.g. Ictaluridae, Centrarchidae; Lee *et al.* 1980; Wootton 1976). Kynard (1979*b*) also observed nearly simultaneous appearance of pumpkinseed sunfish, *Lepomis gibbosus* (Centrarchidae) and extinction of *G. aculeatus* in one lake. Failure to find threespine stickleback after a careful search of coastal western European deposits that lack warm-water predatory fishes but contain specimens of more obvious species with which stickleback occur today would increase our confidence that *G. aculeatus* was absent from the Atlantic basin until recently compared with its antiquity in the Pacific.

THE PALAEOECOLOGICAL DISTRIBUTION OF FOSSIL *GASTEROSTEUS PHENOTYPES*

Gasterosteus aculeatus has a broad ecological distribution and its fossils occur in marine, brackish, lacustrine, and fluvial deposits (Table 15.1). The palaeoecological distribution of fossil threespine stickleback phenotypes is limited, but it is consistent with the distribution of these phenotypes among modern habitats.

Morphology of marine fossil *Gasterosteus*

The only two fossil marine *G. aculeatus* from Sakhalin Island, Russia (Sychevskaya and Grechina 1981) and central California, USA (Bell 1977) are complete morphs with deep lateral plates, long dorsal and pelvic spines, and robust pelvic structures. Except for the southernmost marine populations, where low and partial morphs may occur (Münzing 1963; Black 1977), extant marine *G. aculeatus* share this armour morphology with the fossils. It would be premature on the basis of just two fossil records to conclude that marine *G. aculeatus* has been morphologically static since the Miocene, but geographical homogeneity of extant marine *G. aculeatus* supports this conclusion. Such evolutionary conservatism is to be expected because the biology of marine stickleback and the habitat they occupy both favour range shifts in response to environmental change instead of adaptation (Pease *et al.* 1989). Consequently, the morphology of extant marine *G. aculeatus* probably represents the ancestral condition from which freshwater populations have diverged since the Miocene.

Morphology of freshwater fossil *Gasterosteus*

In contrast to marine *Gasterosteus*, extant freshwater populations are variable for armour structures (reviewed in Bell 1984*a*; Reimchen Chapter 9 this volume), and similar variability occurs among fossils. Hagen and Moodie (1982) showed that complete lateral plate morphs dominate, and low morphs are absent, in extant freshwater populations from regions with warm summers and cold winters, including central Europe and the eastern coasts of Asia and North America. A fossil specimen from Siberia appears to be a partial morph (Sychevskaya and Grechina 1981), and specimens from eastern North America are complete morphs (McAllister *et al.* 1981, 1989). All articulated freshwater *Gasterosteus* fossils from south-western North America, where the low morph predominates today in lakes and sluggish streams (Baumgartner and Bell 1984), are low morphs (Bell 1973*a*,*b*; Bell *et al.* 1985*b*; Fierstine *et al.* unpubl. data). Data are limited, but the distribution of lateral plate morphs in fossil threespine stickleback suggests that the existing relationship between lateral plate morph and climate is very old.

Absence of predatory fishes often results in extreme armour reduction in extant *G. aculeatus* (Bell 1984*a*, 1987; Reimchen page 246 this volume), and this tendency is exhibited in fossil threespine stickleback. *Gasterosteus doryssus* generally lacks lateral plates and usually has reduced dorsal spines and pelvic structures (Mural 1973; Bell 1974; Bell *et al.* 1985*b*; see below). Although birds ate *G. doryssus*, predatory fishes were insignificant in its habitat (see below). The Ridge Basin (southern California, USA) stickleback is represented by only two specimens, which appear to exhibit extreme armour reduction (David 1945; Bell 1973*b*), and no predatory fishes are known from the deposit in which it occurs (Welton and Link 1982). Stickleback from the Pleistocene Soboba Formation, southern California, are low morphs with moderate armour ($\leq$ 5 lateral plates, a robust pelvic girdle, moderately serrated spines) but reduced dorsal spine number (Bell 1973*b*). Although predatory fishes were not known from the poorly sampled Soboba Formation, presence of stickleback bones in putative bird pellets (Bell 1973*b*) indicates that the Soboba stickleback was eaten by birds (Wilson 1987). Most other freshwater stickleback fossils are isolated bones from fluvial deposits, and therefore their associations with other species may be unreliable. However, the available evidence indicates that absence of fish predation in fresh water has resulted in selection for armour reduction in *Gasterosteus* since the Miocene.

The palaeoecological distribution of *Gasterosteus* suggests a peculiar kind of evolutionary conservatism for at least 10 Ma. During this period, the species complex has inhabited a broad variety of habitats. To the extent that they are represented in the fossil record, ancient habitat–phenotype associations conform to existing ones. However, persistent habitat–phenotype associations do not reflect evolutionary stasis. Rather, they have been maintained by a steady state between repeated colonization of a fixed array of habitats, to each of which there has been a consistent evolutionary response, followed by extinction owing to habitat change (Bell 1987, 1988; Bell and Foster page 14; McPhail page 436 this volume).

THE PALAEOBIOLOGY OF *GASTEROSTEUS DORYSSUS*

The vast majority of fossil threespine stickleback available for study represent *G. doryssus* (Fig. 15.2) from late middle Miocene lacustrine diatomaceous shales of the middle member of the Truckee Formation, Nevada, USA. Fossil threespine stickleback from other deposits number less than a few hundred mostly fragmentary specimens, but more than 8000 specimens of *G. doryssus* have been collected. Specimens are often very well

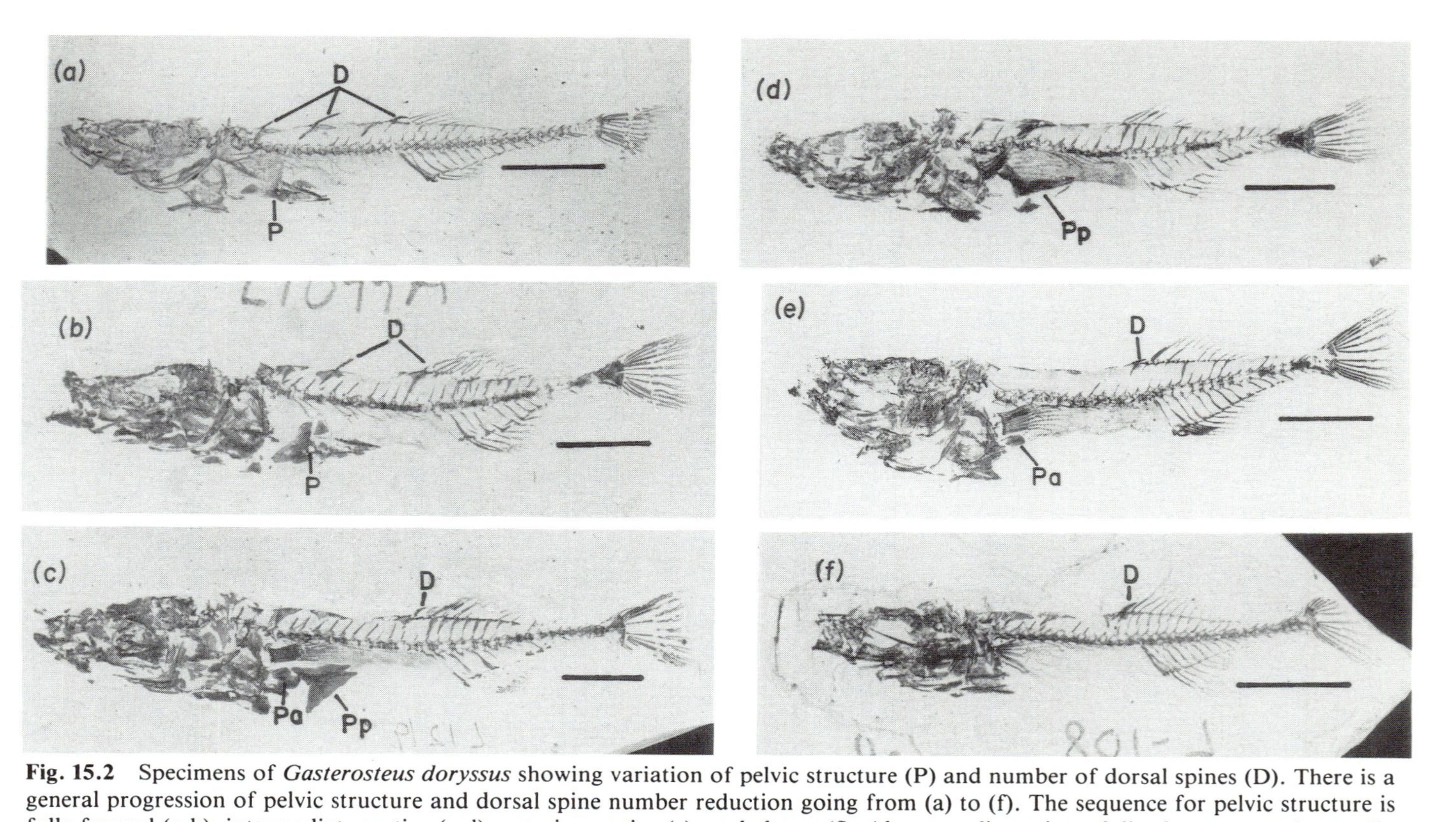

Fig. 15.2 Specimens of *Gasterosteus doryssus* showing variation of pelvic structure (P) and number of dorsal spines (D). There is a general progression of pelvic structure and dorsal spine number reduction going from (a) to (f). The sequence for pelvic structure is fully formed (a,b), intermediate vestige (c,d), anterior vestige (e), and absent (f). Also note disruption of distal segments of some fin rays, e.g. the dorsal fin in (b). Other abbreviations: Pa, anterior pelvic vestige; Pp, posterior pelvic vestige. Scale bars are 1 cm long. All specimens are from the University of Michigan Museum of Paleontology, UMMP74773.

preserved and can be collected in large numbers with exceptionally fine stratigraphic precision. Properties of middle Truckee Formation rocks, of the fossil assemblage, and of fossil *G. doryssus* permit palaeoecological inferences that are critical for evolutionary interpretations. Similarities between pelvic reduction in extant threespine stickleback and *G. doryssus* (Bell 1987, 1988) also provide critical support for interpretation of stratigraphic variation of pelvic girdle phenotypes in the fossils. Finally, specimens from a thoroughly sampled section exhibit variation for six characters that appear to exhibit temporal patterns (Bell *et al.* 1985*b*).

Gasterosteus doryssus was described almost simultaneously in 1907 by Jordan as *Merriamella doryssa* and by Hay as *Gasterosteus williamsoni lepotosomus*. Jordan (1908) accepted Hay's (1907) assignment of this form to *Gasterosteus* but established seniority for his nominal species. He also perpetuated Hay's erroneous belief that *G. doryssus* came from the Pleistocene Lahontan beds, which was corrected by La Rivers (1953). *G. doryssus* was not studied again until 1973, when Mural proposed *G. apodus* as a new species. Bell (1974) immediately synonymized *G. apodus* (see above) with *G. doryssus* and described its variation.

All Truckee Formation fossils discussed below have been deposited in the University of Michigan Museum of Paleontology (UMMP). Specimens

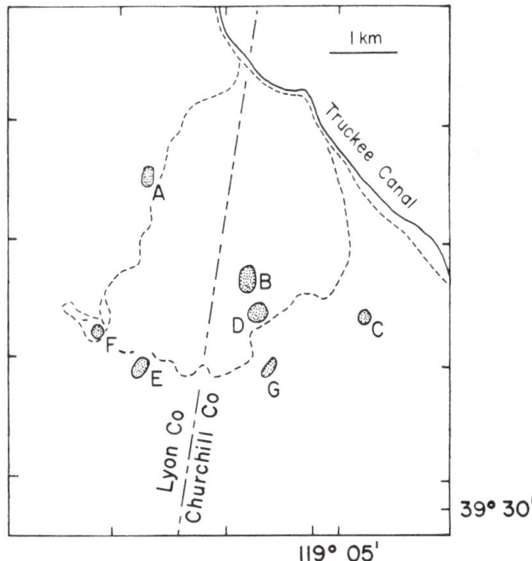

Fig. 15.3 Locations of quarries (stippled) in the middle member of the Truckee Formation that have produced *G. doryssus* and other fossil fishes. Quarries are referred to by letter in the text. Curved dashed lines are roads. (Based on Two Tips Quadrangle, Nevada, 15 minute series [topographic], US Geological Survey.)

studied by Jordan (1907) and Hay (1907) came from the Truckee Canal, which runs along the base of the mountains, but all subsequent samples of *G. doryssus* come from commercial diatomite quarries near by in the Virginia Range, near Hazen, Nevada (Fig. 15.3). Quarry D has been most intensively sampled and has produced several thousand specimens of *G. doryssus*. Quarries A and C contain relatively short stratigraphic sections that correlate with sections in Quarry D (Brown pers. comm.). Quarry B, which is adjacent to Quarry D and appears to be stratigraphically above it, contains abundant killifish (*Fundulus nevadensis*, Eastman 1917) but no stickleback (Bell unpubl. data). Stratigraphic relations among other quarries are unknown.

Gasterosteus doryssus provides fascinating insights into patterns of evolution over periods of several thousand years. In this section I introduce *G. doryssus* and the 'Lake Truckee' environment in which it lived. Inferences from lithology of the middle Truckee diatomites and the palaeoecology and taphonomy of *G. doryssus* form the basis for evolutionary interpretations of stratigraphic variation in this fossil species.

Palaeoecology of *Gasterosteus doryssus*

Palaeogeography, palaeoclimatology, and dispersal of Miocene sticklebacks in western North America

The distribution of extant freshwater *G. aculeatus* can be explained only by assumption of dispersal through the ocean (e.g. McPhail and Lindsey 1970; Bell 1976*a*, 1984*a*; Bell and Foster page 16; McPhail page 401 this volume). Like other sticklebacks, *G. aculeatus* has a boreal to temperate distribution (Wootton 1976, 1984*a*), and records of Miocene *Gasterosteus* in western North America (Table 15.1) are near the southern limit of its range (Fig. 1.1).

Palaeobotanical evidence indicates that a climatic cooling trend had developed in the Great Basin of western North America by the middle Miocene (G. R. Smith 1981*a*). Based on vascular plants from Quarries A and D and the Hazen Flora (Axelrod 1948), Axelrod (in Bell 1974) inferred that Lake Truckee experienced warm but rarely hot summers and mild winters with occasional light frosts. He concluded that rainfall averaged 38–44 cm annually and was absent in the summer. Presence of encysted diatoms confirms occurrence of a dry season (Ruben 1971). All five fish genera from the Truckee Formation, *Gasterosteus*, *Fundulus*, *Oncorhynchus*, *Gila*, and *Ictalurus* (G. R. Smith 1981*b*; see below), are characteristic of the north temperate zone. Palaeobotanical and palaeoichthyological evidence indicates that Lake Truckee had a temperate climate.

Miocene *Gasterosteus* from deposits near the Pacific coast of North America at about 35 ° north latitude (Bell 1977; Welton and Link 1982; Reynolds 1991) have been tectonically displaced northward more than 300 km relative to Lake Truckee since the Miocene (M. L. Smith 1981). Thus, stickleback records in southern California deposits suggest that climatic cooling in the Miocene permitted *Gasterosteus* to spread south along the Pacific coast to a point well south of Lake Truckee.

During the Miocene, the Sacramento and San Joaquin valleys of California, west of the present Sierra Nevada mountain range, were inundated by the Pacific Ocean, and the Sierra Nevada, a formidable barrier to fish dispersal today, had not yet formed (Axelrod 1957, 1962; Cole and Armentrout 1979; G.R. Smith 1981*b*). Distributions of both extant and fossil fishes provide ample evidence for bidirectional movement of fishes between the Pacific coast of southern California and inland basins near Lake Truckee (Bell 1974; G.R. Smith 1981*a*; M.L. Smith 1981). The sister-species relationship between *Ictalurus hazenensis* from the middle member of the Truckee Formation (Baumgartner 1982) and *I. vespertinus* from Neogene and early Pleistocene deposits of north-western USA (Miller and Smith 1967) suggests that *Gasterosteus* could also have entered Nevada by a route north of California. Consequently, *G. doryssus* could have reached Lake Truckee by any of several routes from the Pacific coast.

Palaeolimnology

Lake Truckee appears to have been a relatively large, productive, and temperate lake that received little input of terrigenous sediment. The minimum size of the lake is set by the 3.7 km distance between correlated horizons in quarries A and C, and additional exposures occur beyond these quarries (Fig. 15.3). Based on diversity of molluscs from the lower member of the Truckee Formation, Yen (1950) also inferred that Lake Truckee was large, that it had muddy and rocky shores, and that its pH or water velocity varied. A fossil shore bird (Charadriidae or Scolapscidae; Bickert pers. comm.; UMMP 74779) from Quarry D also indicates presence of muddy or sandy shores. However, middle Truckee deposits are rich in diatoms and devoid of sandstone turbidites (Brown pers. comm.), indicating little input of terrestrial sediment into offshore depositional environments. Based on the diatom species present, Krebs and Bradbury (1984) inferred that the lake was shallow, with salt-marsh vegetation on shore. Krebs and Bradbury (1984) attributed absence of fluvial sediment from a similar nearby deposit to filtration by near-shore vegetation of water discharged from tributary streams, and perhaps this process also withheld coarse fluvial sediment from laminated diatomites of the Truckee Formation. Gastropods studied by Yen (1950) indicated habitats rich in filamentous algae and aquatic leafy plants, as well as rocky-bottom habitats rich in algae and diatoms. Lake Truckee probably was large and ecologically heterogeneous.

The diatom flora of the middle member of the Truckee Formation indicates that Lake Truckee was saline (Ruben 1971; Krebs and Bradbury 1984). The only two common fish species present, *G. doryssus* and *Fundulus nevadensis*, belong to euryhaline groups that tolerate saline water. However, a frog, *Rana pipiens*, from a nearby depositional basin of about the same age as Truckee Formation lake beds (La Rivers 1953), and salamander bones in a fossil bird pellet (Wake pers. comm.) from Quarry D, both demonstrate proximity to Lake Truckee of adjacent freshwater habitat. Presence of the freshwater diatom, *Aulacoseira* (Krebs and Bradbury 1984), in Quarry D and of *I. hazenensis*, which belongs to a clade that is normally restricted to fluvial habitats (Baumgartner 1982), indicate existence of fresh waters tributary to Lake Truckee. Evidence from such diverse organisms as diatoms, fishes, and amphibians all indicates that Lake Truckee was saline but that freshwater habitats occurred nearby.

Predation on *Gasterosteus doryssus*

Predation plays a crucial role in evolution of stickleback armour (Bell 1984*a*; Reimchen Chapter 9 this volume), and armour structures of *G. doryssus* vary stratigraphically. Thus, it is important to consider possible predators on *G. doryssus*. Insects often occur in lacustrine fossil deposits (Gray 1988) and in modern saline lakes (De Deckker 1988), but surprisingly, not a single aquatic or terrestrial insect has been found in the extensively sampled middle Truckee Formation beds where *G. doryssus* occurs. Absence of flying insects, which must have flown onto the lake and drowned (Wilson 1980, 1988), demonstrates that this deposit did not preserve insects. Consequently, there is no fossil evidence concerning predatory aquatic insects from Lake Truckee.

Fry and egg cannibalism is common in *G. aculeatus* (Foster *et al.* 1988; Foster 1988, page 395 this volume; Hyatt and Ringler 1989*a*, *b*; Whoriskey and FitzGerald page 202 this volume), but egg predation should have occurred in *G. doryssus* only if it was a benthic feeder (Foster 1988; Ridgway and McPhail 1988). There is no indication of fry cannibalism from stomach contents.

Four other fish species occur in the middle member of the Truckee Formation. A killifish, *Fundulus nevadensis*, is locally abundant (Quarry B, top of D), but it occurs with *G. doryssus* infrequently and rarely is large enough (Bell unpubl. data) to have preyed on stickleback larger than about 1 cm standard length (SL, *sensu* Hubbs and Lagler 1970). Single specimens of a catfish, *Ictalurus hazenensis*, from Quarry F (Baumgartner 1982) and of an undescribed trout (cf. *Oncorhynchus mykiss*, G. R. Smith pers. comm.) from Quarry D were probably transported *post mortem* from fluvial habitats into Lake Truckee. Displaced scales on the trout suggest *post mortem* transport (Elder and Smith 1988), and the catfish belongs to a group whose extant members are restricted to streams (Baumgartner 1982). These species evidently were not significant stickleback predators.

However, a large undescribed minnow, *Gila* sp., that might have eaten *G. doryssus* occurs in Quarry E and possibly in an older (11.2 Ma; Krebs and Bradbury 1984) nearby deposit at Clark Siding, Nevada. Fossil plants are abundant in these quarries, and the diatomites appear to have been bioturbated (Krebs and Bradbury 1984), both of which findings indicate deposition near shore (Wilson 1980, 1988). Near-shore environments may have had reduced salinity from stream discharge, permitting sympatry of the stickleback and minnow. The northern squawfish, *Ptychocheilus oregonensis*, a large minnow from north-western North America, preys on *G. aculeatus* (e.g. Hagen and Gilbertson 1972; Wydoski and Whitney 1979). Therefore the large *Gila* from the Truckee Formation might also have taken *G. doryssus* (G. R. Smith pers. comm.). However, Truckee Formation deposits have been sampled sufficiently to conclude that this large minnow and other potential piscivorous fishes were absent or too rare in the open lake to have been a significant cause of mortality for *G. doryssus* there.

Ruben (1971) reported a fossil snake that contained three fish, one of which he tentatively identified as *G. doryssus*. This specimen seems to have come from the vicinity of Quarry F, and another small snake was collected recently from Quarry A (Bell unpubl. data; UMMP74786). Garter snakes, *Thamnophis couchi*, are selective predators on threespine stickleback that favour retention of a moderate number of lateral plates (five per side), but spines do not appear to impede garter snake predation (Bell and Haglund 1978). Snakes ate *G. doryssus*, but it is unclear whether predation by snakes influenced the evolution of their armour.

Bickert (pers. comm.) reported six bird species from Quarry D and an adjacent exposure (north of Quarry D, east of Quarry B). A shore bird (Charadriiformes or Scolopacidae; UMMP74779) belongs to a group that is not piscivorous, and a small songbird (Passeriformes; UMMP74781) was too small to have been a significant stickleback predator. Another songbird was large and could have preyed on stickleback in shallow water (Whoriskey and FitzGerald 1985*b*). A duck (*Anas* sp.; UMMP74783) also occurred, and ducks may take stickleback (Table 9.1). More importantly, however, a grebe (*Podiceps* sp.; UMMP74778) and a cormorant (*Phalacrocorax* sp.; UMMP74780, 74784), piscivorous birds that commonly prey on threespine stickleback (Reimchen page 243 this volume, Table 9.1), occurred in Lake Truckee. Numerous small fish bones in the abdominal cavity of one cormorant (UMMP74784) show that it ate fish, but none of the bones is diagnostic for *Gasterosteus*. However, bone aggregations containing remains of multiple stickleback specimens (Fig. 15.4) occur in the same deposits as articulated *G. doryssus*. Wilson (1987) interpreted similar aggregations from Eocene lacustrine deposits as regurgitated bird pellets, and their presence in the middle member of the Truckee Formation, where the only piscivores are birds, supports his interpretation. Bird predation is a significant cause of selective mortality in extant lacustrine threespine stickleback populations (Reimchen 1980, 1983, 1988, page 250 this volume), and bird predation should have favoured retention of armour in *G. doryssus*.

In summary, piscivorous birds are the only known stickleback predators in the middle member of the Truckee Formation. Although there is no record

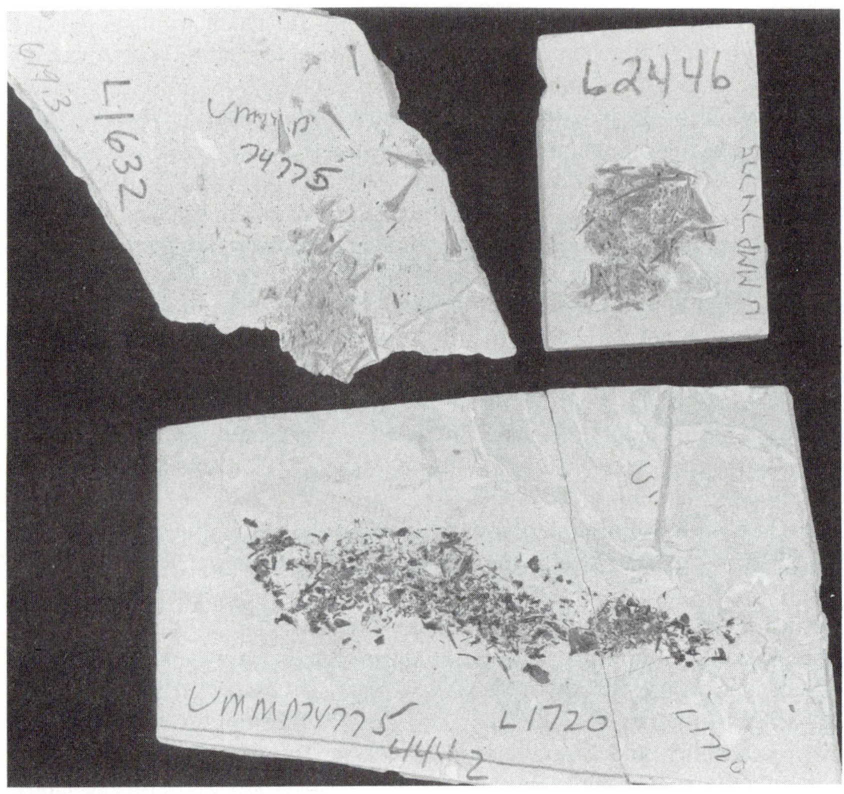

Fig. 15.4 Fossil bird pellets containing spines and other bones of *Gasterosteus doryssus* from the Truckee Formation. The elongated triangular objects in the two top specimens are stickleback spines.

of insects that might have eaten stickleback, modern saline lakes contain them unless their salinity exceeds that of seawater, and large benthic predatory insects should have occurred in near-shore habitats (De Deckker 1988). Except for a large *Gila* that may have taken *G. doryssus* near shore, piscivorous fish appear to have been allopatric with stickleback or too small to have posed a significant predation threat to them. In Reimchen's (page 251 his volume) terminology, Lake Truckee was probably a bird–insect lake, where selection should favour reduction of armour. However, it was a moderately large lake, in which armour reduction might not be expected to be as pronounced as it actually was (Fig. 9.2).

Palaeoautecology of *Gasterosteus doryssus*

There is no evidence that *G. doryssus* had any fundamental ecological differences from extant threespine stickleback. Its size, morphology, inferred habitat, reproductive biology, food use, and social behaviour probably did not deviate from those of modern populations.

Lateral plates are almost always absent in *G. doryssus* (unpubl. data), and the pelvic girdle and dorsal spines are usually reduced (see below). Weak expression of armour in this stickleback is appropriate for the inferred predation regime in Lake Truckee.

The abdomens of rare specimens of *G. doryssus* contain numerous opaque masses that resemble eggs (Fig. 15.5). These specimens suggest that female *G. doryssus* produced large clutches of large eggs, as in extant threespine stickleback (Baker page 164 this volume). It follows from this observation that other aspects of the reproductive biology of extant *G. aculeatus* (e.g. courtship, paternal care) occurred in this Miocene stickleback. This conclusion is consistent with presence of similar reproductive biology in *G. wheatlandi* and other stickleback genera (McLennan *et al.* 1988).

At most horizons in Quarry D, mean SL of *G. doryssus* appears to have been slightly greater (unweighted $\bar{X} = 47.28$, SD = 4.06) than that of most extant threespine stickleback populations (Appendix 6.1). Larger specimens of *G. doryssus* often reached 75 mm SL, and at one horizon (i.e. sample 10 of Bell *et al.* 1985*b*), they reached 100 mm SL, near the maximum for threespine stickleback (Bell 1984*b*). McPhail (1977) attributed extremes in SL to intense fish predation, but the *G. doryssus* samples with relatively large mean SL often had greatly reduced pelvic structure and dorsal spine number, and predatory fishes were absent. Thus, there must be reasons other than intense fish predation for large SL in *Gasterosteus* (Bell 1984*a*,*b*).

The SL of specimens pooled among six time-averaged samples is strongly unimodal (see Fig. 1 of Bell 1984*b*), suggesting that individuals usually died at one time of year and that one age class dominated the death assemblage (see Wilson 1984 for rationale). It also suggests that *G. doryssus* had a generation time of 1 yr, which is common in extant *Gasterosteus* (Baker page 157 this volume). However, it is impossible to estimate clutch size from SL in *G. doryssus* because the relationship between clutch size and

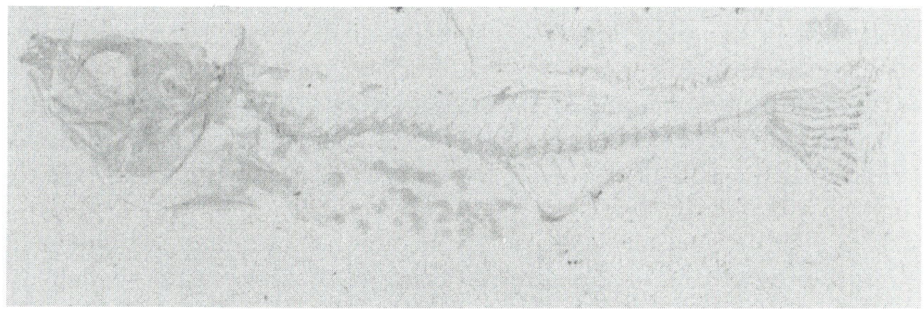

Fig. 15.5 Specimen of *Gasterosteus doryssus* with abdominal masses that appear to represent eggs. Note displaced caudal fin rays.

SL varies greatly among populations (Fig. 6.4).

Intact plant material is rare in Quarry D, where *G. doryssus* abounds, suggesting that the deposit formed well off shore (Wilson 1980, 1988). Consistent lamination of Quarry D rock indicates that bottom water at this site was anoxic for enough of the year to exclude benthic invertebrates (Anderson *et al.* 1985; Anderson and Dean 1988). Sedimentation in Quarry D averaged about 0.35 mm yr^{-1} (Bell and Haglund 1982), indicating sufficient primary productivity to support abundant zooplankton for stickleback prey. However, the few *G. doryssus* stomachs with prey contained benthic ostracods (family Cyprididae; Forester, pers. comm.). Mass mortality layers of *G. doryssus* comprised specimens with a limited size range, indicating that the mass mortality layers represent schools that died catastrophically above the depositional site (Bell *et al.* 1989). Feeding habits of *G. doryssus* cannot be inferred from morphology because gill rakers, which might indicate prey type (Gross and Anderson 1984; McPhail page 418 this volume), are rarely preserved, and body form, which might indicate prey type (McPhail 1984, page 419 this volume; Lavin and McPhail 1985; Baumgartner *et al.* 1988), has not been studied. Although information is limited and somewhat conflicting, it appears that *G. doryssus* was a schooling planktivore that fed in open water.

In summary, *G. doryssus* seems to have been an ecologically modern threespine stickleback. Birds appear to have been its only major vertebrate predator, and it had greatly reduced armour and relatively large body size. It schooled in open water where it presumably ate zooplankton, but it also ate benthic prey. Females appear to have produced large clutches of eggs, suggesting that reproductive biology was much like that of extant threespine stickleback.

Fine-scale chronology in the middle member of the Truckee Formation

Sedimentation of diatomites

A reliable chronological framework is needed to infer rates and patterns of change in biostratigraphic sequences (MacLeod 1991). An important attribute of the middle member of the Truckee Formation is that it appears to contain annual laminations, or varves, each consisting of a couplet of light and dark layers. The varves have been used to convert stratigraphic distance to years (Bell and Haglund 1982; Bell *et al.* 1985*b*), so it is important to carefully examine the evidence that the laminations are actually annual.

The palaeoecological and palaeolimnological information presented above indicates that Lake Truckee had a temperate climate with dry summer conditions, and that it was large, shallow, productive, and saline. Diatom growth and deposition should have been seasonal and could have produced varves. Although Lake Truckee was shallow, diatomites of the middle member of the Truckee Formation are continuously laminated in all quarries except Quarry E. Bell *et al.* (1989) reviewed the evidence that each lamination represents an annual cycle of sedimentation. Although such laminations in lacustrine biogenic sediments are generally interpreted as varves (Birks and Birks 1980; Anderson *et al.* 1985; Anderson and Dean 1988), Anderson *et al.* (1985) proposed stringent criteria for recognition of varves. Thus, there is some uncertainty that the Truckee Formation diatomites are varved.

Alternative explanations for laminations in Truckee Formation diatomites seem unlikely. Storms can produce laminated sediments (Smith and Elder 1985), but the preponderance of biogenic silicate and absence of clastic particles of fluvial origin (i.e. sand) in Quarry D rocks are inconsistent with this interpretation (Kelts and Hsü 1978). Furthermore, distal segments of fin rays in *G. doryssus* are sometimes displaced (Fig. 15.2(b), 15.5), indicating slow (i.e. $>$ 8 wk at low temperature) burial (Smith and Elder 1985; Elder and Smith 1988). Stickleback preservation and absence of larger clastic particles indicate that middle Truckee Formation diatomites were not produced by storm events.

Two sets of couplets could form each year in a lacustrine diatomite if autumn cooling eliminated the vertical density gradient caused by temperature differences, allowing wind-driven overturn to bring deep, nutrient-rich water to the surface to stimulate a secondary diatom bloom. The vertical density gradient in saline lakes is caused largely by salinity, and surface cooling will have little effect on the gradient (Wetzel 1975; Anderson *et al.* 1985). Because Lake Truckee was saline (Ruben 1971; Krebs and Bradbury 1984), it should not have experienced autumn overturn nor have produced a fall lamination.

In summary, palaeoecological and palaeoclimatic inferences, dominance of biogenic silicate in the sediment, absence of clastic material of fluvial origin, and fish taphonomy all suggest that the dark members of sedimentary couplets in the middle member of the Truckee Formation resulted from winter fluvial discharge that was filtered through marginal swamps before entering the open lake. The light members of couplets represent summer diatom blooms. Although there must be some uncertainty (Anderson *et al.* 1985), laminations in the middle member of the Truckee Formation are probably varves.

Chronology

Varves in Truckee Formation diatomites can be used for a very-fine-scale chronology of stratigraphic variation in *G. doryssus*. Mean varve thickness in Quarry D is 0.347 ± 0.034 mm (Bell and Haglund 1982), at the lower end of the range expected for lacustrine diatomites (Schindel 1980). Although the boundaries between laminations are usually too indistinct for their individual thicknesses to be measured, they can be counted over a measured distance to determine local mean thicknesses.

Failure of varves to form or their loss by erosion could cause chronological errors. None of the sections in which *G. doryssus* occurs shows evidence of erosion or even brief interruption of sedimentation. The Quarry D section is about 55 m thick, consisting of laminated diatomite with occasional crystalline (previously interpreted as sandstone, which is absent) and glassy volcanic ashes (Brown pers. comm.). Ashes form about 4 per cent of the section (Bell and Haglund 1982). Compressional deformation of the thicker (≥ 5 cm) ash layers may distort stratigraphic distances within a few centimetres of the ash, but thick ash layers are rare and this deformation does not affect microstratigraphic measurement elsewhere in the section. Fossilized mud cracks, unconformities indicating erosion, fluvial clastics, and structures representing stream channels are all absent. Although a few varves may be obscured or lost locally (Bell *et al.* 1985*b*), there is no evidence in sampled sections of the middle member of the Truckee Formation of significant depositional gaps. Sedimentary gaps characterize the fossil record (Schindel 1980, 1982*a*; Sadler 1981) and distort rates and patterns of stratigraphic change (Gingerich 1983; Jones 1988; MacLeod 1991; but see Kraus and Bown 1986), but the laminated diatomite of the Truckee Formation is nearly stratigraphically complete (Schindel 1982*a*) at a time scale of 1 yr.

Even under these ideal conditions, chronological precision is limited by the precision of field measurements. Errors in stratigraphic distance resulted from the summation of rounding errors in numerous individual measurements made during quarrying. Daily remeasurement of quarried sections in the field revealed maximum deviations of up to about 3 cm (*c.* 86.5 yr) from original measurements, but daily recalibration to original measurements precluded accumulation of greater errors. However, stratigraphic precision of less than 0.5 cm (*c.* 14.4 yr) was possible within short stratigraphic intervals under favourable conditions.

It is possible to recognize specimens of *G. doryssus* that died the same year because each varve consists of a couplet of light and dark laminations (Bell *et al.* 1989). The delicate structure of articulated fish fossils precludes exposure and redeposition of specimens on a later varve. Thus, specimens of *G. doryssus* from one varve must have died during the same year. Bell *et al.* (1989) argued that mass-mortality layers of stickleback on single varves were members of a school, and thus represent a population. Most fossil samples, including *G. doryssus* samples, accumulated over many generations (i.e. they are time averaged), and their character variances and correlations contain separate components owing to intrapopulation variation and to change through time (Bell *et al.* 1987, 1989; Bookstein 1988). Analysis of samples from single varves, especially large mass-mortality samples, allows estimation of intrapopulation and time-averaging components of variance and correlation in *G. doryssus* (Bell *et al.* 1987, 1989).

The middle member of the Truckee Formation provides almost ideal conditions for analysis of stratigraphic variation. There is no evidence for discontinuous deposition, and it is possible to convert stratigraphic distance to time with an exceptional level of accuracy. Even if there are missing time intervals, they are probably brief and more or less randomly distributed; if the chronology is distorted, it is uniformly distorted. At worst, stratigraphic change in *G. doryssus* can be measured against a clock that ticks at a stochastically constant rate; at best, varves can be counted to convert stratigraphic distance to years or generations. Assuming the worst case, events separated by about 6 cm (173 yr) can be resolved. Few biostratigraphic sequences have been resolved with the precision possible for *G. doryssus* (Schindel 1982*b*; Behrensmeyer and Schindel 1983; but see Wilson 1984; McCune 1987*a*, 1990), and this precision approaches the time scale of evolutionary studies using laboratory and extant natural populations (Schindel 1982*b*; Gingerich 1983).

Patterns of stratigraphic variation in *Gasterosteus doryssus*

Inconsistencies between the descriptions of *G. doryssus* in 1907 by Jordan and Hay represent morphological variation. Bell (1974) described the range of variation of six characters, including the number of dorsal spines and pelvic girdle structure, in this form, and Bell and Haglund (1982) concluded that some of these characters vary stratigraphically. Bell *et al.* (1985*b*) published the most detailed assessment of stratigraphic variation in *G. doryssus*, and this study forms the basis for the summary below. Stratigraphic change does not necessarily represent evolution or a specific evolutionary mechanism. Interpretation of stratigraphic variation in *G. doryssus* is discussed after the variation is described.

The patterns of stratigraphic variation

Bell *et al.* (1985*b*) made 26 samples separated by roughly 5000 yr intervals from a section in Quarry D representing about 110 000 yr of deposition. Pelvic structure, standard length (SL), and number of dorsal spines, predorsal pterygiophores, and dorsal and anal fin rays were scored. Sample sizes and the number of samples were large enough for reasonably precise estimates of character means and for detection of patterns of change through time. Most pairs of characters are significantly correlated among samples, even when adjusted for stratigraphic trends, but on average these correlations explain only about one-quarter of the variation. Phenotypic composition of samples varied through time, but changes in mean values usually failed to deviate from expectations based on random-walk models (Bell *et al.* 1985*b*; Bookstein 1988). Mean values for all traits except pelvic structure had a significant rank-order correlation with time, but small steps appeared to be superimposed on these overall trends (see also Bell and Legendre 1987). Only the number of dorsal spines and dorsal fin rays differed between the first and last samples, because of reversals of apparent trends near the ends of other series. Except for SL, all traits appear to change between samples 20 and 21. *Gasterosteus doryssus* exhibited extensive chronological variation and was not in stasis (i.e. no significant change through time).

Stratigraphic variation of pelvic structure and dorsal spine number are the most interesting characters because of their significance for defence against predatory vertebrates (e.g. Hoogland *et al.* 1957; Gross 1977, 1978;

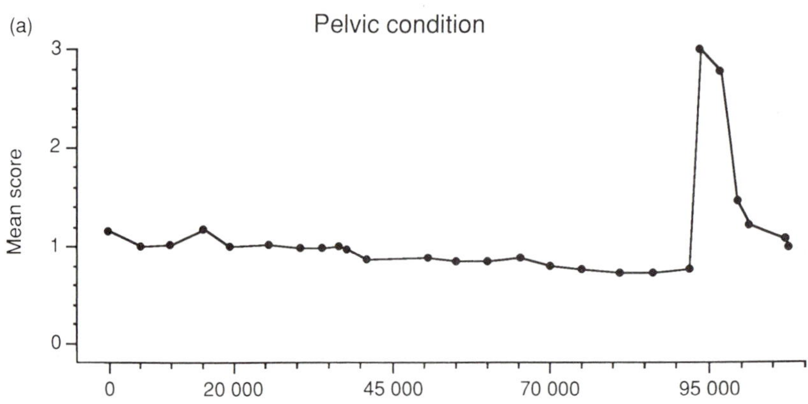

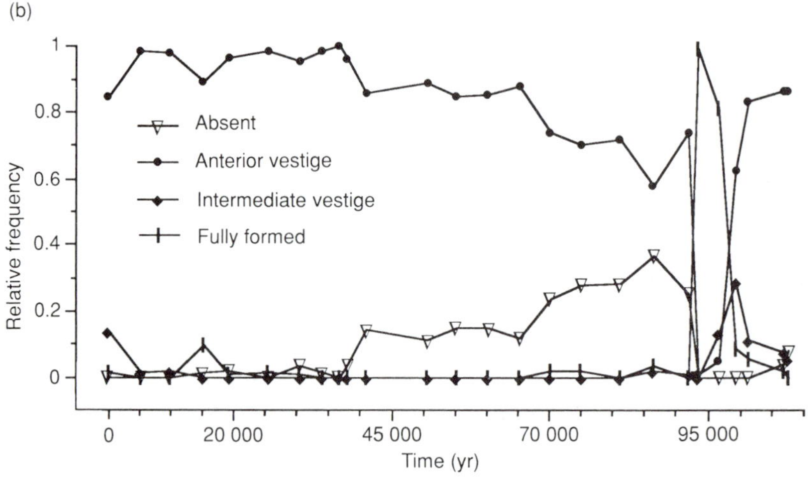

Fig. 15.6 Temporal variation of pelvic girdle phenotypes of *Gasterosteus doryssus*: (a) mean phenotype based on transformation of phenotypes to scores (fully formed, 3; intermediate vestige, 2; anterior vestige, 1; absent, 0), and (b) phenotype frequencies. Examples of these phenotypes are shown in Fig. 15.2. (Based on data in Bell *et al.* 1985*b*).

Reimchen 1980, 1983, 1991*a*, Chapter 9 this volume). Dorsal spine numbers were counted, but it was necessary to place pelvic phenotypes in four ordered classes (Fig. 15.2) which were assigned numerical values. Mean dorsal spine numbers and pelvic scores were computed and analysed in relation to time. Mean pelvic (Fig. 15.6) structure experienced a nearly linear decline at a very low rate, forming a significant rank-order correlation with time for the earliest 20 samples (92 000 yr). Mean pelvic score increased dramatically

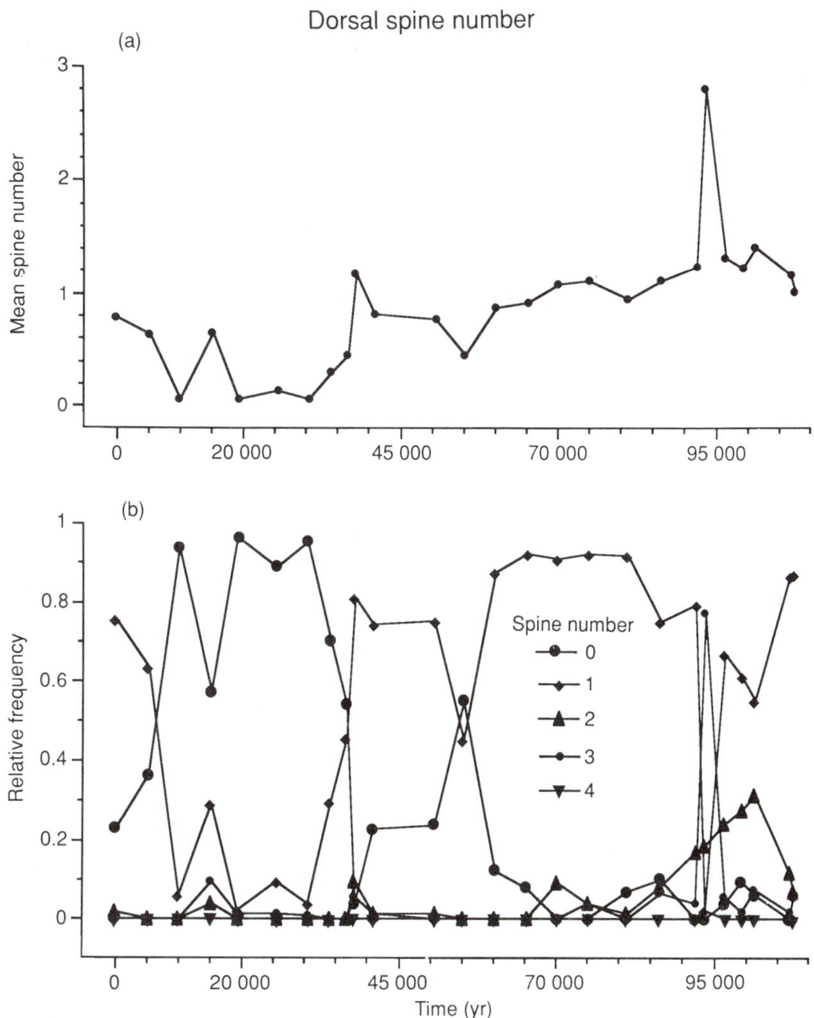

Fig. 15.7 Temporal variation of dorsal spine number phenotypes of *Gasterosteus doryssus*: (a) mean phenotype, and (b) phenotype frequencies. Examples of these phenotypes are shown in Fig. 15.2. (Based on data in Bell *et al.* 1985*b*.)

to its maximum value between samples 20 and 21, and then experienced a rapid decline. The pattern for dorsal spine number (Fig. 15.7) is less striking because of the generally irregular course of change in dorsal spine number, but shows an overall increasing trend. Mean dorsal spine number has a sharp increase between samples 20 and 21, followed by a more irregular but rapid decline. Transient increases in the frequency of intermediate phenotypes for both pelvic structure and dorsal spine number as their means declined after sample 21 fit the pattern expected for evolution of polygenic threshold traits (Bell *et al.* 1985*b*).

Bell and Legendre (1987) reanalysed the Quarry D samples using a chronological clustering technique designed to detect discontinuities in the composition of ordered sets of samples (Legendre *et al.* 1985). Analysis of the composition of samples for multivariate morphotypes that reflect mostly dorsal spine number and pelvic structure indicated discontinuities between samples 20 and 21, as had been indicated by analysis of the single-character plots, and a second discontinuity between samples 14 and 15, which was not obvious in the single-character sequences. Principal coordinate analysis illustrated the discontinuous morphological displacement between samples 20 and 21, and a more progressive return by the end of the sequence to the position of sample 20 in the morphospace defined by the first two principal coordinate axes.

Resolution of the displacement between samples 20 and 21

The stratigraphic sampling interval used by Bell *et al.* (1985*b*) produced what Eldredge and Gould (1972) called a 'geologically instantaneous' transition. This transition might form one of three patterns using finer temporal resolution: (1) truly instantaneous (at any time scale) replacement of low-spined (i.e. with small ovoid pelvic vestiges, zero or one dorsal spine) by spinier sticklebacks, (2) replacement of low-spined by spinier fish after a period of co-occurrence, and (3) gradual replacement of low-spined by spinier fish involving a sequence of intermediates progressively trending toward spinier phenotypes.

A sample of 123 scorable specimens was collected with ± 0.5 cm stratigraphic precision from a 27 cm section between samples 20 and 21 containing the transition from low-spined to spiny sticklebacks (Bell unpubl. data). Approximately 1086 varves occur in this 27 cm section. About 50 scorable low-spined specimens occur alone for the first 646 yr, 7 spiny and 12 low-spined specimens are interspersed within a 122 yr interval, and about 48 spiny specimens occur alone for the final 318 yr. (Number of specimens is approximate because they differ for dorsal spine and pelvic phenotypes.) Although these results are tentative, clearly spiny and low-spined stickleback coexisted, but did not form a transitional sequence in which intermediate phenotypes (i.e. large pelvic vestiges with separate anterior and posterior elements [e.g. Fig. 15.2(c),(d)]; two dorsal spines) dominated numerically.

Thus, this transition does not represent an evolutionary process, but was an ecological event in which a low-spined biological species was replaced by a spiny one after a brief period of coexistence. Causes of such replacement events are difficult to infer (Kitchell 1985), and there is no evidence for the cause of this one. During 14 000 yr after this replacement event, reduced pelvic girdle and dorsal spine number phenotypes 're-evolved'. Demonstration that the transition from low-spined to spiny phenotypes was a species replacement event justifies separate analyses of the first 20 and last six samples of the sequence presented by Bell *et al.* (1985*b*).

Interpretation of stratigraphic variation in *Gasterosteus doryssus*

The punctuated equilibria model (Eldredge and Gould 1972) has stimulated numerous analyses of biostratigraphic sequences, but it is extremely difficult to show that they have yielded a faithful rendition of intrapopulation (intrademic) evolution. Simulation studies place bounds on the expectations for intrapopulation evolution (Kirkpatrick 1982; Petry 1982; Newman *et al.* 1985; Lande 1986; Milligan 1986), but they are sensitive to simplifying assumptions that may limit their applicability to real population lineages. The dearth of reliable empirical information on stratigraphic variation of population lineages is a serious deficiency for the development of evolutionary theory.

Criteria for recognition of intrapopulation evolution in the fossil record

The ideal biostratigraphic sequence to study intrapopulation evolution would have to meet several stringent criteria. It must include many samples, each of which is a large random sample from one generation of a single deme. Samples must be made at reasonably uniform and reliably estimated time intervals. It must be possible to resolve the time between samples down to a few hundred generations because the simulation studies cited above suggest that very high rates of evolution are possible. It must be reasonable to assume that much of the observed phenotypic variation reflects genetic variation, rather than phenotypic plasticity or ontogenetic variation. The samples must represent populations with ancestor–descendant relationships; gene flow, geographic variation, and species replacement should either be recognizable or be unlikely causes for stratigraphic variation. Unfortunately, these criteria are rarely met individually and never in combination in existing biostratigraphic sequences. Which of these criteria does the *G. doryssus* sequence described above meet, and to what extent does failure to meet them limit conclusions?

Some of these criteria can be dealt with briefly. Sample sizes vary among samples and characters in the Quarry D stickleback sequence (Appendix 1 of Bell *et. al.* 1985*b*), but they are large enough for means to be estimated reliably. There are enough (i.e. 26) samples for trends to be inferred using rank-order correlation or regression analysis. Samples were made at roughly

5000 yr intervals, and the problems of inaccurate chronology (MacLeod 1991) and irregular sampling time interval (Gingerich 1983) are insignificant. Characters that vary in *G. doryssus* appear to have a significant component of genetic variation in extant stickleback populations (Lindsey 1962; Blouw and Boyd 1992). The fossil specimens used exceed the size at which the characters studied have completed development (Bell and Harris 1985; unpubl. obs.). Thus, some of the criteria for analysis of patterns of intrapopulation evolution are easily met by the *G. doryssus* sequence; others discussed below warrant more careful consideration.

The problem of time averaging in fossil samples

Samples from extant populations comprise specimens from a short time interval and a limited area, but even they may represent several generations and multiple differentiated subpopulations. These pooling problems are greater in fossil samples, even those collected from a narrow stratigraphic range and a restricted area, because of the long time during which fossil samples typically accumulate. The character mean of a time-averaged sample (i.e. comprising specimens deposited over a long time interval) will be weighted according to the relative abundance of specimens from different times and places, but it is still a good indicator of the average value of the character in the source area during that time interval.

However, temporal and spatial pooling may significantly increase sample variances (Bell *et al.* 1987) and correlations between characters (Bookstein 1988; Bell *et al.* 1989). Terrestrial mammals, coastal marine molluscs, and oceanic microfossils, which constitute the majority of available biostratigraphic sequences, often occur in poorly stratified deposits that limit temporal resolution (e.g. Schindel 1980; 1982*a*,*b*; Sadler 1981; Behrensmeyer and Schindel 1983; Jones 1988) or are rare enough that specimens from different stratigraphic horizons must be pooled for statistical characterization. In either case, each sample is time averaged and may include specimens that lived thousands of generations apart. There is little evidence concerning evolutionary rates at this time scale (Gingerich 1983), but theoretical studies are consistent with high enough rates of evolution (Kirkpatrick 1982; Petry 1982; Newman *et al.* 1985; Lande 1986; Milligan 1986; Lynch 1990) to produce significant effects on character variance and correlations within a few thousand generations. Indeed, stratification of Eocene land mammal samples produced evidence that time averaging increases character variance (Kelly and Gingerich 1991).

Individual samples of *G. doryssus* accumulated over an average of 1335.5 yr (SD = 526.8 yr). Bell *et al.* (1987) examined dorsal fin ray number in temporally stratified samples and two mass-mortality layers of *G. doryssus*. Neither sample stratification nor comparison of time-averaged samples with mass-mortality layers or extant populations indicated that time averaging inflated variance of dorsal fin ray number in *G. doryssus* samples. However, although dorsal spine number and pelvic structure were correlated among time-averaged samples, they were not significantly associated among individuals within mass-mortality layers (Bell *et al.* 1989). This result indicates that the statistical properties of *G. doryssus* samples are affected by time averaging. Nevertheless, samples of *G. doryssus* approach the ideal of a sample from one generation much more closely than do samples of other fossil taxa.

The problem of spatial pooling in fossil samples

Post mortem transport, gene flow, geographic variation, and presence of multiple species are forms of spatial variation that can also increase character variation and correlation within fossil samples. Taphonomic analysis is usually applied to palaeoecology, but it can also be used to help assess the effects of spatial variation on fossil samples (Bell *et al.* 1987, 1989; Behrensmeyer and Kidwell 1988; Kidwell and Behrensmeyer 1988). The durability of many fossil taxa facilitates *post mortem* transport and redeposition after burial, which can mix individuals from different times and places or sort them by size or hydrodynamic properties. These problems appear to be minimal in *G. doryssus* because *post mortem* transport, either before or after initial burial, would disarticulate delicate fossil fish (Wilson 1980, 1988; Elder 1985; Smith and Elder 1985), and specimens of *G. doryssus* are generally well preserved. Thus, *post mortem* transport probably had little effect on *G. doryssus* samples, but other causes of spatial pooling are possible.

Biostratigraphic sequences have often been implicitly treated as samples from sets of populations with ancestor–descendant relationships. Gingerich (1979) named this treatment 'stratophenetics' and formalized the method, but it has serious pitfalls (Schaeffer *et al.* 1972; Englemann and Wiley 1977). Successive fossil samples may represent separate populations or species, between which differences have accumulated over an unknown period of time since their latest common ancestor. This problem is the Achilles' heel of evolutionary palaeobiology (but see Prothero and Lazarus 1980).

Intralacustrine and parapatric lake–stream species pairs are rare, and both tend to differ in characters that probably could be detected if they had been present in *G. doryssus* (Schluter and McPhail 1992; McPhail page 418 this volume). However, a low-spined biological species was replaced by a spiny one in the Quarry D sequence, and four of the first seven samples in this sequence with otherwise low-spined phenotypes include low frequencies of specimens with fully formed pelvic girdles and three dorsal spines. Two biological species of *Gasterosteus* were present within the Lake Truckee basin, but they were distinctive for spine number and pelvic structure. It seems unlikely that mixtures of species within samples or a progression of slightly divergent species produced the stratigraphic trends observed in *G. doryssus*, because character distributions are usually unimodal within samples (Bell *et al.* 1985*b*). Interspecific differences appear to affect only the dramatic transition between samples 20 and 21 and variation within a few of the early samples in the Quarry D sequence.

Geographical variation within a population poses a similar problem for inference of evolutionary change in the fossil record (Gould and Eldredge 1977; Bell and Haglund 1982). Stratigraphic variation could represent either evolutionary change or changes in the geographical locations of differentiated subpopulations. It may be possible to eliminate geographical variation as an explanation by showing that it is too limited in related extant taxa to account for stratigraphic variation (Koch 1986). However, geographical variation cannot be dismissed this way for the *G. doryssus* sequence, because small-scale spatial variation is common within extant threespine stickleback populations (e.g. Reimchen 1980; Bell and Richkind 1981; Baumgartner and Bell 1984; Baumgartner 1986*b*). In addition, palaeoecological and palaeolimnological information reviewed above indicates that there were ecologically contrasting aquatic habitats in the Lake Truckee basin in which selection might have favoured divergence (Endler 1977). Information from the fossil record itself will be needed to determine whether stratigraphic variation in *G. doryssus* could represent geographical variation instead of evolution.

Despite the potential for intrapopulation spatial variation to contribute to stratigraphic variation in *G. doryssus*, palaeontological evidence indicates that observed variation represents intrapopulation evolution within two successive populations (i.e. samples 1–20 and 21–26). The unimodal phenotypic distributions of *G. doryssus* samples and the extremely gradual trend for pelvic structure during the first 92 000 yr of the sequence do not indicate that a shifting geographic distribution strongly influenced stratigraphic variation (Bell *et al.* 1985*b*).

Current studies are consistent with this assessment (unpubl. data). Samples made at four corresponding horizons in quarries D and A either are indistinguishable or differ slightly in the frequencies of similar phenotypic classes. Similarly, the contents of bird pellets, which may have been transported a great distance between ingestion and regurgitation, and articulated specimens, which must have been deposited close to the place of death (Wilson 1980, 1988; Elder 1985; Smith and Elder 1985), have similar spine phenotypes in the same samples. These results indicate that the Quarry D sequence reported by Bell *et al.* (1985*b*) is representative of a widespread stickleback population in Lake Truckee. Although it is extremely difficult to show that variation in a biostratigraphic sequence represents intrapopulation evolution, and the evidence presented here is only circumstantial, there is no reason to believe that stratigraphic change in the *G. doryssus* sequence does not represent intrapopulation evolution.

The problem of random walks

Bookstein (1986, 1988; Bookstein and Reyment 1992) noted that symmetrical random walks produce sequences of points that resemble patterns normally attributed to a deterministic process (i.e. natural selection). He argued that it is necessary to reject the random walk hypothesis before making mechanistic inferences from biostratigraphic sequences. He proposed the use of a range test to test the hypothesis that the maximum departure of a character mean from its original value was either too large or too small to be produced by a random walk with the observed distribution of step sizes between successive sample means. Rejection of the null hypothesis would indicate presence of either anagenesis (large range) or stasis (small range). However, Bookstein (1988) noted that existence of an expectation for a stratigraphic trend based on independent evidence (e.g. selection in an extant analogue) could justify inference of a deterministic causal mechanism for a biostratigraphic sequence that did not deviate from random walk expectations.

Using this range statistic, Bookstein (1988) concluded that the range of excursion for sequences of the six characters of *G. doryssus* analysed by Bell *et al.* (1985*b*) generally occurred within the envelope of maximum range expected for random walks. The one clear exception was standard length (SL), which appeared to be in stasis. Bell *et al.* (1985*b*) had inferred the presence of a significant tendency for trend reversal superimposed on an overall rank-order correlation with time for this character. Bookstein's inference of stasis is interesting because SL in *G. doryssus* appears to have been continuously near the upper limit of SL in extant *G. aculeatus* populations (Baker page 157 this volume).

Bookstein (1988) allowed that existence of the stratigraphic trend for pelvic reduction, combined with palaeoecological conditions that favour reduction in extant *G. aculeatus* (i.e. absence of predatory fishes; see above), permits the inference that the trend was not due to a random walk, even though the range test on it was not falsified. Although the requirement of an external expectation for stratigraphic change was met only by pelvic structure, one could extend this expectation to reduction of dorsal spine number because of the functional interaction of dorsal spines with the pelvis (Reimchen 1983, 1991, page 248 this volume) and the co-occurrence of pelvic reduction and decreased dorsal spine number in some extant populations of *Gasterosteus* (Bell 1974; Reimchen 1980). Development of similar expectations for other variable characters of *G. doryssus* is limited by the difficulty of inferring palaeoecological conditions. In general, use of an independent expectation for stratigraphic change in biostratigraphic sequences is impracticable because the necessary information either on palaeoecology or on selection in extant analogues will be unreliable or non-existent.

Application of Bookstein's (1986, 1988) range test to biostratigraphic sequences is problematical. If the variance of the step size between successive samples is increased or decreased by natural selection, the range required to deviate significantly from the range expected for a random walk would be affected (Rohlf pers. comm.). Sequences influenced by combinations of drift and selection are unlikely to exceed the critical limits of the range test (Fenster and Sorhannus 1991). Similarly, sequences resulting from a combination of stabilizing and directional selection, or selection in varying directions (e.g. Felsenstein 1988), should often fail to deviate significantly from the expectations for a random walk. Only consistent directional change (i.e. orthogenesis) or very strict prolonged stabilizing selection will produce sequences that are inconsistent with a random walk. Although strict stasis has been proposed as a dominant mode in evolution (e.g. Eldredge and Gould 1972; Stanley 1979), orthogenesis has long since been abandoned (e.g. Simpson 1944), and there is ample reason to expect inconsistent operation of selection on a population. Therefore, falsification of a hypothesis based on a random walk model to explain a biostratigraphic sequence provides strong support for operation of natural selection, but failure to falsify the hypothesis is ambiguous.

Conclusion

Although there will always be an element of uncertainty for biostratigraphic sequences, the *G. doryssus* sequence meets the criteria for analysis of intrapopulation evolution in the fossil record reasonably well. Stratigraphic variation in *G. doryssus* appears to represent intrapopulation evolution of heritable traits, with little influence of geographical variation. Although two species were present, they are easy to distinguish. The chronology for the sequence is precise and accurate, and samples are evenly spaced. *G. doryssus* seems to exhibit trends within a stratigraphic section representing about 110 000 yr. These trends appear to be gradual, without imposition of stasis by the intrinsic constraints sometimes implied under the punctuated equilibria model (e.g. Gould 1980; Maderson *et al.* 1982). Although SL seems to change slightly, it may be constrained below an upper limit that characterizes SL in extant threespine stickleback. Selection cannot be excluded as the constraint on maximum SL in the *G. aculeatus* complex.

Stratigraphic change in *G. doryssus* generally results from progressive shifts among adjacent samples for phenotype frequency distributions. Some of the phenotypic changes observed represent major morphological transitions for *Gasterosteus* and cannot be dismissed as trivial random departures from long-term stasis. In particular, the magnitude of reduction of both pelvic structure and the number of dorsal spines in *G. doryssus* rarely occurs among extant populations of *G. aculeatus* (e.g. Moodie and Reimchen 1976*a*; Gross 1978; Bell 1984*a*, 1987, unpubl. obs.; Bell *et al.* 1985*a*). The major discontinuity in which pelvic structure and dorsal spine number increase instantaneously (after sample 20, Bell *et al.* 1985*b*) represents replacement of one biological species by another, not intrapopulation evolution. Reduction of pelvic structure and dorsal spine number following this species replacement is rapid, but represents phyletic gradualism that results in functionally significant morphological evolution within a population lineage. The 're-evolution' of reduced pelvic structure and dorsal spine number after the species replacement is added evidence that conditions in Lake Truckee favoured loss of armour.

The major exception to strictly gradual evolution involves pelvic reduction in the spiny species that appears after sample 20. Loss of the pelvic spine is almost always accompanied by reduction of the pelvic girdle to separate anterior and posterior vestiges and loss of the ascending branch (Bell 1987, unpubl. data). Consistent synchrony of these three morphological changes omits phenotypes that occur as intermediates in morphological series from extant populations with pelvic reduction (Bell 1987). Evolutionary reduction of the pelvic girdle saltates past the phenotype in which the pelvic girdle is intact but lacks spines. Evolution of *G. doryssus* is consistent with neo-Darwinian mechanisms (Charlesworth *et al.* 1982), but with the added important implication that developmental processes sometimes bias expression of phenotypic variation during evolution.

CONCLUSIONS

Taken alone, the fossil record of the *G. aculeatus* complex is not generally exceptional. It consists mostly of small numbers of specimens from about 20 deposits, and the vast majority of fossil stickleback specimens come from a single deposit from the Truckee Formation. It is an excellent example of intrapopulation evolution, and it adds precision for interpretation of the tempo of evolutionary divergence in extant threespine stickleback.

Phylogenetic pattern

Variation among extant and fossil stickleback samples demonstrates that the species complex has been diversifying since the Miocene without sustained divergence. Marine populations appear to have been phenotypically stable. The most highly divergent extant freshwater populations occur in habitats that must have been colonized by marine stickleback postglacially. Thus, divergence from the morphology of their common marine ancestor postdates deglaciation. Variation among fossil samples, however, indicates that phenotypic differentiation of freshwater threespine stickleback occurred as long ago as 10 million years. Thus, it is paradoxical that *G. wheatlandi* is the only old distinctive species to diverge from the *G. aculeatus* species complex. This combination of repeated divergence from a stable common ancestor and absence of numerous, old, distinctive species produces a phylogenetic topology called a raceme (Williams 1992, see also Bell and Foster page 14 this volume). The phylogenetic raceme for *G. aculeatus* comprises a phenotypically stable marine ancestor from which many freshwater isolates have diverged rapidly and then gone extinct (Bell 1987, 1988; Bell and Foster page 14; McPhail page 401 this volume). Phenotypic diversity of ninespine stickleback, *Pungitius pungitius*, in many postglacial freshwater habitats indicates that their phylogeny, too, may form a raceme, but their fossil record is too sparse to tell. Phylogenetic racemes may be common but less conspicuous in taxa other than *G. aculeatus* (Williams 1992). The fossil record of *G. aculeatus* is necessary for recognition of its raceme.

Chronology and zoogeography

Buth and Haglund's (page 82 this volume; Haglund *et al.* 1992*a*) allozyme analysis indicates that Atlantic and Pacific basin threespine stickleback clades diverged very recently. A set of Japanese populations forms the sister group of all other *G. aculeatus*, but the genetic distance between other Pacific basin populations and those in the Atlantic is low, suggesting recent derivation of Atlantic basin populations from a Pacific basin ancestor. However, all five stickleback genera and *G. wheatlandi* occur in the Atlantic basin, suggesting that generic radiation of the Gasterosteidae occurred in the Atlantic and that *G. aculeatus* arose there. The threespine stickleback occurred in the Pacific at least 11 million years (Ma) ago, and possibly as long ago as 16 Ma, but it did not appear in the Atlantic basin until about 1.9 Ma ago (Table 15.1). Although a greater effort should be made to search European deposits for fossil stickleback, the fossil data indicate that *G. aculeatus* arose in the Pacific and recently spread to the Atlantic.

Evolution of a Miocene population

Eldredge and Gould's (1972) punctuated equilibria model postulated that evolutionary change is restricted to relatively brief periods during speciation. Many biostratigraphic sequences have been studied to assess the validity of punctuated equilibria, but most studies have flaws that limit their utility to distinguish between punctuated equilibria and phyletic gradualism. The Quarry D sequence of *G. doryssus* (Bell *et al.* 1985*b*) is almost free of these flaws. The pattern of change does not deviate fundamentally from expectations of neo-Darwinian theory; genetic drift and selection could produce observed patterns. Although evolutionary rates were variable, there is no evidence that evolutionary stasis was imposed by intrinsic properties (e.g. developmental mechanisms) or that significant evolutionary saltations occurred, as postulated in later expressions of punctuated equilibria (e.g. Gould 1980; Maderson *et al.* 1982).

The tempo of evolution in threespine stickleback populations

Restriction of the more divergent threespine stickleback populations to postglacial habitats imposes an upper limit of about 5000 to 25 000 yr on their age. However, habitat age imposes only an upper limit on population age, and it is uncertain from studies of postglacial divergence how fast threespine stickleback really evolve. The Quarry D sequence of *G. doryssus* (Bell *et al.* 1985*b*) appears to represent two separate population lineages, one occurring for the first 92 000 yr of the sequence and another for the final 14 000 yr. Evolution of armour phenotypes in the earlier lineage is relatively slow, but the earliest samples in the sequence already had extreme spine and pelvic reduction. In contrast, when the second lineage appeared, it possessed the primitive states for both of these traits. In this respect, arrival of this spiny species in Lake Truckee is analogous to postglacial colonization of lakes by marine or anadromous threespine stickleback.

In the second lineage, mean dorsal spine number changed by about 1.5 spines within 3200 yr but by only about 0.3 of a spine in the remaining 11 000 yr of the sequence. The relatively brief 3200 yr period in which the greatest change occurred was too long to resolve the pattern of evolution. Pelvic reduction in the second lineage evolved more slowly than dorsal spine number. About 90 per cent of the reduction in mean pelvic score occurred during the first half of the sequence (i.e. 7800 yr) for this lineage. The magnitude of pelvic girdle and dorsal spine number reduction observed during this sequence is greater than that seen in most extant postglacial populations (Bell 1974, 1987; Reimchen 1980). Thus, the tempo of change in *G. doryssus* suggests that the upper limit placed by time since deglaciation on the age of extant populations may be too long to estimate rates of divergence.

ACKNOWLEDGEMENTS

I am especially indebted to F.H. Brown for the K/Ar date for Truckee Formation stickleback and for numerous geological insights. D.J. Long kindly shared with me results of his unpublished study of fossil *Gasterosteus* from California. J. Bickert identified the Truckee Formation birds. C.C. Swift and S. Funder brought published stickleback records from California and Greenland, respectively, to my attention. I thank W.N. Krebs and R.E. Reynolds for information on the age and palaeoecology of Miocene deposits in south-western USA, J.G. Fleagle and D.W. Krause for discussion of the Neogene fossil record, F.J. Rohlf and F.L. Bookstein for explanations of random walk statistics, and T. Hrbek, A. Meyer, and K.T. Shao for translation or extraction of information from papers published in various languages. S.C. Nemtzov read the manuscript. My research on sticklebacks has been supported by grants from the National Science Foundation of the USA (currently by BSR-8905758), Earthwatch, and the National Geographic Society. My thanks to Earthwatch volunteers, friends, and students who broke rocks in the hot Nevada sun to collect fossil stickleback.

16

Evolutionary inference: the value of viewing evolution through stickleback-tinted glasses

Susan A. Foster and Michael A. Bell

The threespine stickleback species complex possesses a number of attributes that make it particularly suitable for evolutionary studies. This stickleback has undergone remarkable phenotypic diversification in freshwater habitats since the deglaciation of north-western North America, Scotland, and other less extensively studied regions (reviewed in Bell and Foster page 16 this volume). In some instances, speciation has occurred since the last deglaciation (McPhail Chapter 14 this volume). In addition, there exist many seemingly older, freshwater populations, and the marine, estuarine, and anadromous ancestors (hereafter referred to as marine ancestors) of the freshwater populations are extant in both the Atlantic and Pacific basins (Bell and Foster page 2). Extensive postglacial diversity, coupled with a limited, though long and informative, fossil record, provides a chronological dimension that is lacking in other species. The diversity of phenotypes, population ages, and interrelationships makes possible interpopulation comparisons that will shed light upon a variety of issues in evolutionary biology and that would be impossible in species with more obscure intraspecific phylogeny.

Research on the evolution of morphological phenotypes has in many respects progressed beyond that on other aspects of the phenotype in threespine stickleback. This advance is undoubtedly due to the relative ease with which morphological character states can be sampled. Because much of the special value of the threespine stickleback for evolutionary research lies in the extensive population diversification and well-documented racemic phylogeny of the species complex (Bell 1987, 1988; Williams 1992; Bell and Foster page 13 this volume), the ability to compare large numbers of individuals from many populations with relative ease facilitates evolutionary inference considerably. Early research on population biology (e.g. Hagen 1967; McPhail 1969; Hagen and McPhail 1970) was of critical importance because it highlighted evolutionary explanations and generated questions regarding the phylogenetic history of the species complex. Although most research has concentrated on the interpretation of morphological differentiation, efforts are now being made to interpret population differences in behaviour and life history in an evolutionary context as well (Chapters by Baker, Huntingford *et al.*, and Foster this volume). Evolutionary research on physiology is in a still earlier stage of development, but there is excellent potential for comparative studies and for research in physiological ecology (Chapters by Guderley, Wootton this volume).

The threespine stickleback species complex possesses a number of attributes other than its unusual phylogenetic history that enhance its value for evolutionary research. Not only is this stickleback readily reared in the laboratory, making genetic and physiological research possible, it is also abundant and easily collected and observed in the field. Finally, as amply demonstrated by the chapters in this volume, there already exists substantial information on many aspects of the biology of threespine stickleback.

A primary purpose of this closing chapter is to highlight the particular value of the threespine stickleback complex for evolutionary research. This topic is discussed to varying degrees in earlier chapters, and most of the authors have suggested important future research directions in their fields. We will attempt to bring together these ideas in an effort to identify major questions that must be addressed if this species complex is to be exploited fully as a source of evolutionary insights. We close with a brief comparison of this diversification with better-studied adaptive radiations, and with a plea for the conservation of the populations that constitute this remarkable species complex.

ADAPTATION AND THE EVOLUTION OF FRESHWATER POPULATIONS

The chapters in this volume provide convincing evidence of a prominent role for natural selection in the diversification of postglacial freshwater populations of threespine stickleback. In large part, this inference has been possible because of the unusual phylogenetic history (racemic phylogeny) of the group. One advantage of this history for phylogenetic inference is that many of the freshwater populations have been independently derived from the marine ancestor and have thus been exposed independently to similar selection regimes (e.g. Bell and Foster page 20, McPhail page 401 this volume). In effect, numerous 'replicate experiments' have been established, permitting examination of correlations between iterative apomorphies and environmental variables with more certainty of statistical independence among the populations than is usually possible. The value of the postglacial freshwater diversification for correlative studies of adaptation is also increased by the low level of gene flow between populations, a condition that should facilitate adaptation to local environments (e.g. Mayr 1942; Endler 1973, 1977; Slatkin 1985, 1987). Similarly, the recency of the diversification reduces the likelihood that historical events such as dispersal or stream capture have affected many populations.

There are, of course, pitfalls inherent in the use of correlative studies for adaptive inference (e.g. Gould 1984; Endler 1986). Especially when only a few populations are sampled, correlations with some environmental variables are likely to be detected by chance alone (e.g. Clarke 1975). This problem can be alleviated if a large number of populations can be sampled (Endler 1986). Geographical complexity in the pattern of environmental variation also reduces the probability of covariation among environmental variables, thereby reducing the likelihood that phenotypic variation will be ascribed to the wrong environmental factor (e.g. Endler 1977). Both requirements can be met readily in many regions in which freshwater stickleback have diversified since the last glacial recession. Finally, it is essential that the variation in the traits being studied have known heritability to avoid confusing environmental induction of phenotypically plastic traits with selection on heritable variation (e.g. Endler 1986). Because stickleback can readily be reared in the laboratory, phenotype heritabilities can be readily assessed. Despite this potential, our knowledge of trait heritabilities is extremely limited except for a handful of morphological (Lindsey 1962; Hagen 1973; Hagen and Gilbertson 1973*a*; McPhail 1977) and life history (Snyder and Dingle 1989; Snyder 1991*b*) traits.

Even when all the above conditions are met, correlative studies can, at best, suggest a causal relationship between an environmental variable and phenotypic variation. Other methods are required to demonstrate a causal relationship (Endler 1986). At present, evidence of adaptive variation in behaviour and life history is primarily correlational. Evidence for adaptive diversification of several morphological phenotypes is far stronger.

Morphological phenotypes

Variation in armour structure has been particularly well studied in an evolutionary context, beginning with the seminal work of Hagen and his colleagues. Hagen and Gilbertson (1972) originally detected an association between the presence of predatory fishes in lakes and a modal number of seven bony lateral plates on each side of the abdomen. Because they also demonstrated that plate number was heritable (Hagen and Gilbertson 1973*b*) and that predation by fish favours an increase in the frequency of seven-plated stickleback in natural populations (Hagen and Gilbertson 1973*a*) and in the laboratory (Moodie *et al.* 1973), they provided convincing evidence that selection by predatory fish maintains this plate number phenotype in natural populations. Reimchen (1983) subsequently showed that specific lateral plates interact functionally with the dorsal spines and pelvis, but the functional significance of the most anterior plates, which are present in seven-plated individuals, is currently being investigated (Reimchen pers. comm.). Such combined approaches are sufficient to meet Endler's (1986) criteria for demonstration of selection in natural populations.

The structure of the pelvic girdle also varies in relation to environmental variables. Predation regime (Reimchen 1980) and calcium concentration (Giles 1983*a*) have been proposed as factors influencing evolution of pelvic structure. The pelvic girdle and spines are generally robust and play an important part in the deterrence of predation by vertebrates (e.g. Hagen and Gilbertson 1972; Gross 1978), but in rare cases the pelvic spines and girdle may be partially or totally lost (Bell 1987). Pelvic structure is highly heritable in the related ninespine stickleback, *Pungitius pungitius* (Blouw and Boyd 1992), and probably has a strong genetic basis in *Gasterosteus* as well.

Reimchen's work on the functional interactions of the pelvic girdle and lateral plates (Reimchen 1983), and on the relation between gape size of trout, *Oncorhynchus clarki*, and the distance between the tips of the dorsal and pelvic spines (Reimchen 1991, page 271 this volume), demonstrates the functional importance of the pelvis in deterring ingestion and mastication by predatory fishes. In contrast, Reist's (1980*b*) study of brook stickleback, *Culaea inconstans*, with variable pelvic structure indicated that individuals with pelvic reduction were less susceptible to predation by dytiscid beetles than were those with the primitive fully formed pelvis. Analysis of pelvic reduction has the added dimension of a fossil record, which suggests that it is an ancient phenomenon but can occur very rapidly (Bell *et al.* 1985*b*; Bell page 459 this volume).

Consideration of the functional morphology of this armour complex reveals why this combination of the pelvic girdle and seven lateral plates is maintained by fish predation. Threespine stickleback possess two large pelvic spines and three dorsal spines, the anterior two of which are large and serrated. Four of the seven lateral plates retained in populations exposed to vertebrate predation buttress the large, locking dorsal spines, while the pelvic spines, which also lock in an erect position, are supported on the pelvic girdle. Together the lateral plates and pelvic girdle enclose the abdomen of the stickleback in a flexible segmented armour complex that is an effective deterrent to many piscivorous vertebrates (Hoogland *et al.* 1957; Reimchen 1983 page 248 this volume).

These studies provide compelling evidence that this armour complex is maintained in populations as a consequence of selection imposed by predatory vertebrates (Reimchen Chapter 9 this volume). There is evidence that both lateral plate number and pelvic girdle expression are associated with vertebrate predation across a diversity of populations in geographically disparate locations. Both traits are known to be heritable, and selective predation by a piscivorous fish has been shown to favour the seven-plate phenotype. Finally, full expression of the pelvic girdle and lateral plates that support the dorsal spines is essential for the armour complex to act as an effective deterrent to predatory vertebrates.

In light of this evidence, variation in lateral plate number and pelvic girdle expression seems extremely unlikely to represent a 'non-adaptive product of history' (Gould and Lewontin 1979; Gould and Vrba 1982). Although other factors, such as the presence of abundant predatory invertebrates (Reimchen 1980; Reist 1980*b*) and low calcium levels (Giles 1983*a*; Bell *et al.* 1985*a*, in press), may favour the loss of these structures, selection by vertebrate predators is clearly influential in maintaining the pelvic girdle and a high frequency of specimens with seven lateral plates per side.

Evidence of adaptive diversification in morphology is not limited to armour structures. Hagen and McPhail and their colleagues have provided evidence that population differences in morphological characters including mouth size, body shape, and the number and length of gill rakers are inherited (Hagen 1973; Lavin and McPhail 1987). They have also provided evidence that trait differences are strongly associated with specific ecological factors (Hagen and Gilbertson 1972; Lavin and McPhail 1985, 1986; McPhail Chapter 14 this volume). As in the case of the armour complex, similar associations are found over a wide geographic range, throughout which the different phenotypes are interspersed in a complex geographic pattern. Therefore divergence in these traits is also likely to be a consequence of evolution in response to local selection pressures, and similarities among geographically distant populations seem to be the products of parallel evolution. A similar combination of field, genetic, and laboratory research was used to explain the evolution of black nuptial coloration in populations of stickleback in the Chehalis River drainage in western Washington (McPhail 1969; Hagen and Moodie 1979; Hagen *et al.* 1980).

Non-morphological phenotypes

It is only recently that these methods have been used to infer the adaptive values of specific behavioural phenotypes. Because data on behavioural phenotypes are more difficult to collect in the field and laboratory than are morphological data, comparatively few populations have been studied. Nevertheless, these more limited comparisons provide compelling evidence of ecotypic differentiation in aggressive (Bakker Chapter 12 this volume), foraging (Hart and Gill Chapter 8, McPhail Chapter 14 this volume), antipredator (Huntingford *et al.* Chapter 10 this volume), and reproductive (Foster Chapter 13 this volume) behaviour. Efforts to assess the extent to which population differences in behaviour are mediated by genetic differences are few, but population differences in both the aggressive and the antipredator behaviours that have been examined have proven to be genetically based (Bakker Chapter 12, Huntingford *et al.* Chapter 10 this volume). Although the adaptive values of many of the behavioural patterns that have been studied seem evident (e.g. antipredator behaviours are best developed in populations exposed to relatively high levels of predation by vertebrates), in no instance has it been shown directly that particular behavioural traits contribute to fitness.

In summary, although the data are limited, evidence available to date suggests that behavioural phenotypes have undergone adaptive diversification akin to that observed for morphological phenotypes. Clearly, additional research on behavioural phenotypes, including direct measures of natural selection on behaviour and assessments of the genetic underpinnings of population differences in behaviour, would do much to advance our understanding of the extent to which natural selection has influenced the evolution of behavioural phenotypes of a vertebrate in nature.

One approach to understanding the role of adaptation in behavioural diversification involves testing game theory models of the evolution of behaviour (e.g. Maynard Smith 1982) and optimal foraging models (e.g. Stephens and Krebs 1986) through interpopulation comparisons. Many of these models predict the evolution of different behavioural phenotypes under different ecological and social conditions. Judicious comparisons of populations exposed to appropriate selection regimes can permit tests of these evolutionary models (e.g. Riechert 1986). This approach has not been taken in research on stickleback, but should be.

The comparative approach to understanding adaptation is even more poorly developed in the study of life history evolution in stickleback (Baker Chapter 6 this volume), and has rarely been used to elucidate the adaptive values of variable physiological traits (Guderley Chapter 4 this volume) or of population differences in patterns of somatic and gonadal allocation (Wootton page 115 this volume). Baker's chapter provides clear evidence of the potential of this approach for understanding the ways in which natural selection has moulded life history characteristics. He also stresses the importance of uniform methodology if data from different studies are to be of use in comparisons among populations. This concern applies not only to the study of life history data, but also to research on any phenotype.

Similarly, our understanding of adaptation is only as good as our understanding of the selection regimes to which populations are exposed. We have detailed information on the ecology of few of the habitats in which stickleback live. Equally, the measures of selection pressures that we use are not necessarily accurate (Reimchen page 273 this volume) or complete (e.g. Arnold and Wade 1984). Our understanding of the causes of diversification among populations of threespine stickleback clearly would benefit from a more complete ecological assessment of the habitats in which stickleback live and from a clearer perception of how the fitness of specific phenotypes is affected by specific ecological processes.

By arguing for a prominent role of natural selection in moulding many phenotypic components in the postglacial diversification of threespine stickleback, we are not uncritically invoking the adaptationist programme (Gould and Lewontin 1979). We feel that the research summarized above and in the chapters in this volume provides ample evidence for adaptive diversification in many aspects of phenotype. In making this argument we do not intend to exclude effects of genetic drift (including founder effects and population bottlenecks), ontogenetic constraints, or allometry as factors in population differentiation (Bell 1981, 1987). We wish only to reiterate the point made by McPhail (page 435 this volume) and Reimchen (page 276 this volume) that a remarkable proportion of the diversification in postglacial populations of threespine stickleback has been shown to be adaptive.

PATTERNS OF EVOLUTIONARY CHANGE

The direction of evolutionary change

The radiation of threespine stickleback that has taken place in freshwater habitats in recently deglaciated regions is of particular value for evolutionary research because plesiomorphic character states can be inferred with remarkable certainty. In the case of morphological traits this inference is relatively straightforward. Virtually all marine and anadromous stickleback, the ancestors of freshwater stickleback, have a complete set of lateral plates (complete morphs), a fully developed pelvic girdle, three dorsal spines, and paired pelvic spines (Bell and Foster page 5 this volume). Furthermore, marine stickleback appear to have possessed this morphology for at least the last 10 million years (Bell 1977, page 441 this volume). There can be little question that these are plesiomorphic character states relative to the postglacial freshwater diversification.

What is particularly unusual, however, is the ability to infer plesiomorphy for behavioural and life history traits. The fossil record rarely can provide evidence of stasis in either kind of trait (but see Boucot 1990), and it is plausible that behavioural or life history phenotypes have changed in marine populations since the last glacial maximum. However, the cladogram for populations of *G. aculeatus* (Haglund *et al.* 1992*a*; Buth and Haglund page 78 this volume) suggests that the behavioural and life history phenotypes of marine *G. aculeatus* are also ancient. A set of Japanese populations forms the sister group to all other *G. aculeatus*, which in turn are divided into separate monophyletic groups in the Atlantic and Pacific basins. The fossil record shows that *G. aculeatus* has been in the Pacific basin for at least 10 million years and possibly 16 million years, but the earliest record of threespine stickleback in the Atlantic basin is about 1.9 million years (Bell page 444 this volume). Consequently, character states shared between Pacific and Atlantic marine populations are presumably primitive character states relative to the postglacial diversifications in either basin.

The quality of such inference with respect to morphological, behavioural, or life history phenotypes is dependent on the accuracy of the assumption of relative uniformity of character states in marine and anadromous populations. The evidence of morphological uniformity is strong. Of the many reports on the morphology of marine threespine stickleback (reviewed in Bell 1984*a*), there are only a few reports of unusual morphological phenotypes in marine and anadromous populations (Münzing 1963; Black 1977; Borg 1985). In contrast, we are glaringly deficient in our knowledge of variation in other aspects of phenotype in marine populations.

The ability to infer the direction of evolutionary change enables much more accurate adaptive inference. For example, adaptive explanations for the loss of diversionary displays in freshwater populations are very different from those that might explain iterative evolution of novel diversionary display behaviour in freshwater populations (Foster 1988, page 394 this volume). Knowledge of character polarity can insure that the appropriate explanation is provided. Inference of character polarity can also facilitate tests of specific evolutionary hypotheses, such as Kaneshiro's (1976) speciation hypothesis (see Foster page 393 this volume).

Perhaps one of the most intriguing benefits of the ability to identify derived character states is that it should be possible to determine whether the independent evolution of similar, derived phenotypes is the consequence of the same or different genetic and developmental modifications. For example, Wilkens' (1971) interpopulation crosses of cave fish, *Anoptichthys* (= *Astyanax*) *mexicanus*, suggested that reduction of the eyes had evolved independently and depended on different genes in different populations. Such comparisons could provide a unique test of two alternative hypotheses using populations that have diversified under natural conditions. The first holds that uniform selection on closely related lineages tends to produce the same genetic changes because it favours the same shared alleles (e.g. Muller 1939). The second holds that uniform selection on closely related lineages (even populations drawn from the same source population) can act as a diversifying force in evolution (Cohan 1984; Cohan and Hoffman 1989). These alternatives can potentially be discriminated through judicious comparisons of populations in an adaptive diversification with the unusual characteristics of the threespine stickleback postglacial diversification.

Similarly, analysis of the role of changes in the timing of developmental processes in evolution (heterochrony) requires placement of phenotypic differences in a phylogenetic context (Fink 1982). Interpopulation comparisons have already demonstrated a tendency for simplification or reduction of phenotypic states of freshwater populations relative to their marine ancestors (Bell 1981, 1987). Undoubtedly, additional insights could be derived from more detailed comparative analyses of changes in developmental programmes.

Molecular methods in the inference of evolutionary pattern

Uncertainties about character polarity in endemic radiations of *G. aculeatus* would benefit from construction of regional population-level cladograms based on molecular data. Phenotypic characters are unsuitable for this purpose because similar phenotypes are often the products of iterative evolution and are therefore homoplasies. Furthermore, if phenotypic integration is as important as we think it is (Foster *et al.* 1992; Bell and Foster page 22 this volume), use of any phenotypic data to construct a phylogenetic framework for inference of character polarity would be circular (Brooks and

McLennan 1991). At present, however, the appropriate combination of methodology and molecules to resolve cladistic relations among postglacial populations is unavailable.

Postglacial populations are too young for a sufficient number of apomorphic allozymes to have been produced by mutational processes. Withler and McPhail (1985) conducted an allozyme survey of *G. aculeatus* populations in south-western British Columbia. Although their data provided evidence of lower polymorphism and greater population differentiation in freshwater populations than in anadromous populations, the electromorphs of freshwater populations included only one allele that was absent from anadromous populations (present in 11 of 40 freshwater populations). Thus, their data provided no evidence of sufficient, appropriate electromorph variation for use in reconstruction of phylogenetic relationships among these populations.

Similarly, restriction fragment length polymorphisms (RFLPs) appear to evolve too slowly to resolve cladistic relations among populations that have been isolated postglacially (Avise *et al.* 1987). Gach and Reimchen (1989) distinguished two groups of freshwater threespine stickleback populations in the Queen Charlotte Islands using RFLP analysis, but finer phylogenetic resolution was impossible. They seem to have detected the presence of an older group that had persisted in a glacial refugium, as well as populations resulting from postglacial colonization of freshwater.

Analysis of the control region (i.e. 'D-loop') of the mitochondrial genome using the polymerase chain reaction (PCR) and direct sequencing does not appear promising either. 'Universal primers' for the D-loop (Kocher *et al.* 1989; Meyer *et al.* 1990), which amplify this region in most species, do not appear to work in *G. aculeatus* (Ortí pers. comm.), but other effective primers for this region in threespine stickleback probably will be developed. The more fundamental problem, however, is that postglacial populations may be too young for a sufficient number of nucleotide substitutions to have occurred within the D-loop to produce a cladogram. For example, Meyer *et al.* (1990) analysed an 803 base-pair region that includes the most variable region of the D-loop in 32 cichlid species endemic to Lake Victoria, East Africa. They found a total of only 15 positions in the sequence that varied between at least two species, and estimated the age of the group at less than 200 000 yr. However, this age is an order of magnitude greater than that of postglacial threespine stickleback, in which only a few substitutions should be expected. If other fast-evolving sequences are identified, it might become possible to infer population-level cladograms for postglacial threespine stickleback populations, but the amount of sequence necessary to make such inferences for populations that have diverged so recently may make it impracticable using currently available methods.

Molecular methods for phylogenetic inference are developing at an astonishing pace, and a suitable method for cladistic analyses of postglacial populations may be developed soon. For example, a method called RAPD (Williams *et al.* 1990) uses the entire genome and appeared to be promising for cladistic analyses. However, this method evidently is unreliable because binding of primer to the DNA sample is sensitive to a variety of experimental conditions (Eanes pers. comm.), and failure of a primer to bind will alter the result of this analysis.

Although it seems likely that the methodological impediments to inference of local interpopulation relations are temporary, there may be a more fundamental problem. If freshwater stickleback populations of a region were derived by means of sequential branching within a clade as lakes were colonized, the time between successive branch points may have been so brief that few base substitutions occurred between branches (i.e. nodes in a cladogram). Even if enough base substitutions occurred between successive branch points to permit a cladistic analysis, subsequent substitutions within the same DNA sequence might obscure synapomorphic substitutions needed for group formation in a cladogram. Alternatively, freshwater populations of a region might have been derived independently from a common marine ancestor. In either case, the resulting cladogram will be totally unresolved, a polytomy in which each population branches from a common point representing the hypothetical ancestor. Thus, even large quantities of molecular data may not produce a regional phylogenetic framework for inference of character polarities. This outcome would necessitate reliance on marine populations alone for inference of character polarities. Until appropriate methodologies are developed, these issues must remain unresolved.

The tempo of evolutionary change

There are serious problems inherent in the quantitative comparison of evolutionary rates for phenotypic traits (Simpson 1944). Differences in phenotypic complexity among taxa complicate estimation of evolutionary rates (Schopf *et al.* 1975), and imprecise estimates of the time since divergence between populations limit the reliability of rate estimates. Even if time since divergence could be estimated reliably, and a valid common phenotypic scale could be developed (e.g. variance units of properly scaled data), computed rates of evolution are inversely related to the time interval over which they are measured (Gingerich 1983).

When evolutionary rates are calculated using extant radiations, inferences are made based on 'molecular clock' models (Zuckerkandl and Pauling 1965) or on maximum habitat age (e.g. Carson 1970; Fryer and Iles 1972; Grant 1986). Molecular clock calibrations are proving remarkably variable, however, and are therefore of less value in the inference of evolutionary rates than was originally hoped (e.g. Avise and Aquadro 1982; Vawter and Brown 1986; Harrison 1991). Use of maximum habitat age is also problematic, because colonization could have occurred well after the habitat first became available for colonization and evolutionary rates may have varied over the postcolonization interval. For example, the fossil record suggests that time since deglaciation under-estimates the time necessary for morphological divergence of postglacial threespine stickleback populations, and that evolutionary rates may have slowed significantly in postglacial populations as they approached an adaptive peak (Bell page 470 this volume).

Despite these serious methodological limitations, it appears that phenotypic evolution and speciation occur more rapidly in *G. aculeatus* than is usually assumed (e.g. Simpson 1944; Stanley 1979; McPhail page 433 this volume). Postglacial freshwater populations of threespine stickleback that have been studied range in age from about 22 000 yr in parts of Cook Inlet, Alaska, USA (Reger pers. comm.) down to about 11 000 yr in British Columbia, Canada (McPhail page 401 this volume) and 8000 yr in the Outer Hebrides, Scotland, UK (Campbell and Williamson 1979). Thus, the maximum age of initiation of the postglacial radiation of *G. aculeatus* is much less than those cited in other radiations (Grant 1986). Consequently, it is not clear whether *G. aculeatus* evolves unusually rapidly or whether rapid evolution is a common phenomenon that becomes detectable only with the unusually fine time scale available for postglacial stickleback populations (Gingerich 1983). Differentiation of other boreal fishes, including *Pungitius pungitius* (McPhail 1963), *Myoxocephalus* (Johnson 1964), and a variety of other northern freshwater fishes (Smith and Todd 1984; Mina 1991), suggests that the high rates of phenotypic evolution in *Gasterosteus* may not be as unusual as often assumed (e.g. McPhail page 433 this volume).

THE EVOLUTION OF CORRELATED CHARACTERS

Character correlations can result from pleiotropy, linkage disequilibrium, common mechanisms of developmental or physiological regulation, forced association by natural selection, or a common response to environmental conditions (e.g. Olson and Miller 1958; Thorpe 1976; Sokal 1978). Although genetic correlations among traits undoubtedly constrain the evolutionary responses of the traits to selection in the short term, there can be little doubt that genetic correlations can evolve, as do the traits themselves (Riska *et al.* 1989; reviewed in Barton and Turelli 1989).

The threespine stickleback radiation could provide unusually detailed insights into the ways in which selection in natural environments affects genetic correlations. Because many of the populations in recently deglaciated regions are thought to have been derived from a relatively uniform ancestor within a short time span, such populations can be treated as replicate lines drawn from a common ancestral gene pool. Consequently, comparison of laboratory estimates of genetic correlations in threespine stickleback populations could provide tests of a variety of hypotheses concerning the ways in which genetic correlations evolve in natural populations and about the causes of character correlations.

Although the value of this approach to the study of correlated characters in threespine stickleback has not been widely recognized, Bakker's review (Chapter 12 this volume) emphasized this potential. For example, Bakker demonstrated that unexpected correlations among behavioural and life history traits are partly a consequence of common influences of the hormones of the pituitary-gonadal axis. Additionally he pointed out that population comparisons can be used to examine the ways in which selection on life history, for example, can affect reproductive behaviour through common hormonal controls.

PHENOTYPIC INTEGRATION

When subsets of a relatively uniform ancestral population are independently subjected to similar novel selection regimes, the outcome may be a set of populations that are phenotypically similar in a diverse array of characters (e.g. Muller 1939). Alternatively, it is possible that the populations will diverge in at least some character states (Cohan 1984; Cohan and Hoffman 1989), possibly in a compensatory manner. For example, it is possible that behavioural and morphological defences will always be positively correlated with one another and with the intensity of predation. Alternatively, populations that evolve effective antipredator behaviour may be more likely to exhibit armour reductions than are those that evolve no such behaviour. Phenotypic integration encompasses both possibilities, in so far as both reflect interactive responses of multiple characters to specific selection regimes.

The postglacial diversification of populations of threespine stickleback provides an excellent system in which to examine patterns of phenotypic integration, as exemplified by the limnetic–benthic contrast (Foster *et al.* 1992; Bell and Foster page 21, McPhail page 418 this volume). In this case, limnetic forms appear to possess a set of common phenotypic traits that differ from those that typify benthic forms (Fig. 1.8). There is no evidence of compensation. However, if more kinds of phenotypes are evaluated simultaneously in more populations, evidence of compensatory relationships among some aspects of the phenotype across populations might emerge.

Comparisons among postglacial freshwater populations can be used to test specific hypotheses about the patterns and causes of phenotypic integration. For example, Thorpe (1976) and Sokal (1978) suggested that characters with strong intrapopulation correlations should also have strong interpopulation correlations. If this proposal is correct, populations with strong intrapopulation correlations but weak interpopulation correlations (or strong correlations with opposite sign) would presumably reflect selection against the correlation. This hypothesis has rarely been tested (Riska 1985). Francis *et al.* (1986) found that in threespine stickleback populations, four pairs of morphological characters with high intrapopulation correlations also had high interpopulation correlations, as expected. However, three, the correlation of body depth with eye length, head length, and standard length, are strong within ($r \geq 0.85$) and weaker among populations ($r \leq 0.65$), indicating that any correlated response of these characters has been broken in the course of postglacial differentiation of these populations.

Clearly, comparison of a diversity of characters across populations of freshwater and marine stickleback in recently deglaciated areas has the potential to provide substantial insights into the patterns and mechanisms of phenotypic integration among populations independently subjected to similar selection regimes. It should be possible to learn a great deal about the evolution of complexes of functionally integrated traits, especially when such research is combined with that on the genetic, developmental, and physiological bases of the differences in phenotype.

CLOSING THOUGHTS

Diversification of the threespine stickleback contrasts sharply with other radiations of freshwater fishes (Echelle and Kornfield 1984). Most other endemic radiations of freshwater fishes comprise numerous sympatric species. McPhail (Chapter 14 this volume) reports only three types of sympatry or parapatry in *Gasterosteus*: (1) benthic–limnetic lacustrine, (2) stream–lake, and (3) anadromous–stream-resident species pairs. However, there is never more than a pair of species, one of which concentrates on plankton and the other on benthic prey. The only known exception to this rule is sympatry of the 'white stickleback' and the common marine *G. aculeatus* in Nova Scotia, Canada (Blouw and Hagen 1990). Otherwise, trophic divergence appears to be the only means for coexistence of biological species of threespine stickleback.

The exceptional youth of threespine stickleback radiations does not appear to explain the absence of multiple sympatric threespine stickleback species. McCune *et al.* (1984; see also McCune 1987*a*,*b*) reported cyclical occurrence of intralacustrine radiations of semionotid fishes. These radiations occurred in Mesozoic rift lakes during pluvial cycles with periods of 22 000 yr. At least 21 semionotid species evolved during one cycle (P4 cycle of McCune *et al.* 1984) within one early Jurassic lake (McCune 1987*b*). Thus, the age of postglacial radiation does not appear to account for generation of only pairs of sympatric species of the *G. aculeatus* complex.

Dominey (1984*b*) suggested that mating systems that promote sexual selection should facilitate the evolution of reproductive isolation. Although this view is not held universally (reviewed in Greenwood 1991), if it is correct, evolution of isolating mechanisms should be facilitated in threespine stickleback. The elaborate courtship behaviour in which threespine stickleback often engage (Rowland page 313, Foster page 382 this volume) offers ample opportunity for sexual selection. Other commonly cited factors, such as small body size and the potential to become isolated in small habitats, also apply to *G. aculeatus*. It appears that the most likely hypothesis for coexistence of no more than two biological species of *G. aculeatus* anywhere is their limited potential for divergence in use of trophic resources.

Most evolutionary radiations that have been studied in detail have been termed adaptive radiations because of the seemingly major role of natural selection in causing diversification (e.g. Simpson 1953; Carlquist 1980; Grant 1986; Simon 1987). Although exceptions may exist (Gittenberger 1991), radiations tend to be associated with adaptive shifts into novel environments or with novel patterns of resource use, and they are often attributed to underutilized ecological niche space. These are qualities characteristic of most isolated archipelagos because many taxonomic groups are poor at dispersing across ocean expanses. These qualities also characterize freshwater habitats created by glacial recession that are now occupied by threespine stickleback (e.g. Bell 1984*a*; McPhail and Lindsey 1970, 1986; Bell and Foster page 17). Depauperate faunas combined with novel ecological conditions seem to have favoured rapid evolutionary responses of marine threespine stickleback to differences in selection regimes among habitats, resulting in population differentiation.

In closing, we would like to make a plea for the conservation of threespine stickleback populations. Threespine stickleback are most abundant and diverse in far northern boreal habitats that have been relatively undisturbed by human activities. Thus, the fascinating endemic species and phenotypically divergent populations, such as those reported by McPhail and Reimchen in this volume, remain largely untouched. However, serious losses have occurred in more heavily disturbed regions of western Europe and North America, and this history should serve as a warning of the vulnerability of this fascinating fish to human activity.

The 'unarmoured threespine stickleback', *G. a. williamsoni*, was once abundant and widespread in coastal streams around Los Angeles, California, USA, but is now represented only by relict populations in a few small headwater habitats that are threatened by urbanization and introduction of exotic aquatic species (Ono and Williams 1983). Less unusual threespine stickleback have disappeared throughout much of southern California (Swift pers. comm.). Similarly, threespine stickleback have declined and disappeared in European habitats where they once abounded (Lelek 1987), and have all but disappeared from some Dutch habitats that supplied specimens for classical ethological experiments (van Alphen pers. comm.). This decline appears to reflect a general decline in water quality and construction of barriers to entry by anadromous stickleback.

In boreal habitats, where the greatest concentration of phenotypic diversity occurs, regional habitat degradation owing to timber exploitation and point disturbances, especially from fisheries management, represent serious threats to stickleback. For example, planning for management of natural resources in the Tongass National Forest, Alaska, USA, considers at length the impact of timber exploitation on salmonid fisheries and the value of enhancing salmon stocks, but ignores threespine stickleback entirely (US Department of Agriculture, Forest Service 1979).

Use of ichthyocides to eliminate stickleback and other 'trash fish' to improve game fish production was a common practice throughout western North America. Although this practice was generally unsuccessful, the stickleback that repopulated treated habitats are of uncertain ancestry. However, even without poisoning, sport fish introduction can by itself do serious damage. Kynard (1979*b*) reported simultaneous increase of an introduced sunfish, *Lepomis gibbosus*, and decline of threespine stickleback. Similarly, trout, *Oncorhynchus mykiss*, introductions to small lakes around Cook Inlet, Alaska, USA, have seemingly resulted in virtual extinction of some threespine stickleback populations with extreme pelvic reduction (Bell, unpubl. data). Because of the high degree of endemism in northern lakes, loss of even a single population may represent a significant decline in diversity. Therefore, some morphologically divergent endemic populations in British Columbia, Canada, have received special attention (Moodie 1984; Reimchen 1984; McAllister *et al.* 1985).

Threespine stickleback are often abundant and seem to represent an inexhaustible resource. Nevertheless, experience has proven that they are easily eliminated by human activities and therefore deserve serious conservation efforts. This volume amply demonstrates their value as a source of information on evolutionary processes. If populations continue to be extinguished at the current rate, we will have lost a remarkable source of information on pattern and mechanism in evolutionary change. We cannot afford this loss.

ACKNOWLEDGEMENTS

We thank all the contributors to this volume for providing us with so many insights and ideas to summarize. Our research has been supported by a USA National Institute of Mental Health Research Service Award to S.A.F. and by USA National Science Foundation grants (BSR 86-00114, BSR 88-905758 to M.A.B. and BSR 91-08132 to S.A.F.).

References

Adalsteinsson, H. (1979). Size and food of arctic char *Salvelinus alpinus* and stickleback *Gasterosteus aculeatus* in Lake Myvatn. *Oikos*, **32**, 228–231.

Ahsan, S.N. and Hoar, W.S. (1963). Some effects of gonadotropic hormones on the threespine stickleback, *Gasterosteus aculeatus. Canadian Journal of Zoology*, **41**, 1045–1053.

Alexander, R. McN. (1967). *Functional design in fishes*. Hutchinson, London.

Ali, M.Y. and Lindsey, C.C. (1974). Heritable and temperature-induced variation in the medaka, *Oryzias latipes. Canadian Journal of Zoology*, **52**, 959–976.

Allen, J.R.M. and Wootton, R.J. (1982*a*). Effect of food on the growth of carcase, liver and ovary in female *Gasterosteus aculeatus* L. *Journal of Fish Biology*, **21**, 537–547.

Allen, J.R.M. and Wootton, R.J. (1982*b*). Age, growth and rate of food consumption in an upland population of the three-spined stickleback *Gasterosteus aculeatus* L. *Journal of Fish Biology*, **21**, 95–105.

Allen, J.R.M. and Wootton, R.J. (1982*c*). The effect of ration and temperature on the growth of the three-spined stickleback, *Gasterosteus aculeatus* L. *Journal of Fish Biology*, **20**, 409–422.

Allen, J.R.M. and Wootton, R.J. (1983). Rate of food consumption in a population of threespine sticklebacks, *Gasterosteus aculeatus*, estimated from the faecal production. *Environmental Biology of Fishes*, **8**, 157–162.

Allen, J.R.M. and Wootton, R.J. (1984). Temporal patterns of diet and rate of food consumption of the three-spined stickleback (*Gasterosteus aculeatus*) in Llyn Frongoch, an upland Welsh lake. *Freshwater Biology*, **14**, 335–346.

Allis, E.P. (1909). The cranial anatomy of the mail-cheeked fishes. *Zoologica*, **57**, 112–219.

Anderson, E. (1936). The species problem in *Iris. Annals of the Missouri Botanical Garden*, **23**, 457–509.

Anderson, R.Y. and Dean, W.E. (1988). Lacustrine varve formation through time. *Palaeogeography, Palaeoclimatology, Palaeoecology*, **62**, 215–235.

Anderson, R.Y., Dean, W.E., Bradbury, J.P. and Love, D. (1985). Meromictic lakes and varved lake sediments in North America. *US Geological Survey Bulletin*, **1607**, 1–19.

Andersson, E., Mayer, I. and Borg, B. (1988). Inhibitory effect of 11-ketoandrostenedione and androstenedione on spermatogenesis in the three-spined stickleback, *Gasterosteus aculeatus* L. *Journal of Fish Biology*, **33**, 835–840.

Andersson, E., Borg, B. and Leeuw, R. de (1989). Characterization of gonadotropin-releasing hormone binding sites in the pituitary of the three-spined stickleback, *Gasterosteus aculeatus. General and Comparative Endocrinology*, **76**, 41–45.

Andersson, M. (1982). Sexual selection, natural selection and quality advertisement. *Biological Journal of the Linnean Society*, **17**, 375–393.

Andersson, M. (1986). Evolution of condition-dependent sex ornaments and mating preferences: sexual selection based on viability differences. *Evolution*, **40**, 804–816.

Ando, S., Yamazaki, F., Hatano, M. and Zama, K. (1986). Deterioration of chum salmon (*Oncorhynchus keta*) muscle during spawning migration. III: Changes in protein composition and protease activity of juvenile chum salmon muscle upon treatment with sex steroids. *Comparative Biochemistry and Physiology*, **83B**, 325–330.

Andrusak, H. and Northcote, T.G. (1971). Segregation between cutthroat trout (*Salmo clarki*) and Dolly Varden (*Salvenius malma*) in small coastal British Columbian lakes. *Journal of the Fisheries Research Board of Canada*, **28**, 1259–1268.

Aneer, G. (1973). Biometric characteristics of the three-spined stickleback (*Gasterosteus aculeatus* L.) from the northern Baltic proper. *Zoologica Scripta*, **2**, 157–162.

Anker, G.C. (1974). Morphology and kinetics of the head of the stickleback, *Gasterosteus aculeatus. Transactions of the Zoological Society of London*, **32**, 311–416.

Anker, G.C. (1978). Analysis of respiration and feeding movements in the three-spined stickleback, *Gasterosteus aculeatus* L. *Netherlands Journal of Zoology*, **28**, 485–523.

Anonymous (1830). Some accounts of *Gasterosteus aculeatus. Magazine of Natural History*, **3**, 329–332.

Appellöf, A. (1894). Über einige reultate der Kreuzbefruchtung bei Knochenfischen. *Bergens Museums Aarborg*, **1894–1895**, 1–17.

Arderon, W. (1746). Abstract of a letter from Mr. Wm. Arderon, F.R.S. to Mr. Henry Baker, F.R.S. containing some observations made on the bansticle, or pricklebag, alias prickleback, and also on fish in general. *Philosophical Transactions of the Royal Society*, **44**, 424–428.

Armstrong, E.A. (1949*a*). Diversionary display. Part 1. Connotation and terminology. *Ibis*, **91**, 88–97.

Armstrong, E.A. (1949*b*). Diversionary display. Part 2. The nature and origin of distraction display. *Ibis*, **91**, 179–188.

Armstrong, J.E. (1981). Post-Vashon Wisconsin glaciation, Fraser Lowland, British Columbia. *Geological Survey of Canada Bulletin*, **322**, 1–34.

Armstrong, R.H. (1971). Age, food, and migration of sea-run cutthroat trout, *Salmo clarki*, at Eva Lake, southeast Alaska. *Transactions of the American Fisheries Society*, **100**, 302–306.

Arnold, S.J. (1977). Polymorphism and geographic variation in the feeding behavior of the garter snake, *Thamnophis elegans. Science*, **197**, 676–678.

Arnold, S.J. and Bennett, A.F. (1984). Behavioural variation in natural populations. III. Antipredator displays in the garter snake *Thalamnus radix. Animal Behaviour*, **32**, 1108–1118.

Arnold, S.J. and Wade, M.J. (1984). On the measurement of natural and sexual selection: applications. *Evolution*, **38**, 720–734.

Arnoult, J. (1986). Gasterosteidae. In *Check-list of the freshwater fishes of Africa*, Vol. 2 (ed. J. Daget, J.-P. Gosse and D.F.E. Thys van den Audenaerde), pp. 280. Institute Royal des Sciences Naturelles de Belgique, Brussels.

Assem, J. van den (1967). Territorality in the three-spined stickleback, *Gasterosteus aculeatus* L.: an experimental study in intra-specific competition. *Behaviour Supplement*, **16**, 1–164.

Assem, J. van den and Molen, J.N. van der (1969). Waning of the aggressive response in the three-spined stickleback upon constant exposure to a conspecific. I. A preliminary analysis of the phenomenon. *Behaviour*, **36**, 286–334.

Assem, J. van den and Sevenster, P. (ed.) (1985). First international symposium on stickleback behaviour. *Behaviour*, **93**, 1–277.

Audet, C., FitzGerald, G. J. and Guderley, H. (1985*a*). Salinity preferences of four sympatric sticklebacks (Gasterosteidae) during their reproductive season. *Copeia*, **1985**, 209–213.

Audet, C., FitzGerald, G. J. and Guderley, H. (1985*b*). Prolactin and cortisol control of salinity preferences in *Gasterosteus aculeatus* and *Apeltes quadracus*. *Behaviour*, **93**, 36–55.

Audet, C., FitzGerald, G. J. and Guderley, H. (1986*a*). Environmental control of salinity preferences in four sympatric sticklebacks *Gasterosteus aculeatus*, *Gasterosteus wheatlandi*, *Pungitius pungitius*, and *Apeltes quadracus*. *Journal of Fish Biology*, **28**, 725–739.

Audet, C., FitzGerald, G. J. and Guderley, H. (1986*b*). Photoperiod effects on plasma cortisone levels in *Gasterosteus aculeatus*. *General and Comparative Endocrinology*, **61**, 76–81.

Austad, S. N. (1984). A classification of alternative reproductive behaviors and methods for field-testing ESS models. *American Zoologist*, **12**, 309–319.

Austad, S. N. (1989). Game theory and the evolution of animal contests. *Trends in Ecology and Evolution*, **4**, 2–3.

Avise, J. C. (1974). Systematic value of electrophoretic data. *Systematic Zoology*, **23**, 465–481.

Avise, J. C. (1976). Genetics of plate morphology in an unusual population of threespine sticklebacks (*Gasterosteus aculeatus*). *Genetical Research*, **27**, 33–46.

Avise, J. C. and Aquadro, C. F. (1982). A comparative summary of genetic distances in the vertebrates: patterns and correlations. *Evolutionary Biology*, **15**, 151–185.

Avise, J. C. and Smith, M. H. (1977). Gene frequency comparisons between sunfish (Centrarchidae) populations at various stages of evolutionary divergence. *Systematic Zoology*, **26**, 319–335.

Avise, J. C., Smith, M. H. and Ayala, F. J. (1975). Adaptive differentiation with little genic change between two native California minnows. *Evolution*, **29**, 411–426.

Avise, J. C., Arnold, J., Ball, R. M., Bermingham, E., Lamb, T., Nigel, J. E., Reeb, C. A. and Saunders, N. C. (1987). Intraspecific phylogeography: the mitochondrial DNA bridge between population genetics and systematics. *Annual Review of Ecology and Systematics*, **18**, 489–522.

Axelrod, D. I. (1948). Climate and evolution in western North America during Middle Pliocene time. *Evolution*, **2**, 127–144.

Axelrod, D. I. (1957). Late Tertiary floras and the Sierra Nevadan uplift. *Bulletin of the Geological Society of America*, **68**, 19–45.

Axelrod, D. I. (1962). Post-Pliocene uplift of the Sierra Nevada, California. *Bulletin of the Geological Society of America*, **73**, 183–198.

Baerends, G. P. (1985). Do the dummy experiments with sticklebacks support the IRM-concept? *Behaviour*, **93**, 258–277.

Baggerman, B. (1957). An experimental study on the timing of breeding and migration in the three-spined stickleback, *Gasterosteus aculeatus* L. *Archives Néerlandaises de Zoologie*, **12**, 105–318.

Baggerman, B. (1960). Salinity preference, thyroid activity and the seaward migration of four species of Pacific salmon (*Oncorhynchus*). *Journal of the Fisheries Research Board of Canada*, **17**, 295–322.

Baggerman, B. (1966). On the endocrine control of reproductive behaviour in the male three-spined stickleback (*Gasterosteus aculeatus*). *Symposium of the Society of Experimental Biology*, **20**, 427–456.

Baggerman, B. (1968). Hormonal control of reproductive and parental behaviour in fishes. In *Perspectives in endocrinology: hormones in the lives of lower vertebrates* (ed. E.J.W. Barrington and C. Barker Jørgensen), pp. 351–404. Academic Press, London.

Baggerman, B. (1972). Photoperiodic responses in the stickleback and their control by a daily rhythm of photosensitivity. *General and Comparative Endocrinology, Supplement*, **3**, 466–476.

Baggerman, B. (1980). Photoperiodic and endogenous control of the annual reproductive cycle in teleost fishes. In *Environmental physiology of fishes* (ed. M.A. Ali), pp. 533–567. Plenum, New York.

Baggerman, B. (1985). The role of biological rhythms in the photoperiod regulation of seasonal breeding in the stickleback *Gasterosteus aculeatus* L. *Netherlands Journal of Zoology*, **35**, 14–31.

Baggerman, B. (1989). On the relationship between gonadal development and response time to photostimulation of sticklebacks living under natural conditions and under constant short-day conditions for long periods of time. *Canadian Journal of Zoology*, **67**, 126–135.

Baker, M.C. (1971). Habitat selection in fourspine sticklebacks (*Apeltes quadracus*). *American Midland Naturalist*, **85**, 239–242.

Bakker, T.C.M. (1985). Two-way selection for aggression in juvenile, female and male sticklebacks (*Gasterosteus aculeatus* L.), with some notes on hormonal factors. *Behaviour*, **93**, 69–81.

Bakker, T.C.M. (1986). Aggressiveness in sticklebacks (*Gasterosteus aculeatus* L.): a behaviour-genetic study. *Behaviour*, **98**, 1–144.

Bakker, T.C.M. and Feuth-de Bruijn, E. (1988). Juvenile territoriality in stickleback *Gasterosteus aculeatus* L. *Animal Behaviour*, **36**, 1556–1558.

Bakker, T.C.M. and Milinski, M. (1991). Sequential female choice and the previous male effect in sticklebacks. *Behavioral Ecology and Sociobiology*, **29**, 205–210.

Bakker, T.C.M. and Sevenster, P. (1983). Determinants of dominance in male sticklebacks (*Gasterosteus aculeatus* L.). *Behaviour*, **86**, 55–71.

Bakker, T.C.M. and Sevenster, P. (1988). Plate morphs of *Gasterosteus aculeatus* Linnaeus (Pisces: Gasterosteidae): comments on terminology. *Copeia*, **1988**, 659–663.

Bakker, T.C.M. and Sevenster, P. (1989). Changes in the sexual tendency accompanying selection for aggressiveness in the three-spined stickleback, *Gasterosteus aculeatus* L. *Journal of Fish Biology*, **34**, 233–243.

Bakker, T.C.M., Feuth-de Bruijn, E. and Sevenster, P. (1988). Population differences in aggressiveness of sticklebacks (*Gasterosteus aculeatus* L.) [abstract]. *Behavior Genetics*, **18**, 706.

Bakker, T.C.M., Feuth-de Bruijn, E. and Sevenster, P. (1989). Asymmetrical effects of prior winning and losing on dominance in sticklebacks (*Gasterosteus aculeatus*). *Ethology*, **82**, 224–229.

Balbi, D. (1990). Mouth morphology and feeding behaviour in two populations of threespine stickleback. Unpublished B.Sc. thesis. University of Leicester.

Ballinger, R.E. (1983). Life history variations. In *Lizard ecology: studies of a model organism* (ed. R.B. Huey, E.R. Pianka, and T.W. Schoener), pp. 241–260. Harvard University Press, Cambridge, Massachusetts.

Banister, K. (1967). The anatomy and classification of the order Gasterosteiformes (Pisces). Unpublished Ph.D. thesis. University of Newcastle-upon-Tyne.

Barel, C.D.N. (1984). Form-relations in the context of constructional morphology:

the eye and suspensorium of lacustrine Cichlidae (Pisces, Teleostei). *Netherlands Journal of Zoology*, **34**, 439–502.

Barlow, G.W. (1968). Ethological units of behavior. In *The central nervous system and fish behavior* (ed. D. Ingle), pp. 217–232. University of Chicago Press, Chicago.

Barlow, G.W. (1974). Extraspecific imposition of social grouping among surgeonfishes (Pisces: Acanthuridae). *Journal of the Zoological Society of London*, **174**, 333–340.

Barton, N.H. and Charlesworth, B. (1984). Genetic revolutions, founder effects, and speciation. *Annual Review of Ecology and Systematics*, **15**, 133–148.

Barton, N.H. and Hewitt, G.M. (1985). Analysis of hybrid zones. *Annual Review of Ecology and Systematics*, **16**, 113–148.

Barton, N.H. and Turelli, M. (1989). Evolutionary quantitative genetics: how little do we know? *Annual Review of Genetics*, **23**, 337–370.

Baskin, J.N. (1974). Survey of the unarmored threespine stickleback (*Gasterosteus aculeatus williamsoni*) in the upper Santa Clara River drainage. *Final Report for the Bureau of Sport Fisheries and Wildlife*, Contract No. 14-16-0001-5387SE, California State Polytechnic University, Pomona.

Baskin, J.N. (1975). Biology and habitat of the unarmored threespine stickleback (*Gasterosteus aculeatus williamsoni*) in the upper Santa Clara River, California. *Final Report for the California State Department of Fish and Game*, Contract AB-27, California State Polytechnic University, Pomona.

Bastock, M. (1967). *Courtship: a zoological study*. Heinemann, London.

Baum, D.A. and Larson, A. (1991). Adaptation reviewed: a phylogenetic methodology for studying character macroevolution. *Systematic Zoology*, **40**, 1–18.

Baumgartner, J.V. (1982). A new fossil ictalurid catfish from the Miocene middle member of the Truckee Formation, Nevada. *Copeia*, **1982**, 38–46.

Baumgartner, J.V. (1986*a*). Phenotypic and genetic aspects of morphological differentiation in the threespine stickleback, *Gasterosteus aculeatus*. Unpublished Ph.D. thesis. State University of New York at Stony Brook.

Baumgartner, J.V. (1986*b*). The genetics of differentiation in a stream population of the threespine stickleback, *Gasterosteus aculeatus*. *Heredity*, **57**, 199–208.

Baumgartner, J.V. (1992). Spatial variation of morphology in a freshwater population of the threespine stickleback, *Gasterosteus aculeatus*. *Canadian Journal of Zoology*, **70**, 1140–1148.

Baumgartner, J.V. and Bell, M.A. (1984). Lateral plate morph variation in California populations of the threespine stickleback, *Gasterosteus aculeatus*. *Evolution*, **38**, 665–674.

Baumgartner, J.V., Bell, M.A. and Weinberg, P.H. (1988). Body form differences between the Enos Lake species pair of threespine sticklebacks (*Gasterosteus aculeatus* complex). *Canadian Journal of Zoology*, **66**, 467–474.

Baxter, R. (1956). Effectiveness of gill nets and sportfish tackle in sampling fish populations. *Quarterly Progress Report: Federal Aid in Fish Restoration*. United States Fish and Wildlife Service and Alaska Game Commission. June 30, 1956.

Becker, P.H., Frank, D. and Walter, U. (1987). Geographical and annual variations of feeding of common terns *Sterna hirundo* on the German North Sea coast. *Journal of Ornithology*, **128**, 457–476.

Behrensmeyer, A.K. and Kidwell, S.M. (1988). Taphonomy's contributions to paleobiology. *Paleobiology*, **11**, 105–119.

Behrensmeyer, A.K. and Schindel, D.E. (1983). Resolving time in paleobiology. *Paleobiology*, **9**, 1–8.

Belanger, G., Guderley, H. and FitzGerald, G. J. (1987). Salinity during embryonic development influences the response to salinity of *Gasterosteus aculeatus* L. (trachurus). *Canadian Journal of Zoology*, **65**, 451-454.

Bell, M. A. (1973*a*). The Pliocene stickleback, *Pungitius haynesi*, a junior synonym of *Gasterosteus aculeatus. Copeia*, **1973**, 588-590.

Bell, M. A. (1973*b*). Pleistocene threespine sticklebacks *Gasterosteus aculeatus* (Pisces) from southern California. *Journal of Paleontology*, **47**, 479-483.

Bell, M. A. (1974). Reduction and loss of the pelvic girdle in *Gasterosteus* (Pisces): a case of parallel evolution. *Natural History Museum of Los Angeles County Contributions in Science*, **257**, 1-36.

Bell, M. A. (1976*a*). Evolution of phenotypic diversity in *Gasterosteus aculeatus* superspecies on the Pacific coast of North America. *Systematic Zoology*, **25**, 211-227.

Bell, M. A. (1976*b*). Reproductive character displacement in threespine sticklebacks. *Evolution*, **30**, 847-848.

Bell, M. A. (1977). A late Miocene marine threespine stickleback, *Gasterosteus aculeatus aculeatus*, and its zoogeographic and evolutionary significance. *Copeia*, **1977**, 277-282.

Bell, M. A. (1978). Fishes of the Santa Clara River system, southern California. *Natural History Museum of Los Angeles County Contributions in Science*, **295**, 1-20.

Bell, M. A. (1979). Low-plate morph of the threespine stickleback breeding in salt water. *Copeia*, **1979**, 529-533.

Bell, M. A. (1981). Lateral plate polymorphism and ontogeny of the complete plate morph of threespine sticklebacks (*Gasterosteus aculeatus*). *Evolution*, **35**, 67-74.

Bell, M. A. (1982). Melanism in a high elevation stream population of *Gasterosteus aculeatus. Copeia*, **1982**, 829-835.

Bell, M. A. (1984*a*). Evolutionary phenetics and genetics: the threespine stickleback, *Gasterosteus aculeatus*, and related species. In *Evolutionary genetics of fishes* (ed. B. J. Turner), pp. 431-528. Plenum, New York.

Bell, M. A. (1984*b*). Gigantism in threespine sticklebacks: implications for causation of body size evolution. *Copeia*, **1984**, 530-534.

Bell, M. A. (1987). Interacting evolutionary constraints in pelvic reduction of threespine sticklebacks, *Gasterosteus aculeatus* (Pisces, Gasterosteidae). *Biological Journal of the Linnean Society*, **31**, 347-382.

Bell, M. A. (1988). Stickleback fishes: bridging the gap between population biology and paleobiology. *Trends in Ecology and Evolution*, **3**, 320-325.

Bell, M. A. and Baumgartner, J. V. (1984). An unusual population of *Gasterosteus aculeatus* from Boston, Massachusetts. *Copeia*, **1984**, 258-262.

Bell, M. A. and Haglund, T. R. (1978). Selective predation of threespine sticklebacks *Gasterosteus aculeatus* by garter snakes. *Evolution*, **32**, 304-319.

Bell, M. A. and Haglund, T. R. (1982). Fine-scale temporal variation of the Miocene stickleback *Gasterosteus doryssus. Paleobiology*, **8**, 282-292.

Bell, M. A. and Harris, E. I. (1985). Developmental osteology of the pelvic complex of *Gasterosteus aculeatus. Copeia*, **1985**, 789-792.

Bell, M. A. and Legendre, P. (1987). Multicharacter chronological clustering in a sequence of fossil sticklebacks. *Systematic Zoology*, **36**, 52-61.

Bell, M. A. and Richkind, K. E. (1981). Clinal variation of lateral plates in threespine stickleback fish. *American Naturalist*, **117**, 113-132.

Bell, M. A., Francis, R. C. and Havens, A. C. (1985*a*). Pelvic reduction and its directional asymmetry in threespine sticklebacks from the Cook Inlet region, Alaska. *Copeia*, **1985**, 437–444.

Bell, M.A., Baumgartner, J.V. and Olson, E.C. (1985*b*). Patterns of temporal change in single morphological characters of a Miocene stickleback fish. *Paleobiology*, **11**, 258–271.

Bell, M.A., Sadagursky, M.S. and Baumgartner, J.V. (1987). Utility of lacustrine deposits for the study of variation within fossil samples. *Palaios*, **2**, 455–466.

Bell, M.A., Wells, C.E. and Marshall, J.A. (1989). Mass-mortality layers of fossil stickleback fish: catastrophic kills of polymorphic schools. *Evolution*, **43**, 607–619.

Bell, M.A., Ortí, G., Walker, J.A. and Koenings, J.A. (in press). Evolution of pelvic reduction in threespine stickleback fish: a test of competing hypotheses. *Evolution*.

Belles-Isles, J.-C. and FitzGerald, G.J. (1991). Filial cannibalism in sticklebacks: a male reproductive strategy? *Ethology, Ecology and Evolution*, **3**, 49–62.

Belles-Isles, J.-C. and FitzGerald, G.J. (in press). A fitness advantage of cannibalism in female sticklebacks *Gasterosteus aculeatus* L. *Ethology, Ecology and Evolution*.

Belles-Isles, J.-C., Cloutier, D. and FitzGerald, G.J. (1990). Female cannibalism and male courtship tactics in threespine sticklebacks. *Behavioral Ecology and Sociobiology*, **26**, 363–368.

Bengtson, S.A. (1971). Food and feeding of diving ducks breeding at Lake Myvatn, Iceland. *Ornis Fennica*, **48**, 77–92.

Benjamin, M. (1974). Seasonal changes in the prolactin cell of the pituitary gland of the freshwater stickleback, *Gasterosteus aculeatus* form leiurus. *Cell and Tissue Research*, **152**, 93–102.

Benjamin, M. (1980). The response of prolactin, ACTH, and growth hormone cells in the pituitary gland of the three-spined stickleback, *Gasterosteus aculeatus* L. form leiurus, to increased environmental salinities. *Acta Zoologica (Stockholm)*, **61**, 1–7.

Bentzen, P. and McPhail, J.D. (1984). Ecology and evolution of sympatric sticklebacks (*Gasterosteus*): specialization for alternative trophic niches in the Enos Lake species pair. *Canadian Journal of Zoology*, **62**, 2280–2286.

Bentzen, P., Ridgway, M.S. and McPhail, J.D. (1984). Ecology and evolution of sympatric sticklebacks (*Gasterosteus*): spatial segregation and seasonal habitat shifts in the Enos Lake species pair. *Canadian Journal of Zoology*, **62**, 2436–2439.

Benzie, V.L. (1965). Some aspects of the anti-predator responses of two species of sticklebacks. Unpublished Ph.D. thesis. Oxford University.

Berg, L.S. (1949). *Freshwater fishes of the USSR and adjacent countries*. Israel Program for Scientific Translations, Jerusalem.

Berg, L.S. (1965). *Freshwater fishes of the USSR and adjacent countries*. Vol. III. Israel Program for Scientific Translations, Jerusalem.

Bertin, L. (1925). Recherches bionomiques, biométriques et systématiques sur les Epinoches (Gastérostéides). *Annales de L'Institut Océanographique, Monaco*, **2**, 1–204.

Berven, K.A. and Gill, D.E. (1983). Interpreting geographic variation in life-history traits. *American Zoologist*, **23**, 85–97.

Beukema, J.J. (1964). A study of the time pattern of food intake in the three-spined stickleback (*Gasterosteus aculeatus*) by means of a semi-automatic recording apparatus. *Archives Néerlandaises de Zoologie*, **16**, 167–168.

Beukema, J. J. (1968). Predation by the three-spined stickleback (*Gasterosteus aculeatus* L.). *Behaviour*, **31**, 1-126.

Beyer, C., Larsson, K., Perez-Palacios, G. and Morali, G. (1973). Androgen structure and male sexual behavior in the castrated rat. *Hormones and Behavior*, **4**, 99-108.

Biether, M. (1970). Die Chloridzellen de Stichlings. *Zeitschrift für Zellforschung*, **107**, 421-446.

Bigelow, H. B. and Schroeder, W. C. (1953). Fishes of the Gulf of Maine. *Fishery Bulletin, U.S. Fish and Wildlife Service*, **53**, 1-577.

Billard, R., Fostier, A., Weil, C. and Breton, B. (1982). Endocrine control of spermatogenesis in teleost fish. *Canadian Journal of Fisheries and Aquatic Sciences*, **39**, 65-79.

Birks, H. J. B. and Birks, H. H. (1980). *Quaternary paleoecology*. University Park Press, Baltimore.

Björnsson, B. T., Yamauchi, K., Nichioka, R. S., Deftos, L. J. and Bern, H. A. (1987). Effects of hypophysectomy and subsequent hormone replacement therapy on hormonal and osmoregulatory status of coho salmon, *Oncorhynchus kisutch*. *General and Comparative Endocrinology*, **68**, 421-430.

Black, E. A. (1977). Population differentiation of the threespine stickleback (*Gasterosteus aculeatus*). Unpublished M.Sc. thesis. University of British Columbia, Vancouver.

Black, R. (1971). Hatching success in the three-spined stickleback (*Gasterosteus aculeatus*) in relation to changes in behaviour during the parental phase. *Animal Behaviour*, **19**, 532-541.

Black, R. and Wootton, R. J. (1970). Dispersion in a natural population of three-spined sticklebacks. *Canadian Journal of Zoology*, **48**, 1133-1135.

Blaxter, J. H. S. and Hempel, G. (1963). The influence of egg size on herring larvae (*Clupea harengus* L.). *Journal du Conseil Permanent International pour l'Exploration de la Mer*, **28**, 211-240.

Blouw, D. M. and Boyd, G. J. (1992). Inheritance of reduction, loss, and asymmetry of the pelvis of *Pungitius pungitius* (ninespine stickleback). *Heredity*, **68**, 33-42.

Blouw, D. M. and Hagen, D. W. (1981). Ecology of the fourspine stickleback, *Apeltes quadracus*, with respect to a polymorphism for dorsal spine number. *Canadian Journal of Zoology*, **59**, 1677-1692.

Blouw, D. M. and Hagen, D. W. (1984*a*). The adaptive significance of dorsal spine variation in the fourspine stickleback, *Apeltes quadracus*. I. Geographic variation in spine number. *Canadian Journal of Zoology*, **62**, 1329-1339.

Blouw, D. M. and Hagen, D. W. (1984*b*). The adaptive significance of dorsal spine variation in the fourspine stickleback, *Apeltes quadracus*. II. Phenotype-environment correlations. *Canadian Journal of Zoology*, **62**, 1340-1350.

Blouw, D. M. and Hagen, D. W. (1984*c*). The adaptive significance of dorsal spine variation in the fourspine stickleback, *Apeltes quadracus*. III. Correlated traits and experimental evidence on predation. *Heredity*, **53**, 371-382.

Blouw, D. M. and Hagen, D. W. (1984*d*). The adaptive significance of dorsal spine variation in the fourspine stickleback, *Apeltes quadracus*. IV. Phenotypic covariation with closely related species. *Heredity*, **53**, 383-396.

Blouw, D. M. and Hagen, D. W. (1990). Breeding ecology and evidence of reproductive isolation of a widespread stickleback fish (Gasterosteidae) in Nova Scotia, Canada. *Biological Journal of the Linnean Society*, **39**, 195-217.

Blueweiss, L., Fox, H., Kudzma, V., Nakashima, D., Peters, R. and Sams, S. (1978).

Relationships between body size and some life history parameters. *Oecologia*, **37**, 257–272.

Boag, P.T. and Grant, P.R. (1981). Intense natural selection in a population of Darwin's finches (Geospizinae) in the Galapagos. *Science*, **214**, 82–85.

Bock, F. (1928). Die Hypophyse des Stichlings (*Gasterosteus aculeatus* L.) unter besonderer Berucksichtigung der jahrescyklischen Verahderungen. *Zeitschrift für wissenschaftliche Zoologie*, **131**, 645–710.

Bodaly, R.A. (1979). Morphological and ecological divergence within the lake whitefish (*Coregonus clupeaformis*) species complex in Yukon Territory. *Journal of the Fisheries Research Board of Canada*, **36**, 1214–1222.

Bolduc, F. and FitzGerald, G.J. (1989). The role of selected environmental factors and sex ratio upon egg production in threespine sticklebacks, *Gasterosteus aculeatus*. *Canadian Journal of Zoology*, **67**, 2013–2020.

Bolger, T. and Connolly, P.L. (1989). The selection of suitable indices for the measurement and analysis of fish condition. *Journal of Fish Biology*, **34**, 171–182.

Bolin, R.L. (1936). The systematic position of *Indostomus paradoxus* Prashad and Mukerji, a fresh water fish from Burma. *Journal of the Washington Academy of Sciences*, **26**, 420–423.

Bollens, S.M. and Frost, B.W. (1989). Predator-induced diel vertical migration in a planktonic copepod. *Journal of Plankton Research*, **11**, 1047–1065.

Bolton, J.P., Collie, N.L., Kawauchi, H. and Hirano, T. (1987). Osmoregulatory actions of growth hormone in rainbow trout (*Salmo gairdneri*). *Journal of Endocrinology*, **112**, 63–68.

Bone, Q. and Marshall, N.B. (1982). *Biology of fishes*. Blackie, Glasgow.

Bookstein, F.L. (1986). Random walk and the existence of evolutionary rates. *Paleobiology*, **13**, 446–464.

Bookstein, F.L. (1988). Random walk and the biometrics of morphological characters. *Evolutionary Biology*, **23**, 369–398.

Bookstein, F.L. and Reyment, R.A. (1992). Random walk and quantitative stratigraphic sequences. *Terra Nova*, **4**, 147–151.

Bookstein, F.L., Chernoff, B., Elder, R., Humphries, J., Smith, G. and Strauss, R. (1985). *Morphometrics in evolutionary biology*. Special publication 15, The Academy of Natural Sciences, Philadelphia.

Borg, B. (1981). Effects of methyltestosterone on spermatogenesis and secondary sexual characters in the three-spined stickleback (*Gasterosteus aculeatus* L.). *General and Comparative Endocrinology*, **44**, 177–180.

Borg, B. (1982*a*). Seasonal effects of photoperiod and temperature on spermatogenesis and male secondary sexual characteristics in the three-spined stickleback, *Gasterosteus aculeatus* L. *Canadian Journal of Zoology*, **60**, 3377–3386.

Borg, B. (1982*b*). Extraretinal photoreception involved in photoperiod effects on reproduction in male three-spined sticklebacks, *Gasterosteus aculeatus*. *General and Comparative Endocrinology*, **47**, 84–87.

Borg, B. (1985). Field studies on three-spined sticklebacks in the Baltic. *Behaviour*, **93**, 153–157.

Borg, B. (1987). Stimulation of reproductive behaviour by aromatizable and non-aromatizable androgens in the male three-spined stickleback, *Gasterosteus aculeatus*. In *Proceedings of the fifth congress of European ichthyologists, Stockholm 1985*, (ed. S.O. Kullander and B. Fernholm), pp. 269–271. Swedish Museum of Natural History, Stockholm.

Borg, B. and Ekström, P. (1981). Gonadal effects of melatonin in the three-spined stickleback, *Gasterosteus aculeatus* L. during different seasons and photoperiods. *Reproduction, Nutrition et Développement*, **21**, 919–927.

Borg, B. and Veen, T. van (1982). Seasonal effects of photoperiod and temperature on the ovary of the three-spined stickleback, *Gasterosteus aculeatus* L. *Canadian Journal of Zoology*, **60**, 3387–3393.

Borg, B., Reschke, M., Peute, J. and Hurk, R. van den (1985). Effects of castration and androgen-treatment on the pituitary and testes of the three-spined stickleback, *Gasterosteus aculeatus* L., in the breeding season. *Acta Zoologica (Stockholm)*, **66**, 47–54.

Borg, B., Paulson, G. and Peute, J. (1986). Stimulatory effects of methyltestosterone on pituitary gonadotropic cells and testes Leydig cells of the three-spined stickleback, *Gasterosteus aculeatus* L., in winter. *General and Comparative Endocrinology*, **62**, 54–61.

Borg, B., Timmers, R. J. M. and Lambert, J. G. D. (1987*a*). Aromatase activity in the brain of the three-spined stickleback, *Gasterosteus aculeatus*. I. Distribution and effects of season and photoperiod. *Experimental Biology*, **47**, 63–68.

Borg, B., Timmers, R. J. M. and Lambert, J. G. D. (1987*b*). Aromatase activity in the brain of the three-spined stickleback, *Gasterosteus aculeatus*. II. Effects of castration. *Experimental Biology*, **47**, 69–71.

Borg, B., Peute, J., Reschke, M. and Hurk, R. van den (1987*c*). Effects of photoperiod and temperature on testes, renal epithelium, and pituitary gonadotropic cells of the threespine stickleback, *Gasterosteus aculeatus* L. *Canadian Journal of Zoology*, **65**, 14–19.

Borg, B., Peute, J. and Paulson, G. (1988). Seasonal changes in the gonadotropic cells of the male threespine stickleback, *Gasterosteus aculeatus* L. *Canadian Journal of Zoology*, **66**, 1961–1967.

Borg, B., Schoonen, W. G. E. J. and Lambert, J. G. D. (1989*a*). Steroid metabolism in the testes of the breeding and non-breeding threespine stickleback, *Gasterosteus aculeatus*. *General and Comparative Endocrinology*, **73**, 40–45.

Borg, B., Andersson, E., Mayer, I. and Lambert, J. G. D. (1989*b*). Aromatase activity in the brain of the three-spined stickleback, *Gasterosteus aculeatus* L. III. Effects of castration under different conditions and of replacement with different androgens. *Experimental Biology*, **48**, 149–152.

Borg, B., Andersson, E., Mayer, I., Zandbergen, M. A. and Peute, J. (1989*c*). Effects of castration on pituitary gonadotropic cells of the male three-spined stickleback, *Gasterosteus aculeatus* L., under long photoperiod in winter: indications for a positive feedback. *General and Comparative Endocrinology*, **76**, 12–18.

Borland, M. A. (1986). Size-assortative mating in threespine sticklebacks from two sites on the Salmon River, British Columbia. Unpublished M.Sc. thesis. University of British Columbia, Vancouver.

Boucot, A. J. (1990). *Evolutionary paleobiology of behavior and coevolution*. Elsevier, Englewood Cliffs, New Jersey.

Boulé, V. and FitzGerald, G. J. (1989). Effects of constant and fluctuating temperatures on egg production in the threespine stickleback (*Gasterosteus aculeatus*). *Canadian Journal of Zoology*, **67**, 1599–1602.

Bowne, P. S. (1985). The systematic position of Gasterosteiformes. Unpublished Ph.D. thesis. University of Alberta, Edmonton.

Brafield, A. E. (1985). Laboratory studies of energy budgets. In *Fish energetics: new perspectives* (ed. P. Tytler and P. Calow), pp. 257–281. Croom Helm, London.

Brafield, A. E. and Llewellyn, M. J. (1982). *Animal energetics*. Blackie, Glasgow.

Breck, J.E. and Gitter, M.J. (1983). Effect of fish size on the reactive distance of bluegill (*Lepomis macrochirus*) sunfish. *Canadian Journal of Fisheries and Aquatic Sciences*, **40**, 162–167.

Breden, F., Scott, M.A. and Michel, E. (1987). Genetic differentiation for anti-predator behaviour in the Trinidad guppy, *Poecilia reticulata. Animal Behaviour*, **35**, 618–620.

Brett, J.R. (1979). Environmental factors and growth. In *Fish physiology*, Vol. VIII (ed. W.S. Hoar, D.J. Randall and J.R. Brett), pp. 599–675. Academic Press, London.

Brett, J.R. and Groves, T.D.D. (1979). Physiological energetics. In *Fish physiology*, Vol. VIII (ed. W.S. Hoar, D.J. Randall and J.R. Brett), pp. 279–352. Academic Press, London.

Brewer, G.J. (1970). *An introduction to isozyme techniques*. Academic Press, New York.

Brewer, K.J. and McKeown, B.A. (1980). Prolactin regulation in the coho salmon, *Oncorhynchus kisutch. Journal of Comparative Physiology*, **140**, 217–225.

Bridge, T.W. and Boulenger, G.A. (1904). Fishes. In *The Cambridge natural history*, Vol. VII (ed. S.F. Harmer and A.E. Shipley), pp. 141–760. MacMillan, London.

Broadhurst, P.L. and Jinks, J.L. (1974). What genetical architecture can tell us about the natural selection of behavioural traits. In *The genetics of behaviour* (ed. J.H.F. van Abeelen), pp. 43–63. North-Holland, Amsterdam.

Brodie, E.D., Jr., Nussbaum, R.A. and DiGiovanni, M. (1984). Anti-predator adaptations of Asian salamanders (Salamandridae). *Herpetologica*, **40**, 56–68.

Brooks, D.R. and McLennan, D.A. (1991). *Phylogeny, ecology and behavior: a research program in comparative biology*. University of Chicago Press, Chicago.

Brown, W.L., Jr. (1957). Centrifugal speciation. *Quarterly Review of Biology*, **32**, 247–277.

Brown, W.L., Jr. (1958). General adaptation and evolution. *Systematic Zoology*, **7**, 157–167.

Brush, A.H. and Reisman, H.M. (1965). The carotenoid pigments in the three-spined stickleback, *Gasterosteus aculeatus. Comparative Biochemistry and Physiology*, **14**, 121–125.

Bull, J.L. (1964). *Birds of the New York area*. Harper and Row, New York.

Burghardt, G.M. and Gittleman, J.L. (1990). Comparative behavior and phylogenetic analyses: new wine, old bottles. In *Interpretation and explanation in the study of animal behavior*. Vol. 2. *Explanation, evolution and adaptation* (ed. M. Bekoff and D. Jamieson), pp. 192–225. Westview, Boulder, Colorado.

Burt, A., Kramer, D.L., Nakatsuru, K., and Spry, C. (1988). The tempo of reproduction in *Hyphessobrycon pulchripinnis* (Characidae), with a discussion on the biology of 'multiple spawning' in fishes. *Environmental Biology of Fishes*, **22**, 15–27.

Bush, G.L. (1975). Modes of animal speciation. *Annual Review of Ecology and Systematics*, **7**, 339–364.

Buth, D.G. (1982). Locus assignments for general muscle proteins of darters (Etheostomatini). *Copeia*, **1982**, 217–219.

Buth, D.G. (1983). Duplicate isozyme loci in fishes: origins, distribution, phyletic consequences, and locus nomenclature. In *Isozymes: current topics in biological and medical research*, Vol. 10 (ed. M.C. Rattazzi, J.G. Scandalios, and G.S. Whitt), pp. 381–400. Liss, New York.

Buth, D.G. (1984). The application of electrophoretic data in systematic studies. *Annual Review of Ecology and Systematics*, **15**, 502–522.

Buth, D.G. (1990). Genetic principles and the interpretation of electrophoretic data. In *Electrophoretic and isoelectric focusing techniques in fisheries management* (ed. D.H. Whitmore), pp. 1–21. CRC, Boca Raton, Florida.

Buth, D.G. and Mayden, R.L. (1981). Taxonomic status and relationships among populations of *Notropis pilsbryi* and *Notropis zonatus* (Cypriniformes: Cyprinidae) as shown by the glucosephosphate isomerase, lactate dehydrogenase, and phosphoglucomutase enzyme systems. *Copeia*, **1981**, 583–590.

Buth, D.G. and Murphy, R.W. (1990). Appendix 1: Enzyme staining formulas. In *Molecular systematics* (ed. D.M. Hillis and C. Moritz), pp. 99–126. Sinauer, Sunderland, Massachusetts.

Buth, D.G., Crabtree, C.B., Orton, R.D. and Rainboth, W. J. (1984). Genetic differentiation between the freshwater subspecies of *Gasterosteus aculeatus* in southern California. *Biochemical Systematics and Ecology*, **12**, 423–432.

Buth, D.G., Dowling, T.E. and Gold, J.R. (1991). Molecular and cytological investigations (chapter 4). In *Cyprinid fishes: systematics, biology, and exploitation* (ed. I.J. Winfield and J.S. Nelson), pp. 83–126. Chapman and Hall, London.

Butlin, R. (1989). Reinforcement of premating isolation. In *Speciation and its consequences* (ed. D. Otte and J. A. Endler), pp. 158–179. Sinauer, Sunderland, Massachusetts.

Calow, P. (1985). Adaptive aspects of energy allocation. In *Fish energetics: new perspectives* (ed. P. Tytler and P. Calow), pp. 13–31. Croom Helm, London.

Campbell, R.N. (1979). Sticklebacks (*Gasterosteus aculeatus* L. and *Pungitius pungitius* L.) in the Outer Hebrides, Scotland. *Hebridean Naturalist*, **3**, 8–15.

Campbell, R.N. (1984). Morphological variation in the three-spined stickleback (*Gasterosteus aculeatus*) in Scotland. *Behaviour*, **93**, 161–168.

Campbell, R.N. and Williamson, R.B. (1979). The fishes of inland waters in the Outer Hebrides. *Proceedings of the Royal Society of Edinburgh*, **77B**, 377–393.

Campeau, S., Guderley, H. and FitzGerald, G.J. (1984). Salinity tolerances and preferences of fry of two species of sympatric sticklebacks: possible mechanisms of habitat segregation. *Canadian Journal of Zoology*, **62**, 1048–1051.

Carl, G.C. (1953). Limnobiology of Cowichan Lake, British Columbia. *Journal of the Fisheries Research Board of Canada*, **9**, 417–449.

Carlisle, T.R. (1982). Brood success in variable environments: implications for parental care allocation. *Animal Behaviour*, **30**, 824–836.

Carlquist, S. (1980). *Hawaii, a natural history*. Pacific Tropical Botanical Gardens, Honolulu.

Carson, H.L. (1970). Chromosome tracers of the origin of species. *Science*, **168**, 1414–1418.

Carson, H.L. (1975). The genetics of speciation at the diploid level. *American Naturalist*, **109**, 73–92.

Carson, H.L. (1982). Speciation as a major reorganization of polygenic balances. In *Mechanisms of speciation* (ed. C. Barigozzi), pp. 411–433. Liss, New York.

Carson, H.L. (1987). The genetic system, the deme, and the origin of the species. *Annual Review of Genetics*, **21**, 405–423.

Carson, H.L. and Templeton, A.R. (1984). Genetic revolutions in relation to speciation phenomena: the founding of new populations. *Annual Review of Ecology and Systematics*, **15**, 97–131.

Casteel, R.W. and Hutchison, J.H. (1973). *Orthodon* (Actinopterygii, Cyprinidae) from the Pliocene and Pleistocene of California. *Copeia*, **1973**, 358–361.

Castonguay, M. and FitzGerald, G.J. (1990). The ecology of the calanoid copepod, *Eurytemora affinis*, in salt marsh tide pools. *Hydrobiologica*, **202**, 125–133.

Caswell, H. (1983). Phenotypic plasticity in life-history traits: demographic effects and evolutionary consequences. *American Zoologist*, **23**, 35–46.

Chanin, P. (1985). *The natural history of otters*. Croom Helm, London.

Chappell, L.H. (1969). The parasites of the three-spined stickleback *Gasterosteus aculeatus* L. from a Yorkshire pond. I. Seasonal variation of parasite fauna. *Journal of Fish Biology*, **1**, 137–152.

Chappill, J.A. (1989). Quantitative characters in phylogenetic analysis. *Cladistics*, **5**, 217–234.

Charlesworth, B., Lande, R. and Slatkin, M. (1982). A neo-Darwinian commentary on macroevolution. *Evolution*, **36**, 474–498.

Chauvin-Muckensturm, B. (1979). Les réactions agressives des Épinoches en lumières colorées. *Comptes Rendus hebdomadaires des séances de l'Académie des Sciences, Paris, Série D, Sciences naturelles*, **289**, 1065–1067.

Chellappa, S. and Huntingford, F.A. (1989). Depletion of energy reserves during reproductive aggression in male three-spined stickleback, *Gasterosteus aculeatus* L. *Journal of Fish Biology*, **35**, 315–316.

Chellappa, S., Huntingford, F.A., Strang, R.H.C. and Thomson, R.Y. (1989). Annual variation in energy reserves in male three-spined stickleback, *Gasterosteus aculeatus* L. (Pisces, Gasterosteidae). *Journal of Fish Biology*, **35**, 275–286.

Chen, T.R. and Reisman, H.M. (1970). A comparative chromosome study of the North American species of sticklebacks (Teleostei: Gasterosteidae). *Cytogenetics*, **9**, 321–332.

Cheverud, J.M. (1988). A comparison of genetic and phenotypic correlations. *Evolution*, **42**, 958–968.

Clague, J.J. (1981). *Late Quaternary geology and geochronology of British Columbia. Part 2: Summary and discussion of radiocarbon-dated Quaternary history*. Geological Survey of Canada, Paper No. 80–34 Ottawa. Ministry of Supply and Services, Quebec.

Clarke, B.C. (1975). The contribution of ecological genetics to evolutionary theory: detecting the direct effects of natural selection on particular polymorphic loci. *Genetics* (suppl.), **79**, 101–113.

Clutton-Brock, T.H. (ed.) (1988). *Reproductive success*. University of Chicago Press.

Coad, B.W. (1981). A bibliography of the sticklebacks (Gasterosteidae: Osteichthyes). *Syllogeus*, **35**, 1–142.

Coad, B.W. (1983). Plate morphs in freshwater samples of *Gasterosteus aculeatus* from Arctic and Atlantic Canada: complementary comments on a recent contribution. *Canadian Journal of Zoology*, **61**, 1174–1177.

Coad, B. W. and Power, G. (1973*a*). Observations on the ecology and phenotypic variation of the threespine stickleback, *Gasterosteus aculeatus* L., 1758, and the blackspotted stickleback, *G. wheatlandi* Putnam, 1867 (Osteichthyes: Gasterosteidae) in Armory Cove, Quebec. *Canadian Field-Naturalist*, **87**, 113–122.

Coad, B.W. and Power, G. (1973*b*). Observations on the ecology of lacustrine populations of the threespine stickleback (*Gasterosteus aculeatus* L., 1758) in the Matamek River system, Quebec. Le *Naturaliste Canadien*, **100**, 437–445.

Cohan, F.M. (1984). Can uniform selection retard random genetic divergence between isolated conspecific populations? *Evolution*, **39**, 495–504.

Cohan, F.M. and Hoffman, A.A. (1989). Uniform selection as a diversifying force in evolution: evidence from *Drosophila. American Naturalist*, **134**, 613–637.

Cole, M.R. and Armentrout, J.M. (1979). Neogene paleogeography of the western United States. In *Cenozoic paleogeography of the western United States* (ed. J.M. Armentrout, M.R. Cole, and H. Terbest, Jr.), pp. 297–323. Pacific Coast Paleogeography Symposium 3, Pacific Section of the Society of Economic Paleontologists and Mineralogists, Los Angeles.

Cole, S.J. (1978). Studies on the energy budget of two freshwater teleosts. Unpublished M.Sc. thesis. University of Wales, Aberystwyth.

Colebrook, J.M. (1960). Some observations of zooplankton swarms in Windermere. *Journal of Animal Ecology*, **29**, 241–242.

Colgan, P.W. (1986). Motivational basis of fish behaviour. In *The behaviour of teleost fishes* (1st edn) (ed. T.J. Pitcher), pp. 23–46. Croom Helm, London.

Collias, N.E. (1990). Statistical evidence for aggressive response to red by male three-spined sticklebacks. *Animal Behaviour*, **39**, 401–403.

Collie, N.L., Bolton, J.P., Kawauchi, H. and Hirano, T. (1989). Survival of salmonids in seawater and the time-frame of growth hormone action. *Fish Physiology and Biochemistry*, **7**, 315–321.

Constantz, G.D. (1985). Allopaternal care in the tessellated darter, *Etheostoma olmstedi* (Pisces: Percidae). *Environmental Biology of Fishes*, **14**, 175–183.

Convey, P. (1988). Competition for perches between larval damselflies: the influence of perch use on feeding efficiency, growth rate and predator avoidance. *Freshwater Biology*, **18**, 15–28.

Coope, G.R., Shotton, F.W. and Strachan, I. (1961). A late Pleistocene fauna and flora from Upton Warren, Worcestershire. *Philosophical Transactions of the Royal Society*, **244**, 379–421.

Cope, E.D. (1872). Observations on the systematic relations of the fishes. *Proceedings of the American Association for the Advancement of Science,* **20**, 317–343.

Copeman, D.J. (1977). Population differences in rainbow smelt, *Osmerus mordax*: multivariate analysis of mensural and meristic data. *Journal of the Fisheries Research Board of Canada*, **34**, 1220–1229.

Courtenay, S.C. and Keenleyside, M.H.A. (1983). Nest site selection in the fourspine stickleback, *Apeltes quadracus* (Mitchell). *Canadian Journal of Zoology*, **61**, 1443–1447.

Covens, M., Covens, L., Ollevier, F. and DeLoof, A. (1987). A comparative study of some properties of vitellogenin (Vg) and yolk proteins in a number of freshwater and marine teleost fishes. *Comparative Biochemistry and Physiology*, **B88**, 75–80.

Covens, M., Stynen, D., Ollevier, F. and DeLoof, A. (1988). Concanavalin A reactivity of vitellogenin and yolk proteins of the threespine stickleback, *Gasterosteus aculeatus* (Teleostei). *Comparative Biochemistry and Physiology*, **90B**, 227–233.

Cowen, R.K., Chiarella, L.A., Gomez, C.J. and Bell, M. A. (1991). Offshore distribution, size, age and lateral plate variation of late larval/early juvenile sticklebacks (*Gasterosteus*) off the Atlantic coast of New Jersey and New York. *Canadian Journal of Fisheries and Aquatic Sciences*, **48**, 1679–1684.

Cox, C.R. and LeBoeuf, B.J. (1977). Female incitation of male competition: a mechanism in sexual selection. *American Naturalist*, **111**, 317–335.

Coyne, J. A. and Orr, H. A. (1989). Patterns of speciation in *Drosophila*. *Evolution*, **43**, 362–381.

Craig, D. and FitzGerald, G. J. (1982). Reproductive tactics of four sympatric sticklebacks (Gasterosteidae). *Environmental Biology of Fishes*, **7**, 369–375.

Craig-Bennett, M. A. (1931). The reproductive cycle of the three-spined stickleback, *Gasterosteus aculeatus*, Linn. *Philosophical Transactions of the Royal Society*, **B219**, 197–279.

Craik, J. C. A. and Harvey, S. M. (1986). Phosphorus metabolism and water uptake during final maturation of ovaries of teleosts with pelagic and demersal eggs. *Marine Biology*, **90**, 285–289.

Creed, R. (ed.) (1971). *Ecological genetics and evolution*. Blackwell, Oxford.

Crim, L. W., Peter, R. E. and Billard, R. (1981). Onset of gonadotropic hormone accumulation in the immature trout pituitary gland in response to estrogen or aromatizable androgen steroid hormones. *General and Comparative Endocrinology*, **44**, 372–381.

Crivelli, A. J. and Britton, R. H. (1987). Life history adaptations of *Gasterosteus aculeatus* in a Mediterranean wetland. *Environmental Biology of Fishes*, **18**, 109–125.

Cronly-Dillon, J. and Sharma, S. C. (1968). Effect of season and sex on the photopic spectral sensitivity of the three-spined stickleback. *Journal of Experimental Biology*, **49**, 679–687.

Crow, R. T. and Liley, N. R. (1979). A sexual pheromone in the guppy, *Poecilia reticulata* (Peters). *Canadian Journal of Zoology*, **57**, 184–188.

Cruz, A. and Wiley, J. W. (1989). The decline of an adaptation in the absence of a presumed selection pressure. *Evolution*, **43**, 55–62.

Cui, Y. and Wootton, R. J. (1988*a*). Bioenergetics of growth of a cyprinid, *Phoxinus phoxinus*: the effect of ration, temperature and body size on food consumption, faecal production and nitrogenous excretion. *Journal of Fish Biology*, **33**, 431–443.

Cui, Y. and Wootton, R. J. (1988*b*). Effects of ration, temperature and body size on the body composition, energy content and condition of the minnow *Phoxinus phoxinus* (L.). *Journal of Fish Biology*, **32**, 749–764.

Cullen, E. (1960). Experiment on the effect of social isolation on reproductive behaviour in the three-spined stickleback. *Animal Behaviour*, **8**, 235.

Cummins, K. C. and Wuycheck, D. D. (1971). Caloric equivalents for investigations in ecological energetics. *Internationale Vereingung für Theoretische und Angewandte Limnologie*, **18**, 1–158.

Curio, E. (1976). *The ethology of predation*. Springer-Verlag, Berlin.

Cvancara, A. M., Clayton, L., Bickley, W. B., Jr., Jacob, A. F., Ashworth, A. C., Brophy, J. A., Shay, C. T., Delorme, L. D. and Lammers, G. E. (1971). Paleolimnology of late Quaternary deposits: Siebold Site, North Dakota. *Science*, **171**, 172–174.

Daniel, W. (1985). Fragen zum Wanderverhalten des dreistachligen Stichlings (*Gasterosteus aculeatus* L.). *Faunistische-Ökologische Mitteilungen*, **5**, 419–429.

Darwin, C. (1859). *The origin of species* (facsimile of the 1st edn). Harvard University Press, Cambridge, Massachusetts.

Darwin, C. (1872). *The origin of species* (6th edn). Murray, London.

David, L. R. (1945). A Neogene stickleback from the Ridge Formation of California. *Journal of Paleontology*, **19**, 315–318.

Dawkins, R. (1980). Good strategy or evolutionary stable strategy? In *Sociobiology beyond nature/nurture* (ed. G. W. Barlow and J. Silverberg), pp. 331–367. Westview, Boulder, Colorado.

Dawson, J.W. (1871). The post-Pliocene geology of Canada. *Canadian Naturalist* (new series), **6**, 19–42, 166–187, 241–259, 369–416.

Dawson, J.W. (1872). Notes on the post-Pliocene geology of Canada; with especial reference to the conditions of accumulation of the deposits and the marine life of the period. Mitchell and Wilson, Montreal. [reprint of Dawson 1871]

Dawson, J.W. (1894). *The Canadian ice age*. Scientific, New York.

De Deckker, P. (1988). Biological and sedimentary facies of Australian salt lakes. *Palaeogeography, Palaeoclimatology, Palaeoecology*, **62**, 237–270.

Delbeek, J.C. and Williams, D.D. (1987*a*). Food resource partitioning between sympatric populations of brackish water sticklebacks. *Journal of Animal Ecology*, **56**, 949–967.

Delbeek, J.C. and Williams, D.D. (1987*b*). Morphological differences among females of four species of stickleback (Gasterosteidae) from New Brunswick and their possible ecological significance. *Canadian Journal of Zoology*, **65**, 289–295.

Delbeek, J.C. and Williams, D.D. (1988). Feeding selectivity of four species of sympatric sticklebacks in brackish-water habitats in eastern Canada. *Journal of Fish Biology*, **32**, 41–62.

Dingle, H. (1988). Quantitative genetics of life history evolution in a migrant insect. In *Population genetics and evolution* (ed. G. de Jong), pp. 83–93. Springer-Verlag, Berlin.

Dobson, F.S. (1985). The use of phylogeny in behavior and ecology. *Evolution*, **39**, 1384–1388.

Dobzhansky, Th. (1937). *Genetics and the origin of species* (1st edn). Columbia University Press, New York.

Dobzhansky, Th. (1940). Speciation as a stage in evolutionary divergence. *American Naturalist*, **74**, 312–321.

Dobzhansky, Th. (1951). *Genetics and the origin of species* (3rd edn). Columbia University Press, New York.

Dominey, W.J. (1984*a*). Alternative mating tactics and evolutionary stable strategies. *American Zoologist*, **24**, 385–396.

Dominey, W.J. (1984*b*). Effects of sexual selection and life history on speciation: species flocks of African cichlids and Hawaiian *Drosophila*. In *Evolution of fish species flocks* (ed. A.A. Echelle and I. Kornfield), pp. 231–249. University of Maine, Orono.

Donoghue, M.J. (1989). Phylogenies and the analysis of evolutionary sequences, with examples from seed plants. *Evolution*, **43**, 1137–1156.

Douglas, R.H., Eva, J. and Guttridge, N. (1988). Size constancy in goldfish (*Carassius auratus*). *Behavioural Brain Research*, **30**, 37–42.

Douglas, S.D. and Reimchen, T.E. (1988). Habitat characteristics and population estimate of breeding Red-throated Loon, *Gavia stellata*, on the Queen Charlotte Islands. *Canadian Field-Naturalist*, **102**, 679–684.

Dowdey, T.G. and Brodie, E.D., Jr. (1989). Antipredator strategies of salamanders: individual and geographical variation in responses of *Eurycea bislineata* to snakes. *Animal Behaviour*, **38**, 707–711.

Downing, J.A. (1986). A regression technique for the estimation of epiphytic invertebrate populations. *Freshwater Biology*, **16**, 161–173.

Drickamer, L. and Vessey, S. (1992). *Animal behavior: mechanisms, ecology and evolution* (3rd edn). Brown, Dubuque, Iowa.

Duarte, C.M. and Alcaraz, M. (1989). To produce many small or few large eggs: a size-independent reproductive tactic of fish. *Oecologia*, **80**, 401–404.

Dufresne, F., FitzGerald, G.J. and Lachance, S. (1990). Age and size related differences in reproductive costs in threespine sticklebacks. *Behavioral Ecology*, **1**, 140–147.

Dunbar, R.I.M. (1982). Intraspecific variations in mating strategy. In *Perspectives in ethology*, Vol. 5 (ed. P.P.G. Bateson and P.H. Klopfer), pp. 385–431. Plenum, New York.

Dunham, A.E. and Miles, D.B. (1985). Patterns of covariation in life history traits of squamate reptiles: the effects of size and phylogeny reconsidered. *American Naturalist*, **126**, 231–257.

Dunham, A.E., Miles, D.B. and Reznick, D.N. (1988). Life history patterns in squamate reptiles. In *Biology of the Reptilia*, Vol. 16B (ed. C. Gans and R.B. Huey), pp. 442–511. Liss, New York.

Eastman, C.R. (1917). Fossil fishes in the collection of the United States National Museum. *Proceedings of the US National Museum*, **52**, 235–304.

Eastman, R. (1969). *The kingfisher*. Collins, London.

Eberhard, W.G. (1982). Behavioral characters for the higher classification of orb-weaving spiders. *Evolution*, **36**, 1067–1095.

Echelle, A.A. and Kornfield, I. (ed.) (1984). *Evolution of fish species flocks*. University of Maine, Orono.

Edmunds, M. (1974). *Defence in animals: a survey of anti-predator defences*. Longman, New York.

Eggers, D.M. (1982). Planktivore preference by prey size. *Ecology*, **63**, 381–390.

Eggers, D.M., Bartoo, N.W., Rickard, N.A., Wissmar, R.C., Burgner, R.L., and Devol, A.H. (1978). The Lake Washington ecosystem: the perspective from the fish community production and forage base. *Journal of the Fisheries Research Board of Canada*, **35**, 1553–1571.

Ehrenbaum, E. (1904). Eier und Larven von Fischen der deutschen Bucht. III. Fische mit festsitzenden Eiern. *Wissenschaftliche Meeresuntersuchungen Herausgegben von der Kommission zur Wissenschaftliche Untersuchung der Deutschen mere in Kiel und der Biologischen Anstalt auf Helgoland*, **6**, 127–200.

Ehrlich, P.R. and Raven, P.H. (1969). Differentiation of populations. *Science*, **165**, 1228–1232.

Ekström, P. (1984). Central neural connections of the pineal organ and retina in the teleost *Gasterosteus aculeatus* L. *Journal of Comparative Neurology*, **226**, 321–335.

Ekström, P. and Meissel, H. (1989). Signal processing in a simple vertebrate photoreceptor system: the teleost pineal organ. *Physiologia Bohemoslovaca*, **38**, 311–326.

Ekström, P. and Veen, T. van (1983). Central connections of the pineal organ in the three-spined stickleback, *Gasterosteus aculeatus* L. (Teleostei). *Cell and Tissue Research*, **232**, 141–155.

Ekström, P., Borg, B., and Veen, T. van (1983). Ontogenetic development of the pineal organ, parapineal organ, and retina of the three-spined stickleback, *Gasterosteus aculeatus*, L. *Cell and Tissue Research*, **233**, 593–609.

Ekström, P., Honkanen, T., and Ebbesson, S.O.E. (1988). FMRFamide-like immunoreactive neurons of the nervus terminalis of the teleosts innervate both retina and pineal organ. *Brain Research*, **460**, 68–75.

Elder, R.L. (1985). Principles of aquatic taphonomy with examples from the fossil record. Unpublished Ph.D. thesis. University of Michigan, Ann Arbor.

Elder, R.L. and Smith, G.R. (1988). Fish taphonomy and environmental inference in paleolimnology. *Palaeogeography, Palaeoclimatology, Palaeoecology*, **62**, 577–592.

Eldredge, N. and Gould, S. J. (1972). Punctuated equilibria: an alternative to phyletic gradualism. In *Models in paleobiology* (ed. T. J. M. Schopf), pp. 82–115. Freeman Cooper, San Francisco.

Elgar, M. A. and Heaphy, L. J. (1989). Covariation between clutch size, egg weight and egg shape: comparative evidence for chelonians. *Journal of Zoology, London*, **219**, 137–152.

Elliott, J. M. (1976). The energetics of feeding, metabolism and growth of brown trout (*Salmo trutta* L.) in relation to body weight, water temperature and ration size. *Journal of Animal Ecology*, **45**, 923–948.

Elliott, J. M. (1979). Energetics of freshwater teleosts. *Symposia of the Zoological Society of London*, **44**, 29–61.

Elliott, J. M. (1981). Some aspects of thermal stress in freshwater teleosts. In *Stress and fish* (ed. A. D. Pickering), pp. 209–245. Academic Press, London.

Endler, J. A. (1973). Gene flow and population differentiation. *Science*, **179**, 243–250.

Endler, J. A. (1977). *Geographic variation, speciation and clines*. Monographs in Population Biology, No. 10. Princeton University Press, Princeton, New Jersey.

Endler, J. A. (1978). A predator's view of animal color patterns. *Evolutionary Biology*, **11**, 319–364.

Endler, J. A. (1980). Natural selection on color patterns in *Poecilia reticulata. Evolution*, **34**, 76–91.

Endler, J. A. (1982). Problems in distinguishing historical from ecological factors in biogeography. *American Zoologist*, **22**, 441–452.

Endler, J. A. (1986). *Natural selection in the wild*, Monographs in Population Biology, No. 21. Princeton University Press, Princeton, New Jersey.

Engel, L. J. (1971). *Annual progress report for evaluation of sport fish stocking on the Kenai Peninsula–Cook Inlet area, July 1, 1970 to June 30, 1971*. Project F-9-3, Job g-ll-F, pp. 1–34, Alaska Department of Fish and Game, Juneau.

Englemann, G. F. and Wiley, E. O. (1977). The place of ancestor–descendant relationships in phylogeny reconstruction. *Systematic Zoology*, **26**, 1–11.

Erlinge, S. (1968). Food studies on captive otters *Lutra lutra* L. *Oikos*, **23**, 327–335.

Erlinge, S. and Jensen, B. (1981). The diet of otters *Lutra lutra* in Denmark. *Natura Jutlandica*, **19**, 161–165.

Evers, M. (1878). Zur Charakteristik des Stichlings (*Gasterosteus aculeatus*). *Jahresberichte des Naturwissenschaftlichen Vereins in Elberfeld nebst wissenschaftlichen Beilagen*, **5**, 26–46.

Ewert, J.-P. (1980). *Neuro-ethology*. Springer-Verlag, New York.

Falconer, D. S. (1981). *Introduction to quantitative genetics* (2nd edn). Longman, London.

Falconer, D. S. (1989). *Introduction to quantitative genetics* (3rd edn). Longman, Harlow.

Faris, A. A. (1986). Some effects of acid water on the biology of *Gasterosteus aculeatus* (Pisces). Unpublished Ph.D. thesis. University of Wales, Aberystwyth.

Faris, A. A. and Wootton, R. J. (1987). Effects of water pH and salinity on the the survival of eggs and larvae of the euryhaline teleost, *Gasterosteus aculeatus* L. *Environmental Pollution*, **48**, 49–59.

Farr, J. A. and Travis, J. (1986). Fertility advertisement by female sailfin mollies, *Poecilia latipinna* (Pisces: Poeciliidae). *Copeia*, **1986**, 467–472.

Farris, J. S. (1972). Estimating phylogenetic trees from distance matrices. *American Naturalist*, **106**, 645–668.

Fatio, V. (1882). *Faunes des vertebrés de la Suisse*, Vol. 4 *Histoire naturelle des poissons*. Georg, Genève.

Feldmeth, C.R. and Baskin, J.N. (1976). Thermal and respiratory studies with reference to temperature and oxygen tolerance for the unarmored stickleback, *Gasterosteus aculeatus williamsoni* Hubbs. *Bulletin of the Southern California Academy of Science*, **75**, 127–131.

Felsenstein, J. (1978). The number of evolutionary trees. *Systematic Zoology*, **27**, 27–33.

Felsenstein, J. (1981). Skepticism towards Santa Rosalia, or why are there so few kinds of animals? *Evolution*, **35**, 124–138.

Felsenstein, J. (1985). Phylogenies and the comparative method. *American Naturalist*, **125**, 1–15.

Felsenstein, J. (1988). Phylogenies and quantitative characters. *Annual Review of Ecology and Systematics*, **19**, 445–471.

Fenster, E. J. and Sorhannus, U. (1991). On the measurement of morphological rates of evolution: a review. *Evolutionary Biology*, **25**, 375–410.

Feuth-de Bruijn, E. and Sevenster, P. (1983). Parental reactions to young in sticklebacks (*Gasterosteus aculeatus* L.). *Behaviour*, **83**, 186–203.

Fink, W.L. (1982). The conceptual relationship between ontogeny and phylogeny. *Paleobiology*, **8**, 254–264.

Fisher, S.E., Shaklee, J.B., Ferris, S.D., and Whitt, G. S. (1980). Evolution of five multilocus isozyme systems in the chordates. *Genetica*, **52/53**, 73–85.

Fitch, W.M. (1971). Toward defining the course of evolution: minimum change for a specific tree topology. *Systematic Zoology*, **20**, 406–416.

FitzGerald, G. J. (1983). The reproductive ecology and behaviour of three sympatric sticklebacks (Gasterosteidae) in a saltmarsh. *Biology of Behaviour*, **8**, 67–79.

FitzGerald, G. J. (1992*a*). Filial cannibalism in fishes: why do parents eat their offspring? *Trends in Ecology and Evolution*, **7**, 7–10.

FitzGerald, G.J. (1992*b*). Egg cannibalism by sticklebacks: spite or selfishness? *Behavioral Ecology and Sociobiology*, **30**, 201–206.

FitzGerald, G.J. and Dutil, J.D. (1981). Evidence for differential predation on an estuarine stickleback community. *Canadian Journal of Zoology*, **59**, 2394–2395.

FitzGerald, G.J. and Havre, N. van (1985). Flight, fright and shoaling in sticklebacks (Gasterosteidae). *Biology of Behaviour*, **10**, 321–331.

FitzGerald, G. J. and Havre, N. van (1987). The adaptive significance of cannibalism in sticklebacks (Gasterosteidae: Pisces). *Behavioral Ecology and Sociobiology*, **20**, 125–128.

FitzGerald, G. J. and Kedney, G. I. (1987). Aggression, fighting and territoriality in sticklebacks: three different phenomena? *Biology of Behaviour*, **12**, 186–195.

FitzGerald, G.J. and Whoriskey, F.G. (1985). The effects of interspecific interactions upon male reproductive success in two sympatric sticklebacks, *Gasterosteus aculeatus* and *G. wheatlandi*. *Behaviour*, **93**, 112–126.

FitzGerald, G.J. and Whoriskey, F.G. (1992). Empirical studies of cannibalism in fish. In *Cannibalism: ecology and evolution among diverse taxa* (ed. M.A. Elgar and B. J. Crespi), pp. 238–255. Oxford University Press.

FitzGerald, G. J. and Wootton, R. J. (1986). Behavioural ecology of sticklebacks. In *The behaviour of teleost fishes* (1st edn) (ed. T.J. Pitcher), pp. 409–432. Croom Helm, London.

FitzGerald, G.J., Gaudreault, A., and Havre, N. van (1986). Decision making by parental sticklebacks *Gasterosteus aculeatus* in a variable environment. In

Behavioural ecology and population biology (ed. L.C. Drickamer), pp. 71–75. Privat, Toulouse.

FitzGerald, G.J., Guderley, H., and Picard, P. (1989). Hidden reproductive costs in the three-spined stickleback (*Gasterosteus aculeatus*). *Experimental Biology*, **48**, 295–300.

FitzGerald, G.J., Whoriskey, F.G., Morrissette, J., and Harding, M. (1992). Habitat scale, female cannibalism, and male reproductive success in threespine stickleback *Gasterosteus aculeatus. Behavioral Ecology*, **3**, 141–147.

Fivizzani, A.J. and Meir, A.H. (1976). Circadian temporal synergism of cortisol and prolactin influences gonadal growth in *Fundulus grandis abstract. American Zoologist*, **16**, 259.

Fivizzani, A.J. and Meir, A.H. (1978). Temporal synergism of cortisol and prolactin influences salinity preference of gulf killifish, *Fundulus grandis. Canadian Journal of Zoology*, **56**, 2597–2602.

Fleming, I.A. and Ng, S. (1987). Evaluation of techniques for fixing, preserving, and measuring salmon eggs. *Canadian Journal of Fisheries and Aquatic Sciences*, **44**, 1957–1962.

Fletcher, D.A. (1984). Effects of food supply on egg quality and quantity in fishes. Unpublished Ph.D. thesis. University of Wales, Aberystwyth.

Follenius, E. (1968). Cytologie et cytophysiologie des cellules interstitielles de l'épinoche: *Gasterosteus aculeatus* L.: étude au microscope électronique. *General and Comparative Endocrinology*, **11**, 198–219.

Foote, C.J., Wood, C.C., and Withler, R.E. (1989). Biochemical genetic comparison of sockeye salmon and kokanee, the anadromous and nonanadromous forms of *Oncorhynchus nerka. Canadian Journal of Fisheries and Aquatic Sciences*, **46**, 149–158.

Ford, E.B. (1975). *Ecological genetics* (4th edn). Wiley, New York.

Foster, S.A. (1985*a*). Group foraging in a coral reef fish: a mechanism for gaining access to defended resources. *Animal Behaviour*, **33**, 782–792.

Foster, S.A. (1985*b*). Size-dependent territory defense by a damselfish. *Oecologia*, **67**, 499–505.

Foster, S.A. (1987). Acquisition of a defended resource: a benefit of group foraging for the neotropical wrasse, *Thalassoma lucasanum. Environmental Biology of Fishes*, **19**, 215–222.

Foster, S.A. (1988). Diversionary displays of paternal stickleback: defenses against cannibalistic groups. *Behavioral Ecology and Sociobiology*, **22**, 335–340.

Foster, S.A. (1990). Courting disaster in cannibal territory. *Natural History*, November, pp. 52–61.

Foster, S. [A.] and Ploch, S. (1990). Determinants of variation in antipredator behavior of territorial male threespine stickleback in the wild. *Ethology*, **84**, 281–294.

Foster, S.A., Baker, J.A., and Bell, M.A. (1992). Phenotypic integration of life history and morphology: an example from the three-spined stickleback, *Gasterosteus aculeatus* L. *Journal of Fish Biology*, **41**, 21–35 (Supplement).

Foster, S.A., Garcia, V. B., and Town, M.Y. (1988). Cannibalism as the cause of an ontogenetic shift in habitat use by fry of the threespine stickleback. *Oecologia*, **74**, 577–585.

Francis, R.C. (1983). Experiential effects on agonistic behavior in the paradise fish, *Macropodus opercularis. Behaviour*, **85**, 292–313.

Francis, R.C., Havens, A.C., and Bell, M.A. (1985). Unusual lateral plate variation of threespine sticklebacks (*Gasterosteus aculeatus*) from Knik Lake, Alaska. *Copeia*, **1985**, 619-624.

Francis, R.C., Baumgartner, J.V., Havens, A.C., and Bell, M.A. (1986). Historical and ecological sources of variation among lake populations of threespine sticklebacks, *Gasterosteus aculeatus*, near Cook Inlet, Alaska. *Canadian Journal of Zoology*, **64**, 2257-2265.

Fraser, D.F. and Gilliam, J.F. (1987). Feeding under predation hazard: response of the guppy and Hart's rivulus from sites with contrasting predation hazard. *Behavioral Ecology and Sociobiology*, **21**, 203-209.

Fraser, D.F. and Huntingford, F.A. (1986). Feeding and avoiding predation hazard: the behavioural response of the prey. *Ethology*, **73**, 56-68.

Fredskild, B. (1983). The Holocene vegetational development of the Godthåbsfjord area, West Greenland. *Meddelelser om Grønland* (*Geoscience*) **10**, 1-28.

Fredskild, B. (1985). The Holocene vegetational development of Tugtuligssuaq and Qeqertat, Northwest Greenland. *Meddelelser om Grønland (Geoscience)*, **14**, 1-20.

Fredskild, B. and Røen, U. (1982). Macrofossils of an interglacial peat deposit at Kap Kobenhaun, North Greenland. *Boreas*, **11**, 181-185.

Fretwell, S.D. and Lucas, H.L. (1970). On territorial behaviour and other factors influencing habitat distribution in birds. *Acta Biotheoretica*, **19**, 16-36.

Frey, D.F. and Miller, R.J. (1972). The establishment of dominance relationships in the blue gourami, *Trichogaster trichopterus* (Pallus). *Behaviour*, **42**, 8-62.

Fries, G. (1965). Langen-, Gewichts- und Eiverhaltnisse beim dreistacheligen Stichling (*Gasterosteus aculeatus* L.). *Zeitschrift für Fisch Hilfswiss*, **13**, 171-180.

Frost, W.E. (1954). The food of pike, *Esox lucius* L., in Windermere. *Journal of Animal Ecology*, **23**, 339-360.

Fry, F.E.J. (1971). The effect of environmental factors on the physiology of fish. In *Fish physiology*, Vol. VI (ed. W.S. Hoar and D.J. Randall), pp. 1-98. Academic Press, London.

Fryer, G. and Iles, T.D. (1972). *The cichlid fishes of the Great Lakes of Africa*. Oliver and Boyd, Edinburgh.

Futuyma, D.J. and Mayer, G.C. (1980). Non-allopatric speciation in animals. *Systematic Zoology*, **29**, 254-271.

Gach, M.A. and Reimchen, T.E. (1989). Mitochondrial DNA patterns among endemic stickleback from the Queen Charlotte Islands: a preliminary survey. *Canadian Journal of Zoology*, **67**, 1324-1328.

Gaillard, J.-M., Pontier, D., Allaine, D., Lebreton, J.D., Trouvilliez, J., and Clobert, J. (1989). An analysis of demographic tactics in birds and mammals. *Oikos*, **56**, 59-76.

Gale, W.F. (1986). Indeterminant fecundity and spawning behavior of captive red shiners—fractional, crevice spawners. *Transactions of the American Fisheries Society*, **115**, 429-437.

Gardiner, B.G. (1966). *Catalogue of Canadian fossil fishes*. Contribution No. 68, Life Sciences, Royal Ontario Museum, University of Toronto, pp. 1-154.

Garrett, G.P. (1982). Variation in the reproductive traits of the Pecos pupfish, *Cyprinodon pecosensis*. *American Midland Naturalist*, **108**, 355-363.

Garside, E.T., Heinze, D.G., and Barbour, S.E. (1977). Thermal preference in relation to salinity in the threespine stickleback, *Gasterosteus aculeatus* L., with an interpretation of its significance. *Canadian Journal of Zoology*, **55**, 590-594.

Gaudant, J. (1979). L'ichthyofaune tiglienne de Tegelen (Pays-Bas): signification paléoécologique et paléoclimatique. *Scripta Geologica*, **50**, 1-16.

Gaudant, J. (1981). Contribution de la paléoichthyologie continentale à la reconstitution des paléoenvironments cénozoíques d'Europe occidentale: approche systématique, paléoécologique, paléogéograpique et paléoclimatologique. Thesis Université P. et M. Curie, Paris Mémoires des Sciences de la Terre, No. 81-53.

Gaudreault, A. and FitzGerald, G.J. (1985). Field observations of intraspecific and interspecific aggression among sticklebacks (Gasterosteidae). *Behaviour*, **94**, 203-211.

Gaudreault, A., Miller, T., Montgomery, W.L., and FitzGerald, G.J. (1986). Interspecific interactions and diet of sympatric juvenile brook charr, *Salvelinus fontinalis*, and adult ninespine sticklebacks, *Pungitius pungitius*. *Journal of Fish Biology*, **28**, 133-140.

George, D.G. (1981). Zooplankton patchiness. *Annual Report of the Freshwater Biological Association*, **49**, 32-44.

George, D.G. and Edwards, R.W. (1976). The effect of wind on the distribution of chlorophyll a and crustacean plankton in a shallow eutrophic reservoir. *Journal of Applied Ecology*, **13**, 667-690.

Gerell, R. (1968). Food habits of the mink, *Mustela vison* Screb. in Sweden. *Viltrevy*, **5**, 120-194. [not seen]

Gerking, S.H. and Lee, R.M. (1980). Reproductive performance of the desert pupfish (*Cyprinodon n. nevadensis*) in relation to salinity. *Environmental Biology of Fishes*, **5**, 375-378.

Gerrish, N. and Bristow, J.M. (1979). Macroinvertebrate association with aquatic macrophytes and artificial substrates. *Journal of Great Lakes Research*, **5**, 69-72.

Gibson, R.M. (1980). Optimal prey-size selection by three-spined sticklebacks (*Gasterosteus aculeatus*): a test of the apparent-size hypothesis. *Zeitschrift für Tierpsychologie*, **52**, 291-307.

Gilbertson, L.G. (1980). Variation and natural selection in an Alaskan population of the threespine stickleback (*Gasterosteus aculeatus* L.). Unpublished Ph.D. thesis. University of Washington, Seattle.

Giles, N. (1981). Summer diet of the grey heron. *Scottish Birds*, **11**, 153-159.

Giles, N. (1983*a*). The possible role of environmental calcium levels during the evolution of phenotypic diversity in Outer Hebridean populations of the three-spined stickleback. *Journal of Zoology, London*, **199**, 535-544.

Giles, N. (1983*b*). Behavioural effects of the parasite *Schistocephalus solidus* (Cestoda) on an intermediate host, the threespine stickleback, *Gasterosteus aculeatus* L. *Animal Behaviour*, **31**, 1192-1194.

Giles, N. (1984*a*). Implications of parental care of offspring for the anti-predator behaviour of adult male and female three-spined sticklebacks, *Gasterosteus aculeatus* L. In *Fish reproduction: strategies and tactics* (ed. G.W. Potts and R.J. Wootton), pp. 275-289. Academic Press, London.

Giles, N. (1984*b*). Development of the overhead fright response in wild and predator-naive three-spined sticklebacks, *Gasterosteus aculeatus* L. *Animal Behaviour*, **32**, 276-279.

Giles, N. (1987*a*). A comparison of the behavioural responses of parasitised and non-parasitised three-spined sticklebacks, *Gasterosteus aculeatus* L., to progressive hypoxia. *Journal of Fish Biology*, **30**, 631-638.

Giles, N. (1987*b*). Predation risk and reduced foraging activity in fish: experiments with parasitised and non-parasitised three-spined sticklebacks, *Gasterosteus aculeatus* L. *Journal of Fish Biology*, **31**, 37-44.

Giles, N. (1987*c*). Population biology of the three-spined sticklebacks, *Gasterosteus aculeatus*, in Scotland. *Journal of Zoology, London*, **212**, 255-265.

Giles, N. and Huntingford, F.A. (1984). Predation risk and inter-population variation in anti-predator behaviour in the three-spined stickleback, *Gasterosteus aculeatus* L. *Animal Behaviour*, **32**, 264–275.

Giles, N. and Huntingford, F.A. (1985). Variability in breeding biology of three-spined sticklebacks (*Gasterosteus aculeatus*): problems with measuring population differences in aggression. *Behaviour*, **93**, 57–68.

Gill, T. (1884). On the mutual relationships of the hemibranchiate fishes. *Proceedings of the Academy of Natural Sciences of Philadelphia*, **14**, 154–166.

Gilliam, J.F. and Fraser, D.F. (1987). Habitat selection under predation: test of a model with foraging minnows. *Ecology*, **68**, 1856–1862.

Gingerich, P.D. (1979). The stratophenetic approach to phylogeny reconstruction in vertebrate paleontology. In *Phylogenetic analysis and paleontology* (ed. J. Cracraft and N. Eldredge), pp. 41–77. Columbia University Press, New York.

Gingerich, P.D. (1983). Rates of evolution: effect of time and temporal scaling. *Science*, **222**, 159–161.

Gingerich, P.D. (1990). Stratophenetics. In *Paleobiology: a synthesis* (ed. D.E.G. Briggs and P.R. Crowther), pp. 437–442. Blackwell, Oxford.

Gittenberger, E. (1991). What about non-adaptive radiation? *Biological Journal of the Linnean Society*, **43**, 263–272.

Glebe, B.D. and Leggett, W.C. (1981). Latitudinal differences in energy allocation and use during the freshwater migrations of American shad (*Alosa sapidissima*) and their life history consequences. *Canadian Journal of Fisheries and Aquatic Science*, **38**, 806–820.

Godin, J.-G.J. and Sproul, C.D. (1988). Risk taking in parasitized sticklebacks under threat of predation: effects of energetic need and food availability. *Canadian Journal of Zoology*, **66**, 2360–2367.

Goldschmidt, T. and Bakker, T.C.M. (1990). Determinants of reproductive success of male sticklebacks in the field and in the laboratory. *Netherlands Journal of Zoology*, **40**, 664–687.

Goldschmidt, T., Foster, S.A., and Sevenster, P. Internest distance and sneaking in three-spined stickleback. *Animal Behaviour*, **44**, 793–795.

Goldthwaite, R.O., Coss, R.G. and Owings, D.H. (1990). Evolutionary dissipation of an antisnake system: differential behaviour by California and Arctic ground squirrels in above- and below-ground contexts. *Behaviour*, **112**, 246–269.

Goodey, W. and Liley, N.R. (1986). The influence of early experience on escape behaviour in the guppy. *Canadian Journal of Zoology*, **64**, 885–888.

Goodrich, E.S. (1909). Vertebrata craniata. I. Cyclostomes and fishes. In *A treatise on zoology*, Part 9 (ed. R. Lankester), pp. 1–518. Black, London.

Goolish, E.M. (1991). Aerobic and anaerobic scaling in fish. *Biological Review*, **66**, 33–56.

Gosline, W.A. (1971). *Functional morphology and classification of Teleostean fishes*. University Press of Hawaii, Honolulu.

Gotceitas, V. (1990). Foraging and predator avoidance: a test of a patch choice model with juvenile bluegill sunfish. *Oecologia*, **83**, 346–351.

Gould, S.J. (1980). Is a new and general theory of evolution emerging? *Paleobiology*, **6**, 119–130.

Gould, S.J. (1984). Covariance sets and ordered geographic variation in *Cerion* from Aruba, Bonaire and Curaçao, a way of studying nonadaptation. *Systematic Zoology*, **33**, 217–237.

Gould, S.J. (1987). *Time's arrow, times's cycle: myth and metaphor in the discovery of geological time*. Harvard University Press, Cambridge, Massachusetts.

Gould, S.J. (1989). *Wonderful life: the Burgess shale and the nature of history*. Hutchinson Radius, London.

Gould, S.J. and Eldredge, N. (1977). The meaning of punctuated equilibria: the tempo and mode of evolution reconsidered. *Paleobiology*, **3**, 115–151.

Gould, S.J. and Lewontin, R.C. (1979). The spandrels of San Marco and the Panglossian paradigm: a critique of the adaptationist programme. *Proceedings of the Royal Society*, **B205**, 581–598.

Gould, S.J. and Vrba, E.S. (1982). Exaptation—a missing term in the science of form. *Paleobiology*, **8**, 4–15.

Goyens, J. and Sevenster, P. (1976). Influence du facteur héréditaire et de la densité de population sur l'ontogénèse de l'agressivité chez l'épinoche (*Gasterosteus aculeatus* L.). *Netherlands Journal of Zoology*, **26**, 427–431.

Grant, B.R. and Grant, P.R. (1989*a*). Sympâtric speciation and Darwin's finches. In *Speciation and its consequences* (ed. D. Otte and J.A. Endler), pp. 433–457. Sinauer, Sunderland, Massachusetts.

Grant, B.R. and Grant, P.R. (1989*b*). Natural selection in a population of Darwin's finches. *American Naturalist*, **133**, 377–393.

Grant, P.R. (1986). *Ecology and evolution of Darwin's finches*. Princeton University Press, Princeton, New Jersey.

Gray, J. (1988). Evolution of the freshwater ecosystem: the fossil record. *Palaeogeography, Palaeoclimatology, Palaeoecology*, **62**, 1–214.

Greenbank, J. and Nelson, P.R. (1959). Life history of the threespine stickleback *Gasterosteus aculeatus* in Karluk Lake and Bare Lake, Kodiak Island, Alaska. *Fishery Bulletin, U.S. Fish and Wildlife Service*, **59**, 537–559.

Greene, H.W. (1986). Natural history and evolutionary biology. In *Predator-prey relationships* (ed. M.E. Feder and G.V. Lauder), pp. 99–108. University of Chicago Press, Chicago.

Greene, H.W. (1988). Antipredator mechanisms in reptiles. In *Biology of the reptiles*, Vol. XVI (ed. C. Gans and R.B. Huey), pp. 1–152. Liss, New York.

Greenwood, P.H. (1965). The cichlid fishes of Lake Nabugabo, Uganda. *Bulletin of the British Museum of Natural History (Zoology)*, **7**, 315–357.

Greenwood, P.H. (1974). The cichlid fishes of Lake Victoria, East Africa: the biology and evolution of a species flock. *Bulletin of the British Museum of Natural History (Zoology) Supplement*, **6**, 1–134.

Greenwood, P.H. (1991). Speciation. In *Cichlid fishes: behaviour, ecology and evolution* (ed. M.H.A. Keenleyside), pp. 88–128. Chapman and Hall, London.

Greenwood, P.H., Rosen, D.E., Weitzman, S.H., and Myers, G.S. (1966). Phyletic studies of teleostean fishes, with a provisional classification of living forms. *Bulletin of the American Museum of Natural History*, **131**, 341–455.

Grier, J.W. (1984). *Biology of animal behavior*. Times Mirror/Mosby, St. Louis, Missouri.

Griswold, B.L. and Smith, L.L., Jr. (1972). Early survival and growth of the ninespine stickleback, *Pungitius pungitius*. *Transactions of the American Fisheries Society*, **101**, 350–352.

Griswold, B.L. and Smith, L.L., Jr. (1973). The life history and trophic relationship of the ninespine stickleback, *Pungitius pungitius* in the Apostle Islands area of Lake Superior. *Fishery Bulletin, U.S. National Marine Fisheries Service*, **71**, 1039–1060.

Gross, H.P. (1977). Adaptive trends of environmentally sensitive traits in the three-spined stickleback, *Gasterosteus aculeatus* L. *Zeitschrift für zoologische Systematik und Evolutionsforschung*, **15**, 252–278.

Gross, H.P. (1978). Natural selection by predators on the defensive apparatus of the three-spined stickleback, *Gasterosteus aculeatus* L. *Canadian Journal of Zoology*, **56**, 398–413.

Gross, H.P. and Anderson, J.M. (1984). Geographic variation in the gill rakers and diet of European threespine sticklebacks, *Gasterosteus aculeatus*. *Copeia*, **1984**, 87–97.

Gross, H.P. and Franck, D. (1979). Sexual selection in sticklebacks: resource prediction [abstract]. *Second Biennial Conference on the Ethology and Behavioral Ecology of Fishes, Normal, Illinois*.

Gross, M.R. (1984). Sunfish, salmon, and the evolution of alternative reproductive strategies and tactics in fishes. In *Fish reproduction: strategies and tactics* (ed. G. Potts and R.J. Wootton), pp. 55–75. Academic Press, London.

Gross, M.R. and Charnov, E.L. (1980). Alternative male life histories in bluegill sunfish. *Proceedings of the National Academy of Sciences USA*, **77**, 6937–6940.

Gross, M.R., Coleman, R.C., and McDowall, R. (1988). Aquatic productivity and the evolution of diadromous fish migration. *Science*, **239**, 1291–1293.

Guderley, H. and Blier, P. (1988). Thermal acclimation in fish: conservative and labile properties of swimming muscle. *Canadian Journal of Zoology*, **66**, 1105–1115.

Guderley, H. and Foley, L. (1990). Anatomic and metabolic responses to thermal acclimation in the nine-spine stickleback, *Pungitius pungitius*. *Fish Physiology and Biochemistry*, **8**, 465–473.

Guiton, P. (1960). On the control of behaviour during the reproductive cycle of *Gasterosteus aculeatus*. *Behaviour*, **15**, 163–184.

Gutz, M. (1970). Experimentelle Untersuchungen zur Salzadaptation verschiedener Rassen des Dreistchligen Stichlings (*Gasterosteus aculeatus* L.). *Internationale Revue der Gesamten Hydrobiologie*, **55**, 845–894.

Hagen, D.W. (1967). Isolating mechanisms in threespine sticklebacks (*Gasterosteus*). *Journal of the Fisheries Research Board of Canada*, **24**, 1637–1692.

Hagen, D.W. (1973). Inheritance of numbers of lateral plates and gill rakers in *Gasterosteus aculeatus*. *Heredity*, **30**, 303–312.

Hagen, D.W. and Blouw, D.M. (1983). Heritability of dorsal spines in the fourspine stickleback. *Heredity*, **50**, 275–281.

Hagen, D.W. and Gilbertson, L.G. (1972). Geographic variation and environmental selection in *Gasterosteus aculeatus* L. in the Pacific northwest, America. *Evolution*, **26**, 32–51.

Hagen, D.W. and Gilbertson, L.G. (1973*a*). The genetics of plate morphs in freshwater threespine sticklebacks. *Heredity*, **31**, 75–84.

Hagen, D.W. and Gilbertson, L.G. (1973*b*). Selective predation and the intensity of selection acting upon the lateral plates of threespine sticklebacks. *Heredity*, **30**, 273–287.

Hagen, D.W. and McPhail, J.D. (1970). The species problem within *Gasterosteus aculeatus* on the Pacific coast of North America. *Journal of the Fisheries Research Board of Canada*, **27**, 147–155.

Hagen, D.W. and Moodie, G.E.E. (1979). Polymorphism for breeding colors in *Gasterosteus aculeatus*. I. Their genetics and geographic distribution. *Evolution*, **33**, 641–648.

Hagen, D.W. and Moodie, G.E.E. (1982). Polymorphism for plate morphs in *Gasterosteus aculeatus* on the east coast of Canada and an hypothesis for their global distribution. *Canadian Journal of Zoology*, **60**, 1032–1042.

Hagen, D.W., Moodie, G.E.E., and Moodie, P.F. (1980). Polymorphism for breeding colors in *Gasterosteus aculeatus*. II. Reproductive success as a result of convergence for threat display. *Evolution*, **34**, 1050–1059.

Haglund, T.R. (1981). Differential reproduction among the lateral plate phenotypes of *Gasterosteus aculeatus*, the threespine stickleback. Unpublished Ph.D. thesis. University of California, Los Angeles.

Haglund, T.R. and Buth, D.G. (1988). Allozymes of the unarmored threespine stickleback (*Gasterosteus aculeatus williamsoni*) and identification of the Shay Creek population. *Isozyme Bulletin*, **21**, 196.

Haglund, T.R., Buth, D.G., and Blouw, D.M. (1990). Allozyme variation and the recognition of the 'white stickleback'. *Biochemical Systematics and Ecology*, **18**, 559–563.

Haglund, T.R., Buth, D.G., and Lawson, R. (1992*a*). Allozyme variation and phylogenetic relationships of Asian, North American, and European populations of the threespine stickleback. *Copeia*, **1992**, 432–443.

Haglund, T.R., Buth, D.G., and Lawson, R. (1992*b*). Allozyme variation and phylogenetic relationships of Asian, North American, and European populations of the ninespine stickleback *Pungitius pungitius*. In *Systematics, historical ecology, and North American fishes* (ed. R.L. Mayden), pp. 438–452. Stanford University Press, Palo Alto, California.

Hairston, N.G., Li, K.T., and Easter, S.S., Jr. (1982). Fish vision and the detection of planktonic prey. *Science*, **218**, 1240–1242.

Hall, M.F. (1956). A comparative study of the reproductive behaviour of the sticklebacks (Gasterosteidae). Unpublished Ph.D. thesis. Oxford University.

Halliburton, R. and Gall, G.A.E. (1981). Disruptive selection and assortative mating in *Tribolium castaneum*. *Evolution*, **35**, 829–843.

Hamilton, W.D. (1979). Wingless and fighting males in fig wasps and other insects. In *Sexual selection and reproductive competition in insects* (ed. M.S. Blum and N.A. Blum), pp. 167–220. Academic Press, New York.

Hamilton, W.D. and Zuk, M. (1982). Heritable true fitness and bright birds: a role for parasites? *Science*, **218**, 384–387.

Hancock, A. (1852). Observations on the nidification of *Gasterosteus aculeatus and Gasterosteus spinachia*. *Annals and Magazine of Natural History*, **2** (10), 241–248.

Hangelin, C. and Vuorinen, I. (1988). Food selection in juvenile three-spined sticklebacks studied in relation to size, abundance and biomass of prey. *Hydrobiologia*, **157**, 169–177.

Hardy, J.D., Jr. (1978). *Development of fishes of the Mid-Atlantic Bight. An atlas of egg, larval and juvenile stages*. Vol. 2. *Anguillidae through Syngnathidae*. Biology Survey Program, Fish and Wildlife Service, US Department of the Interior, Washington, DC.

Harley, C.B. (1981). Learning the evolutionarily stable strategy. *Journal of Theoretical Biology*, **89**, 611–633.

Harold, F.M. (1986). *The vital force: a study of bioenergetics*. Freeman, New York.

Harrison, R.G. (1991). Molecular changes at speciation. *Annual Review of Ecology and Systematics*, **22**, 281–308.

Harrison, R.G., Rand, D.M., and Wheeler, W.C. (1987). Mitochondrial DNA variation in field crickets across a narrow hybrid zone. *Molecular Biology and Evolution*, **4**, 144–158.

Hart, M. 't (1978). A study of a short term behaviour cycle: creeping through in the three-spined stickleback (*Gasterosteus aculeatus* L.). *Behaviour*, **67**, 1–66.

Hart, P. J. B. (1986). Foraging in teleost fishes. In *The behaviour of teleost fishes* (1st edn) (ed. T. J. Pitcher), pp. 211–235. Croom Helm, London.

Hart, P. J. B. (1989). Predicting resource utilization: the utility of optimal foraging models. *Journal of Fish Biology*, **35** (Supp. A), 271–277.

Hart, P. J. B. and Ison, S. (1991). The influence of prey size and abundance, and individual phenotype on prey choice by the three-spined stickleback, *Gasterosteus aculeatus* L. *Journal of Fish Biology*, **38**, 359–372.

Hartigan, J. A. (1973). Minimum mutation fits to a given tree. *Biometrics*, **29**, 53–65.

Hartley, P. H. T. (1948). Food and feeding relationships in a community of fresh-water fishes. *Journal of Animal Ecology*, **17**, 1–14.

Harvey, P. H. and Mace, G. M. (1983). Comparisons between taxa and adaptive trends. In *Current problems in sociobiology* (ed. King's College Sociobiology Group), pp. 343–361. Cambridge University Press, Cambridge.

Harvey, P. H. and Pagel, M. D. (1991). *The comparative method in evolutionary biology*. Oxford University Press, Oxford.

Hasegawa, S., Hirano, T., Ogasawara, T., Iwata., M., Akiyama, T., and Arai, S. (1987). Osmoregulatory ability of chum salmon, *Oncorhynchus keta*, reared in fresh water for prolonged periods. *Fish Physiology and Biochemistry*, **4**, 101–110.

Hasler, A. D., Scholz, A. T., and Horrall, R. M. (1978). Olfactory imprinting and homing in salmon. *American Scientist*, **200**, 68–74.

Havens, A. C., Sweet, D. E., Baer, C. L., and Bradley, T. J. (1984). *Investigation of threespine stickleback abundance in landlocked Matanuska-Susitna Valley lakes*. Alaska Department of Fish and Game, Juneau.

Havre, N. van and FitzGerald, G. J. (1988). Shoaling and kin recognition in the threespine stickleback (*Gasterosteus aculeatus* L.). *Biology of Behaviour*, **13**, 190–201.

Hay, D. E. (1969). Mate selection in the threespine stickleback (*Gasterosteus*). Unpublished M.Sc. thesis. University of British Columbia, Vancouver.

Hay, D. E. and McPhail, J. D. (1975). Mate selection in three-spine sticklebacks (*Gasterosteus*). *Canadian Journal of Zoology*, **53**, 441–450.

Hay, O. P. (1907). A new fossil stickleback fish from Nevada. *Proceedings of the US National Museum*, **32**, 271–273.

Hedgepeth, J. W. (1957). Estuaries and lagoons II. Biological aspects. In *Treatise on marine ecology and paleoecology* (ed. J. W. Hedgepeth), Memoirs of the Geological Society of America, Vol. 67, No. 1, pp. 693-729. The Geological Society of America, New York.

Hedrick, A. and Riechert, S. E. (1990). Genetically-based variation between two spider populations in foraging behavior. *Oecologia*, **80**, 533–539.

Heinroth, O. (1911). Beiträge zur Biologie, namentlich Ethologie und Psychologie der Anatiden, pp. 589–702. *Verhanalung des V Internationalen Ornithologen Kongresses, Berlin*.

Heins, D. C. and Baker, J. A. (1987). Analysis of factors associated with intraspecific variation in propagule size of a stream-dwelling fish. In *Community and evolutionary ecology of North American stream fishes* (ed. W. J. Matthews and D. C. Heins), pp. 223–231. University of Oklahoma Press, Norman.

Heins, D. C. and Baker, J. A. (1988). Egg sizes in fishes: do mature oocytes accurately demonstrate size statistics of ripe ova? *Copeia*, **1988**, 238–240.

Heins, D. C. and Rabito, F. G., Jr. (1986). Spawning performance in North

American minnows: direct evidence of the occurrence of multiple clutches in the genus *Notropis*. *Journal of Fish Biology*, **28**, 343–357.

Heller, R. and Milinski, M. (1979). Optimal foraging of sticklebacks on swarming prey. *Animal Behaviour*, **27**, 1127–1141.

Henricson, J. and Nyman, L. (1976). The ecological and genetic segregation of two sympatric species of dwarf char (*Salvelinus alpinus* (L.) species complex). *Report of the Institute of Freshwater Research, Drottningolm*, **55**, 15–37.

Herter, K. (1930). Weitre Dressurversuche an Fischen. *Zeitschrift für Vergleichende Physiologie*, **11**, 730–748.

Herzog, H.A. and Schwartz, J.M. (1990). Geographical variation in anti-predator behaviour of neonate garter snakes, *Thamnophis sirtalis*. *Animal Behaviour*, **40**, 597–601.

Heuts, M.J. (1947*a*). Experimental studies on adaptive evolution in *Gasterosteus aculeatus* L. *Evolution*, **1**, 89–102.

Heuts, M.J. (1947*b*). The phenotypical variability of *Gasterosteus aculeatus* (L.) populations in Belgium: its bearing on the general geographical variability of the species. *Verhandelingen van de Koninklijke Vlaamse Acadamie Voor Wetenschappen Letteren en Schone Kunsten van Belgie Klasse der Wetenschappen*, **9**, 1–63.

Hirai, K., Tanaka, S., and Kato, F. (1973). *The ecology of the landlocked three-spined stickleback (trachurus form),* Gasterosteus aculeatus *L., in the Ono Basin, Japan*. Board of Education, Ono City, Fukui Prefectural, Japan.

Hirano, T. and Mayer-Gostan, N. (1978). Endocrine control of osmoregulation in fish. In *Eighth international symposium on comparative endocrinology* (ed. P.J. Gaillard and H.H. Boer), pp. 94–103. Springer-Verlag, New York.

Hislop, J.R.G. (1984). A comparison of the reproductive tactics and strategies of cod, haddock, whiting and Norway pout in the North Sea. In *Fish reproduction: strategies and tactics* (ed. G.W. Potts and R.J. Wootton), pp. 311–329. Academic Press, London.

Hoar, W.S. (1962*a*). Hormones and the reproductive behaviour of the male three-spined stickleback (*Gasterosteus aculeatus*). *Animal Behaviour*, **10**, 247–266.

Hoar, W.S. (1962*b*). Reproductive behavior of fish. *General and Comparative Endocrinology Supplement*, **1**, 206–216.

Hoffman, A. (1989). *Arguments on evolution*. Oxford University Press, Oxford.

Hollis, K.L. (1990). The role of Pavlovian conditioning in territorial aggression and reproduction. In *Contemporary issues in comparative psychology* (ed. D.A. Dewsbury), pp. 197–219. Sinauer, Sunderland, Massachusetts.

Honma, Y. and Tamura, E. (1984). Anatomical and behavioral differences among threespine sticklebacks: the marine form, the landlocked form and their hybrids. *Acta Zoologica (Stockholm)*, **65**, 79–87.

Honma, Y., Teshigawara, H., and Chiba, A. (1976). Change in the cells of the adenohypophysis associated with the diadramous migration of the threespine stickleback, *Gasterosteus aculeatus* L. *Archives Histology Japan*, **39**, 1–14.

Honma, Y., Chiba, A., and Tamura, E. (1986). Fine structure of the sterile testis of hybrid threespine stickleback between marine and landlocked forms. *Japanese Journal of Ichthyology*, **33**, 262–268.

Hoogland, R.D. (1951). On the fixing-mechanism in the spines of *Gasterosteus aculeatus* L. *Proceedings of the Koninklijke Nederlandse Acadamie van Wetenschappen*, **C54**, 171–180.

Hoogland, R. [D.], Morris, D. and Tinbergen, N. (1957). The spines of sticklebacks

(*Gasterosteus* and *Pygosteus*) as a means of defence against predators (*Perca* and *Esox*). *Behaviour*, **10**, 205–236.

Houde, A.E. (1988). Genetic difference in female choice between two guppy populations. *Animal Behaviour*, **36**, 510–516.

Houde, A.E. and Endler, J.A. (1990). Correlated evolution of female mating preferences and male color patterns in the guppy *Poecilia reticulata*. *Science*, **248**, 1405–1408.

Houston, A.I. and McNamara, J.M. (1988). The ideal free distribution when competitive abilities differ: an approach based on statistical mechanics. *Animal Behaviour*, **36**, 166–174.

Howes, D.E. (1983). Late Quaternary sediments and geomorphic history of northern Vancouver Island, B.C. *Canadian Journal of Earth Sciences*, **20**, 57–65.

Hoyle, J.A. and Keast, A. (1987). The effect of prey morphology and size on handling time in a piscivore, the largemouth bass (*Micropterus salmoides*). *Canadian Journal of Zoology*, **65**, 1972–1977.

Hubbs, C.L. (1929). The Atlantic American species of the fish genus *Gasterosteus*. *Occasional Papers of the Museum of Zoology, University of Michigan*, **200**, 1–9.

Hubbs, C.L. (1955). Hybridization between fish species in nature. *Systematic Zoology*, **4**, 1–20.

Hubbs, C.L. and Lagler, K.F. (1970). *Fishes of the Great Lakes Region*. University of Michigan Press, Ann Arbor.

Hudon, J. and Guderley, H. (1984). An electrophoretic study of the phylogenetic relationships among four species of sticklebacks (Pisces: Gasterosteidae). *Canadian Journal of Zoology*, **62**, 2313–2316.

Huntingford, F.A. (1976*a*). The relationship between anti-predator behaviour and aggression among conspecifics in the three-spined stickleback, *Gasterosteus aculeatus*. *Animal Behaviour*, **24**, 245–260.

Huntingford, F.A. (1976*b*). A comparison of the reactions of sticklebacks in different reproductive conditions towards conspecifics and predators. *Animal Behaviour*, **24**, 694–697.

Huntingford, F.A. (1976*c*). An investigation of the territorial behaviour of the three-spined stickleback (*Gasterosteus aculeatus*) using principal components analysis. *Animal Behaviour*, **24**, 822–834.

Huntingford, F.A. (1977). Inter- and intraspecific aggression in male sticklebacks. *Copeia*, **1977**, 158–159.

Huntingford, F.A. (1979). Pre-breeding aggression in male and female three-spined sticklebacks (*Gasterosteus aculeatus*). *Aggressive Behavior*, **5**, 51–58.

Huntingford, F.A. (1981). Further evidence for an association between lateral scute number and aggressiveness in the three-spined stickleback, *Gasterosteus aculeatus*. *Copeia*, **1981**, 717–720.

Huntingford, F.A. (1982). Do inter- and intraspecific aggression vary in relation to predation pressure in sticklebacks? *Animal Behaviour*, **30**, 909–916.

Huntingford, F.A. (1984). *The study of animal behaviour*. Chapman and Hall, London.

Huntingford, F.A. and Giles, N. (1987). Individual variation in anti-predator responses in the three-spined stickleback (*Gasterosteus aculeatus* L.). *Ethology*, **74**, 205–210.

Huntingford, F.A. and Wright, P.J. (1989). How sticklebacks learn to avoid dangerous feeding patches. *Behavioural Processes*, **19**, 181–189.

Huntingford, F.A. and Wright, P.J. (in press). Inherited population differences in avoidance conditioning in three-spined sticklebacks. *Behaviour.*

Hyatt, K.D. and Ringler, N.H. (1989*a*). Role of nest raiding and egg predation in regulating population density of threespine sticklebacks (*Gasterosteus aculeatus*) in a coastal British Columbia lake. *Canadian Journal of Fisheries and Aquatic Science*, **46**, 372–383.

Hyatt, K.D. and Ringler, N.H. (1989*b*). Egg cannibalism and the reproductive strategies of threespine sticklebacks (*Gasterosteus aculeatus*) in a coastal British Columbia lake. *Canadian Journal of Zoology*, **67**, 2036–2046.

Hynes, H.B.N. (1950). The food of freshwater sticklebacks (*Gasterosteus aculeatus* and *Pygosteus pungitius*) with a review of the methods used in studies of the food of fishes. *Journal of Animal Ecology*, **19**, 36–58.

Ibrahim, A.A. (1988). Diet choice, foraging behaviour and the effect of predators on feeding in the threespined stickleback (*Gasterosteus aculeatus*). Unpublished Ph.D. thesis. University of Glasgow.

Ibrahim, A.A. and Huntingford, F.A. (1988). Foraging efficiency in relation to within-species variation in morphology in three-spined sticklebacks, *Gasterosteus aculeatus. Journal of Fish Biology*, **33**, 823–824.

Ibrahim, A.A. and Huntingford, F.A. (1989*a*). Laboratory and field studies of the effect of predation risk on foraging in three-spined sticklebacks (*Gasterosteus aculeatus*). *Behaviour*, **109**, 46–57.

Ibrahim, A.A. and Huntingford, F.A. (1989*b*). Laboratory and field studies on diet choice in three-spined sticklebacks (*Gasterosteus aculeatus* L.) in relation to profitability and visual features of prey. *Journal of Fish Biology*, **34**, 245–257.

Ibrahim, A.A. and Huntingford, F.A. (1989*c*). The role of visual cues in prey selection in three-spined sticklebacks (*Gasterosteus aculeatus*). *Ethology*, **81**, 265–272.

Ida, H. (1976). Removal of the family Hypoptychidae from the suborder Ammodytoidei, order Perciformes, to the suborder Gasterosteoidei, order Syngnathiformes. *Japanese Journal of Ichthyology*, **23**, 33–42.

Idler, D.R. and Ng, T.B. (1983). Teleost gonadotropins: isolation, biochemistry, and function. In *Fish physiology*, Vol. IX: *Reproduction; part A: Endocrine tissues and hormones* (ed. W.S. Hoar, D.J. Randall and E.M. Donaldson), pp. 187–221. Academic Press, New York.

Iersel, J.J.A. van (1953). An analysis of the parental behaviour of the male three-spined stickleback (*Gasterosteus aculeatus* L.). *Behaviour Supplement*, **3**, 1–159.

Iersel, J.J.A. van (1958). Some aspects of territorial behaviour of the male three-spined stickleback. *Archives Néerlandaises de Zoologie*, **13**, 383–400.

Igarishi, K. (1964). Observations on the development of the scutes in the landlocked form of three-spined stickleback, *Gasterosteus aculeatus aculeatus* Linnaeus. *Bulletin of the Japanese Society of Scientific Fisheries*, **30**, 95–103.

Igarishi, K. (1970*a*). Formation of the scutes in the marine form of the three-spined stickleback, *Gasterosteus aculeatus aculeatus* (L.). *Annotationes Zoologicae Japonenses*, **43**, 34–42.

Igarishi, K. (1970*b*). On the variation of the scute in the three-spined stickleback, *Gasterosteus aculeatus aculeatus* (Linnaeus) from Nasu Area, Tochigi-Ken. *Annotationes Zoologicae Japonenses*, **43**, 43–49.

Ikeda, K. (1933). Effect of castration on the secondary sexual characters of anadromous three-spined sticklebacks, *Gasterosteus aculeatus* (L.). *Japanese Journal of Zoology*, **5**, 135–157.

Ikeda, K. (1934). On the variation and inheritance of lateral shields in the three-spined stickleback. *Japanese Journal of Genetics*, **9**, 104–106.

Immelmann, K. (1975). Ecological significance of imprinting and early learning. *Annual Review of Ecology and Systematics*, **6**, 15–37.

International Union of Biochemistry, Nomenclature Committee (1984). *Enzyme nomenclature, 1984*. Academic Press, Orlando, Florida.

Ivlev, V.W. (1961). *Experimental ecology of the feeding of fishes*. Yale University Press, New Haven, Connecticut.

Jakobsen, P.J. and Johnsen, G.H. (1987). The influence of predation on horizontal distribution of zooplankton species. *Freshwater Biology*, **17**, 501–507.

Jakobsen, P.J., Johnsen, G.H., and Larsson, P. (1988). Effects of predation risk and parasitism on the feeding ecology, habitat use, and abundance of lacustrine threespine stickleback (*Gasterosteus aculeatus*). *Canadian Journal of Fisheries and Aquatic Science*, **45**, 426–431.

Jameson, E.W., Jr. (1988). *Vertebrate reproduction*. Wiley, New York.

Jamieson, I. and Colgan, P.W. (1989). Eggs in the nest of males and their effect on mate choice in the three-spined stickleback. *Animal Behaviour*, **38**, 859–865.

Jamieson, I.G., Blouw, D.M., and Colgan, P.W. (1992*a*). Field observations on the reproductive biology of a newly discovered stickleback. *Canadian Journal of Zoology*, **70**, 1057–1063.

Jamieson, I., Blouw, D.M., and Colgan, P.W. (1992*b*). Parental care as a constraint on male mating success in fishes: a comparative study of threespine and white sticklebacks. *Canadian Journal of Zoology*, **70**, 956–962.

Jefferson, G.T. (1991). The Camp Cady local fuana: stratigraphy and paleontology of the Lake Manix basin. *San Bernardino County Museum Association Quarterly*, **38**, 93–99.

Jenkins, D., Walker, J.G.K., and McCowan, D. (1979). Analyses of otter (*Lutra lutra*) faeces from Deeside, N.E. Scotland. *Journal of Zoology*, **187**, 235–244.

Jenni, D.A. (1972). Effects of conspecifics and vegetation on nest site selection in *Gasterosteus aculeatus*. *Behaviour*, **42**, 97–118.

Jobling, M. (1981). The influences of feeding on the metabolic rate of fishes: a short review. *Journal of Fish Biology*, **18**, 385–400.

Jobling, M. (1983). Towards an explanation of specific dynamic action (SDA). *Journal of Fish Biology*, **23**, 549–555.

Jobling, M. (1985). Growth. In *Fish energetics: new perspectives* (ed. P. Tytler and P. Calow), pp. 213–230. Croom Helm, London.

Johnson, L. (1964). Marine-glacial relicts of the Canadian Arctic islands. *Systematic Zoology*, **13**, 76–91.

Johnston, I.A. and Dunn, J. (1987). Temperature acclimation and metabolism in ectotherms with particular reference to teleost fish. In *Temperature and animal cells*, Society for Experimental Biology Symposium, No. 41 (ed. K. Bowler and B.J. Bowler), pp. 67–93.

Jones, D.H. and John, A.W.G. (1978). The three-spined stickleback, *Gasterosteus aculeatus* L. from the north Atlantic. *Journal of Fish Biology*, **13**, 231–236.

Jones, J.S. (1988). Gaps in fossil teeth: saltations or sampling errors? *Trends in Ecology and Evolution*, **3**, 208–213.

Jones, J.W. and Hynes, H.B.N. (1950). The age and growth of *Gasterosteus aculeatus, Pygosteus pungitius* and *Spinachia vulgaris*, as shown by their otoliths. *Journal of Animal Ecology*, **19**, 59–73.

Jordan, C.M. and Garside, E.T. (1972). Upper lethal temperatures of three-spine stickleback *Gasterosteus aculeatus* (L.) in relation to thermal and osmotic acclimation, ambient salinity and size. *Canadian Journal of Zoology*, **50**, 1405–1411.

Jordan, D.S. (1907). The fossil fishes of California, with supplementary notes on other species of extinct fishes. *University of California Publications, Department of Geology Bulletin*, **5**, 95–144.

Jordan, D.S. (1908). Note on a fossil stickleback fish from Nevada. *Smithsonian Miscellaneous Collections*, **52**, 117–118.

Jürss, K., Bittorf, T., Vökler, T., and Wacke, R. (1982). Experimental studies on biochemical and physiological differences between the three morphs of the three-spined stickleback, *Gasterosteus aculeatus* L. I. Gill Na/K-ATPase, muscle alanine aminotransferase and muscle aspartate aminotransferase activites. *Zoologische Jahrbücher Abteilung für Allgemine Zoologie und Physiologie der Tiere*, **86**, 267–272.

Jürss, K., Bittorf, T., Vökler, T., and Wacke, R. (1983). Experimental studies on biochemical and physiological differences between the three morphs of the three-spined stickleback, *Gasterosteus aculeatus* L. II. Alanine aminotransferase, aspartate aminotransferase and glutamate dehydrogenase activities of the liver. *Zoologische Jahrbücher Abteilung für Allgemine Zoologie und Physiologie der Tiere*, **87**, 1–7.

Jürss, K., Bittorf, T., Vökler, T., and Wacke, R. (1985). Experimental studies on biochemical and physiological differences between the three morphs of the three-spined stickleback, *Gasterosteus aculeatus* L. III. Liver-somatic index and enzyme activities at different times of the year. *Zoologisch Jahrbücher Abteilung für Allgemine Zoologie und Physiologie der Tiere*, **89**, 441–451.

Kaneshiro, K.Y. (1976). Ethological isolation and phylogeny in the *Planitibia* subgroup of Hawaiian *Drosophila. Evolution*, **30**, 740–745.

Kaneshiro, K.Y. (1980). Sexual isolation, speciation and the direction of evolution. *Evolution*, **34**, 437–444.

Kean-Howie, J.C., Pearre, S. and Dickie, L.M. (1988). Experimental predation by sticklebacks on larval mackerel and protection of fish larvae by zooplankton alternative prey. *Journal of Experimental Marine Biology and Ecology*, **124**, 239–259.

Kedney, G.I., Boulé, V., and FitzGerald, G.J. (1987). The reproductive ecology of threespine sticklebacks breeding in fresh and brackish water. In *Common strategies of anadromous and catadromous fishes*, American Fisheries Society Symposium No. 1 (ed. M.J. Dadswell, R.J. Klauda, C.M. Moffitt, R.L. Saunders, R.A. Rulifson and J.E. Cooper), pp. 151–161. American Fisheries Society, Bethesda, Maryland.

Keenleyside, M.H.A. (1955). Some aspects of schooling behaviour of fish. *Behaviour*, **8**, 183–248.

Keenleyside, M.H.A. (1979). *Diversity and adaptation in fish behaviour*. Springer-Verlag, Berlin.

Kelly, J. and Gingerich, P.D. (1991). Effects of time accumulation on metric variability in fossil samples [abstract]. *Journal of Vertebrate Paleontology Supplement*, **11**, 39A–40A.

Kelts, K. and Hsü, K.J. (1978). Freshwater carbonate sedimentation. In *Lakes: chemistry, geology, and physics* (ed. A. Lerman), pp. 295–323. Springer-Verlag, New York.

Kerfoot, W.C. (1975). The divergence of adjacent populations. *Ecology*, **56**, 1298–1313.

Ketele, A.G. and Verheyen, R.F. (1985). Competition for space between the three-spined stickleback, *Gasterosteus aculeatus* L. f. leiura, and the nine-spined stickleback, *Pungitius pungitius* (L.). *Behaviour*, **93**, 127–138.

Kidwell, S.M. and Behrensmeyer, A.K. (1988). Overview: ecological and evolutionary implications of taphonomic processes. *Palaeogeography, Palaeoclimatology, Palaeoecology*, **63**, 1–13.

Kilarski, W.L. and Kozlowska, M. (1983). Ultrastructural characteristics of the teleostean muscle fibers and their nerve endings: the stickleback (*Gasterosteus aculeatus* L.). *Zeitschrift für Mikroskopisch-anatomische Forschung Leipzig*, **97**, 1022–1036.

King, P.A. and Goldstein, L. (1983). Organic osmolytes and cell volume regulation in fish. *Molecular Physiology*, **4**, 53–66.

Kingsolver, J.G. and Schemske, D.W. (1991). Path analysis of selection. *Trends in Ecology and Evolution*, **9**, 276–280.

Kirkpatrick, M. (1982). Quantum evolution and punctuated equilibria in continuous genetic characters. *American Naturalist*, **119**, 833–848.

Kitchell, J.A. (1985). Evolutionary paleoecology: recent contributions to evolutionary theory. *Paleobiology*, **11**, 91–104.

Kleerekoper, H. (1969). *Olfaction in fishes*. Indiana University Press, Bloomington.

Klinger, S.A., Magnuson, J.J., and Gallepp, G.W. (1982). Survival mechanisms of the central mudminnow (*Umbra limi*), fathead minnow (*Pimephales promelas*) and brook stickleback (*Culaea inconstans*) for low oxygen in winter. *Environmental Biology of Fishes*, **7**, 113–120.

Kobayasi, H. (1962). Morphological and genetical observations in hybrids of some teleost fishes. *Journal of Hokkaido Gakugei University*, Section 2B, **13** (Supplement) 1–112.

Koch, H.J. and Heuts, M.J. (1942). Influence de l'hormone thyrodienne sur la régulation osmotique chez *Gasterosteus aculeatus* L. forme *gymnurus*. *Annales de la Societé Royale Zoologique de Belgique*, **73**, 165–172.

Koch, P.E. (1986). Clinal geographic variation in mammals: implications for the study of chronoclines. *Paleobiology*, **12**, 269–281.

Kocher, T.D., Thomas, W.K., Meyer, A., Edwards, S.V., Paabo, S., Villablanca, F.X., and Wilson, A.C. (1989). Dynamics of mitochondrial DNA evolution in animals: amplification and sequencing with conserved primers. *Proceedings of the National Academy of Sciences*, **86**, 6196–6200.

Kodric-Brown, A. (1986). Satellites and sneakers: opportunistic male breeding tactics in pupfish (*Cyprinodon pecosensis*). *Behavioral Ecology and Sociobiology*, **19**, 425–432.

Kodric-Brown, A. and Brown, J.H. (1984). Truth in advertising: the kinds of traits favored by sexual selection. *American Naturalist*, **124**, 305–322.

Kornfield, I.L., Beland, K.F., Moring, J.R., and Kircheis, F.W. (1981). Genetic similiarity among endemic Artic char (*Salvelinus alpinus*) and implications for their management. *Canadian Journal of Fisheries and Aquatic Science*, **38**, 32–39.

Kraus, M.J. and Bown, T.M. (1986). Paleosols and time resolution in alluvial stratigraphy. In *Paleosols: their origin, classification, and interpretation* (ed. V.P. Wright), pp. 180–207. Blackwell, London.

Krebs, J.R. (1978). Optimal foraging: decision rules for predators. In *Behavioural ecology, an evolutionary approach* (1st edn) (ed. J.R. Krebs and N.B. Davies), pp. 23–63. Sinauer, Sunderland, Massachusetts.

Krebs, J.R. and Cowie, R.J. (1976). Foraging strategies in birds. *Ardea*, **63**, 98–115.

Krebs, J.R. and Davies, N.B. (1987). *An introduction to behavioural ecology* (2nd edn). Blackwell, Oxford.

Krebs, J.R., Kacelnik, A., and Taylor, P. (1978). Test of optimal sampling by foraging great tits. *Nature*, **275**, 27–31.

Krebs, W.N. and Bradbury, J.P. (1984). Field trip guide to non-marine diatomites near Reno, Nevada. In *Geological uses of diatoms, Short course*, pp. 1–4. National Geological Society of America meeting, Reno, Nevada, November 9, 1984.

Kronnie, G. te, Tatarczuch, L., Raamsdonk, W. van, and Kilarski, W. (1983). Muscle fibre types in the myotome of stickleback, *Gasterosteus aculeatus* L.: a histochemical, immunohistochemical and ultrastructural study. *Journal of Fish Biology*, **22**, 303–316.

Krupp, F. and Coad, B.W. (1985). Notes on a population of threespine stickleback, *Gasterosteus aculeatus*, from Syria (Pisces: Osteichthyes: Gasterosteidae). *Senckenbergiana Biologica*, **66**, 35–39.

Kuntz, A. and Radcliffe, L. (1917). Notes on the embryology and larval development of twelve teleostean fishes. *Fishery Bulletin, U.S. Fish and Wildlife Service*, **35**, 87–134.

Kynard, B.E. (1972). Male breeding behavior and lateral plate phenotypes in the threespine stickleback (*Gasterosteus aculeatus* L.). Unpublished Ph.D. thesis. University of Washington, Seattle.

Kynard, B.E. (1978*a*). Breeding behavior of a lacustrine population of threespine sticklebacks (*Gasterosteus aculeatus* L.). *Behaviour*, **67**, 178–207.

Kynard, B.E. (1978*b*). Nest desertion of male *Gasterosteus aculeatus. Copeia*, **1978**, 702–703.

Kynard, B.E. (1979*a*). Nest habitat preference of low plate number morphs in threespine sticklebacks (*Gasterosteus aculeatus*). *Copeia*, **1979**, 525–528.

Kynard, B.E. (1979*b*). Population decline and change in frequencies of lateral plates in threespine sticklebacks (*Gasterosteus aculeatus*). *Copeia*, **1979**, 635–638.

LaBarbera, M. (1989). Analyzing body size as a factor in ecology and evolution. *Annual Review of Ecology and Systematics*, **20**, 97–117.

Lachance, S. (1990). L'investissement parental chez l'épinoch (Gasterosteidae) dans un milieu imprévisible. Unpublished M.Sc. thesis. Laval University, Québec.

Lachance, S. and FitzGerald, G.J. (1992). Parental care tactics of three-spined sticklebacks living in a harsh environment. *Behavioral Ecology*, **3**, 360–366.

Lachance, S., Magnan, P., and FitzGerald, G.J. (1987). Temperature preferences of three sympatric sticklebacks (*Gasterosteidae*). *Canadian Journal of Zoology*, **67**, 1573–1576.

Lagomarsino, I.V., Francis, R.C., and Barlow, G.W. (1988). The lack of correlation between size of egg and size of hatchling in the Midas cichlid, *Cichlasoma citrinellum. Copeia*, **1988**, 1086–1089.

Lam, T.J. and Hoar, W.S. (1967). Seasonal effects of prolactin on freshwater osmoregulation of the marine form (*trachurus*) of the stickleback *Gasterosteus aculeatus. Canadian Journal of Zoology*, **45**, 509–516.

Lam, T.J. and Leatherland, J.F. (1969). Effect of prolactin on freshwater survival of the marine form (trachurus) of the threespine stickleback, *Gasterosteus aculeatus* in early winter. *General and Comparative Endocrinology*, **12**, 385–387.

Lam, T.J. and Leatherland, J.F. (1970). Effect of hormones on survival of the marine form (trachurus) of the threespine stickleback (*Gasterosteus aculeatus* L.) in deionized water. *Comparative Biochemistry and Physiology*, **33**, 295–303.

Lam, T.J., Nagahama, Y., Chan, K., and Hoar, W.S. (1978). Overripe eggs and postovulatory corpora lutea in the threespine stickleback, *Gasterosteus aculeatus* L., form trachurus. *Canadian Journal of Zoology*, **56**, 2029–2036.

Lam, T.J., Chan, K., and Hoar, W.S. (1979). Effect of progesterone and estadiol-

17β on ovarian fluid secretion in the threespine stickleback, *Gasterosteus aculeatus* L., form trachurus. *Canadian Journal of Zoology*, **57**, 468–471.

Lande, R. (1982). Rapid origin of sexual isolation and character divergence in a cline. *Evolution*, **36**, 213–223.

Lande, R. (1986). The dynamics of peak shifts and the pattern of morphological evolution. *Paleobiology*, **12**, 343–354.

Lande, R. and Arnold, S.J. (1983). The measurement of selection on correlated characters. *Evolution*, **37**, 1210–1226.

Lane, M. (1981). *The fish: the story of the stickleback*. Dial Press, New York.

La Rivers, I. (1953). A lower Pliocene frog from western Nevada. *Journal of Paleontology*, **27**, 77–81.

Larson, G.L. (1976). Social behavior and feeding ability of two phenotypes of *Gasterosteus aculeatus* in relation to their spatial and trophic segregation in a temperate lake. *Canadian Journal of Zoology*, **54**, 107–121.

Lauder, G.V. (1986). Homology, analogy, and the evolution of behavior. In *Evolution of animal behavior: paleontological and field approaches* (ed. M.H. Nitecki and J.A. Kitchell), pp. 9–40. Oxford University Press, New York.

Lauder, G.V. and Liem, K.F. (1982). Symposium summary: evolutionary patterns in actinopterygian fishes. *American Zoologist*, **22**, 343–345.

Lauder, G.V. and Liem, K.F. (1983). The evolution and interrelationships of the acanthopterygian fishes. *Bulletin of the Museum of Comparative Zoology*, **150**, 95–197.

Lavin, P.A. and McPhail, J.D. (1985). The evolution of freshwater diversity in threespine stickleback (*Gasterosteus aculeatus*): site-specific differentiation of trophic morphology. *Canadian Journal of Zoology*, **63**, 2632–2638.

Lavin, P.A. and McPhail, J.D. (1986). Adaptive divergence of trophic phenotype among freshwater populations of threespine stickleback (*Gasterosteus aculeatus*). *Canadian Journal of Fisheries and Aquatic Sciences*, **43**, 2455–2463.

Lavin, P.A. and McPhail, J.D. (1987). Morphological divergence and the organization of trophic characters among lacustrine populations of the threespine stickleback (*Gasterosteus aculeatus*). *Canadian Journal of Fisheries and Aquatic Sciences*, **44**, 1820–1829.

Lavin, P.A. and McPhail, J.D. (in press). Parapatric lake and stream dwelling threespine sticklebacks on northern Vancouver Island: disjunct distribution or parallel evolution? *Canadian Journal of Zoology*.

Leatherland, J.F. (1970*a*). Seasonal variation in the structure and ultrastructure of the pituitary of the marine form (*trachurus*) of the threespine stickleback, *Gasterosteus aculeatus* L. I. Rostral pars distalis. *Zeitschrift für Zellforschung*, **104**, 301–317.

Leatherland, J.F. (1970*b*). Seasonal variation in the structure and ultrastructure of the pituitary in the marine form (*trachurus*) of the threespine stickleback, *Gasterosteus aculeatus* L. II. Proximal pars distalis and neuro-intermediate lobe. *Zeitschrift für Zellforschung*, **104**, 318–336.

Leatherland, J.F. and Lam, T.J. (1969). Prolactin and survival in deionized water of the marine form (*trachurus*) of the threespine stickleback, *Gasterosteus aculeatus* L. *Canadian Journal of Zoology*, **47**, 989–995.

Lee, D.S., Gilbert, C.R., Hocutt, C.H., Jenkins, R.E., McAllister, D.E., and Stauffer, J.R., Jr. (1980). *Atlas of North American freshwater fishes*. North Carolina Biological Survey, Publication No. 1980–12. North Carolina State Museum of Natural History, Raleigh.

Leeuw, R. de, Wurht, Y.A., Zandbergen, M.A., Peute, J., and Goos, H.J.T. (1986). The effects of aromatizable androgens, nonaromatizable androgens, and estrogens on gonadotropin release in castrated African catfish, *Clarias gariepinus* (Burchell). *Cell and Tissue Research*, **243**, 587–594.

Legendre, P., Dallot, S., and Legendre, L. (1985). Succesion of species within a community: chronological clustering, with applications to marine and freshwater zooplankton. *American Naturalist*, **125**, 257–288.

Leggett, W.C. and Carscadden, J.E. (1978). Latitudinal variation in reproductive characteristics of American Shad (*Alosa sapidissima*): evidence for population specific life history strategies. *Journal of the Fisheries Research Board of Canada*, **35**, 1469–1478.

Leiner, M. (1929). Ökologische Studien an *Gasterosteus aculeatus. Zeitschrift für Morphologie und Ökologie der Tiere*, **14**, 360–399.

Leiner, M. (1930). Fortsetzung der Ökologischen Studien an *Gasterosteus aculeatus. Zeitschrift für Morphologie und Ökologie der Tiere*, **16**, 499–540.

Leiner, M. (1931). Ökologisches von *Gasterosteus aculeatus* L. *Zoologischer Anzeiger*, **93**, 317–333.

Leiner, M. (1934). Beiträge zur ontogenetischen Entwicklung der drei europaischen Stichlingsarten und ihrer Kreuzungsprodukte. *Zeitschrift für Wissenschaft Zoologie*, **145**, 366–388.

Leiner, M. (1940). Kurze mitteilung uber den brutpflegenstinkt von stichlingsbastarden. *Zeitschrift für Tierpsychologie*, **4**, 167–169.

Leiner, M. (1960). The propagation of the sticklebacks and similar fishes. *Tropical Fish Hobbyist*, **9**, (4), 39–49.

Lelek, A. (1987). *The freshwater fishes of Europe*, Vol. 9, *Threatened fishes of Europe*. AULA-Verlag, Wiesbaden.

Lemmetyinen, R. (1973). Feeding ecology of *Sterna paradisaea* Pontopp. and *S. hirundo* L. in the archipelago of southwestern Finland. *Annales Zoologici Fennici*, **10**, 507–525.

Lester, R.J.G. (1971). The influence of *Schistocephalus plerocercoids on the respiration of Gasterosteus* and a possible resulting effect on the behaviour of the fish. *Canadian Journal of Zoology*, **49**, 361–366.

Levinton, J.S. (1988). *Genetics, paleontology, and macroevolution*. Cambridge University Press, Cambridge.

Li, S.K. and Owings, D.H. (1978*a*). Sexual selection in the three-spined stickleback. I. Normative observations. *Zeitschrift für Tierpsychologie*, **46**, 359–371.

Li, S.K. and Owings, D.H. (1978*b*). Sexual selection in the three-spined stickleback. II. Nest raiding during the courtship phase. *Behaviour*, **64**, 298–304.

Liley, N.R. (1966). Ethological isolating mechanisms in four sympatric species of poeciliid fishes. *Behaviour Supplement*, **13**, 1–197.

Liley, N.R. (1969). Hormones and reproductive behavior in fishes. In *Fish physiology*, Vol. III (ed. W.S. Hoar and D.J. Randall), pp. 73–116. Academic Press, New York.

Lima, S.L. and Dill, L.M. (1990). Behavioural decisions made under the risk of predation: a review and a prospectus. *Canadian Journal of Zoology*, **68**, 619–640.

Limbaugh, C. (1962). Life history and ecological notes on the tubenose, *Aulorynchus flavidus*, a hemibranch fish of western North America. *Copeia*, **1962**, 549–555.

Lindsey, C.C. (1962). Experimental study of meristic variation in a population of threespine stickleback, *Gasterosteus aculeatus. Canadian Journal of Zoology*, **40**, 271–312.

Lindsey, C.C. (1963). Sympatric occurrence of two species of humpback whitefish in Squanga Lake, Yukon Territory. *Journal of the Fisheries Research Board of Canada*, **20**, 749–767.

Lindsey, C.C. (1978). Form, function and locomotory habits in fish. In *Fish physiology*, Vol. 7 (ed. W.S. Hoar and D.J. Randall), pp. 1–100. Academic Press, London.

Lindsey, C.C. and McPhail, J.D. (1986). Zoogeography of fishes of the Yukon and Mackenzie basins. In *The zoogeography of North American freshwater fishes* (ed. C.H. Hocutt and E.O. Wiley), pp. 639–674. Wiley, New York.

Linnaeus, C. (1758). *Systema naturae per regna tria naturae, secundum classes, ordines, genera, species com characteribus, differentiis, synonymis, locis.* Editio decima, reformata, Tom. I. Laurentii Salvii, Holmiae, Stockholm.

Littlejohn, M.J. (1981). Reproductive isolation: a critical review. In *Evolution and speciation: essays in honour of M.J.D. White* (ed. W.R. Atchley and D.S. Woodruff), pp. 298–334. Cambridge University Press, Cambridge.

Littlejohn, M.J. and Watson, G.F. (1985). Hybrid zones and homogamy in Australian frogs. *Annual Review of Ecology and Systematics*, **16**, 85–112.

Liu, H.-T. and Wang, N.-C. (1974). A new *Pungitius* from the Nihowan Formation of North China. *Vertebrata Palasiatica*, **12**, 89–98.

Lodge, D.M. (1985). Macrophyte–gastropod associations: observations and experiments on macrophyte choice by gastropods. *Freshwater Biology*, **15**, 695–708.

Loiselle, P.V. and Barlow, G.W. (1978). Do fishes lek like birds? In *Contrasts in behaviour: adaptations in aquatic and terrestrial environments* (ed. E.S. Reese and F.J. Lighter), pp. 31–76. Wiley, New York.

Lonneberg, E. (1938). Kleine Notizen über Carotenoide bei verschiedenen Tierarten. *Arkiv foer Zoologi*, **30**, 1–10.

Lorenz, K. (1950). The comparative method in the study of innate behavior patterns. *Symposia of the Society for Experimental Biology*, **4**, 221–286.

Losey, G.S., Jr. and Sevenster, P. (1991). Can threespine sticklebacks learn when to display? I. Punished displays. *Ethology*, **87**, 45–58.

Loughry, W.J. (1988). Population differences in how black-tailed prairie dogs deal with snakes. *Behavioral Ecology and Sociobiology*, **22**, 61–67.

Loughry, W.J. (1989). Discrimination of snakes by two populations of black-tailed prairie dogs. *Journal of Mammalogy*, **70**, 627–630.

Lowe, G.D. (1978). The measurement by direct calorimetry of the energy lost as heat by a polychaete *Neanthes* (=*Nereis*) *virens*. Unpublished Ph.D. thesis. University of London.

Lynch, M. (1990). The rate of morphological evolution in mammals from the standpoint of the neutral expectation. *American Naturalist*, **136**, 727–741.

Lythgoe, J.N. (1979). *The ecology of vision*. Clarendon, Oxford.

McAllister, D.E. (1960). The twospine stickleback, *Gasterosteus wheatlandi*, new to Canadian freshwater fish fauna. *Canadian Field-Naturalist*, **74**, 177–178.

McAllister, D.E., Cumbaa, S.L., and Harington, C.R. (1981). Pleistocene fishes (*Coregonus*, *Osmerus*, *Microgadus*, *Gasterosteus*) from Green Creek, Ontario, Canada. *Canadian Journal of Earth Sciences*, **18**, 1356–1364.

McAllister, D.E., Parker, B.J., and McKee, P.M. (1985). Rare, endangered and extinct fishes in Canada. *Syllogeus*, **54**, 1–192.

McAllister, D.E., Harington, C.R., Cumbaa, S.L., and Renaud, C.B. (1989). Paleoenvironmental and biogeographic analyses of fossil fishes in peri-Champlain Sea deposits in eastern Canada. In *The late Quaternary development of the*

Champlain Sea basin (ed. N.R. Gadd), pp. 241-258. Geological Association of Canada, Special Paper 35.

Macan, T.T. and Kitching, A. (1972). Some experiments with artificial substrata. *Verhandlungen der Internationalen Vereiningung für theoretische und angewandte Limnologie*, **18**, 213-20.

McCart, P.J. (1970). Evidence for the existence of sibling species of pygmy whitefish (*Prosopium coulteri*) in three Alaskan lakes. In *Biology of coregonid fishes* (ed. C.C. Lindsey and C.S. Woods), pp. 81-98. University of Manitoba Press, Winnipeg.

McClearn, G.E. and DeFries, J.C. (1973). *Introduction to behavioral genetics*. Freeman, San Francisco.

McCormick, S.D. and Bern, H.A. (1989). In vitro stimulation of Na^+/K^+ ATPase activity and ouabain binding by cortisol in coho salmon gill. *American Journal of Physiology*, **256**, R707-R715.

McCune, A.R. (1987*a*). Lakes as laboratories of evolution: endemic fishes and environmental cyclicity. *Palaios*, **2**, 446-454.

McCune, A.R. (1987*b*). Toward a phylogeny of a fossil species flock: semionotid fishes from a lake deposit in the early Jurassic Towaco formation, Newark Basin. *Bulletin of the Peabody Museum of Natural History, Yale University*, **43**, 1-108.

McCune, A.R. (1990). Environmental novelty and atavism in the *Semionotus* complex: relaxed selection during colonization of an expanding lake. *Evolution*, **44**, 71-85.

McCune, A.R., Thomson, K.S., and Olsen, P.E. (1984). Semionotid fishes from the Mesozoic great lakes of North America. In *Evolution of fish species flocks* (ed. A.A. Echelle and I. Kornfield), pp. 27-44. University of Maine, Orono.

McDonald, A.L., Heimstra, N.W., and Damkot, D.K. (1968). Social modification of agonistic behaviour in fish. *Animal Behaviour*, **16**, 437-441.

McDowall, R.M. (1988). *Diadromy in fishes: migrations between freshwater and marine environments*. Croom Helm, London.

McFarland, D.J. (1974). Time-sharing as a behavioural phenomenon. In *Advances in the study of behaviour*, Vol. 5 (ed. D.S. Lehrman, J.S. Rosenblatt, R.A. Hinde and E. Shaw), pp. 201-225. Academic Press, New York.

McFarland, D.J. (1985). *Animal behavior: psychobiology, ethology, and evolution*. Benjamin/Cummings, Menlo Park, California.

McGibbon, S. (1977). Investigation into the effects of salinity on the oxygen consumption and monovalent cation concentration of *Gasterosteus aculeatus* L. Unpublished B.Sc. thesis. University of Wales, Aberystwyth.

McInerney, J.E. (1969). Reproductive behavior of the blackspotted stickleback *Gasterosteus wheatlandi*. *Journal of the Fisheries Research Board of Canada*, **26**, 2061-2075.

McKaye, K.R. (1984). Behavioural aspects of cichlid reproductive strategies: patterns of territoriality and brood defence in Central American substratum spawners and African mouth brooders. In *Fish reproduction: strategies and tactics* (ed. G.W. Potts and R.J. Wootton), pp. 245-273. Academic Press, London.

McKenzie, J.A. (1974). The parental behavior of the male brook stickleback *Culaea inconstans* (Kirtland). *Canadian Journal of Zoology*, **52**, 649-652.

McKenzie, J.A. and Keenleyside, M.H.A. (1970). Reproductive behavior of ninespine sticklebacks (*Pungitius pungitius* (L.)) in South Bay, Manitoulin Island, Ontario. *Canadian Journal of Zoology*, **48**, 55-61.

McLean, E.B. and Godin, J.-G.J. (1989). Distance to cover and fleeing from predators in fish with different amounts of defensive armour. *Oikos*, **55**, 281-290.

MacLean, J. (1980). Ecological genetics of threespine sticklebacks in Heisholt Lake. *Canadian Journal of Zoology*, **58**, 2026–2039.

McLennan, D.A. (1989). Phylogenetic analysis of behavioural evolution: a case study using Gasterosteid fishes. Unpublished M.Sc. thesis. University of British Columbia, Vancouver.

McLennan, D.A. (1991). Integrating phylogeny and experimental ethology: from pattern to process. *Evolution*, **45**, 1773–1789.

McLennan, D.A. and McPhail, J.D. (1989*a*). Experimental investigations of the evolutionary significance of sexually dimorphic nuptial colouration in *Gasterosteus aculeatus* (L.): temporal changes in the structure of the male mosaic signal. *Canadian Journal of Zoology*, **67**, 1767–1777.

McLennan, D.A. and McPhail, J.D. (1989*b*). Experimental investigations of the evolutionary significance of sexually dimorphic nuptial colouration in *Gasterosteus aculeatus* (L.): the relationship between male colour and male behaviour. *Canadian Journal of Zoology*, **67**, 1778–1782.

McLennan, D.A. and McPhail, J.D. (1990). Experimental investigations of the evolutionary significance of sexually dimorphic nuptial colouration in *Gasterosteus aculeatus* (L.): the relationship between male colour and female behaviour. *Canadian Journal of Zoology*, **68**, 482–492.

McLennan, D.A., Brooks, D.R., and McPhail, J.D. (1988). The benefits of communication between comparative ethology and phylogenetic systematics: a case study using gasterosteid fishes. *Canadian Journal of Zoology*, **66**, 2177–2190.

MacLeod, N. (1991). Punctuated anagenesis and the importance of stratigraphy to paleobiology. *Paleobiology*, **17**, 167–188.

McPhail, J.D. (1963). Geographic variation in North American ninespine sticklebacks, *Pungitius pungitius. Journal of the Fisheries Research Board of Canada*, **20**, 27–44.

McPhail, J.D. (1969). Predation and the evolution of a stickleback (*Gasterosteus*). *Journal of the Fisheries Research Board of Canada*, **26**, 3183–3208.

McPhail, J.D. (1977). Inherited interpopulation differences in size at first reproduction in threespine stickleback, *Gasterosteus aculeatus* L. *Heredity*, **38**, 53–60.

McPhail, J.D. (1984). Ecology and evolution of sympatric sticklebacks (*Gasterosteus*): morphological and genetic evidence for a species pair in Enos Lake, British Columbia. *Canadian Journal of Zoology*, **62**, 1402–1408.

McPhail, J.D. (1992). Ecology and evolution of sympatric sticklebacks (*Gasterosteus*): evidence for a species pair in Paxton Lake, Texada Island, British Columbia. *Canadian Journal of Zoology*, **70**, 361–369.

McPhail, J.D. (1993). Ecology and evolution of sympatric sticklebacks (*Gasterosteus*): origin of sympatric pairs. *Canadian Journal of Zoology*, in press.

McPhail, J.D. and Hay, D.E. (1983). Differences in male courtship in freshwater and marine sticklebacks (*Gasterosteus aculeatus*). *Canadian Journal of Zoology*, **61**, 292–297.

McPhail, J.D. and Lindsey, C.C. (1970). Freshwater fishes of northwestern Canada and Alaska. *Bulletin of the Fisheries Research Board of Canada*, **173**, 1–381.

McPhail, J.D. and Lindsey, C.C. (1986). Zoogeography of the freshwater fishes of Cascadia. In *The zoogeography of North American freshwater fishes* (ed. C. Hocutt and E.O. Wiley), pp. 615–637. Wiley, New York.

McPhail, J.D. and Peacock, S.D. (1983). Some effects of the cestode (*Schistocephalus solidus*) on reproduction in the threespine stickleback (*Gasterosteus aculeatus*): evolutionary aspects of a host-parasite interaction. *Canadian Journal of Zoology*, **61**, 901–908.

McQuinn, I., FitzGerald, G. J., and Powles, H. (1983). Environmental effects on embryos and larval survival of the Isle Verte stock of Atlantic herring. *Le Naturaliste Canadien*, **110**, 343–353.

Maddison, W. P. (1990). A method for testing the correlated evolution of two binary characters: are gains or losses concentrated on certain branches of a phylogenetic tree? *Evolution*, **44**, 539–557.

Maddison, W. P. and Maddison, D. R. (1987). *MacClade*, version 2.1. An interactive, graphic program for analyzing phylogenies and studying character evolution. Published privately at the Museum of Comparative Zoology, Harvard University, Cambridge, Massachusetts.

Maddison, W. P., Donoghue, M. J., and Maddison, D. R. (1984). Outgroup analysis and parsimony. *Systematic Zoology*, **33**, 83–103.

Maderson, P. F. A., Alberch, P., Goodwin, B. C., Gould, S. J., Hoffman, A., Murray, J. D., Raup, D. M., de Ricqlés, A., Seilacher, A., Wagner, G. P., and Wake, D. B. (1982). The role of development in macroevolutionary change, group report. In *Evolution and development* (ed. J. T. Bonner), pp. 279–312. Springer-Verlag, Berlin.

Madsen, F. J. (1957). The food of diving ducks in Danish fjords. *Danish Review of Game Biology*, **3**, 19–83.

Magnuson, J. J., Paszkowski, C. A., Rahel, F. J., and Tonn, W. M. (1988). Fish ecology in severe environments of small isolated lakes in northern Wisconsin. In *Freshwater wetlands and wildlife*, DOE Symposium Series No. 61 (ed. R. R. Sharitz and J. W. Gibbons). USDOE Office of Scientific and Technical Information, Oak Ridge, Tennessee.

Magurran, A. E. (1986). Predator inspection behaviour in minnow shoals: differences between populations and individuals. *Behavioral Ecology and Sociobiology*, **19**, 267–273.

Magurran, A. E. (1990). The inheritance and development of minnow anti-predator behaviour. *Animal Behaviour*, **39**, 828–834.

Magurran, A. E. and Pitcher, T. J. (1987). Provenance, shoal size and the sociobiology of predator evasion behaviour in minnow shoals. *Proceeding of the Royal Society*, **B229**, 439–465.

Magurran, A. E. and Seghers, B. H. (1990). Population differences in predator recognition and attack cone avoidance in the guppy *Poecilia reticulata*. *Animal Behaviour*, **40**, 443–452.

Maitland, P. S. (1965). The feeding relationships of salmon, trout, minnows, stone loach and three-spined stickleback in the River Endrick, Scotland. *Journal of Animal Ecology*, **34**, 109–133.

Mangel, M. and Clark, C. W. (1988). *Dynamic modeling in behavioral ecology*. Princeton University Press, Princeton, New Jersey.

Mann, R. H. K. (1971). The populations, growth, and production of fish in four small streams in southern England. *Journal of Animal Ecology*, **40**, 155–190.

Mann, R. H. K. and Mills, C. A. (1979). Demographic aspects of fish fecundity. *Symposia of the Zoological Society of London*, **44**, 161–177.

Mann, R. H. K., Mills, C. A., and Crisp, D. T. (1984). Geographical variation in the life-history tactics of some species of freshwater fish. In *Fish reproduction: strategies and tactics* (ed. G. W. Potts and R. J. Wootton), pp. 171–186. Academic Press, London.

Manzer, J. I. (1976). Distribution, food, and feeding of the threespine stickleback, *Gasterosteus aculeatus*, in Great Central Lake, Vancouver Island, with comments on competition for food with juvenile sockeye salmon, *Oncorhynchus nerka*. *Fishery Bulletin*, *U.S. National Marine Fisheries Service*, **74**, 647–668.

Marconato, A. and Bisazza, A. (1986). Males whose nests contain eggs are preferred by female *Cottus gobio* L. Pisces: Cottidae). *Animal Behaviour*, **34**, 1580–1582.

Marsh, E. (1986). Effects of egg size on offspring fitness and maternal fecundity in the orangethroat darter, *Etheostoma spectabile* (Pisces: Percidae). *Copeia*, **1986**, 18–30.

Marshall, C.R. (1990). Confidence intervals on stratigraphic ranges. *Paleobiology*, **16**, 1–10.

Marteinsdottir, G. and Able, K.W. (1988). Geographic variation in egg size among populations of the mummichog, *Fundulus heteroclitus* (Pisces: Fundulidae). *Copeia*, **1988**, 471–478.

Maskell, M., Parkin, D.T., and Verspoor, E. (1977). Apostatic selection by sticklebacks upon a dimorphic prey. *Heredity*, **39**, 83–89.

Masuda, H., Amaoka, K., Araga, C., Uyeno, T., and Yoshino, T. (eds) (1984). *The fishes of the Japanese archipelago*. Tokai University Press, Tokyo.

Matej, V.E. (1980). The ultrastructure of chloride cells of the gill epithelium in the stickleback, *Gasterosteus aculeatus* L. and crucian carp, *Carassius carassius*. *Tsitologiya*, **22**, 1387–1391.

Matej, V.E., Kharazova, A.D., and Vinogradov, G.A. (1981). The response of *Gasterosteus aculeatus* L. gill epithelium chloride cells to changes of pH and salinity in the environment. *Tsitologiya*, **23**, 159–165.

Mathews, W.H., Fyles, J.G., and Nasmith, H.W. (1970). Postglacial crustal movements in southwestern British Columbia and adjacent Washington state. *Canadian Journal of Earth Sciences*, **7**, 690–702.

Mayer, I., Borg, B., and Schulz, R. (1990*a*). Seasonal changes in and effect of castration/androgen replacement on the plasma levels of five androgens in the male three-spined stickleback, *Gasterosteus aculeatus* L. *General and Comparative Endocrinology*, **79**, 23–30.

Mayer, I., Borg, B., and Schulz, R. (1990*b*). Conversion of 11-ketoandrostenedione to 11-ketotestosterone by blood cells of six fish species. *General and Comparative Endocrinology*, **77**, 70–74.

Maynard Smith, J. (1966). Sympatric speciation. *American Naturalist*, **100**, 637–650.

Maynard Smith, J. (1982). *Evolution and the theory of games*. Cambridge University Press, Cambridge.

Maynard Smith, J. (1989). *Evolutionary genetics*. Oxford University Press, Oxford.

Maynard Smith, J., Burian, R., Kauffman, S., Alberch, P., Campbell, J., Goodwin, B., Lande, R., Raup, D., and Wolpert, L. (1985). Developmental constraints and evolution. *Quarterly Review of Biology*, **60**, 265–287.

Mayr, E. (1942). *Systematics and the origin of species*. Columbia University Press, New York.

Mayr, E. (1954). Change of genetic environment and evolution. In *Evolution as a process* (ed. J. Huxley, A. C. Hardy and E.B. Ford), pp. 157–180. Allen and Unwin, London.

Mayr, E. (1963). *Animal species and evolution*. Harvard University Press, Cambridge, Massachusetts.

Mayr, E. (1982). Processes of speciation. In *Mechanisms of speciation* (ed. C. Barigozzi), pp. 1–19. Liss, New York.

Mayr, E. (1983). How to carry out the adaptationist program? *American Naturalist*, **121**, 324–334.

Mayr, E. and Ashlock, P. D. (1991). *Principles of systematic zoology*. McGraw-Hill, New York.

Meakins, R. H. (1974). A quantitative approach to the effects of the plerocercoid of *Schistocephalus solidus* Muller 1776 on the ovarian maturation of the three-spined stickleback *Gasterosteus aculeatus* L. *Zeitschrift für Parasitenkunde*, **44**, 73–79.

Meakins, R. H. (1975). The effects of activity and season on the respiration of the three-spined stickleback, *Gasterosteus aculeatus* L. *Comparative Physiology and Biochemistry*, **51A**, 155–157.

Meier, A. H. and Fivizzani, A. J. (1980). Physiology of migration. In *Animal migration, orientation and navigation* (ed. S. A. Gauthreaux, Jr.), pp. 225–282. Academic Press, New York.

Meier, A. H., Trobec, T. N., Joseph, M. M., and John, T. M. (1971). Temporal synergism of prolactin and adrenal steroids in the regulation of fat stores. *Proceedings of the Society of Experimental Biology and Medicine*, **137**, 408–415.

Meyer, A., Kocher, T. D., Basasibwaki, P., and Wilson, A. C. (1990). Monophyletic origin of Lake Victoria cichlid fishes suggested by mitochondrial DNA sequences. *Nature*, **347**, 550–553.

Michod, R. E. (1979). Evolution of life histories in response to age-specific mortality factors. *American Naturalist*, **113**, 531–550.

Milinski, M. (1977*a*). Experiments on the selection by predators against spatial oddity of their prey. *Zeitschrift für Tierpsychologie*, **43**, 311–325.

Milinski, M. (1977*b*). Do all members of a swarm suffer the same predation? *Zeitschrift für Tierpsychologie*, **45**, 373–388.

Milinski, M. (1979). An evolutionarily stable feeding strategy in sticklebacks. *Zeitschrift für Tierpsychologie*, **51**, 36–40.

Milinski, M. (1982). Optimal foraging: the influence of intraspecific competition on diet selection. *Behavioral Ecology and Sociobiology*, **11**, 109–115.

Milinski, M. (1984*a*). Parasites determine a predator's optimal feeding strategy. *Behavioral Ecology and Sociobiology*, **15**, 35–37.

Milinski, M. (1984*b*). Competitive resource sharing: an experimental test of a learning rule for ESSs. *Animal Behaviour*, **32**, 233–242.

Milinski, M. (1984*c*). A predator's cost of overcoming the confusion-effect of swarming prey. *Animal Behaviour*, **32**, 1157–1162.

Milinski, M. (1985). Risk of predation of parasitized sticklebacks (*Gasterosteus aculeatus* L.) under competition for food. *Behaviour*, **93**, 203–216.

Milinski, M. (1986). A review of competitive resource sharing under constraints in sticklebacks. *Journal of Fish Biology*, **29**, (Supp. A), 1–14.

Milinski, M. (1988). Games fish play: making decisions as a social forager. *Trends in Ecology and Evolution*, **3**, 325–330.

Milinski, M. (1990). Parasitism and host decision-making. In *Parasites and host behaviour* (ed. C. J. Barnard and J. M. Behnke), pp. 95–116. Taylor and Francis, London.

Milinski, M. and Bakker, T. C. M. (1990). Female sticklebacks use male coloration in mate choice and hence avoid parasitized males. *Nature*, **344**, 330–333.

Milinski, M. and Heller, R. (1978). Influence of a predator on the optimal foraging behaviour of sticklebacks (*Gasterosteus aculeatus* L.). *Nature*, **275**, 642–644.

Milinski, M. and Regelmann, K. (1985). Fading short-term memory for patch quality in sticklebacks. *Animal Behaviour*, **33**, 678–680.

Miller, R. R. and Hubbs, C. L. (1969). Systematics of *Gasterosteus aculeatus*, with particular reference to intergradation and introgression along the Pacific coast of North America: a commentary on a recent contribution. *Copeia*, **1969**, 52–69.

Miller, R. R. and Smith, G. R. (1967). New fossil fishes from Plio-Pleistocene Lake Idaho. *Occasional Papers of the Museum of Zoology, University of Michigan*, **654**, 1–24.

Milligan, B. G. (1986). Punctuated evolution induced by ecological change. *American Naturalist*, **127**, 522–532.

Mina, M. V. (1991). *Microevolution of fishes: evolutionary aspects of phenotypic diversity*. Balkema, Rotterdam.

Mittelbach, G. G. (1981*a*). Foraging efficiency and body size: a study of optimal diet and habitat use by bluegills. *Ecology*, **62**, 1370–1386.

Mittelbach, G. G. (1981*b*). Patterns of invertebrate size and abundance in aquatic habitats. *Canadian Journal of Fisheries and Aquatic Sciences*, **38**, 896–904.

Moenkhaus, W. J. (1911). Cross fertilization among fishes. *Proceedings of the Indiana Academy of Sciences*, **1910–1911**, 353–393.

Molenda, E. and Fiedler, K. (1971). Die Wirkung von Prolactin auf das Verhalten von Stichlings-♂♂ (*Gasterosteus aculeatus* L.). *Zeitschrift für Tierpsychologie*, **28**, 463–474.

Mommsen, T. P., French, C. J., and Hochachka, P. W. (1980). Sites and patterns of amino acid and protein mobilization during the spawning migration of salmon. *Canadian Journal of Zoology*, **58**, 1785–1799.

Monod, T. (1968). Le complexe urophore des poissons téléostéens. *Mémoires de L'Institut Fondamental D'Afrique Noire*, No. 81. Ifan-Dakar.

Moodie, G. E. E. (1972*a*). Predation, natural selection and adaptation in an unusual threespine stickleback. *Heredity*, **28**, 155–167.

Moodie, G. E. E. (1972*b*). Morphology, life history and ecology of an unusual stickleback (*Gasterosteus aculeatus*) in the Queen Charlotte Islands, Canada. *Canadian Journal of Zoology*, **50**, 721–732.

Moodie, G. E. E. (1982). Why asymmetric mating preferences may not show the direction of evolution. *Evolution*, **36**, 1096–1097.

Moodie, G. E. E. (1984). Status of the giant (Mayer Lake) stickleback, *Gasterosteus* sp., on the Queen Charlotte Islands, British Columbia. *Canadian Field-Naturalist*, **98**, 115–119.

Moodie, G. E. E. and Reimchen, T. E. (1976*a*). Phenetic variation and habitat differences in *Gasterosteus* populations of the Queen Charlotte Islands. *Systematic Zoology*, **25**, 49–61.

Moodie, G. E. E. and Reimchen, T. E. (1976*b*). Glacial refugia, endemism and stickleback populations of the Queen Charlotte Islands. *Canadian Field-Naturalist*, **90**, 471–474.

Moodie, G. E. E., McPhail, J. D., and Hagen, D. W. (1973). Experimental demonstration of selective predation in *Gasterosteus aculeatus*. *Behaviour*, **47**, 95–105.

Moore, J. and Gotelli, M. J. (1990). Phylogenetic perspective on the evolution of altered host behaviour: a critical look at the manipulation hypothesis. In *Parasites and host behaviour* (ed. C. J. Barnard and J. M. Behnke), pp. 193–229. Taylor and Francis, London.

Mori, S. (1984). Sexual dimorphism of the landlocked three-spined stickleback *Gasterosteus aculeatus microcephalus* from Japan. *Japanese Journal of Ichthyology*, **30**, 419–425.

Mori, S. (1985). Reproductive behaviour of the landlocked three-spined stickleback, *Gasterosteus aculeatus microcephalus*, in Japan. I. The year-long prolongation of the breeding period in waterbodies with springs. *Behaviour*, **93**, 21–35.

Mori, S. (1987*a*). Geographical variations in freshwater populations of the three-spined stickleback, *Gasterosteus aculeatus*, in Japan. *Japanese Journal of Ichthyology*, **34**, 33–46.

Mori, S. (1987*b*). Divergence in reproductive ecology of the three-spined stickleback, *Gasterosteus aculeatus. Japanese Journal of Ichthyology*, **34**, 165–175.

Mori, S. (1988). The upright nesting behaviours on a vertical shore-wall in the three-spined stickleback, *Gasterosteus aculeatus* (*leiurus* form). *Journal of Ethology*, **6**, 59–62.

Mori, S. (1990*a*). Two morphological types in the reproductive stock of three-spined stickleback, *Gasterosteus aculeatus*, in Lake Harutori, Hokkaido Island. *Environmental Biology of Fishes*, **27**, 21–31.

Mori, S. (1990*b*). The breeding system of the three-spined stickleback, *Gasterosteus aculeatus* (*leiurus* form), with reference to spatial and temporal pattern of nesting activity. Unpublished Ph.D. thesis. Kyoto University, Japan.

Mori, S. and Nagoshi, M. (1987). Growth and maturity size of the three-spined stickleback *Gasterosteus aculeatus* in rearing pool. *Bulletin of the Faculty of Fisheries, Mie University*, **14**, 1–10.

Morizot, D.C. and Schmidt, M.E. (1990). Starch gel electrophoresis and histochemical visualization of proteins. In *Electrophoretic and isoelectric focusing techniques in fisheries management* (ed. D.H. Whitmore), pp. 23–80. CRC, Boca Raton, Florida.

Morris, D. (1956). The function and causation of courtship ceremonies in animals (with special reference to fish). In: *Foundation Signer-Polignac: Colloque Internationale sur l'Instinct, Paris, 1954*, pp. 261–286.

Morris, D. (1958). The reproductive behaviour of the ten-spined stickleback (*Pygosteus pungitius* L.). *Behaviour Supplement*, **6**, 1–154.

Morrow, J.E. (1980). *The freshwater fishes of Alaska*. Alaska Northwest, Anchorage.

Mourier, J.-P. (1972). Étude de la cytodifférenciation du rein de l'épinoche femelle après traitement par la méthyltestostérone. *Zeitschrift für Zellforschung*, **123**, 96–111.

Mourier, J.-P. (1976*a*). Ultrastructural modifications of renal cells in the three-spined stickleback (*Gasterosteus aculeatus* L.) after castration. *Cell and Tissue Research*, **168**, 527–548.

Mourier, J.-P. (1976*b*). Effects of an antiandrogen, cyproterone acetate, on the kidney of the three-spined stickleback (*Gasterosteus aculeatus* L.). *Cell and Tissue Research*, **173**, 357–366.

Mourier, J.-P. (1979). Incorporation of ^{3}H-thymidine in the nephron of *Gasterosteus aculeatus* L. and its stimulation by methyltestoterone. *Cell and Tissue Research*, **201**, 249–262.

Mousseau, T.A. and Roff, D.A. (1987). Natural selection and the heritability of fitness components. *Heredity*, **59**, 181–197.

Muckensturm, B. (1969). La signification de la livrée nuptiale de l'épinoche. *Revue de Comportement de l'Animal*, **3**, 39–64.

Mullem, P.J. van and Vlugt, J.C. van der (1964). On the age, growth and migration of an anadromous stickleback, *Gasterosteus aculeatus* L., investigated in mixed populations. *Archives Néerlandaises de Zoologie*, **16**, 111–139.

Muller, H.J. (1939). Reversibility in evolution considered from the standpoint of genetics. *Biological Reviews of the Cambridge Philosophical Society*, **14**, 261–279.

Munro, A.D. and Pitcher, T.J. (1983). Hormones and agonistic behaviour in teleosts. In *Control processes in fish physiology* (ed. J.C. Rankin, T.J. Pitcher and R.T. Duggan), pp. 155–175. Croom Helm, London.

Munro, J. A. and Clemens, W. A. (1937). The American merganser in British Columbia and its relation to the fish population. *Bulletin of the Biological Board of Canada*, **55**, 1-50.

Münzing, J. (1959). Biologie, Variabilität und Genetik von *Gasterosteus aculeatus* L. (Pisces). Untersuchungen im Elbegebiet. *Internationale Revue der Gesamten Hydrobiologie*, **44**, 317-382.

Münzing, J. (1963). The evolution of variation and distributional patterns in European populations of the three-spined stickleback, *Gasterosteus aculeatus*. *Evolution*, **17**, 320-332.

Mural, R. J. (1973). The Pliocene sticklebacks of Nevada with a partial osteology of the Gasterosteidae. *Copeia*, **1973**, 721-735.

Muramoto, J., Igarishi, K., Itoh, M., and Makino, S. (1969). A study of the chromosomes and enzymatic patterns of sticklebacks of Japan. *Proceedings of the Japan Academy*, **45**, 803-807.

Murphy, R. W., Sites, J. W., Jr., Buth, D. G., and Haufler, C. H. (1990). Proteins. I: Isozyme electrophoresis. In *Molecular systematics* (ed. D. M. Hillis and C. Mortiz), pp. 45-126. Sinauer, Sunderland, Massachusetts.

Myers, G. S. (1930). The killifish of San Ignacio and the stickleback of San Ramon, lower California. *Proceedings of the California Academy of Sciences, 4th Series*, **19**, 95-104.

Narver, D. W. (1969). Phenotypic variation in three-spine sticklebacks (*Gasterosteus aculeatus*) of the Chignik River system, Alaska. *Journal of the Fisheries Research Board of Canada*, **26**, 405-412.

Nei, M. (1972). Genetic distance between populations. *American Naturalist*, **106**, 283-292.

Nei, M. (1978). Estimation of average homozygosity and genetic distance from a small number of individuals. *Genetics*, **89**, 583-590.

Neill, S. R. S. and Cullen, J. M. (1974). Experiments on whether schooling by their prey affects the hunting behaviour of cephalopods and fish predators. *Journal of Zoology, London*, **172**, 549-569.

Nelson, J. S. (1969). Geographic variation in the brook stickleback, *Culaea inconstans*, and notes on nomenclature and distribution. *Journal of the Fisheries Research Board of Canada*, **26**, 2431-2447.

Nelson, J. S. (1971). Comparison of pectoral and pelvic skeletons and of some other bones and their phylogenetic implications in the Aulorhynchidae and Gasterosteidae (Pisces). *Journal of the Fisheries Research Board of Canada*, **28**, 427-442.

Nelson, J. S. (1977). Evidence of a genetic basis for absence of the pelvic skeleton in brook stickleback, *Culaea inconstans*, and notes on the geographical distribution and origin of the loss. *Journal of the Fisheries Research Board of Canada*, **34**, 1314-1320.

Nelson, J. S. (1984). *Fishes of the world* (2nd edn). Wiley-Interscience, New York.

Nevo, E. (1978). Genetic variation in natural populations: patterns and theory. *Theoretical Population Biology*, **13**, 121-177.

Newman, C. M., Cohen, J. E., and Kipnis, C. (1985). Neo-Darwinian evolution implies punctuated equilibria. *Nature*, **315**, 400-401.

Newman, H. H. (1915). Development and heredity in heterogenic teleost hybrids. *Journal of Experimental Zoology*, **18**, 511-576.

Ng, T. B. and Idler, D. R. (1983). Yolk formation and differentiation in teleost fishes. In *Fish physiology*, Vol. 9, *Reproduction*, Part A: *Endocrine tissues and hormones* (ed. W. S. Hoar, D. J. Randall and E. M. Donaldson), pp. 373-404. Academic Press, New York.

Nichols, J.T. and Breder, C.M., Jr. (1927). The marine fishes of New York and southern New England. *Zoologica, New York*, **9**, 1-192.

Nicol, J.A.C. (1989). *The eyes of fishes*. Clarendon, Oxford.

Nicoll, C.S., Walker Wilson, S., Nishioka, R. and Bern, H.A. (1981). Blood and pituitary prolactin levels in tilapia (*Sarotherodon mossamibicus*: Teleostei) from different salinities as measured by a homologous radioimmunoassay. *General and Comparative Endocrinology*, **44**, 365-373.

Nikol'skii, G.V. (1961). *Special ichthyology*. Israel Program for Scientific Translations, Jerusalem.

Nikolsky, G.V. (1963). The ecology of fishes. Academic Press, London.

Nilsson, N. and Northcote, T.G. (1981). Rainbow trout (*Salmo gairdneri*) and cutthroat trout (*S. clarki*) interactions in coastal British Columbian lakes. *Canadian Journal of Fisheries and Aquatic Sciences*, **38**, 1228-1246.

Nilsson, S.G. and Nilsson, I. (1976). Numbers, food consumption, and fish predation by birds in Lake Mockeln, southern Sweden. *Ornis Scandinavica*, **7**, 61-70.

Noaillac-Depeyre, J. and Gas, N. (1973). Mise en évidence d'une zone adaptée au transport des ions dans l'intestin de la carpe commune (*Cyprinus carpio* L.). *Comptes Rendus Hebdomadaires des Séances de* l'Académie des Sciences, Paris, **276**, 773-776.

Noakes, D.L.G. (1986). When to feed: decision making in sticklebacks, *Gasterosteus aculeatus. Environmental Biology of Fishes*, **16**, 95-104.

Nordlie, F.G. and Walsh, S.J. (1989). Adaptive radiation in osmotic regulatory patterns among three species of Cyprinodontids (Teleostei: Atherinomorpha). *Physiological Zoology*, **62**, 1203-1218.

Obrhelova, N. (1977). Fischfauna des Holstein-Interglazials von Tonisberg bei Krefeld (BRD). *Casopis pro Mineralogii a Geologii, Roc*, **22**, 173-188.

O'Brien, W.J., Slade, N.A., and Vinyard, G.L. (1976). Apparent size as the determinant of prey selection by bluegill sunfish (*Lepomis macrochirus*). *Ecology*, **57**, 1204-1310.

O'Brien, W.J., Browman, H.I., and Evans, B.I. (1990). Search strategies of foraging animals. *American Scientist*, **78**, 152-160.

Ogilvie, R.T. (1989). Disjunct vascular flora of northwestern Vancouver Island in relation to Queen Charlotte Islands' endemism and Pacific coast refugia. In *The outer shores* (ed. G.G.E. Scudder and N.A. Gessler), pp. 125-127. Queen Charlotte Islands Museum Press, Skidegate, British Columbia.

O'Hara, K. and Penczak, T. (1987). Production of the three-spined stickleback, *Gasterosteus aculeatus* L., in the River Weaver, England. *Freshwater Biology*, **18**, 353-360.

Ohguchi, O. (1981). Prey density and selection against oddity by three-spined sticklebacks. *Advances in Ethology*, **23**, 1-79.

Okada, Y. (1960). Studies on the freshwater fishes of Japan. *Journal of the Faculty of Fisheries, Prefectural University of Mie*, **4**, 1-860.

Ollason, J.G. (1987). Artificial design in natural history: why it is so easy to understand animal behaviour. *Alternatives, perspectives in ethology*, Vol. 7 (ed. P.P.G. Bateson and P.H. Klopfer). pp. 233-257. Plenum, New York.

Ollevier, F. and Covens, M. (1983). Vitellogenins in *Gasterosteus aculeatus. Annales de la Societé Royale Zoologique de Belgique*, **113**, 327-334.

Olson, E.C. and Miller, R.L. (1958). *Morphological integration*. University of Chicago Press, Chicago.

Olson, S.L., Swift, C.C., and Mokhiber, C. (1979). An attempt to determine the prey of the great auk *Pinguinus impennis. Auk*, **96**, 790-792.

Ono, R. D. and Williams J. D. (1983). *Vanishing fishes of North America*. Stonewall, Washington.

Owings, D. H. and Coss, R. G. (1977). Snake mobbing by Californian ground squirrels: adaptive variation in ontogeny. *Behaviour*, **62**, 50–69.

Paepke, H.-J. (1968). Über eine mischpopulation des dreistachligen stichlings (*Gasterosteus aculeatus* L.) aus der umgegend von Potsdam. *Veröffentlichungen des Bezirksheimatmuseums Potsdam (Beiträge Tierwelt Mark V)*, **16**, 23–34.

Paepke, H.-J. (1970). Studien zur ökologie, variabilität und populationsstruktur des dreistachligen und neunstachligen stichlings. II: Die variabilität der lateralbeschildung von *Gasterosteus aculeatus* L. in einer brandenbergischen intergradationzone und ihre zoogeographisch-historischen hintergrunde. *Veröffentlichungen des Bezirksheimatmuseums Potsdam (Beiträge Tierwelt Mark VII)*, **21**, 5–48.

Paepke, H.-J. (1971). Studien zur ökologie, variabilität und populationsstruktur des dreistachligen und neunstachligen stichlings. III: Grössen-, alters- und geschlechtsverhaltnisse in brandenbergischen population von *Gasterosteus aculeatus* L. *Veröffentlichungen des Bezirksheimatmuseums Potsdam (Beiträge Tierwelt Mark VII)*, **23/24**, 5–22.

Paepke, H.-J. (1983). *Die Stichlinge*. Ziemsen Verlag, Wittenberg.

Page, E. B. (1963). Ordered hypotheses for multiple treatments: a significance test for linear ranks. *Journal of the American Statistical Association*, **58**, 216–230.

Parker, G. A. (1984). Evolutionarily stable strategies. In *Behavioural ecology: an evolutionary approach* (ed. J. R. Krebs and N. B. Davies) (2nd edn), pp. 30–61. Blackwell, Oxford.

Parker, H. and Holm, H. (1990). Patterns of nutrient and energy expenditure in female common eiders nesting in the High Arctic. *Auk*, **107**, 660–668.

Parsons, P. A. (1972). Genetic determination on behavior (mice and men). In *Genetics, environment and behavior* (ed. L. Ehrman, S. Omenn, and E. Caspari), pp. 75–103. Academic Press, New York.

Pascoe, D. and Mattey, D. (1977). Dietary stress in parasitized and non-parasitized sticklebacks. *Zeitschrift für Parasitenkunde*, **51**, 179–186.

Pasmanik, M. and Callard, G. V. (1988). A high abundance androgen receptor in goldfish brain: characteristics and seasonal changes. *Endocrinology*, **123**, 1162–1171.

Paterson, H. E. H. (1978). More evidence against speciation by reinforcement. *South African Journal of Science*, **74**, 369–371.

Paterson, H. E. H. (1981). The continuing search for the unknown and unknowable: a critique of contemporary ideas on speciation. *South African Journal of Science*, **77**, 113–119.

Paterson, H. E. H. (1982). Perspective on speciation by reinforcement. *South African Journal of Science*, **78**, 53–57.

Paterson, H. E. H. (1985). The recognition concept of species. In *Species and speciation* (ed. E. S. Vrba), pp. 21–29. Transvaal Museum Monograph No. 4, Pretoria.

Pearcy, W. G. and Richards, S. W. (1962). Distribution and ecology of fishes of the Mystic River estuary, Connecticut. *Ecology*, **43**, 248–259.

Pease, C. M., Lande, R., and Bull, J. J. (1989). A model of population growth, dispersal and evolution in a changing environment. *Ecology*, **70**, 1657–1664.

Peeke, H. V. S. (1969). Habituation of conspecific aggression in the three-spined stickleback (*Gasterosteus aculeatus* L.). *Behaviour*, **35**, 137–156.

Peeke, H. V. S. (1983). Habituation, sensitization, and redirection of aggression and feeding behavior in the three-spined stickleback (*Gasterosteus aculeatus* L.). *Journal of Comparative Psychology*, **97**, 43–51.

Peeke, H.V.S. and Peeke, S.C. (1973). Habituation in fish with special reference to intraspecific aggressive behavior. In *Habituation*, Vol. 1, *Behavioral studies* (ed. H.V.S. Peeke and M.J. Herz), pp. 59–83. Academic Press, London.

Peeke, H.V.S. and Veno, A. (1973). Stimulus specificity of habituated aggression in the stickleback (*Gasterosteus aculeatus*). *Behavioral Biology*, **8**, 427–432.

Peeke, H.V.S. and Veno, A. (1976). Response independent habituation of territorial aggression in the three-spined stickleback (*Gasterosteus aculeatus*). *Zeitschrift für Tierpsychologie*, **40**, 53–58.

Peeke, H.V.S., Wyers, E.J., and Herz, M.J. (1969). Waning of the aggressive response to male models in the three-spined stickleback (*Gasterosteus aculeatus* L.). *Animal Behaviour*, **17**, 224–228.

Peeke, H.V.S., Figler, M.H., and Blankenship, N. (1979). Retention and recovery of habituated territorial aggressive behavior in the three-spined stickleback (*Gasterosteus aculeatus* L.): the roles of time and nest construction. *Behaviour*, **69**, 171–182.

Pelkwijk, J.J. ter and Tinbergen, N. (1937). Eine reizbiologische Analyse einiger Verhaltensweisen von *Gasterosteus aculeatus* L. *Zeitschrift für Tierpsychologie*, **1**, 193–200.

Penczak, T. (1965). Morphological variation of the stickleback (*Gasterosteus aculeatus* L.) in Poland. *Zoologica Poloniae*, **15**, 3–49.

Penczak, T. (1968). *Gasterosteus aculeatus* L. as food of *Mergus merganser* L. *Przeglad Zologiczny*, **12**, 357–358. (in Polish).

Penczak, T. (1981). Ecological fish production in two small lowland rivers in Poland. *Oecologia*, **48**, 107–111.

Pennycuick, L. (1971). Quantitative effects of three species of parasites on a population of three-spined sticklebacks, *Gasterosteus aculeatus*. *Journal of Zoology, London*, **165**, 143–162.

Perlmutter, A. (1963). Observations on the fishes of the genus *Gasterosteus* in the waters of Long Island, New York. *Copeia*, **1963**, 168–173.

Peter, R.E. (1983). The brain and neurohormones in teleost reproduction. In *Fish physiology*, Vol. IX, Part A: *Endocrine tissues and hormones* (ed. W.S. Hoar, D.J. Randall, and E.M. Donaldson), pp. 97–135. Academic Press, New York.

Peterson, C.W. (1989). Females prefer mating males in the carmine triplefin, *Axoclinus carminalis*, a paternal brood guarder. *Environmental Biology of Fishes*, **26**, 213–221.

Petry, D. (1982). The pattern of phyletic speciation. *Paleobiology*, **8**, 56–66.

Pianka, E.R. and Parker, W.S. (1975). Age-specific reproductive tactics. *American Naturalist*, **109**, 453–464.

Picard, P., Jr., Dodson, J.J., and FitzGerald, G.J. (1990). Habitat segregation among the age groups of *Gasterosteus aculeatus* (Pisces: Gasterosteidae) in the middle St. Lawrence Estuary, Canada. *Canadian Journal of Zoology*, **68**, 1202–1208.

Pietsch, T.W. (1978). Evolutionary relationships of the sea moths (Teleostei: Pegasidae) with a classification of gasterosteiform families. *Copeia*, **1978**, 517–529.

Pimm, S.L. (1978). Sympatric speciation: a simulation model. *Biological Journal of the Linnean Society*, **11**, 131–139.

Pitcher, T.J. (1986). Functions of shoaling behaviour in teleosts. In *The behaviour of teleost fishes* (1st edn) (ed. T.J. Pitcher), pp. 294–337. Croom Helm, London.

Pitcher, T.J., Magurran. A.E., and Allan, J.R. (1983). Shifts of behaviour with shoal size in cyprinids. *Proceedings of the British Freshwater Fisheries Conference*, **3**, 220–228.

Pitcher, T. J., Green, D., and Magurran, A. E. (1986). Dicing with death: predator inspection behaviour in minnow shoals. *Journal of Fish Biology*, **28**, 439–448.

Plafker, G. and Addicott, W. O. (1976). Glaciomoraine deposits of Miocene through Holocene age in the Yagataga Formation along the Gulf of Alaska margin, Alaska. In *Recent and ancient sedimentary environments in Alaska* (ed. T. P. Miller), pp. Q1–Q23. Alaska Geological Society, Anchorage.

Policansky, D. (1983). Size, age and demography of metamorphosis and sexual maturation in fishes. *American Zoologist*, **23**, 57–63.

Pottinger, T. G. (1988). Seasonal variation in specific plasma-and target-tissue binding of androgens, relative to plasma steroid levels in the brown trout. *General and Comparative Endocrinology*, **70**, 334–344.

Potts, G. W. (1984). Parental behaviour in temperate marine teleosts with special reference to the development of nest structures. In *Fish reproduction: strategies and tactics* (ed. G. W. Potts and R. J. Wootton), pp. 223–244. Academic Press, London.

Poulin, R. and FitzGerald, G. J. (1987). The potential of parasitism in the structuring of a salt marsh stickleback community. *Canadian Journal of Zoology*, **65**, 2793–2798.

Poulin, R. and FitzGerald, G. J. (1988). Water temperature, vertical distribution and risk of ectoparasitism in juvenile sticklebacks. *Canadian Journal of Zoology*, **66**, 2002–2005.

Poulin, R. and FitzGerald, G. J. (1989*a*). Early life histories of three sympatric sticklebacks in a salt-marsh. *Journal of Fish Biology*, **34**, 207–221.

Poulin, R. and FitzGerald, G. J. (1989*b*). Risk of parasitism and microhabitat selection in juvenile sticklebacks. *Canadian Journal of Zoology*, **67**, 14–18.

Poulin, R. and FitzGerald, G. J. (1989*c*). Shoaling as an anti-parasite mechanism in juvenile sticklebacks (*Gasterosteus*). *Behavioral Ecology and Sociobiology*, **24**, 251–255.

Power, G. (1965). Notes on the cold-blooded vertebrates of the Nabisipi River region, County Duplessis, Quebec. *Canadian Field-Naturalist*, **79**, 49–64.

Prashad, B. and Mukerji, D. D. (1929). The fish of Indawgyi Lake and the streams of the Myitkyina district (Upper Burma). *Records of the Indian Museum*, **31**, 161–213.

Pressley, P. H. (1981). Parental effort and the evolution of nest-guarding tactics in the threespine stickleback, *Gasterosteus aculeatus* L. Evolution, **35**, 282–295.

Priede, I. G. (1985). Metabolic scope in fishes. In *Fish energetics: new perspectives* (ed. P. Tytler and P. Calow), pp. 33–64. Croom Helm, London.

Prothero, D. R. and Lazarus, D. B. (1980). Planktonic microfossils and the recognition of ancestors. *Systematic Zoology*, **29**, 119–129.

Prunet, P., Beouf, G., and Houdebine, L. M. (1985). Plasma and pituitary prolactin levels in rainbow trout during adaptation to different salinities. *Journal of Experimental Zoology*, **235**, 187–196.

Quast, J. C. (1965). Osteological characteristics and affinities of the hexagrammid fishes, with a synopsis. *Proceedings of the California Academy of Sciences*, **31**, 563–600.

Quinn, T. P. and Light, J. T. (1989). Occurrence of threespine sticklebacks (*Gasterosteus aculeatus*) in the open North Pacific Ocean: migration or drift? *Canadian Journal of Zoology*, **67**, 2850–2852.

Rad, O. (1980). Breeding distribution and habitat selection of Red-breasted Mergansers *Mergus serra* in fresh water in western Norway. *Wildfowl*, **31**, 53–56.

Rafinski, J., Banbura, J., and Przybylski, M. (1989). Genetic differentiation of freshwater and marine sticklebacks (*Gasterosteus aculeatus*) of eastern Europe.

Zeitschrift für Zoologische Systematik und Evolutionsforschung, **27**, 33–43.

Rainboth, W.J. and Whitt, G.S. (1974). Analysis of evolutionary relationships among shiners of the subgenus *Luxilus* (Teleostei, Cypriniformes, *Notropis*) with the lactate dehydrogenase and malate dehydrogenase isozyme systems. *Comparative Biochemistry and Physiology*, **49B**, 241–252.

Rajasilta, M. (1980). Food consumption of the three-spined stickleback (*Gasterosteus aculeatus*). *Annales Zoologici Fennici*, **17**, 123–126.

Rand, D.M. and Harrison, R.G. (1989). Ecological genetics of a mosaic hybrid zone: mitochondrial, nuclear and reproductive differentiation of crickets by soil type. *Evolution*, **43**, 432–449.

Rasmussen, E. (1973). Systematics and ecology of the Isefjord marine fauna (Denmark). *Ophelia*, **11**, 1–507.

Raunich, L., Callegarini, C., and Cucchi, C. (1971). Studio di alcuni parametri biochimici (emoglobine, latticoideidrogenasi) in popolazioni di *Gasterosteus aculeatus* dell'Italia e della Svezia. *Bollettino de Zoologia*, **38**, 557.

Raunich, L., Callegarini, C., and Cucchi, C. (1972). Ecological aspects of hemoglobin polymorphism in *Gasterosteus aculeatus* (Teleostea). In *Fifth European biology symposium* (ed. B. Battaglia), pp. 153–162. Piccin Editore, Padua.

Raven, P. (1986). The size of minnow prey in the diet of young kingfishers *Alcedo atthis*. *Bird Study*, **33**, 6–11.

Rawlinson, S.E. and Bell, M.A. (1982). A stickleback fish (*Pungitius*) from the Neogene Sterling Formation, Kenai Peninsula, Alaska. *Journal of Paleontology*, **56**, 583–588.

Reebs, S.G., Whoriskey, F.G., and FitzGerald, G.J. (1984). Diel patterns of fanning activity, egg respiration and the nocturnal behavior of male three-spined sticklebacks, *Gasterosteus aculeatus* L. (f. *trachurus*). *Canadian Journal of Zoology*, **62**, 329–334.

Regan, C.T. (1909). The species of threespined sticklebacks (*Gasterosteus*). *Annals and Magazine of Natural History*, **4**, 435–437.

Regan, C.T. (1913). The osteology and classification of the teleostean fishes of the order Scleroparei. *Annals and Magazine of Natural History*, Series **8**, (11), 170–184.

Regelmann, K. (1984). Competitive resource sharing: a simulation model. *Animal Behaviour*, **32**, 226–232.

Reiffers, C. (1984). Reproduction et comportement des femelles de trois espèces d'épinoches (Gasterosteidae) sympatriques en milieu naturel. Unpublished M.Sc. thesis. Laval University, Québec.

Reimchen, T.E. (1979). Substratum heterogeneity, crypsis and colour polymorphism in an intertidal snail. *Canadian Journal of Zoology*, **57**, 1070–1085.

Reimchen, T.E. (1980). Spine deficiency and polymorphism in a population of *Gasterosteus aculeatus*: an adaptation to predators? *Canadian Journal of Zoology*, **58**, 1232–1244.

Reimchen, T.E. (1982). Incidence and intensity of *Cyathocephalus truncatus* and *Schistocephalus solidus* infection in *Gasterosteus aculeatus*. *Canadian Journal of Zoology*, **60**, 1091–1095.

Reimchen, T.E. (1983). Structural relationships between spines and lateral plates in threespine stickleback (*Gasterosteus aculeatus*). *Evolution*, **37**, 931–946.

Reimchen, T.E. (1984). Status of un-armoured and spine-deficient populations (Charlotte unarmoured stickleback) of threespine stickleback, *Gasterosteus* sp., on the Queen Charlotte Islands, British Columbia. *Canadian Field-Naturalist*, **98**, 120–126.

Reimchen, T.E. (1988). Inefficient predators and prey injuries in a population of giant stickleback. *Canadian Journal of Zoology*, **66**, 2036–2044.

Reimchen, T.E. (1989). Loss of nuptial color in threespine sticklebacks (*Gasterosteus aculeatus*). *Evolution*, **43**, 450–460.

Reimchen, T.E. (1990). Size-structured mortality in a threespine stickleback (*Gasterosteus aculeatus*)-cutthroat trout (*Oncorhynchus clarki*) community. *Canadian Journal of Fisheries and Aquatic Sciences*, **47**, 1194–1205.

Reimchen, T.E. (1991*a*). Trout foraging failures and the evolution of body size in stickleback. *Copeia*, **1991**, 1098–1104.

Reimchen, T.E. (1991*b*). Evolutionary attributes of headfirst prey manipulation and swallowing in piscivores. *Canadian Journal of Zoology*, **69**, 2912–2916.

Reimchen, T.E. (1992*a*). Extended longevity in a large-bodied *Gasterosteus* population. *Canadian Field-Naturalist*, **106**, 122–125.

Reimchen, T.E. (1992*b*). Injuries and survival of *Gasterosteus* from attacks by a toothed predator (*Oncorhynchus*) and some implications for the evolution of lateral plates. *Evolution*, **46**, 1224–1230.

Reimchen, T.E. and Douglas, S.D. (1980). Observations of loons (*Gavia immer* and *G. stellata*) at a bog lake on the Queen Charlotte Islands. *Canadian Field-Naturalist*, **94**, 398–404.

Reimchen, T.E. and Douglas, S.D. (1984*a*). Seasonal and diurnal abundance of aquatic birds on the Drizzle Lake Reserve, Queen Charlotte Islands, British Columbia. *Canadian Field-Naturalist*, **98**, 22–28.

Reimchen, T.E. and Douglas, S.D. (1984*b*). Feeding schedule and daily food consumption of red-throated loon (*Gavia stellata*) over the pre-fledging period. *Auk*, **101**, 593–599.

Reimchen, T.E. and Nelson, J.S. (1987). Habitat and morphological correlates to vertebral number as shown in a teleost, *Gasterosteus aculeatus*. *Copeia*, **1987**, 868–874.

Reimchen, T.E., Stinson, E.M., and Nelson, J.S. (1985). Multivariate differentiation of parapatric and allopatric populations of threespine stickleback in the Sangan River watershed, Queen Charlotte Islands. *Canadian Journal of Zoology*, **63**, 2944–2951.

Reisman, H.M. (1963). Reproductive behavior of *Apeltes quadracus*, including some comparisons with other gasterosteid fishes. *Copeia*, **1963**, 191–192.

Reisman, H.M. (1968*a*). Reproductive isolating mechanisms of the blackspotted stickleback, *Gasterosteus wheatlandi*. *Journal of the Fisheries Research Board of Canada*, **25**, 2703–2706.

Reisman, H.M. (1968*b*). Effects of social stimuli on the secondary sex characters of male three-spined sticklebacks, *Gasterosteus aculeatus*. *Copeia*, **1968**, 816–826.

Reisman, H.M. and Cade, T.J. (1967). Physiological and behavioral aspects of reproduction in the brook stickleback, *Culaea inconstans*. *American Midland Naturalist*, **77**, 257–295.

Reiss, M.J. (1984). Courtship and reproduction in the three-spined stickleback. *Journal of Biological Education*, **18**, 197–200.

Reist, J.D. (1980*a*). Selective predation upon pelvic phenotypes of brook stickleback, *Culaea inconstans*. *Canadian Journal of Zoology*, **58**, 1245–1252.

Reist, J.D. (1980*b*). Predation upon pelvic phenotypes of brook stickleback, *Culaea inconstans*, by selected invertebrates. *Canadian Journal of Zoology*, **58**, 1253–1258.

Reist, J.D. (1981). Variation in frequencies of pelvic phenotypes of the brook stickleback, *Culaea inconstans*, in Redwater drainage, Alberta. *Canadian Field-Naturalist*, **95**, 178–182.

Reist, J. D. (1983). Behavioural variation in pelvic phenotypes of brook stickleback, *Culaea inconstans*, in response to predation by northern pike, *Esox lucius*. *Environmental Biology of Fishes*, **8**, 255–267.

Reynolds, R. E. (1991). Hemingfordian/Barstovian land mammal age faunas in the central Mojave Desert, exclusive of the Barstow fossil beds. *San Bernardino County Museum Association Quarterly*, **38**, 88–90.

Reynolds, R. E. and Reynolds, R. L. (1991). Late Pleistocene faunas of Lake Thomson. *San Bernardino County Museum Association Quarterly*, **38**, 114–115.

Reznick, D. N. (1983). The structure of guppy life histories: the trade off between growth and reproduction. *Ecology*, **64**, 862–873.

Reznick, D. N. (1990). Plasticity in age and size at maturity in male guppies (*Poecilia reticulata*): an experimental evaluation of alternative models of development. *Journal of Evolutionary Biology*, **3**, 185–203.

Reznick, D. N. and Endler, J. A. (1982). The impact of predation on life history evolution in Trinidadian guppies (*Poecilia reticulata*). *Evolution*, **36**, 160–177.

Reznick, D. N. and Miles, D. B. (1989). Review of life history patterns in poeciliid fishes. In *Ecology and evolution of livebearing fishes (Poeciliidae)* (ed. G. K. Meffe and F. F. Snelson, Jr.), pp. 125–148. Prentice Hall, Englewood Cliffs, New Jersey.

Reznick, D. A., Bryga, H., and Endler, J. A. (1990). Experimentally induced life-history evolution in a natural population. *Nature*, **346**, 357–359.

Rice, W. R. (1987). Speciation via habitat specialization: the evolution of reproductive isolation as a correlated character. *Evolutionary Ecology*, **1**, 301–315.

Richman, N. H. and Zaugg, W. S. (1987). Effects of cortisol and growth hormone on osmoregulation in pre- and desmoltified coho salmon (*Oncorhynchus kisutch*). *General and Comparative Endocrinology*, **65**, 189–198.

Ricker, W. E. (1975). *Computation and interpretation of biological statistics of fish populations. Bulletin of the Fisheries Research Board of Canada*, **191**, 1–382.

Rico, C., Kuhnlein, U., and FitzGerald, G. J. (1992). Male reproductive tactics in the threespine stickleback: an evaluation by DNA fingerprinting. *Molecular Ecology*, **1**, 79–87.

Ridgway, M. S. and McPhail, J. D. (1984). Ecology and evolution of sympatric sticklebacks (*Gasterosteus*): mate choice and reproductive isolation in the Enos Lake species pair. *Canadian Journal of Zoology*, **62**, 1813–1818.

Ridgway, M. S. and McPhail, J. D. (1987). Rival male effects on courtship behaviour in the Enos Lake species pair of sticklebacks (*Gasterosteus*). *Canadian Journal of Zoology*, **65**, 1951–1955.

Ridgway, M. S. and McPhail, J. D. (1988). Raiding shoal size and a distraction display in male sticklebacks (*Gasterosteus*). *Canadian Journal of Zoology*, **66**, 201–205.

Ridley, M. (1983). *The explanation of organic diversity*. Clarendon, Oxford.

Ridley, M. (1986). *Animal behaviour: a concise introduction*. Blackwell, Oxford.

Ridley, M. and Rechten, C. (1981). Female sticklebacks prefer to spawn with males whose nests contain eggs. *Behaviour*, **76**, 152–161.

Riechert, S. E. (1986). Spider fights as a test of evolutionary game theory. *American Scientist*, **74**, 604–610.

Riechert, S. E. and Hedrick, A. V. (1990). Levels of predation and genetically based anti-predator behaviour in the spider, *Agelenopsis aperta*. *Animal Behaviour*, **40**, 679–687.

Rinkel, G. L. and Hirsch, G. C. (1940). Die Restitution des Eiweiss-Sekretes zum Nestbau beim Stichling *Gasterosteus* in Verbindung mit dem Arbeitsrhythmus der Niere. *Zeitschrift für Zellforschung*, **64**, 649–688.

Riska, B. (1985). Group size factors and geographic variation of morphometric correlation. *Evolution*, **39**, 792–803.

Riska, B., Prout, T., and Turelli, M. (1989). Laboratory estimates of heritabilities and gentic correlations in nature. *Genetics*, **123**, 865–871.

Robertson, D.R., Sweatman, H.P.A., Fletcher, E.A., and Clealand, M.E. (1976). Schooling as a mechanism for circumventing the territoriality of competitors. *Ecology*, **57**, 1208–1220.

Røed, K.H. (1979). The temperature preference of the three-spined stickleback, *Gasterosteus aculeatus* L. (Pisces), collected at different seasons. *Sarsia*, **64**, 137–141.

Roeder, M.A. (1985). Late Wisconsin records of *Gasterosteus aculeatus* (threespine stickleback) and *Gila bicolor mojavensis* (Mojave tui chub) from unnamed Mojave River sediments near Daggett, San Bernardino County, California. In *Geological investigations along Interstate 15, Cajon Pass to Manix Lake* (compiler R.E. Reynolds), pp. 171–174. San Bernardino County Museum, Riverside, California.

Roff, D.A. (1984). The evolution of life history parameters in teleosts. *Canadian Journal of Fisheries and Aquatic Science*, **41**, 989–1000.

Roff, D.A. (1988). The evolution of migration and some life history parameters in marine fishes. *Environmental Biology of Fishes*, **22**, 133–146.

Rogers, J.S. (1972). Measures of genetic distance and genetic similiarity. *Studies in genetics VII*, University of Texas Publication No. 7213, pp. 145–153.

Rohlf, F.J. and Bookstein, F.L. (ed.) (1990). *Proceedings of the Michigan morphometrics workshop*, Special Publication Number 2, The University of Michigan Museum of Zoology, Ann Arbor.

Rohwer, S. (1978). Parent cannibalism of offspring and egg raiding as a courtship strategy. *American Naturalist*, **112**, 429–440.

Rome, L.C., Loughna, P.T., and Goldspink, G. (1985). Temperature acclimation: improved sustained swimming performance in carp at low temperatures. *Science*, **228**, 194–196.

Rosen, D.E., and Parenti, L. (1981). Relationships of *Oryzias*, and the groups of atherinomorph fishes. *American Museum Novitates*, **2719**, 1–25.

Rosenzweig, M.L. (1978). Competitive speciation. *Biological Journal of the Linnean Society*, **10**, 275–289.

Roth, F. (1920). Über den Bau und die Entwicklung des Hautzpanzers von *Gasterosteus aculeatus*. *Anatomischer Anzeiger*, **52**, 513–534.

Rothschild, M. (1975). Remarks on carotenoids in the evolution of signals. In *Coevolution of animals and plants* (ed. L.E. Gilbert and P.H. Raven), pp. 20–47. University of Texas Press, Austin.

Rouse, E.F., Coppenger, C.J., and Barnes, P.R. (1977). The effect of an androgen inhibitor on behavior and testicular morphology in the stickleback *Gasterosteus aculeatus*. *Hormones and Behavior*, **9**, 8–18.

Rowland, W.J. (1970). Behavior of three sympatric sticklebacks and its role in their reproductive isolation. Unpublished Ph.D. thesis. State University of New York at Stony Brook.

Rowland, W.J. (1974*a*). Reproductive behavior of the four-spine stickleback *Apeltes quadracus*. *Copeia*, **1974**, 183–194.

Rowland, W.J. (1974*b*). Ground nest construction in the four-spine stickleback, *Apeltes quadracus*. *Copeia*, **1974**, 788–789.

Rowland, W.J. (1982*a*). The effects of male nuptial coloration on stickleback aggression: a re-examination. *Behaviour*, **80**, 118–126.

Rowland, W.J. (1982*b*). Mate choice by male sticklebacks, *Gasterosteus aculeatus*. *Animal Behaviour*, **30**, 1093–1098.

Rowland, W. J. (1983*a*). Interspecific aggression and dominance in *Gasterosteus. Environmental Biology of Fishes*, **8**, 269–277.

Rowland, W. J. (1983*b*). Interspecific aggression in sticklebacks: *Gasterosteus aculeatus* displaces *Apeltes quadracus. Copeia*, **1983**, 541–544.

Rowland, W. J. (1984). The relationships among nuptial coloration, aggression, and courtship of male three-spined sticklebacks, *Gasterosteus aculeatus. Canadian Journal of Zoology*, **62**, 999–1004.

Rowland, W. J. (1988). Aggression versus courtship in threespine sticklebacks and the role of habituation to neighbours. *Animal Behaviour*, **36**, 348–357.

Rowland, W. J. (1989*a*). The effects of body size, aggression and nuptial coloration on competition for territories in male threespine sticklebacks, *Gasterosteus aculeatus. Animal Behaviour*, **37**, 282–289.

Rowland, W. J. (1989*b*). The ethological basis of mate choice in male threespine sticklebacks, *Gasterosteus aculeatus. Animal Behaviour*, **38**, 112–120.

Rowland, W. J. (1989*c*). Mate choice and the supernormality effect in female sticklebacks. *Behavioral Ecology and Sociobiology*, **24**, 433–438.

Rowland, W. J. and Sevenster, P. (1985). Sign stimuli in the threespine stickleback (*Gasterosteus aculeatus*): a re-examination and extension of some classic experiments. *Behaviour*, **93**, 241–257.

Rowland, W. J., Baube, C. L., and Horan, T. T. (1991). Signalling of sexual receptivity by pigmentation pattern in female sticklebacks. *Animal Behaviour*, **42**, 243–249.

Ruben, J. A. (1971). A Pliocene colubrid snake (Reptilia: Colubridae) from west-central Nevada. *Paleobios*, **13**, 1–19.

Ruiter, A. J. H. de (1980). Changes in glomerular structure after sexual maturation and seawater adaptation in males of the euryhaline teleost *Gasterosteus aculeatus* L. *Cell and Tissue Research*, **206**, 1–20.

Ruiter, A. J. H. de (1981). Testosterone-dependent changes *in vivo* and *in vitro* in the structure of the renal glomeruli of the teleost *Gasterosteus aculeatus* L. *Cell and Tissue Research*, **219**, 253–266.

Ruiter, A. J. H. de and Mein, C. G. (1982). Testosterone-dependent transformation of nephronic tubule cells into serous and mucous gland cells in stickleback kidneys *in vivo* and *in vitro. General and Comparative Endocrinology*, **47**, 70–83.

Ruiter, A. J. H. de, Schilstra, A. J., and Wendelaar Bonga, S. E. (1984). Androgen actions on intestinal participation in hydromineral balance of the three-spined stickleback *Gasterosteus aculeatus* L. [abstract]. *General and Comparative Endocrinology*, **53**, 446.

Ruiter, A. J. H. de, Hoogeveen, Y. L., and Wendelaar Bonga, S. E. (1985). Ultrastructure of intestinal and gall bladder epithelium in the teleost *Gasterosteus aculeatus* L., as related to their osmoregulatory function. *Cell and Tissue Research*, **240**, 191–198.

Ruiter, A. J. H. de, Wendelaar Bonga, S. E., Slijkhuis, H., and Baggerman, B. (1986). The effect of prolactin on fanning behavior in the male three-spined stickleback, *Gasterosteus aculeatus* L. *General and Comparative Endocrinology*, **64**, 273–284.

Ryan, M. J. (1990). Sexual selection, sensory systems, and sensory exploitation. In *Oxford surveys in evolutionary biology* (ed. D. Futuyma and J. Antonovics), Vol. 7, pp. 157–195. Oxford University Press, New York.

Sadler, P. M. (1981). Sediment accumulation rates and the completeness of stratigraphic sections. *Journal of Geology*, **89**, 569–584.

Sakamoto, T., Ogasawara, T., and Hirano, T. (1990). Growth hormone kinetics during adaptation to hyperosmotic environment in rainbow trout. *Journal of Comparative Physiology*, **106**, 1-6.

Sanderson, N. (1989). Can gene flow prevent reinforcement? *Evolution*, **43**, 1223-1235.

Sargent, R.C. (1982). Territory quality, male quality, courtship intrusions, and female nest-choice in the threespine stickleback, *Gasterosteus aculeatus. Animal Behaviour*, **30**, 364-374.

Sargent, R.C. (1985). Territoriality and reproductive tradeoffs in the threespine stickleback, *Gasterosteus aculeatus. Behaviour*, **93**, 217-226.

Sargent, R.C. (1992). Ecology of filial cannibalism in fish: theoretical perspectives. In *Cannibalism: ecology and evolution among diverse taxa* (ed. M.A. Elger and B.J. Crespi), pp. 38-62. Oxford University Press.

Sargent, R.C. and Gebler, J.B. (1980). Effects of nest site concealment on hatching success, reproductive success, and paternal behavior of the threespine stickleback, *Gasterosteus aculeatus. Behavioral Ecology and Sociobiology*, **7**, 137-142.

Sargent, R.C. and Gross, M.R. (1986). Williams' principle: an explanation of parental care in teleost fishes. In *The behaviour of teleost fishes* (1st edn) (ed. T.J. Pitcher), pp. 275-293. Croom Helm, London.

Sargent, R.C., Bell, M.A., Krueger, W.H., and Baumgartner, J.V. (1984). A lateral plate cline, sexual dimorphism, and phenotypic variation in the black-spotted stickleback, *Gasterosteus wheatlandi. Canadian Journal of Zoology*, **62**, 368-376.

Sargent, R.C., Gross, M.R., and Berghe, E.P. van den (1986). Male mate choice in fishes. *Animal Behaviour*, **34**, 545-550.

Savage, D.E. and Russell, D.E. (1983). *Mammalian paleofaunas of the world.* Addison Wesley, London.

Savino, J.F. and Stein, R.A. (1989). Behavioural interactions between fish predators and their prey: effects of plant density. *Animal Behaviour*, **37**, 311-321.

Schaeffer, B., Hecht, M., and Eldredge, N. (1972). Phylogeny and paleontology. *Evolutionary Biology*, **6**, 31-46.

Schindel, D.E. (1980). Microstratigraphic sampling and the limits of paleontological resolution. *Paleobiology*, **6**, 408-426.

Schindel, D.E. (1982*a*). Resolution analysis: a new approach to the gaps in the fossil record. *Paleobiology*, **8**, 340-353.

Schindel, D.E. (1982*b*). The gaps in the fossil record. *Nature*, **297**, 282-284.

Schluter, D. and McPhail, J.D. (1992). Ecological character displacement and speciation in sticklebacks. *American Naturalist*, **140**, 85-108.

Schluter, D., Price, T.D., and Grant, P.R. (1985). Ecological character displacement in Darwin's finches. *Science*, **227**, 1056-1058.

Schmidt-Nielsen, B. (1977). Volume regulation of muscle fibres in the killifish, *Fundulus heteroclitus. Journal of Experimental Zoology*, **199**, 411-418.

Schneider, L. (1969). Experimentelle Untersuchungen über den Einfluss von Tageslänge und Temperatur auf die Gonadenreifung beim Dreistachligen Stichling (*Gasterosteus aculeatus*). *Oecologia*, **3**, 249-265.

Schopf, T.J.M., Raup, D.M., Gould, S.J., and Simberloff, D.S. (1975). Genomic versus morphologic rates of evolution: influence of morphologic complexity. *Paleobiology*, **1**, 63-70.

Schreibman, M.P. and Margolis-Nunno, H. (1987). Reproductive biology of the terminal nerve (nucleus olfactoretinalis) and other LHRH pathways in teleost fishes. *Annals of the New York Academy of Sciences*, **519**, 60-68.

Schreibman, M.P., Holtzman, S., and Eckhardt, R.A. (1989). Genetic influences on reproductive system development and function: a review. *Fish Physiology and Biochemistry*, **7**, 237–242.

Schtyl'ko, B. (1934). A Neogene fauna of fresh-water fishes from western Siberia. *United Geological Prospecting Service USSR*, **359**, 82–91.

Schütz, E. (1980). Die Wirkung von untergrund und Nestmaterial auf das Nestbauverhalten des Dreistachligen Stichlings (*Gasterosteus aculeatus*). *Behaviour*, **72**, 242–317.

Scott, W.B. and Crossman, E.J. (1964). *Fishes occurring in the fresh waters of insular Newfoundland*. Queen's Printer, Ottawa.

Scott, W.B. and Crossman, E.J. (1973). Freshwater fishes of Canada. *Bulletin of the Fisheries Research Board of Canada*, **184**, 1–966.

Segaar, J. (1961). Telencephalon and behaviour in *Gasterosteus aculeatus* males. *Behaviour*, **18**, 256–287.

Segaar, J., Bruin, J.P.C. de, Meché, A.P. van der, and Meché-Jacobi, M.E. van der. (1983). Influence of chemical receptivity on reproductive behaviour of the male three-spined stickleback (*Gasterosteus aculeatus*): an ethological analysis of cranial nerve functions regarding nest funning and the zigzag dance. *Behaviour*, **86**, 100–166.

Seghers, B.H. (1974). Schooling behaviour in the guppy (*Poecilia reticulata*): an evolutionary response to predation. *Evolution*, **28**, 286–289.

Selander, R.K., Smith, M.H., Yang, S.Y., Johnson, W.E., and Gentry, J.B. (1971). Biochemical polymorphism and systematics in the genus *Peromyscus*. I. Variation in the old-field mouse (*Peromyscus polionotus*). *Studies in Genetics VI*, University of Texas Publication No. 7103, pp. 49–90.

Semler, D.E. (1971). Some aspects of adaptation in a polymorphism for breeding colours in the threespine stickleback (*Gasterosteus aculeatus*). *Journal of Zoology, London*, **165**, 291–302.

Sevenster, P. (1961). A causal analysis of a displacement activity (fanning in *Gasterosteus aculeatus* L.). *Behaviour Supplement*, **9**, 1–170.

Sevenster, P. (1968). Motivation and learning in sticklebacks. In *The central nervous system and fish behavior* (ed. D. Ingle), pp. 233–245. University of Chicago Press, Chicago.

Sevenster, P. (1973). Incompatibility of response and reward. In *Constraints on learning* (ed. R.A. Hinde and J. Stevenson-Hinde), pp. 265–283. Academic Press, London.

Sevenster, P. and Goyens, J. (1975). Expérience sociale et rôle du sexe dans l'ontogénèse de l'agressivité chez l'épinoche (*Gasterosteus aculeatus* L.). *Netherlands Journal of Zoology*, **25**, 195–205.

Sevenster-Bol, A.C.A. (1962). On the causation of drive reduction after a consummatory act (in *Gasterosteus aculeatus* L.). *Archives Néerlandaises de Zoologie*, **15**, 175–236.

Sibly, R.M. and McFarland, D.J. (1976). On the fitness of behavior sequences. *American Naturalist*, **110**, 601–617.

Siciliano, M.J. and Shaw, C.R. (1976). Separation and visualization of enzymes on gels. In *Chromatographic and electrophoretic techniques*, Vol. 2 (ed. I. Smith), pp. 185–209. Heinemann Medical, London.

Sidell, B.D. and Moerland, T.S. (1989). Effects of temperature on muscular function and locomotory performance in a teleost fish. In *Advances in comparative and environmental physiology* (ed. M. Brouwer), Vol. 5, pp. 115–156. Springer-Verlag, Heidelberg.

Sikkel, P.C. (1989). Egg presence and development stage influence spawning-site choice by female garibaldi. *Animal Behaviour*, **38**, 447–456.

Sillen-Tullberg, B. (1988). Evolution of gregariousness in aposematic butterfly larvae: a phylogenetic analysis. *Evolution*, **42**, 293–305.

Simon, C. (1987). Hawaiian evolutionary biology: an introduction. *Trends in Ecology and Evolution*, **2**, 175–178.

Simpson, G.G. (1944). *Tempo and mode in evolution*. Columbia University Press, New York.

Simpson, G.G. (1953). *The major features of evolution*. Columbia University Press, New York.

Sisson, J.E. and Sidell, B.D. (1987). Effect of thermal acclimation on muscle fiber recruitment of swimming striped bass (*Morone saxatilis*). *Physiological Zoology*, **60**, 310–320.

Sjoberg, K. (1985). Foraging activity patterns in the goosander *Mergus merganser* and the red-breasted merganser *Mergus serrator* in relation to patterns of activity in their major prey species. *Oecologia*, **67**, 35–39.

Sjoberg, K. (1989). Time-related predator/prey interactions between birds and fish in a northern Swedish river. *Oecologia*, **80**, 1–10.

Skutch, A.F. (1955). The parental stratagems of birds. *Ibis*, **97**, 118–142.

Slatkin, M. (1985). Gene flow in natural populations. *Annual Review of Ecology and Systematics*, **16**, 393–430.

Slatkin, M. (1987). Gene flow and the geographic structure of natural populations. *Science*, **236**, 787–792.

Slijkhuis, H. (1978). Ultrastructural evidence for two types of gonadotropic cells in the pituitary gland of the male three-spined stickleback, *Gasterosteus aculeatus*. *General and Comparative Endocrinology*, **36**, 639–641.

Slijkhuis, H., Ruiter, A.J.H. de, Baggerman, B., and Wendelaar Bonga, S.E. (1984). Parental fanning behavior and prolactin cell activity in the male three-spined stickleback, *Gasterosteus aculeatus*. *General and Comparative Endocrinology*, **54**, 297–307.

Slobodkin, L.B. and Rapoport, A. (1974). An optimal strategy of evolution. *Quarterly Review of Biology*, **49**, 181–200.

Smith, G.R. (1981*a*). Effects of habitat size on species richness and adult body sizes of desert fishes. In *Fishes in North American deserts* (ed. R.J. Naiman and D.L. Soltz), pp. 125–177. Wiley, New York.

Smith, G.R. (1981*b*). Late Cenozoic freshwater fishes of North America. *Annual Review of Ecology and Systematics*, **12**, 163–193.

Smith, G.R. and Elder, R.L. (1985). Environmental interpretation of burial and preservation of Clarkia fishes. In *Late Cenozoic history of the Pacific Northwest* (ed. C.J. Smiley), pp. 85–93. Pacific Division, American Association for the Advancement of Science, San Francisco.

Smith, G.R. and Todd, T.N. (1984). Evolution of species flocks of fishes in north temperate lakes. In *Evolution of fish species flocks* (ed. A.A. Echelle and I. Kornfield), pp. 45–68. University of Maine, Orono.

Smith, M.L. (1981). Late Cenozoic fishes in the warm deserts of North America: a reinterpretation of desert adaptations. In *Fishes in North American deserts* (ed. R.J. Naiman and D.L. Soltz), pp. 11–38. Wiley, New York.

Smith, R.J.F. (1970). Effects of food availability on aggression and nest building in brook stickleback (*Culaea inconstans*). *Journal of the Fisheries Research Board of Canada*, **27**, 2350–2355.

Smith, R.S. and Whoriskey, F.G. (1988). Multiple clutches: female threespine sticklebacks lose the ability to recognize their own eggs. *Animal Behaviour*, **36**, 1838–1839.

Snyder, R.J. (1984). Seasonal variation in the diet of the threespine stickleback, *Gasterosteus aculeatus*, in Contra Costa County, California. *California Fish and Game*, **70**, 167–172.

Snyder, R.J. (1988). Migration and the evolution of life histories of the threespine stickleback (*Gasterosteus aculeatus* L.). Unpublished Ph.D. thesis. University of California, Davis.

Snyder, R.J. (1990). Clutch size of anadromous and freshwater sticklebacks: a reassessment. *Canadian Journal of Zoology*, **68**, 2027–2030.

Snyder, R.J. (1991*a*). Migration and life histories of the threespine stickleback: evidence for adaptive variation in growth rate between populations. *Environmental Biology of Fishes*, **31**, 381–388.

Snyder, R.J. (1991*b*). Quantitative genetic analysis of life histories of two freshwater populations of the threespine stickleback. *Copeia*, **1991**, 526–529.

Snyder, R.J. and Dingle, H. (1989). Adaptive, genetically based differences in life history between estuary and freshwater threespine sticklebacks (*Gasterosteus aculeatus* L.). *Canadian Journal of Zoology*, **67**, 2448–2454.

Snyder, R.J. and Dingle, H. (1990). Effects of freshwater and marine overwintering environments on life histories of threespine sticklebacks: evidence for adaptive variation between anadromous and resident freshwater populations. *Oecologia*, **84**, 386–390.

Sokal, R.R. (1978). Population differentiation: something new or more of the same? In *Ecological genetics: the interface* (ed. P.F. Brussard), pp. 215–239. Springer-Verlag, New York.

Sokal, R.R. and Rohlf, F.J. (1981). *Biometry* (2nd edn). Freeman, New York.

Somero, G.N. (1986). Protons, osmolytes and fitness of internal milieu for protein function. *American Journal of Physiology*, **251**, R197–R213.

Soofiani, N.M. and Hawkins, A.D. (1982). Energetic costs at different levels of feeding in juvenile cod, *Gadus morhua* L. *Journal of Fish Biology*, **21**, 577–592.

Stamps, J.A. (1988). Conspecific attraction and aggregation in territorial species. *American Naturalist*, **131**, 329–347.

Stanley, B.V. (1983). Effect of food supply on reproductive behaviour of male *Gasterosteus aculeatus*. Unpublished Ph.D. thesis. University of Wales, Aberystwyth.

Stanley, B.V. and Wootton, R.J. (1986). Effects of ration and male density on the territoriality and nest-building of male three-spined sticklebacks (*Gasterosteus aculeatus* L.). *Animal Behaviour*, **34**, 527–535.

Stanley, S.M. (1979). *Macroevolution: pattern and process*. Freeman, San Francisco.

Starks, E.C. (1902). The shoulder girdle and characteristic osteology of the hemibranchiate fishes. *Proceedings of the US National Museum*, **25**, 619–634.

Stearns, S.C. (1976). Life history tactics: a review of the ideas. *Quarterly Review of Biology*, **51**, 3–47.

Stearns, S.C. and Crandall, R.E. (1984). Plasticity for age and size at sexual maturity: a life-history response to unavoidable stress. In *Fish reproduction: strategies and tactics* (ed. G.W. Potts and R.J. Wootton), pp. 13–33. Academic Press, London.

Stearns, S.C. and Koella, J.C. (1986). The evolution of phenotypic plasticity in life-history traits: predictions of reaction norms for age and size at maturity. *Evolution*, **40**, 893–913.

Stephens, D.W. and Krebs, J.R. (1986). *Foraging theory*. Princeton University Press, Princeton, New Jersey.

Stinson, E.M. (1983). Threespine sticklebacks (*Gasterosteus aculeatus*) in Drizzle

Lake and its inlet, Queen Charlotte Islands: ecological and behavioural relationships and their relevance to reproductive isolation. Unpublished M.Sc. thesis. University of Alberta, Edmonton.

Strauss, R.E. (1979). Reliability estimates for Ivlev's electivity index, the foraging ratio, and a proposed linear index of food selection. *Transactions of the American Fisheries Society*, **108**, 344–352.

Stroband, H.W.J. and Debets, F.M.H. (1978). The ultrastructure and renewal of the intestinal epithelium of the juvenile grasscarp *Ctenopharyngodon idella* (Val.). *Cell and Tissue Research*, **187**, 181–200.

Sutherland, W.J. and Parker, G.A. (1985). Distribution of unequal competitors. In *Behavioural ecology: ecological consequences of adaptive behaviour*, (ed. R.M. Sibly and R.H. Smith), pp. 255–273. Blackwell, Oxford.

Suzuki, K., Kawauchi, H., and Nagahama, Y. (1988*a*). Isolation and characterization of two distinct gonadotropins from chum salmon pituitary glands. *General and Comparative Endocrinology*, **71**, 292–301.

Suzuki, K., Kawauchi, H., and Nagahama, Y. (1988*b*). Isolation and characterization of subunits from two distinct salmon gonadotropins. *General and Comparative Endocrinology*, **71**, 302–306.

Svärdson, G. (1961). Young sibling fish species in northwesten Europe. In *Vertebrate speciation* (ed. W.F. Blair), pp. 498–513. University of Texas Press, Austin.

Swain, D.P. (1986). Adaptive significance of variation in vertebral number in fishes: evidence in *Gasterosteus aculeatus* and *Mylocheilus caurinus*. Unpublished Ph.D. thesis. University of British Columbia, Vancouver.

Swain, D.P. (1992*a*). Selective predation for vertebral phenotype in *Gasterosteus aculeatus*: reversal in the direction of selection at different larval sizes. *Evolution*, **46**, 998–1013.

Swain, D.P. (1992*b*). The functional basis of natural selection for vertebral traits of larvae in the stickleback *Gasterosteus aculeatus. Evolution*, **46**, 987–997.

Swain, D.P. and Lindsey, C.C. (1984). Selective predation for vertebral number of young sticklebacks, *Gasterosteus aculeatus. Canadian Journal of Fisheries and Aquatic Sciences*, **41**, 1231–1233.

Swarup, H. (1958). Stages in the development of the stickleback *Gasterosteus aculeatus* (L.). *Journal of Embryology and Experimental Morphology*, **6**, 373–383.

Sweeting, R.M., Wagner, G.F., and McKeown, B.A. (1985). Changes in plasma glucose, amino acid nitrogen and growth hormone during smoltification and seawater adaptation in coho salmon, *Oncorhynchus kisutch. Aquaculture*, **45**, 185–197.

Swift, C.C. (1989). Late Pleistocene freshwater fishes from the Rancho La Brea deposit, southern California. *Bulletin of the Southern California Academy of Sciences*, **88**, 93–102.

Swinnerton, H.H. (1902). A contribution to the morphology of the teleostean head skeleton. *Quarterly Journal of Microscopical Science*, **45**, 503–593.

Sychevskaya, Y.K. and Grechina, N.I. (1981). Fossil sticklebacks of the genus *Gasterosteus* from the Neogene of the Soviet Far East. *Paleontological Journal*, **1**, 71–80. (transl. of *Paleontologicheskii Zhurnal*).

Symons, P.E.K. (1965). Analysis of spine-raising in the male three-spined stickleback. *Behaviour*, **26**, 1–74.

Symons, P.E.K. (1971). Spacing and density in schooling threespine stickleback (*Gasterosteus aculeatus*) and mummichog (*Fundulus heteroclitus*). *Journal of the Fisheries Research Board of Canada*, **28**, 999–1004.

Taylor, E.B. and McPhail, J.D. (1986). Prolonged and burst swimming in anadromous and freshwater threespine stickleback, *Gasterosteus aculeatus*. *Canadian Journal of Zoology*, **64**, 416–420.

Taylor, W.R. (1967). Outline of a method of clearing tissues with pancreatic enzymes and staining bones of small vertebrates. *Turtox News*, **45**, 308–309.

Teichmann, H. (1954). Vergleichende Untersuchungen an der Nase der Fische. *Zeitschrift der Morphologie und Ökologie der Tiere*, **43**, 171–212.

Templeton, A.R. (1981). Mechanisms of speciation–a population genetic approach. *Annual Review of Ecology and Systematics*, **12**, 23–48.

Templeton, A.R. (1987). Species and speciation. *Evolution*, **41**, 233–235.

Thomas, G. (1974). The influence of encountering a food object on the subsequent searching behaviour in *Gasterosteus aculeatus* L. *Animal Behaviour*, **22**, 941–952.

Thomas, G., Kacelnik, A., and Meulen, J. van der (1985). The three-spined stickleback and the two-armed bandit. *Behaviour*, **93**, 227–240.

Thompson, D.B. (1990). Different spatial scales of adaptation in the climbing behavior of *Peromyscus maniculatus*: geographic variation, natural selection and gene flow. *Evolution*, **44**, 952–965.

Thompson, S.C. and Raveling, D.G. (1987). Incubation behavior of emperor geese compared with other geese: interactions of predation, body size and energetics. *Auk*, **104**, 707–716.

Thorman, S. (1983). Food and habitat resource partitioning between three estuarine fish species on the Swedish west coast. *Estuarine, Coastal and Shelf Science*, **17**, 681–692.

Thorman, S. and Wiederholm, A.-M. (1983). Seasonal occurrence and food resource use of an assemblage of nearshore fish species in the Bothnian Sea, Sweden. *Marine Ecology Progress Series*, **10**, 223–229.

Thorman, S. and Wiederholm, A.-M. (1984). Species composition and dietary relationships in a brackish shallow water fish assemblage in the Bothnian Sea, Sweden. *Estuarine, Coastal and Shelf Science*, **19**, 359–371.

Thorman, S. and Wiederholm, A.-M. (1986). Food, habitat and time niches in a coastal fish species assemblage in a brackish water bay in the Bothnian Sea, Sweden. *Journal of Experimental Marine Biology and Ecology*, **95**, 67–86.

Thornhill, R. and Alcock, J. (1983). *The evolution of insect mating systems*. Harvard University Press, Cambridge, Massachusetts.

Thorpe, R.S. (1976). Biometrical analysis of geographic variation and racial affinities. *Biological Reviews of the Cambridge Philosophical Society*, **51**, 407–452.

Threlfall, W. (1968). A mass die-off of threespined sticklebacks (*Gasterosteus aculeatus* L.). *Canadian Journal of Zoology*, **46**, 105–106.

Tierney, J.F., Huntingford, F.A., and Crompton, D. (in press). Relationship between infectivity of *Schistocephalus solidus* and the anti-predator behaviour of three-spined sticklebacks. *Animal Behaviour*.

Tinbergen, N. (1948). Social releasers and the experimental method required for their study. *Wilson Bulletin*, **60**, 6–51.

Tinbergen, N. (1951). *The study of instinct*. Oxford University Press, New York, and Clarendon, Oxford.

Tinbergen, N. (1952). Derived activities: their causation, biological significance, origin, and emancipation during evolution. *Quarterly Review of Biology*, **27**, 1–32.

Tinbergen, N. (1953). *The social behaviour of animals*. Methuen, London.

Tinbergen, N. (1954). The origin and evolution of threat and courtship display. In *Evolution as a process* (ed. A.C. Hardy, J.S. Huxley, and E.B. Ford), pp. 233–250. Allen and Unwin, London.

Tinbergen, N. (1963). On the aims and methods of ethology. *Zeitschrift für Tierpsychologie*, **20**, 410–433.

Tinbergen, N., Broekhuysen, G. J., Feekes, F., Houghton, J. C. W., Krunk, H., and Szulc, E. (1962). Egg shell removal by the Black-headed gull, *Larus ridibundus*: a behavioural component of camouflage. *Behaviour*, **19**, 74–117.

Titschack, E. (1922). Die sekundaren Geschlechtmerkmale von *Gasterosteus aculeatus* L. *Zoologische Jahrbucher (Abteilungen III: Allgemeine Zoologie und Physiologie)*, **39**, 83–148.

Towers, S. R. and Coss, R. G. (1990). Confronting snakes in the burrow: snake-species discrimination and anti-snake tactics of two California ground squirrel populations. *Ethology*, **84**, 177–192.

Trendall, J. T. (1982). Covariation in life history traits in the mosquitofish, *Gambusia affinis*. *American Naturalist*, **119**, 774–783.

Trexler, J. C. (1989). Phenotypic plasticity in poeciliid life histories. In *Ecology and evolution of livebearing fishes (Poeciliidae)* (ed. G. K. Meffe and F. F. Snelson, Jr.), pp. 201–214. Prentice Hall, Englewood Cliffs, New Jersey.

Trivers, R. L. (1972). Parental investment and sexual selection. In *Sexual selection and the descent of man, 1871–1971* (ed. B. Campbell), pp. 136–179. Aldine, Chicago.

Tschanz, B. and Scharf, M. (1971). Nestortwahl und Orientierung zum Nestort beim Dreistachligen Stichling. *Revue Suisse de Zoologie*, **78**, 715–721.

Tugendhat, B. (1960). The normal feeding behaviour of the three-spined stickleback (*Gasterosteus aculeatus* L.). *Behaviour*, **15**, 284–318.

Tulley, J. J. (1985). Studies on the development of anti-predator responses and aggression in the threespined stickleback (*Gasterosteus aculeatus*). Unpublished Ph.D. thesis. University of Glasgow.

Tulley, J. J. and Huntingford, F. A. (1987*a*). Age, experience and the development of adaptive variation in anti-predator responses in three-spined sticklebacks (*Gasterosteus aculeatus*). *Ethology*, **75**, 285–290.

Tulley, J. J. and Huntingford, F. A. (1987*b*). Parental care and the development of adaptive variation in anti-predator responses in sticklebacks. *Animal Behaviour*, **35**, 1570–1572.

Tulley, J. J. and Huntingford, F. A. (1988). Additional information on the relationship between intra-specific aggression and anti-predator behaviour in the three-spined stickleback, *Gasterosteus aculeatus*. *Ethology*, **78**, 219–222.

Tuomi, J., Hakala, T., and Haukioja, E. (1983). Alternative concepts of reproductive effort, costs of reproduction, and selection in life-history evolution. *American Zoologist*, **23**, 25–34.

Turesson, G. (1922). The genotypic response of the plant species to the habitat. *Hereditas*, **3**, 221–350.

Uchida, K. (1934). Life history of *Aulichthys japonicus* Brevoort (Hemibranchii, Pisces). *Japanese Journal of Zoology*, **5**, 4–5.

Ukegbu, A. A. (1986). Life history patterns and reproduction in the three-spined stickleback *Gasterosteus aculeatus*. Unpublished Ph.D. thesis. University of Glasgow.

Ukegbu, A. A. and Huntingford, F. A. (1988). Brood value and life expectancy as determinants of parental investment in male three-spined sticklebacks, *Gasterosteus aculeatus*. *Ethology*, **78**, 72–82.

Ukegbu, A. A. and Huntingford, F. A. (1989). Age structure and growth in Scottish populations of the three-spined stickleback. *Scottish Naturalist*, 1989, 19–43.

Unger, L. M. and Sargent, R. C. (1988). Allopaternal care in the fathead minnow, *Pimephales promelas*: females prefer males with eggs. *Behavioral Ecology and Sociobiology*, **23**, 27–32.

United States Department of Agriculture, Forest Service (1979). *Tongass land management plan: final environmental impact statement*. US Department of Agriculture, Forest Service, Juneau, Alaska.

Van Valen, L. (1965). Morphological variation and width of ecological niche. *American Naturalist*, **99**, 377–389.

Vawter, L. and Brown, W. M. (1986). Nuclear and mitochondrial DNA comparisons reveal extreme rate variation in the molecular clock. *Science*, **234**, 194–196.

Veen, T. van, Ekström, P., Borg, B., and Møller, M. (1980). The pineal complex of the threespined stickleback, *Gasterosteus aculeatus* L.: a light-electron microscopic and fluorescence histochemical investigation. *Cell and Tissue Research*, **209**, 11–28.

Veen, T. van, Ekström, P., Nyberg, L., Borg, B., Vigh-Teichmann, I., and Vigh, B. (1984). Serotonin and opsin immunoreactives in the developing pineal organ of the three-spined stickleback, *Gasterosteus aculeatus* L. *Cell and Tissue Research*, **237**, 559–564.

Vermeij, G. J. (1982). Unsuccessful predation and evolution. *American Naturalist*, **120**, 701–720.

Vermeij, G. J. (1987). *Evolution and escalation*. Princeton University Press, Princeton, New Jersey.

Vermeij, G. J. (1991). Anatomy of an invasion: the trans-Arctic interchange. *Paleobiology*, **17**, 281–307.

Verrell, P. A. and Arnold, S. J. (1989). Behavioral observations of sexual isolation among allopatric populations of the mountain dusky salamander, *Desmognathus ochrophaeus*. *Evolution*, **43**, 745–755.

Vézina, D. and Guderley, H. (1992). Anatomic and enzymatic responses of the threespine stickleback, *Gasterosteus aculeatus*, to thermal acclimation and acclimatization. *Journal of Experimental Zoology*, **258**, 277–287.

Villars, T. A. (1983). Hormones and aggressive behavior in teleost fishes. In *Hormones and aggressive behaviour* (ed. B. B. Svare), pp. 407–433. Plenum Press, New York.

Visser, M. (1982). Prey selection by the three-spined stickleback (*Gasterosteus aculeatus* L.). *Oecologia*, **55**, 395–402.

Vitt, L. J. and Congdon, J. D. (1978). Body shape, reproductive effort, and relative clutch mass in lizards: resolution of a paradox. *American Naturalist*, **112**, 595–608.

Vivien-Roels, B. (1983). The pineal gland and the integration of environmental information: possible role of hydroxy- and methoxyindoles. *Molecular Physiology*, **4**, 331–345.

Vladykov, V. D. and Kott, E. (1979). Satellite species among the holarctic lampreys (Petromyzonidae). *Canadian Journal of Zoology*, **57**, 860–867.

Vliet, W. H. van (1970). *Shore and freshwater fish collections from Newfoundland*, Publications in Zoology No. 3. National Museum of Natural Sciences, Ottawa.

Voights, D. K. (1976). Aquatic invertebrate abundance in relation to changing marsh vegetation. *American Midland Naturalist*, **95**, 313–322.

Vrat, V. (1949). Reproductive behavior and development of eggs of the three-spined stickleback (*Gasterosteus aculeatus*) of California. *Copeia*, **1949**, 252–260.

Vrijenhoek, R. C., Angus, R. A., and Schultz, R. J. (1977). Variation and heterozygosity in sexually vs. clonally reproducing populations of *Poeciliopsis*. *Evolution*, **31**, 767–781.

Wahli, W., David, I., Ruffer, G.U., and Weber, R. (1981). Vitellogenin and the vitellogenin gene family. *Science*, **212**, 298–304.

Wai, E.H. and Hoar, W.S. (1963). The secondary sex characters and reproductive behavior of gonadectomized sticklebacks treated with methyl testosterone. *Canadian Journal of Zoology*, **41**, 611–628.

Wake, D.B. and Larson, A. (1987). Multidimensional analysis of an evolving lineage. *Science*, **238**, 42–48.

Walkey, M. and Meakins, R.H. (1970). An attempt to balance the energy budget of a host–parasite system. *Journal of Fish Biology*, **2**, 361–372.

Wallace, J.C., and Aasjord, D. (1984). An investigation of the consequences of egg size for the culture of Arctic Charr, *Salvelinus alpinus* (L.) *Journal of Fish Biology*, **24**, 427–435.

Wallace, R.A. and Selman, K. (1978). Oogenesis in *Fundulus heteroclitus*. I. Preliminary observations on oocyte maturation *in vivo* and *in vitro*. *Developmental Biology*, **62**, 354–369.

Wallace, R.A. and Selman, K. (1979). Physiological aspects of oogenesis in two species of sticklebacks, *Gasterosteus aculeatus* L. and *Apeltes quadracus* (Mitchill). *Journal of Fish Biology*, **14**, 551–564.

Wallace, R.A. and Selman, K. (1981). Cellular and dynamic aspects of oocyte growth in teleosts. *American Zoologist*, **21**, 325–343.

Wallace, R.A. and Selman, K. (1982). Oocyte growth in the sheephead minnow, uptake of exogenous proteins by vitellogenic oocytes. *Tissue and Cell*, **14**, 555–571.

Walsh, G. and FitzGerald, G.J. (1984). Resource utilization and coexistence of three species of sticklebacks (Gasterosteidae) in tidal salt-marsh pools. *Journal of Fish Biology*, **25**, 405–420.

Wanntorp, H.-E., Brooks, D.R., Nilsson, T., Nylin, S., Ronquist, F., Stearns S.C., and Wedell, N. (1990). Phylogenetic approaches in ecology. *Oikos*, **57**, 119–132.

Ward, G. and FitzGerald, G.J. (1983). Macrobenthic abundance and distribution in tidal pools of a Quebec salt marsh. *Canadian Journal of Zoology*, **61**, 232–241.

Ward, G. and FitzGerald, G.J. (1987). Male aggression and female mate choice in the threespine stickleback, *Gasterosteus aculeatus* L. *Journal of Fish Biology*, **30**, 679–690.

Ward, G. and FitzGerald, G.J. (1988). Effects of sex ratio on male behaviour and reproductive success in a field population of threespine sticklebacks (*Gasterosteus aculeatus*) (Pisces: Gasterosteidae). *Journal of Zoology, London*, **215**, 597–610.

Ware, D.M. (1975). Relation between egg size, growth and natural mortality of larval fish. *Journal of the Fisheries Research Board of Canada*, **32**, 2503–2512.

Ware, D.M. (1978). Bioenergetics of pelagic fish: theoretical change in swimming speed and relation with body size. *Journal of the Fisheries Research Board of Canada*, **35**, 220–228.

Ware, D.M. (1984). Fitness of different reproductive strategies in teleost fishes. In *Fish reproduction: strategies and tactics* (ed. G.W. Potts and R.J. Wootton), pp. 349–366. Academic Press, London.

Warington, R. (1852). Observations on the natural history of the water snail and fish kept in a confined and limited portion of water. *Annals and Magazine of Natural History*, **(2)10**, 273–280.

Warington, R. (1855). Observations on the habits of the stickleback (being a continuation of a previous paper). *Annals and Magazine of Natural History*, **(2)16**, 330–332.

Warner, B.G., Mathewes, R.W., and Clague, J.J. (1982). Ice-free conditions on the

Queen Charlotte Islands, British Columbia, at the height of Late Wisconsin glaciation. *Science*, **218**, 675–677.

Warner, R.R. (1988). Traditionality of mating-site preferences in a coral reef fish. *Nature*, **335**, 719–721.

Warner, R.R. (1990). Resource assessment versus traditionality in mating site determination. *American Naturalist*, **135**, 205–217.

Wasserman, M. and Koepfer, H.R. (1980). Does asymmetrical mating preference show the direction of evolution? *Evolution*, **34**, 1116–1126.

Webb, P.W. (1975). Hydrodynamics and energetics of fish propulsion. *Bulletin of the Fisheries Research Board of Canada*, **190**, 1–159.

Webb, P.W. (1978). Fast-start performance and body form in seven species of teleost fish. *Journal of Experimental Biology*, **74**, 211–226.

Webb, P.W. (1982). Locomotor patterns in the evolution of actinopterygian fishes. *American Zoologist*, **22**, 329–342.

Webb, P.W. (1984). Body form, locomotion and foraging in aquatic vertebrates. *American Zoologist*, **24**, 107–120.

Webb, P.W. and Weihs, D. (ed.) (1983). *Fish biomechanics*. Praeger, New York.

Weeks, H.J. (1985*a*). Ecology of the threespine stickleback (*Gasterosteus aculeatus*): reproduction in rocky tidepools. Unpublished Ph.D. thesis. Cornell University, Ithaca, New York.

Weeks, H.J. (1985*b*). Growth of juvenile threespine sticklebacks (*Gasterosteus aculeatus*) in rocky tidepools. *American Zoologist*, **25**, (4), 53A.

Welton, B.J. and Link, M.H. (1982). Vertebrate paleontology of Ridge basin, southern California. In *Geologic history of Ridge basin, Southern California* (ed. J.C. Crowell and M.H. Link), pp. 205–210. Pacific Section of the Society of Economic Paleontologists and Mineralogists, Los Angeles.

Wendelaar Bonga, S.E. (1976). The effect of prolactin on kidney structure of the euryhaline teleost *Gasterosteus aculeatus* during adaptation to fresh water. *Cell and Tissue Research*, **66**, 319–338.

Wendelaar Bonga, S.E. (1978). The effects of changes in external sodium, calcium and magnesium concentrations on prolactin cells, skin, and plasma electrolytes of *Gasterosteus aculeatus. General and Comparative Endocrinology*, **34**, 265–275.

Wendelaar Bonga, S.E. (1980). Effect of synthetic salmon calcitonin and low ambient calcium on plasma calcium, ultimobranchial cells, Stannius bodies, and prolactin cells in the teleost, *Gasterosteus aculeatus. General and Comparative Endocrinology*, **40**, 99–108.

Wendelaar Bonga, S.E. and Greven, J.A.A. (1978). The relationship between prolactin cell activity, environmental calcium, and plasma calcium in the teleost *Gasterosteus aculeatus*: observations on stanniectomized fish. *General and Comparative Endocrinology*, **36**, 90–101.

Wendelaar Bonga, S.E. and Veenhuis, M. (1974). The effect of prolactin on the number of membrane-associated particles in kidney cells of the euryhaline teleost *Gasterosteus aculeatus* during transfer from seawater to freshwater: a freeze-etch study. *Journal of Cell Science*, **16**, 687–701.

Wendelaar Bonga, S.E., Greven, J.A.A., and Veenhuis, M. (1977). Vascularization, innervation and ultrastructure of the endocrine cell types of Stannius corpuscles in the teleost *Gasterosteus aculeatus. Journal of Morphology*, **153**, 225–244.

Wenderoff, L.R. (1982). Trophic competition between threespine stickleback (*Gasterosteus aculeatus*) and rainbow trout (*Salmo gairdneri*) in three lakes in the Matanuska Valley in southcentral Alaska. Unpublished M.Sc. thesis. Idaho State University, Pocatello.

Werner, E.E. (1974). The fish size, prey size, handling time relation in several sunfishes and some implications. *Journal of the Fisheries Research Board of Canada*, **31**, 1531–1536.

Werner, E.E. and Gilliam, J.F. (1984). The ontogenetic niche and species interactions in size-structured populations. *Annual Review of Ecology and Systematics*, **15**, 393–425.

Werner, E.E. and Hall, D.J. (1974). Optimal foraging and the size selection of prey by the bluegill sunfish (*Lepomis macrochirus*). *Ecology*, **55**, 1042–1052.

Werner, E.E., Mittelbach, G.C., and Hall, D.J. (1981). The role of foraging profitability and experience in habitat use by the bluegill sunfish. *Ecology*, **62**, 116–125.

Werner, E.E., Gilliam, J.F., Hall, D.J., and Mittelbach, G.C. (1983). An experimental test of the effects of predation risk on habitat use in fish. *Ecology*, **64**, 1540–1548.

Wetterer, J.K. (1989). Mechanisms of prey choice by planktivorous fish: perceptual constraints and rules of thumb. *Animal Behaviour*, **37**, 955–967.

Wetzel, R.G. (1975). *Limnology*. Saunders, Philadelphia.

Whitear, M. (1971). The free nerve endings in fish epidermis. *Journal of Zoology, London*, **163**, 231–236.

Whitman, C.O. (1899). *Animal behavior*. Biological Lectures of the Marine Biological Laboratory, Woods Hole.

Whitt, G.S. (1970). Developmental genetics of the lactate dehydrogenase isozymes of fish. *Journal of Experimental Zoology*, **175**, 1–36.

Whoriskey, F.G. (1991). Stickleback distraction displays: sexual or foraging deception against egg cannibalism? *Animal Behaviour*, **41**, 989–995.

Whoriskey, F.G. and FitzGerald, G.J. (1985*a*). Nest sites of the threespine stickleback: can site characteristics alone protect the nest against egg predators and are nest sites a limiting resource? *Canadian Journal of Zoology*, **63**, 1991–1994.

Whoriskey, F.G. and FitzGerald, G.J. (1985*b*). The effects of bird predation on an estuarine stickleback (Pisces: Gasterosteidae) community. *Canadian Journal of Zoology*, **63**, 301–307.

Whoriskey, F.G. and FitzGerald, G.J. (1985*c*). Sex, cannibalism and sticklebacks. *Behavioral Ecology and Sociobiology*, **18**, 15–18.

Whoriskey, F.G. and FitzGerald, G.J. (1987). Intraspecific competition in sticklebacks (*Gasterosteidae: Pisces*): does mother nature concur? *Journal of Animal Ecology*, **56**, 939–947.

Whoriskey, F.G. and FitzGerald, G.J. (1989). Breeding-season habitat use by sticklebacks (Pisces: Gasterosteidae) at Isle Verte, Quebec. *Canadian Journal of Zoology*, **67**, 2126–2130.

Whoriskey, F.G. and Wootton, R.J. (1987). Swimming endurance of threespine sticklebacks, *Gasterosteus aculeatus* L., from the Afon Rheidol, Wales. *Journal of Fish Biology*, **30**, 335–340.

Whoriskey, F.G., Gaudreault, A., Martel, N., Campeau, S. and FitzGerald, G.J. (1985). The activity budget and behaviour patterns of female threespine sticklebacks, *Gasterosteus aculeatus* L., in a Québec tidal saltmarsh. *Le Naturaliste Canadien (Revue d'Ecologie et de Systématique)*, **112**, 113–118.

Whoriskey, F.G., FitzGerald, G.J., and Reebs, S.G. (1986). The breeding-season population structure of three sympatric, territorial sticklebacks (Pisces: Gasterosteidae). *Journal of Fish Biology*, **29**, 635–648.

Wieser, W. and Medgyesey, N. (1990). Aerobic maximum for growth in the larvae and juveniles of a cyprinid fish: *Rutilus rutilus* (L.): implications for energy budgeting in small poikilotherms. *Functional Ecology*, **4**, 233–242.

Wieser, W., Forstner, H., Medgyesey, N., and Hinterleitner, S. (1988). To switch or not to switch: partitioning of energy between growth and activity in larval cyprinids (Cyprinidae: Teleostei). *Functional Ecology*, 2, 499–507.

Wilbur, H.M. and Morin, P.J. (1988). Life history evolution in turtles. In *Biology of the Reptilia*, Vol. 16B (ed. C. Gans and R.B. Huey), pp. 388–439. Liss, New York.

Wiley, E.O. (1981). *Phylogenetics: the theory and practice of phylogenetic systematics*. Wiley, New York.

Wilkens, H. (1971). Genetic interpretation of regressive evolutionary processes: studies on hybrid eyes of two *Astyanax* cave populations. *Evolution*, **25**, 530–544.

Wilkens, H. and Hüppop, K. (1986). Sympatric speciation in cave fishes? *Zeitschrift für Zoologische Systematik und Evolutionsforschung*, **24**, 223–230.

Williams, D.D. and Delbeek, J.C. (1989). Biology of the threespine stickleback, *Gasterosteus aculeatus*, and the blackspotted stickleback, *G. wheatlandi*, during their marine pelagic phase in the Bay of Fundy, Canada. *Environmental Biology of Fishes*, **24**, 33–41.

Williams, G.C. (1964). Measurements of consociation among fishes and comments on the evolution of schooling. *Publications of the Museum of Michigan State University*, **2**, 351–383.

Williams, G.C. (1966). *Adaptation and natural selection*. Princeton University Press, Princeton, New Jersey.

Williams, G.C. (1975). *Sex and evolution*. Princeton University Press, Princeton, New Jersey.

Williams, G.C. (1992). *Natural selection: domains, levels, and applications*. Oxford University Press, Oxford.

Williams, J.G.K., Kubelik, A.R., Livak, K.J., Rafalski, J.A., and Tingey, S.V. (1990). DNA polymorphisms amplified by arbitrary primers are useful as genetic markers. *Nucleic Acids Research*, **18**, 6531–6535.

Wilson, D.S. (1989). The diversification of single gene pools by density- and frequency-dependent selection. In *Speciation and its consequences* (ed. D. Otte and J.A. Endler), pp. 366–385. Sinauer, Sunderland, Massachusetts.

Wilson, M.V.H. (1980). Eocene lake environments: depth and distance-from-shore variation in fish, insect, and plant assemblages. *Palaeogeography, Palaeoclimatology, Palaeoecology*, **32**, 21–44.

Wilson, M.V.H. (1984). Year classes and sexual dimorphism in the Eocene catostomid fish *Amyzon aggregatum*. *Journal of Vertebrate Paleontology*, **3**, 137–142.

Wilson, M.V.H. (1987). Predation as a source of fish fossils in Eocene lake sediments. *Palaios*, **2**, 497–504.

Wilson, M.V.H. (1988). Reconstruction of ancient lake environments using both autochthonous and allochthonous fossils. *Palaeogeography, Palaeoclimatology, Palaeoecology*, **62**, 609–623.

Wilz, K.J. (1970*a*). Self-regulation of motivation in the three-spined stickleback (*Gasterosteus aculeatus*). *Nature*, **226**, 465–466.

Wilz, K.J. (1970*b*). Causal and functional analysis of dorsal pricking and nest activity in the courtship of the three-spined stickleback *Gasterosteus aculeatus*. *Animal Behaviour*, **18**, 115–124.

Wilz, K.J. (1971). Comparative aspects of courtship behavior in the ten-spined stickleback, *Pygosteus pungitius* (L.). *Zeitschrift für Tierpsychologie*, **29**, 1-10.

Wilz, K.J. (1972). Causal relationships between aggression and the sexual and nest behaviours in the three-spined stickleback (*Gasterosteus aculeatus*). *Animal Behaviour*, **20**, 335-340.

Wilz, K.J. (1973). Quantitative differences in the courtship of two populations of three-spined sticklebacks, *Gasterosteus aculeatus*. *Zeitschrift für Tierpsychologie*, **33**, 141-146.

Wilz, K.J. (1975). Cycles of aggression in male three-spined sticklebacks (*Gasterosteus aculeatus*). *Zeitschrift für Tierpsychologie*, **39**, 1-7.

Winemiller, K.O. (1989). Patterns of variation in life history among South American fishes in seasonal environments. *Oecologia*, **81**, 225-241.

Winn, H.E. (1960). Biology of the brook stickleback *Eucalia inconstans* (Kirtland). *American Midland Naturalist*, **63**, 424-438.

Withler, R.E. and McPhail, J.D. (1985). Genetic variability in freshwater and anadromous sticklebacks (*Gasterosteus aculeatus*) of southern British Columbia. *Canadian Journal of Zoology*, **63**, 528-533.

Withler, R.E., McPhail, J.D., and Devlin, R.H. (1986). Electrophoretic polymorphism and sexual dimorphism in the freshwater and anadromous threespine sticklebacks (*Gasterosteus aculeatus*) of the Little Campbell River, British Columbia. *Biochemical Genetics*, **24**, 701-713.

Woodburne, M.O. (1991). The Mojave Desert Province. *San Bernardino County Museum Association Quarterly*, **38**, 60-77.

Wootton, R.J. (1970). Aggression in the early phases of the reproductive cycle of the male three-spined stickleback (*Gasterosteus aculeatus*). *Animal Behaviour*, **18**, 740-746.

Wootton, R.J. (1971*a*). A note on nest-raiding behavior of male sticklebacks. *Canadian Journal of Zoology*, **49**, 960-962.

Wootton, R.J. (1971*b*). Measures of the aggression of parental male three-spined sticklebacks. *Behaviour*, **40**, 228-262.

Wootton, R.J. (1972*a*). The behaviour of the male three-spined stickleback in a natural situation: a quantitative description. *Behaviour*, **41**, 232-241.

Wootton, R.J. (1972*b*). Changes in the aggression of the male three-spined stickleback after fertilization of eggs. *Canadian Journal of Zoology*, **50**, 537-541.

Wootton, R.J. (1973*a*). Fecundity of the three-spined stickleback, *Gasterosteus aculeatus* (L.). *Journal of Fish Biology*, **5**, 683-688.

Wootton, R.J. (1973*b*). The effect of size of food ration on egg production in the female three-spined stickleback, *Gasterosteus aculeatus* L. *Journal of Fish Biology*, **5**, 89-96.

Wootton, R.J. (1974*a*). The inter-spawning interval of the female three-spined stickleback, *Gasterosteus aculeatus*. *Journal of Zoology, London*, **172**, 331-342.

Wootton, R.J. (1974*b*). Changes in the courtship behaviour of female three-spined sticklebacks between spawnings. *Animal Behaviour*, **22**, 850-855.

Wootton, R.J. (1976). *The biology of the sticklebacks*. Academic Press, London.

Wootton, R.J. (1977). Effect of food limitation during the breeding season on the size, body components and egg production of female sticklebacks (*Gasterosteus aculeatus*). *Journal of Animal Ecology*, **46**, 823-834.

Wootton, R.J. (1979). Energy costs of egg production and environmental determinants of fecundity in teleost fishes. *Symposia of the Zoological Society of London*, **44**, 133-159.

Wootton, R.J. (1984*a*). *A functional biology of sticklebacks*. Croom Helm, London, and University of California Press, Berkeley.

Wootton, R.J. (1984*b*). Introduction: tactics and strategies in fish reproduction. In *Fish reproduction: strategies and tactics* (ed. G.W. Potts and R.J. Wootton), pp. 1-12. Academic Press, London.

Wootton, R.J. (1985*a*). Energetics of reproduction. In *Fish energetics: new perspectives* (ed. P. Tytler and P. Calow), pp. 231-254. Johns Hopkins University Press, Baltimore.

Wootton, R.J. (1985*b*). Effects of food and density on the reproductive biology of the threespine stickleback with a hypothesis on population limitations in sticklebacks. *Behaviour*, **93**, 101-111.

Wootton, R.J. (1990). *Ecology of teleost fishes*. Chapman and Hall, London.

Wootton, R.J. and Evans, G.W. (1976). Cost of egg production in the three-spined stickleback (*Gasterosteus aculeatus* L.). *Journal of Fish Biology*, **8**, 385-395.

Wootton, R.J., Evans, G.W., and Mills, L. (1978). Annual cycle in female three-spined sticklebacks (*Gasterosteus aculeatus* L.) from an upland and lowland population. *Journal of Fish Biology*, **12**, 331-343.

Wootton, R.J., Allen, J.R.M., and Cole, S.J. (1980*a*). Energetics of the annual reproductive cycle in female sticklebacks, *Gasterosteus aculeatus* L. *Journal of Fish Biology*, **17**, 387-394.

Wootton, R.J., Allen, J.R.M., and Cole, S.J. (1980*b*). Effect of body weight and temperature on the maximum daily food consumption of *Gasterosteus aculeatus* L. and *Phoxinus phoxinus* (L.): selecting an appropriate model. *Journal of Fish Biology*, **17**, 695-705.

Worgan, J.P. and FitzGerald, G.J. (1981*a*). Diel activity and diet of three sympatric sticklebacks in tidal salt marsh pools. *Canadian Journal of Zoology*, **59**, 2375-2379.

Worgan, J.P. and FitzGerald, G.J. (1981*b*). Habitat segregation in a salt marsh among adult sticklebacks (Gasterosteidae). *Environmental Biology of Fishes*, **6**, 105-109.

Wright, S. (1978). *Evolution and the genetics of populations*, Vol. 4. University of Chicago Press, Chicago.

Wunder, W. (1928). Experimentelle Untersuchungen an Stichlingen (Kämpfe, Nestbau, Laichen, Brutpflege). *Zoologischer Anzeiger Supplement (Verhandlungen Deutsche Zoologische Gesellschaft)*, **3**, 115-127.

Wunder, W. (1930). Experimentelle Untersuchungen an dreistachligen Stichling (*Gasterosteus aculeatus* L.) während der Laichzeit. *Zeitschrift für Morphologie und Ökologie der Tiere*, **16**, 453-498.

Wunder, W. (1934). Gattenwahlversuche bei Stichlingen und Bitterlingen. *Zoologischer Anzeiger Supplement (Verhandlungen Deutsche Zoologische Gesellschaft)*, **7**, 152-158.

Wunder, W. (1957). Die Sinnesorgane der Fische. *Allgemeine Fischereizeitung*, **82**, 1-24.

Wydoski, R.S. and Whitney, R.R. (1979). *Inland fishes of Washington*. University of Washington Press, Seattle.

Yabe, M. (1985). Comparative osteology and myology of the superfamily Cottoidea (Pisces: Scorpaeniformes), and its phylogenetic classification. *Memoirs of the Faculty of Fisheries, Hokkaido University*, **32**, 1-130.

Yamamoto, T. (1966). An electron microscope study of the columnar epithelial cell in the intestine of fresh water teleosts: goldfish (*Carassius auratus*) and rainbow trout (*Salmo irideus*). *Zeitschrift für Zellforschung*, **72**, 66-78.

Yamanaka, M. (1971). The distribution and habits of the landlocked three-spined stickleback (trachurus form), *Gasterosteus aculeatus*, Aizu district of Japan. *Animal and Nature, Japan*, **1**, 7–11.

Yang, S.Y. and Min, M.S. (1990). Genetic variation and systematics of the sticklebacks (Pisces, Gasterosteidae) in Korea. *Korean Journal of Zoology*, **33**, 499–508.

Yen, T.-C. (1950). A molluscan fauna from the type section of the Truckee Formation. *American Journal of Science*, **248**, 180–193.

Zander, C.D., Möller-Buchner, J., and Totzke, H.-D. (1984). The role of sticklebacks in the food web of the Elbe and Eider estuaries (northern Federal Republic of Germany). *Zoologischer Anzeiger Jena*, **212**, 209–222.

Zaret, T.M. (1980). *Predation and freshwater communities*. Yale University Press, New Haven, Connecticut.

Zietara, M.S. (1989). Electrophoretic lactate and malate dehydrogenase polymorphism in populations of the threespine (*Gasterosteus aculeatus*) and the ninespine (*Pungitius pungitius*) sticklebacks from the Vistula River and the Gdansk Gulf. *Biochemical Systematics and Ecology*, **17**, 253–259.

Ziuganov, V.V. (1981). A contribution to the analysis of parallel variation, the fishes of the genera *Gasterosteus* and *Pungitius* (Gasterosteidae) taken as an example. *Zhiurnal Obshchei Biologii SSSR*, **44**, 718–728. [in Russian].

Ziuganov, V.V. (1983). Genetics of osteal plate polymorphism and microevolution of threespine stickleback (*Gasterosteus aculeatus* L.). *Theoretical and Applied Genetics*, **65**, 239–246.

Ziuganov, V.V. (1988). Studies on the mechanism of ethological reproductive isolation of forms (*trachurus, leiurus, semiarmatus*) of threespine stickleback, *Gasterosteus aculeatus* (Gasterosteiformes, Gasterosteidae) from two geographical regions. *Zoologischeskii Zhurnal*, **57**, 719–727.

Ziuganov, V.V. and Bugayev, V.F. (1988). Isolating mechanisms between spawning populations of the threespine stickleback, *Gasterosteus aculeatus*, of Lake Azabachije, Kamchatka. *Voprosy Ikhtiologii*, **2**, 322–325. [English translation].

Ziuganov, V.V., Golovatjuk, G.J., Savvaitova, K.A., and Bugayev, V.F. (1987). Genetically isolated sympatric forms of threespine stickleback, *Gasterosteus aculeatus*, in Lake Azabachije (Kamchatka Peninsula, USSR). *Environmental Biology of Fishes*, **18**, 241–247.

Zorbidi, Zh. Kh. (1977). Diurnal feeding rhythm of the coho salmon *Oncorhynchus kisutch* of Lake Azabach'ye. *Journal of Ichthyology*, **17**, 166–168.

Zuckerkandl, E. and Pauling, L. (1965). Evolutionary divergence and convergence in proteins. In *Evolving genes and proteins* (ed. V. Bryson and H.J. Vogel), pp. 97–166. Academic Press, New York.

Author Index

Only citations that are central to a discussion are indexed, and only the senior author of papers with more than two authors is listed. If an author's papers are prominent throughout a series of pages, the range of pages is shown.

Subject Index